Molecular Cytogenetics and Crop Improvement

Molecular Cytogenetics and Crop Improvement

Dr. Pramod Kumar Pandey
AICRP on Maize, College of Agriculture, Central Agricultural University (Imphal), Kyrdemkulai- 793105, Meghalaya

Dr. Mohd Ashaq
Department of Botany, Govt Degree College, Thannamandi, J&K

Dr. C Bharatkumar
Regional Agricultural Research Station, Vijayapura, at the University of Agricultural Sciences, Dharwad

Dr. Satish D
Genetics and Plant breeding, Department of Biotechnology and Crop improvement, College of Horticulture, University of Horticultural Sciences Bagalkot, Karnataka

Dr. Jasti Srivarsha
Regional Research Station, Nashik affiliated with the National Horticultural Research and Development Foundation

Dr. Sheetanshu Gupta
Department of Biochemistry, G.B. Pant University of Agriculture & Technology, Pantnagar, Uttarakhand

CRC Press
Taylor & Francis Group
Boca Raton London New York

CRC Press is an imprint of the Taylor & Francis Group, an **informa** business

-EPH-
Elite Publishing House

First published 2026
by CRC Press
4 Park Square, Milton Park, Abingdon, Oxon, OX14 4RN

and by CRC Press
2385 NW Executive Center Drive, Suite 320, Boca Raton FL 33431

CRC Press is an imprint of Informa UK Limited

British Library Cataloguing-in-Publication Data
A catalogue record for this book is available from the British Library

Print edition not for sale in India

ISBN: 9781041205692 (hbk)
ISBN: 9781041207399 (pbk)
ISBN: 9781003717775 (ebk)

DOI: 10.1201/9781003717775

Typeset in A Caslon Pro
by Elite Publishing House, Delhi

-EPH-

Contents

Preface

The advancements in molecular cytogenetics have revolutionized the field of crop improvement, offering unprecedented tools for genome manipulation, gene tracking, and chromosomal modifications. In recent decades, the demand for high-yielding, stress-resistant crops has never been more pressing. Climate change, pest outbreaks, and the depletion of natural resources have challenged traditional breeding methods, pushing scientists to seek more precise and efficient approaches. This book, Molecular Cytogenetics for Crop Improvement, is the culmination of the latest research and methodologies in cytogenetics, aimed at addressing these global agricultural challenges.

The preface serves as an introduction to the scientific principles and techniques covered in the book, including the evolution of cytogenetics from classical methods to molecular-based approaches. The first unit introduces readers to the structure and organization of genomes, focusing on both nuclear and organellar genomes. It explains the complex interplay between nuclear DNA and cytoplasmic genomes and the importance of signal transduction in regulating these interactions. This unit also explores the C value paradox and the roles of repetitive sequences, introns, and exons, setting the stage for understanding genome dynamics.

In the second unit, the focus shifts to cytogenetic techniques, including karyotyping, chromosome banding, and chromosome painting. These tools have transformed the ability to visualize chromosomal behavior during cell division, track introgressions, and localize specific genes or genomic segments using FISH and GISH. These techniques provide a critical foundation for the genetic mapping processes discussed in subsequent units.

Unit three delves into the practical applications of cytogenetics in crop improvement, particularly pre-breeding programs. It covers methodologies for locating and mapping genes on chromosomes using deficiency methods and chromosomal interchanges. The use of balanced lethal systems and multiple interchanges is also explored, demonstrating their utility in maintaining genetic diversity and producing inbred lines. This unit also discusses the role of chromosomal duplications, inversions, and translocations in gene transfer and location, providing essential tools for modern plant breeders.

Unit four presents a detailed examination of trisomics and monosomics, their production, breeding behavior, and utility in gene mapping and hybrid seed production. The use of balanced tertiary trisomics in hybrid seed production is a novel approach covered in this section, offering new possibilities for increasing crop productivity. Additionally, the telocentric method of mapping genes is introduced, highlighting its relevance in modern genetic research.

The fifth unit explores cytogenomics, a rapidly advancing field that integrates cytogenetics with genomic data. It emphasizes the tools and techniques used for isolating specific chromosomes and developing molecular maps for gene location. Cytogenomics has expanded the capabilities of traditional cytogenetic methods, allowing for more precise identification of genetic markers and their associated traits.

Finally, unit six addresses the role of polyploidy in crop evolution and breeding. Polyploidy, both auto- and allopolyploidy, has played a significant role in the development of many crop species. This unit covers the behavior of interspecific and intergeneric crosses, barriers to hybridization, and the use of distant hybridization in creating new crop varieties.

Through this book, we aim to provide a comprehensive resource for understanding the molecular cytogenetic techniques that are shaping the future of crop improvement. The integration of molecular biology with classical cytogenetic principles has opened new avenues for breeding crops that are not only higher yielding but also more resilient to environmental stresses. We hope that this book will serve as a valuable reference for researchers, students, and professionals dedicated to advancing agricultural science and ensuring global food security.

Authors

About the Authors

Dr. Pramod Kumar Pandey is working at the College of Agriculture, Kyrdemkulai, Meghalaya, which is part of the Central Agricultural University headquartered in Imphal, Manipur. He was born in the village of Kondari, which is in the Ghazipur District of Uttar Pradesh. He did his undergraduate studies at PG College, Ghazipur, U.P., and obtained his postgraduate degree in Biotechnology from KIIT University, Bhubaneswar, Odisha. He obtained his Ph.D. in Biotechnology from Dibrugarh University, Dibrugarh, Assam. He has authored numerous scholarly research papers, and articles and contributed to various book chapters. He possesses about 10 years of professional and academic expertise in the field of Endophytes. He is currently involved in maize breeding activities and has more than 6 years of experience in the field of maize breeding. He serves as the In-Charge of AICRP on Maize, CAU Brapanai Centre.

Dr. Mohd Ashaq is an Associate Professor of Botany in the Higher Education Department of the Jammu and Kashmir Government presently posted at Govt Degree College Thannamandi, J&K, India. He completed his M.Sc., M.Phil., and Ph. D degrees in Botany with a specialization in Plant Virology from Aligarh Muslim University, Aligarh, India in 1992, 1993, and 1997 respectively. Dr. Ashaq served in Eritrea for 8 years between 2000-2008, finally as Director of Research at Eritrea Institute of Technology, Asmara, State of Eritrea (2006-2008). Dr. Ashaq has 6 books, 8 book chapters, 30 Research papers, 17 patents, and over 50 popular articles to his credit. He has edited over 100 issues of magazines and newsletters for various organizations. Several of his publications on Refugees and volunteerism have been translated into French, Portuguese, and Spanish. He is the commissioned member of 5 of 6 IUCN (International Union of Conservation of Nature) commissions at present. He has presented papers and served as a resource person, invited speaker, and keynote speaker in over 150 conferences/seminars/workshops, etc. He has organized dozens of international and national conferences/seminars, workshops, and FDPs. He is a Reviewer/Editorial board member/ fellow of about three dozen national and international peer-reviewed Scopus/web of Science journals and societies besides 15 development organizations.

Dr. C Bharatkumar is currently a Scientist (Plant Breeding), AICRP on Chickpea, Regional Agricultural Research Station, Vijayapura, at the University of Agricultural Sciences, Dharwad. He completed a Ph.D. (Genetics and Plant Breeding) from the University of Agricultural Sciences, Bangalore. He has 9 years of experience as a scientist. He has been associated with the teaching of undergraduate and postgraduate students. He obtained the Best Scientist Award and Excellence in Research Award. He has published 2 Books, 26 international research papers in national referred journals, 5 book chapters, and 14 popular articles. He has participated in and presented papers in various international and national workshops, seminars, and conferences.

Dr. Satish D. currently working as an Associate Professor of Genetics and Plant Breeding in the Department of Biotechnology and Crop Improvement at the College of Horticulture, University of Horticultural Sciences Bagalkot, Karnataka, India. He graduated in Agricultural Sciences from the College of Agriculture, Raichur. He completed his Master and obtained a doctoral degree in Genetics and Plant Breeding from the University of Agricultural Sciences Dharwad. He has 13 years of teaching and research experience. He has guided nine MSc and three PhD students and served as a member of the advisory committee for more than 60 postgraduate students of different disciplines. He has handled externally funded research projects sanctioned by different funding agencies as PI and Co-PI. He is also associated with the development of 11 varieties of different vegetable crops and four technologies related to horticultural sciences. He has received ICAR sponsored BEST TEACHER award for his excellence in teaching from the University Horticultural Sciences Bagalkot. Published more than 50 research papers in various national and international journals. He has attended national and international conferences and seminars and presented research papers. He has visited UAE, Malaysia, Nepal, and Thailand to attend international conferences. He has good exposure in Agri/Horti startups, innovations, idea generation, evaluation, mentoring of Agri startups, e-marketing platforms, and IPR issues.

Dr. Jasti Srivarsha is working at the Regional Research Station, Nashik, a part of the National Horticultural Research and Development Foundation. She was born in Hyderabad of Telangana. She graduated in horticulture from the College of Horticulture, Rajendranagar under YSR Horticultural University, Andhra Pradesh. She obtained her master's degree in Genetics and Plant Breeding from Vasantrao Naik Marathwada Krishi Vidyapeeth, Parbhani, Maharashtra, and secured two gold medals for academic excellence. She completed her Ph.D. in Genetics and Plant Breeding from Dr. Balasaheb Sawant Konkan Krishi Vidyapeeth, Dapoli, Maharashtra. She has authored numerous scientific publications and book chapters. She has experience in breeding activities of vegetable crops. She is currently involved in onion breeding. She serves as Senior Technical Officer (Plant Breeding) at NHRDF, Nashik.

Dr. Sheetanshu Gupta working in Biochemistry, Molecular Biology, and Biotechnology for the last 13 years, an alumnus of G.B. Pant University of Agriculture & Technology, Pantnagar, Uttarakhand, India, ICAR ARS NET (Plant Biochemistry and Plant Physiology) and

CSIR NET. Dr. Gupta gained valuable experience in the areas of environmental problems, aeroponics and hydroponics, and protein biochemistry. His work showcased his expertise in applying innovative farming techniques and exploring novel approaches to plant protection. Dr. Gupta researched tissue culture hydroponics and mushroom farming, contributing to the advancement of these fields. Dr. Gupta's academic qualifications, research experience, and expertise in biochemistry and biotechnology make him a valuable asset in the field. With his passion for teaching and research, dedication to innovative farming techniques, and student advocacy, Dr. Gupta continues to contribute significantly to the advancement of agricultural and biochemical sciences with more than 18 published books, 150 book chapters, 2 patents, and more than 25 Research/ Review papers.

Chapter - 1

Introduction to Molecular Cytogenetics for Crop Improvement

A. Genetics

Genetics is a central pillar of modern biology, focused on the study of genes, genetic variation, and the mechanisms of heredity in living organisms. As the scientific study of how traits and characteristics are passed from one generation to the next, genetics forms the backbone of biological research and has far-reaching applications in medicine, agriculture, and biotechnology. This field explores the intricate mechanisms by which genetic information is encoded, replicated, and expressed within cells, providing insights into the fundamental processes of life. The science of genetics has evolved significantly since its inception, driven by groundbreaking discoveries and technological advancements that have transformed our understanding of biological inheritance and diversity.

The Foundations of Genetics

The foundations of genetics were laid in the mid-19th century through the pioneering work of Gregor Mendel, an Augustinian monk who conducted meticulous experiments with pea plants. Mendel's work, which went largely unnoticed until the early 20th century, uncovered the basic principles of inheritance that are now known as Mendel's laws. These principles revolutionized our understanding of how traits are passed from parents to offspring and laid the groundwork for classical genetics.

Mendel's Laws

Mendel's experiments focused on discrete traits such as flower color, seed shape, and plant height. By cross-pollinating pea plants and analyzing the resulting offspring, Mendel formulated three fundamental laws of inheritance:

1. **Law of Segregation:**

   » The law of segregation states that each individual possesses two alleles for a particular trait, which separate during gamete formation. This ensures that each gamete receives only one allele, resulting in offspring inheriting one allele from each parent. This principle explains the reappearance of recessive traits in the F2 generation, despite their absence in the F1

DOI: 10.1201/9781003717775-1

generation. It underscores the concept of alleles as discrete units of inheritance and provides a framework for understanding genetic variation. The law of segregation is a cornerstone of genetics, explaining how genetic information is maintained and transmitted across generations.

2. **Law of Independent Assortment:**

   » The law of independent assortment states that alleles of different genes are distributed independently of one another during gamete formation. This results in genetic variation, as different combinations of alleles can be inherited. This principle highlights the importance of genetic diversity in evolution and adaptation, as it explains how multiple traits are inherited simultaneously and independently. The law of independent assortment forms the basis for understanding complex genetic interactions and the inheritance of traits governed by multiple genes. This principle is essential for predicting the outcomes of genetic crosses and understanding the genetic basis of phenotypic diversity.

3. **Law of Dominance:**

   » The law of dominance posits that when two different alleles are present in an individual, one allele may mask the expression of the other. The dominant allele is expressed in the phenotype, while the recessive allele is masked. This principle explains the presence of dominant and recessive traits and provides insights into the expression of genetic traits. The law of dominance is fundamental for predicting phenotypic ratios in genetic crosses and understanding the inheritance patterns of traits governed by dominant and recessive alleles. It also helps in explaining the concept of homozygosity and heterozygosity in genetic studies.

Mendel's work went largely unnoticed until the early 20th century when it was rediscovered by scientists such as Hugo de Vries, Carl Correns, and Erich von Tschermak. This rediscovery marked the beginning of genetics as a scientific discipline and laid the groundwork for subsequent discoveries in the field. Mendel's principles of inheritance provided a framework for understanding the genetic basis of traits and served as a foundation for the development of modern genetics.

The Molecular Revolution

The discovery of DNA as the hereditary material and the elucidation of its double-helical structure by James Watson and Francis Crick in 1953 marked a turning point in the field of genetics. DNA, or deoxyribonucleic acid, is the molecule that carries the genetic instructions for the development, functioning, growth, and reproduction of all known organisms. The understanding of DNA's structure revealed how genetic information is stored and transmitted, paving the way for the field of molecular genetics.

Molecular Genetics

Molecular genetics focuses on the molecular structure and function of genes, exploring how they are regulated and how they influence the phenotype of an organism. This field has expanded to include a variety of techniques that have transformed our understanding of genetic processes:

1. **Gene Cloning:** Gene cloning involves the creation of copies of specific genes or DNA

sequences. This technique allows researchers to study the function of genes and their products, leading to insights into genetic diseases and potential therapeutic interventions. Gene cloning is a fundamental tool in molecular biology, enabling the production of recombinant proteins, such as insulin, and facilitating the study of gene function and regulation. It has revolutionized research in areas such as gene therapy, functional genomics, and biotechnology, providing insights into the molecular mechanisms underlying genetic diseases and offering potential solutions for their treatment.

2. **DNA Sequencing:** DNA sequencing involves determining the precise order of nucleotides within a DNA molecule. This technology has revolutionized genetics by enabling the identification of genetic variations associated with diseases and traits. DNA sequencing provides a comprehensive view of the genome, allowing researchers to identify mutations, study genetic diversity, and explore the evolutionary relationships between species. It is applied in personalized medicine, forensic investigations, and evolutionary biology, offering valuable information on the genetic foundations of diseases and aiding in the creation of specialized treatments.
3. **Gene Expression Analysis:** Gene expression analysis explores the activation or suppression of genes across various tissues and conditions. Analyzing gene expression patterns sheds light on the molecular processes that drive development and disease progression. This approach reveals how gene networks are regulated and the impact of specific genes on overall health, offering valuable information for designing therapeutic treatments and strategies for managing diseases. Gene expression analysis is used to study cancer biology, developmental processes, and response to environmental stimuli, offering a deeper understanding of the complex interactions between genes and their environment.
4. **Genetic Engineering:** Genetic engineering entails altering an organism's genome to introduce or modify specific traits. Advanced methods like CRISPR-Cas9 have made precise genome editing possible, paving the way for innovative research and practical uses in fields such as medicine and agriculture. Through genetic engineering, genetically modified organisms (GMOs) are created, therapeutic proteins are produced, and genetic disorders can be corrected. This technology holds promise for tackling issues like food security, disease prevention, and environmental sustainability, providing solutions for developing crops with improved qualities and addressing the needs of a growing global population.
5. **Genetics in Crop Improvement:** In the context of crop improvement, genetics provides the foundation for understanding the hereditary components that contribute to important agronomic traits. Crop improvement aims to develop new varieties with enhanced characteristics, such as increased yield, disease resistance, and stress tolerance. By manipulating the genetic makeup of crops, breeders can achieve these goals and address the challenges posed by a growing global population and changing environmental conditions.

Key Contributions of Genetics to Crop Improvement

1. **Agronomic Traits:**

   » Genetics allows researchers to identify the genes and genetic pathways that control agronomic traits. By understanding the genetic basis of traits such as yield, disease resistance, and stress tolerance, breeders can develop strategies to improve these characteristics. This is essential for addressing challenges such as climate change, pests, and diseases, as it provides insights into the genetic mechanisms underlying trait expression and adaptation. The identification of genes associated with desirable traits facilitates the development of breeding programs aimed at improving crop productivity, quality, and resilience.

2. **Marker-Assisted Selection (MAS):**

   » Marker-assisted selection (MAS) is a breeding technique that uses molecular markers to select plants with desirable traits. This approach accelerates the breeding process by enabling the identification of individuals with specific genetic markers associated with favorable traits, without the need for extensive phenotypic evaluation. MAS is used to improve traits such as disease resistance, drought tolerance, and nutritional content in crops, enhancing the efficiency and accuracy of breeding programs. It allows breeders to focus on individuals with the desired genetic makeup, reducing the time and resources required for traditional breeding methods and leading to the development of superior crop varieties.

3. **Genetic Diversity and Conservation:**

   » Genetic diversity is crucial for the resilience and adaptability of crops. Genetics provides tools to assess and conserve genetic diversity in crop species and their wild relatives. This diversity serves as a reservoir of genes that can be used to improve crop resilience to pests, diseases, and environmental stresses. The conservation of genetic diversity is essential for ensuring the long-term sustainability of agriculture and food security, as it enables breeders to access a wide range of genetic resources for crop improvement and adaptation to changing environments. Strategies for conserving genetic diversity include the establishment of gene banks, in situ conservation, and the use of traditional and modern breeding techniques to maintain and enhance genetic diversity in crop populations.

4. **Genomic Selection:**

   » Genomic selection involves using genome-wide information to predict the breeding value of individuals. This approach allows breeders to select the best individuals for breeding based on their genetic potential, improving the efficiency and accuracy of breeding programs. Genomic selection is used to enhance traits such as yield, quality, and resilience in crops, accelerating the development of new varieties and reducing the time and resources required for traditional breeding methods. By incorporating genomic information into breeding programs, breeders can make more informed decisions and develop crop varieties that meet the demands of a changing environment and growing population.

5. Biotechnology and Genetic Engineering:

» Genetic engineering enables the introduction of new traits into crops through the insertion of specific genes. This technology has been used to develop genetically modified crops with traits such as herbicide resistance, insect resistance, and improved nutritional content. Biotechnology and genetic engineering have transformed agriculture by providing tools to address challenges such as food security, pest control, and environmental sustainability. They offer solutions for developing crops with enhanced characteristics, such as improved yield, quality, and resilience, and contribute to the development of sustainable agricultural practices that meet the needs of a growing global population.

Integration with Cytogenetics and Molecular Techniques

The combination of genetics with cytogenetics and molecular methods has significantly advanced the creation of enhanced crop varieties. Cytogenetics offers a deeper understanding of chromosomal structure and the organization of genetic material, while molecular techniques provide precise tools for gene manipulation and analysis. Together, these approaches have accelerated progress in crop improvement, allowing for the development of plants with better traits and higher resilience.

Integration in Crop Improvement

1. **Chromosome Mapping:** The integration of cytogenetics with molecular techniques allows for the mapping of genes to specific chromosomal locations. This information is essential for identifying genes associated with desirable traits and for developing strategies to manipulate these genes. Chromosome mapping is used to identify quantitative trait loci (QTLs) associated with agronomic traits, facilitating the development of marker-assisted breeding strategies. It provides insights into the genetic architecture of complex traits and contributes to the development of improved crop varieties by enabling breeders to target specific genomic regions for trait improvement.
2. **Genetic Linkage Analysis:** Linkage analysis involves studying the association between genetic markers and traits. By identifying markers linked to important traits, breeders can select individuals with the desired genetic makeup, improving the efficiency of breeding programs. Genetic linkage analysis is used to identify genes associated with traits such as disease resistance, yield, and quality, providing a framework for understanding the genetic basis of complex traits and informing breeding strategies aimed at improving crop performance. This technique is essential for developing genetic maps that guide the selection and improvement of crop varieties.
3. **Quantitative Trait Loci (QTL) Mapping:** QTL mapping pinpoints specific regions within the genome that are linked to quantitative traits like yield and disease resistance. This valuable data aids in designing marker-assisted breeding approaches and helps identify potential candidate genes for in-depth research. By revealing the genetic foundations of complex traits, QTL mapping supports breeding initiatives focused on developing improved crop varieties. By targeting genomic regions associated with favorable traits, breeders can enhance specific aspects of the genome, leading to the cultivation of crops with superior characteristics and improved overall performance.

4. **Functional Genomics:** Functional genomics explores the roles of genes and their interactions within the genome. By understanding gene function and regulation, researchers can develop strategies to manipulate gene expression and improve crop traits. Functional genomics is used to study the molecular mechanisms underlying development, stress response, and adaptation in crops, providing insights into gene networks and regulatory pathways. This information is essential for developing strategies to enhance crop performance, resilience, and adaptability to changing environmental conditions, and for ensuring the sustainability of agricultural practices.

Genetics, through its integration with cytogenetics and molecular biology, continues to drive advancements in crop improvement. The insights gained from genetic research are essential for developing sustainable agricultural practices and ensuring food security in the face of global challenges. By leveraging the tools and techniques of genetics, breeders can develop crop varieties that meet the demands of a growing global population and changing environmental conditions, contributing to the long-term sustainability and resilience of agriculture.

B. Cytogenetics

Cytogenetics is a specialized branch of genetics that focuses on the study of chromosomes, their structure, function, and behavior during cell division. This field of science acts as a bridge between genetics and cytology (the study of cells), examining how genetic information is organized within the cell nucleus and how it is transmitted from one generation to the next. The significance of cytogenetics lies in its ability to provide crucial insights into chromosomal abnormalities, their causes, and their effects on the phenotype of organisms. The chromosomal basis of genetic information, cytogenetics plays a pivotal role in advancing our knowledge of heredity, genetic diseases, and evolutionary biology. This field has broad applications across medical genetics, evolutionary studies, and agricultural biotechnology.

Historical Development of Cytogenetics

The field of cytogenetics has a rich history, with its roots tracing back to the early 20th century. It emerged from the foundational work of pioneers such as Walther Flemming and Theodor Boveri, who observed chromosomes under a microscope and recognized their role in heredity. The discovery of chromosomes as carriers of genetic information marked a significant milestone in the study of heredity and laid the groundwork for modern cytogenetics.

Key Milestones in Cytogenetics

1. **Chromosome Discovery:** The discovery of chromosomes as discrete structures within the cell nucleus that carry genetic information was a pivotal moment in cytogenetics. Researchers observed that chromosomes undergo distinct changes during cell division, leading to the realization that they play a crucial role in inheritance. This discovery provided the first glimpse into the physical basis of heredity, as chromosomes were identified as the carriers of genetic material. The ability to visualize chromosomes under a microscope allowed scientists to study their behavior during cell division, leading to insights into the mechanisms of genetic transmission and the role of chromosomes in development and evolution.

2. **Chromosomal Basis of Inheritance:** The chromosomal theory of inheritance, proposed by Walter Sutton and Theodor Boveri, posited that genes are located on chromosomes and that the behavior of chromosomes during meiosis accounts for Mendelian inheritance. This theory provided a framework for understanding how traits are passed from parents to offspring and established the chromosome as the unit of heredity. The chromosomal theory of inheritance bridged the gap between Mendelian genetics and cytology, integrating the study of genes with the study of chromosomes. It explained the principles of independent assortment and segregation, as well as the chromosomal basis of genetic linkage and recombination.

   Development of Karyotyping: Karyotyping, the process of visually examining chromosomes under a microscope, became a cornerstone of cytogenetic research. This technique allowed scientists to study the number, size, and shape of chromosomes, leading to the identification of chromosomal disorders. Karyotyping involves the preparation of metaphase chromosome spreads, followed by staining and visualization, enabling researchers to detect numerical and structural chromosomal abnormalities. The development of karyotyping provided a powerful tool for diagnosing genetic disorders, studying chromosomal evolution, and understanding the genetic basis of developmental and reproductive processes. Karyotyping remains a fundamental technique in cytogenetics, with applications in clinical diagnostics, genetic counseling, and research on chromosomal abnormalities.

3. **Advancements in Chromosome Staining Techniques:** The development of chromosome staining techniques, such as Giemsa banding, enabled researchers to identify specific chromosome regions and banding patterns. These techniques allowed for the precise characterization of chromosomal structures and the detection of subtle chromosomal abnormalities. Chromosome banding techniques provided a means to differentiate individual chromosomes and their regions, leading to the identification of structural rearrangements, such as deletions, duplications, and translocations. These advancements have improved the precision of cytogenetic analyses, allowing researchers to investigate the genetic basis of diseases more thoroughly and accurately map genes to specific chromosomal regions.

4. **Introduction of Molecular Cytogenetics:** The integration of molecular biology techniques, such as fluorescence in situ hybridization (FISH), revolutionized cytogenetics by enabling the direct visualization of specific DNA sequences on chromosomes. Molecular cytogenetics provided new insights into chromosomal structure and function, facilitating the study of complex genomes and chromosomal rearrangements. The introduction of molecular cytogenetics marked a significant advancement in the field, as it allowed for the precise localization of genes, the detection of cryptic chromosomal abnormalities, and the study of chromosomal architecture and dynamics. Molecular cytogenetics bridged the gap between traditional cytogenetics and molecular genetics, offering new tools for studying genetic disorders and chromosomal evolution.

Techniques in Cytogenetics

Cytogenetics employs a variety of techniques to study chromosomes and their behavior. These techniques provide valuable insights into the genetic and chromosomal basis of phenotypic traits and disorders. The development and refinement of cytogenetic techniques have enabled researchers to explore the complexities of chromosomal structure, function, and evolution, contributing to our understanding of genetics and disease mechanisms.

Key Techniques in Cytogenetics

1. **Karyotyping:** Karyotyping involves the preparation and visualization of chromosomes in a cell. This technique provides information about the number and structure of chromosomes, allowing researchers to identify chromosomal abnormalities, such as aneuploidy (abnormal number of chromosomes) and structural rearrangements. Karyotyping is performed by arresting cells in metaphase, staining the chromosomes, and examining them under a microscope. This technique is essential for diagnosing genetic disorders, such as Down syndrome, Turner syndrome, and Klinefelter syndrome, as well as for identifying chromosomal changes associated with cancer. Karyotyping remains a fundamental tool in clinical genetics, cytogenetic research, and evolutionary studies.
2. **Fluorescence In Situ Hybridization (FISH):** FISH is a powerful molecular cytogenetic technique that uses fluorescently labeled probes to visualize specific DNA sequences on chromosomes. This technique enables the detection of chromosomal rearrangements, gene mapping, and the identification of specific genetic abnormalities. FISH involves the hybridization of fluorescent probes to complementary DNA sequences on chromosomes, allowing researchers to visualize the location and organization of genes and chromosomal regions. FISH is extensively utilized in clinical diagnostics, cancer research, and prenatal screening, offering crucial insights into the genetic foundations of diseases and chromosomal abnormalities. With its high sensitivity and specificity, it serves as an essential technique for analyzing complex chromosomal rearrangements and the structure of the genome.
3. **Comparative Genomic Hybridization (CGH):** CGH is a technique used to detect genomic imbalances, such as duplications and deletions, by comparing the genomic DNA of a test sample with a reference sample. This method provides a comprehensive view of the genome and is useful for identifying copy number variations associated with genetic disorders. CGH involves labeling the test and reference DNA samples with different fluorescent dyes, hybridizing them to a microarray or metaphase chromosomes, and analyzing the fluorescence ratio to identify regions of genomic gain or loss. CGH is extensively employed in cancer research, prenatal diagnostics, and the investigation of genomic disorders, offering valuable insights into the genetic underpinnings of various diseases and aiding in the advancement of personalized medicine. This technique enables the detection of submicroscopic chromosomal alterations that are often undetectable through traditional karyotyping methods.
4. **Chromosome Painting:** Chromosome painting involves the use of fluorescent probes that hybridize to entire chromosomes or specific chromosomal regions. This technique is used to study chromosomal rearrangements and to visualize complex karyotypes. Chromosome painting provides a comprehensive view of the entire genome, allowing researchers to identify structural changes, such as translocations, inversions, and insertions. It is particularly useful for studying chromosomal evolution, species differentiation, and the identification of cryptic rearrangements in complex genomes. Chromosome painting has applications in

cancer research, evolutionary biology, and comparative genomics, offering insights into the mechanisms of chromosomal change and genome organization.

5. **Spectral Karyotyping (SKY):** Spectral karyotyping is an advanced technique that uses multiple fluorescent dyes to label each chromosome with a unique color. This technique allows for the simultaneous visualization of all chromosomes in a cell, facilitating the detection of complex chromosomal rearrangements. SKY is particularly useful for identifying translocations, insertions, and other structural abnormalities that may not be apparent with traditional karyotyping. SKY provides a high-resolution perspective of the genome, allowing researchers to explore the genetic foundations of diseases, chromosomal evolution, and the consequences of genomic instability. It is widely applied in cancer research, genetic diagnostics, and evolutionary studies, offering valuable insights into the mechanisms of chromosomal rearrangements and their influence on genetic diversity and variation.

Cytogenetics in Agriculture

In agriculture, cytogenetics is crucial for advancing crop improvement by offering techniques to examine and modify plant chromosomal structures. By studying the chromosomal foundations of agronomic traits, breeders can create new crop varieties with superior characteristics. Cytogenetics provides essential knowledge of the genetic architecture of crops, enabling the identification of advantageous traits and the formulation of effective strategies for enhancing crop performance and productivity.

Applications of Cytogenetics in Crop Improvement

1. **Identification of Chromosomal Rearrangements:** Cytogenetics enables breeders to detect chromosomal rearrangements, including duplications, deletions, inversions, and translocations, which can significantly impact a crop's phenotype. By examining a plant's chromosomal structure, breeders can pinpoint valuable traits and select individuals with a more favorable genetic composition. Understanding chromosomal rearrangements is critical for unraveling the genetic basis of traits and developing targeted approaches to manipulate them in crop breeding programs. Cytogenetic analysis also offers insights into the genetic diversity of crop populations, allowing breeders to improve crop performance, increase resilience, and boost adaptability to various environmental conditions.

2. **Polyploidy and Genome Evolution:** Polyploidy, the condition of having more than two sets of chromosomes, is common in many crop species and plays a significant role in genome evolution. Cytogenetics provides tools to study polyploidy and its impact on crop diversity and adaptation. Polyploidy can lead to increased genetic diversity, enhanced stress tolerance, and improved agronomic traits, making it a valuable mechanism for crop improvement. Cytogenetic techniques, such as chromosome counting and karyotyping, enable researchers to study the effects of polyploidy on gene expression, genome stability, and evolutionary processes. The study of polyploidy contributes to our understanding of the mechanisms of speciation and adaptation in plants, offering new opportunities for developing polyploid crops with enhanced characteristics.

3. **Hybridization and Introgression:** Cytogenetics plays a pivotal role in studying hybridization and

introgression, processes through which genes from wild relatives are introduced into cultivated crop varieties. This incorporation boosts genetic diversity and introduces valuable traits like disease resistance and stress tolerance. Hybridization and introgression are fundamental tools for enhancing crops, as they allow breeders to integrate desirable characteristics from wild species into domesticated crops. Cytogenetic techniques, such as FISH and GISH (genomic in situ hybridization), are instrumental in analyzing the genomic composition of hybrids, enabling researchers to monitor trait inheritance and assess the impact of introgression on crop performance. The study of these processes is essential for developing new crop varieties with greater resilience and adaptability to changing environmental conditions.

4. **Chromosome Mapping and Gene Localization:** Cytogenetic techniques, like FISH, are crucial for gene mapping to precise chromosomal locations. This detailed information helps in identifying genes linked to desirable traits and supports the development of strategies to manipulate these genes effectively. Chromosome mapping plays a key role in understanding the genetic framework of complex traits, aiding in the identification of quantitative trait loci (QTLs) and candidate genes for further research. By pinpointing the exact location of genes on chromosomes, breeders can focus on specific genomic regions for trait enhancement, thereby improving the efficiency and accuracy of breeding programs. This approach is invaluable in studying traits such as yield, quality, and stress tolerance, and is fundamental in creating marker-assisted breeding strategies for crop improvement.
5. **Cytogenetic Markers in Breeding:** Cytogenetic markers, such as chromosome banding patterns and specific DNA sequences, are used in breeding programs to track the inheritance of traits and to select individuals with favorable genetic makeup. Cytogenetic markers provide a means to monitor the transmission of genetic material in breeding populations, enabling breeders to identify individuals with the desired genetic characteristics. The use of cytogenetic markers enhances the efficiency of breeding programs, reducing the time and resources required for selection and evaluation. Cytogenetic markers are essential tools for studying the genetic diversity of crop populations, assessing the effects of selection on genetic variation, and developing new crop varieties with improved performance and resilience.

Chromosomal Abnormalities and Their Impact

Cytogenetics provides insights into chromosomal abnormalities and their effects on phenotype. Chromosomal abnormalities can arise from errors in cell division, leading to genetic disorders and phenotypic variation. The study of chromosomal abnormalities is essential for understanding the genetic basis of diseases, developmental disorders, and evolutionary processes.

Types of Chromosomal Abnormalities

1. **Aneuploidy:** Aneuploidy refers to an abnormal number of chromosomes in a cell. It can result in genetic disorders, such as Down syndrome, which is caused by an extra copy of chromosome 21. Aneuploidy can arise from errors in chromosome segregation during meiosis or mitosis, leading to the gain or loss of chromosomes. In crops, aneuploidy can lead to reduced fertility, developmental abnormalities, and altered phenotypic traits. The study of aneuploidy provides insights into the mechanisms of chromosome segregation,

genomic instability, and the effects of chromosomal imbalances on phenotype. Cytogenetic techniques, such as karyotyping and FISH, are used to detect aneuploidy and to study its impact on genetic variation and adaptation.

2. **Structural Rearrangements:** Structural rearrangements, such as deletions, duplications, inversions, and translocations, involve changes in the structure of chromosomes. These rearrangements can disrupt gene function and lead to phenotypic variation. Structural rearrangements can arise from errors in DNA replication, repair, or recombination, leading to changes in chromosomal structure and organization. The study of structural rearrangements provides insights into the mechanisms of genomic rearrangement, gene regulation, and the evolution of chromosomal architecture. Cytogenetic techniques, such as FISH and chromosome painting, are used to detect structural rearrangements and to study their effects on gene expression, genetic diversity, and adaptation.
3. **Copy Number Variations (CNVs):** Copy number variations (CNVs) involve changes in the number of copies of a particular gene or genomic region. CNVs can influence gene expression and contribute to genetic diversity and adaptation. CNVs can arise from errors in DNA replication or recombination, leading to changes in gene dosage and genomic architecture. The study of CNVs provides insights into the mechanisms of gene regulation, genomic plasticity, and the effects of genetic variation on phenotype. Cytogenetic techniques, such as CGH and array-based technologies, are used to detect CNVs and to study their impact on genetic diversity, disease susceptibility, and evolutionary processes.
4. **Chromosome Fusions and Fissions:** Chromosome fusions and fissions involve the joining or splitting of chromosomes. These events can alter the genetic landscape and contribute to genome evolution. Chromosome fusions and fissions can arise from errors in DNA repair or recombination, leading to changes in chromosomal structure and organization. The study of chromosome fusions and fissions provides insights into the mechanisms of chromosomal evolution, speciation, and adaptation. Cytogenetic techniques, such as karyotyping and FISH, are used to detect chromosome fusions and fissions and to study their effects on genetic diversity, genome organization, and evolutionary processes. Cytogenetics provides tools to study these abnormalities and their impact on phenotype, contributing to our understanding of genetic diversity and adaptation in crops. The insights gained from cytogenetic research are essential for developing strategies to manage genetic disorders, improve crop performance, and enhance the resilience of agricultural systems.

Integration with Molecular Techniques

The integration of cytogenetics with molecular techniques has transformed the field, enabling more precise analysis and manipulation of chromosomes. This integration provides a comprehensive view of the genome, facilitating the study of the structural and functional organization of chromosomes and the identification of genomic variations associated with traits.

Integration in Crop Improvement

1. **Molecular Markers and Cytogenetics:** Molecular markers, such as single nucleotide

polymorphisms (SNPs) and simple sequence repeats (SSRs), are used in conjunction with cytogenetic techniques to study the genetic basis of traits and to track the inheritance of markers in breeding programs. The integration of molecular markers with cytogenetics enhances the resolution of genetic analyses, enabling researchers to identify and map genes associated with desirable traits. Molecular markers provide a means to monitor the transmission of genetic material in breeding populations, facilitating the selection of individuals with favorable genetic characteristics. This integration improves the efficiency and accuracy of breeding programs, contributing to the development of crop varieties with enhanced performance and resilience.

2. **Cytogenomic Approaches:** Cytogenomic approaches combine cytogenetics with genomic technologies, such as next-generation sequencing (NGS), to provide a comprehensive view of the genome. This integration allows researchers to study the structural and functional organization of chromosomes and to identify genomic variations associated with traits. Cytogenomic approaches provide insights into the genetic architecture of complex traits, facilitating the identification of quantitative trait loci (QTLs) and candidate genes for further study. The integration of cytogenetics with genomics enhances our understanding of the mechanisms of gene regulation, genome evolution, and adaptation, offering new opportunities for crop improvement and genetic research.
3. **Genome Editing Technologies:** Genome editing technologies, such as CRISPR-Cas9, offer new opportunities for manipulating chromosomes and genes. By precisely editing specific genomic regions, researchers can introduce or modify traits, enhancing crop improvement efforts. The integration of genome editing with cytogenetics provides tools for studying the effects of genetic modifications on chromosomal structure and function, enabling researchers to develop crop varieties with enhanced characteristics. Genome editing technologies offer solutions for addressing challenges such as food security, pest control, and environmental sustainability, contributing to the development of sustainable agricultural practices and the improvement of crop performance.

Cytogenetics, through its integration with molecular biology, continues to play a crucial role in advancing our understanding of genetics and improving crop varieties. The insights gained from cytogenetic research are essential for developing sustainable agricultural practices and ensuring food security in the face of global challenges. By leveraging the tools and techniques of cytogenetics, researchers and breeders can develop crop varieties that meet the demands of a growing global population and changing environmental conditions, contributing to the long-term sustainability and resilience of agriculture.

C. Molecular Cytogenetics

Molecular cytogenetics is an advanced and specialized field that combines the principles of cytogenetics and molecular biology to study the structure and function of chromosomes at the molecular level. This field represents a significant advancement in the study of chromosomes by providing tools to identify and map genes, study chromosomal rearrangements, and analyze complex genomes. Molecular cytogenetics has become an indispensable component of genetic research, offering insights that are critical for both basic science and practical applications, such as crop improvement. It allows researchers to visualize specific DNA sequences on chromosomes, providing a more detailed and comprehensive understanding of genetic material and its organization within the genome. This integration of cytogenetics with molecular biology has revolutionized the study of chromosomes and genomes, facilitating the identification of chromosomal abnormalities and the understanding of genome evolution.

Principles of Molecular Cytogenetics

Molecular cytogenetics involves the use of various molecular techniques to visualize specific DNA sequences on chromosomes and analyze their organization within the genome. These techniques integrate the study of chromosome structure and behavior with molecular biology, enabling researchers to investigate the genetic basis of phenotypic traits and diseases. The primary focus of molecular cytogenetics is to provide detailed insights into chromosomal architecture, identify genetic variations, and explore the functional implications of these variations.

Key Techniques in Molecular Cytogenetics

1. **Fluorescence in Situ Hybridization (FISH):**

    » **Principles:** FISH is a powerful technique that uses fluorescently labeled DNA probes to hybridize specific DNA sequences on chromosomes. This allows for the direct visualization of chromosomal regions of interest under a fluorescence microscope. FISH provides high-resolution mapping of chromosomal regions, enabling researchers to detect and analyze specific genetic sequences, gene loci, and chromosomal rearrangements.

    » **Applications:** FISH is widely used for gene mapping, identifying chromosomal rearrangements, and studying the organization of the genome. It is applicable in both basic research and clinical diagnostics, providing insights into genetic disorders and chromosomal abnormalities. In clinical diagnostics, FISH is used to detect genetic abnormalities associated with diseases such as cancer, Down syndrome, and other chromosomal disorders. In research, FISH is utilized to study the organization and function of complex genomes, explore evolutionary relationships, and investigate the genetic basis of phenotypic traits.

2. **Genomic In Situ Hybridization (GISH):**

    » **Principles:** GISH is a variation of FISH that uses labeled genomic DNA as a probe to differentiate between parental genomes in hybrids and polyploids. This technique is particularly useful for studying interspecific hybrids and polyploid species. GISH allows researchers to visualize the genomic composition of hybrids, identify parental contributions, and analyze the distribution of genetic material between species.

    » **Applications:** GISH is used to analyze the genomic composition of hybrids, identify parental contributions, and study the evolutionary relationships between species. It plays a crucial role in understanding polyploidy and its impact on genetic diversity. In plant breeding, GISH is utilized to study the genomic composition of hybrid species, assess gene flow between species, and investigate the effects of introgression on genetic diversity and adaptation. In evolutionary biology, GISH provides insights into the mechanisms of speciation and the evolutionary dynamics of polyploid genomes.

3. **Comparative Genomic Hybridization (CGH):**

   » **Principles:** CGH is a technique used to detect genomic imbalances, such as duplications and deletions, by comparing the genomic DNA of a test sample with a reference sample. This method provides a comprehensive view of the genome and is useful for identifying copy number variations. CGH involves the hybridization of differentially labeled test and reference DNA samples to a microarray or metaphase chromosomes, allowing researchers to detect genomic regions with altered copy numbers.

   » **Applications:** CGH is used in cancer research, genetic diagnostics, and the study of genomic disorders. It provides insights into the genetic basis of diseases and contributes to the development of personalized medicine. In cancer research, CGH is utilized to identify genomic alterations associated with tumorigenesis, assess tumor heterogeneity, and explore the genetic mechanisms underlying cancer progression. In clinical diagnostics, CGH is used to detect copy number variations associated with genetic disorders, such as developmental delay and congenital anomalies, providing valuable information for patient diagnosis and management.

4. **Spectral Karyotyping (SKY):**

   » **Principles:** SKY involves the use of multiple fluorescent dyes to label chromosomes, allowing for the visualization of each chromosome in a different color. This technique provides a comprehensive view of the entire genome and facilitates the identification of complex chromosomal rearrangements. SKY utilizes a combination of fluorescently labeled probes that hybridize to specific chromosomal regions, enabling researchers to visualize chromosomal aberrations with high resolution and accuracy.

   » **Applications:** SKY is used in cancer research, genetic diagnostics, and evolutionary studies. It provides insights into chromosomal abnormalities and contributes to the understanding of genome evolution. In cancer research, SKY is utilized to detect chromosomal rearrangements associated with cancer development, assess genomic instability, and explore the genetic basis of tumor progression. In evolutionary biology, SKY is used to study chromosomal evolution, investigate the mechanisms of speciation, and explore the genetic diversity of different species. In clinical diagnostics, SKY is used to identify chromosomal abnormalities associated with genetic disorders, providing valuable information for patient diagnosis and treatment.

Molecular Cytogenetics in Crop Improvement

In the context of crop improvement, molecular cytogenetics provides valuable insights into the genetic basis of important agronomic traits. It allows breeders to identify and map genes associated with traits such as disease resistance, yield, and stress tolerance, facilitating the development of improved crop varieties. Molecular cytogenetics offers tools for analyzing complex genomes, understanding the genetic architecture of traits, and exploring the evolutionary dynamics of crop species.

Applications in Crop Improvement

1. **Gene Mapping and Marker-Assisted Selection (MAS):**
   » Molecular cytogenetics provides tools for mapping genes to specific chromosomal locations, allowing breeders to identify genes associated with desirable traits. This information is essential for developing marker-assisted breeding strategies, which use molecular markers to select individuals with favorable genetic makeup. Gene mapping provides insights into the genetic architecture of complex traits, facilitating the identification of quantitative trait loci (QTLs) and candidate genes for further study. The integration of molecular cytogenetics with MAS enhances the efficiency and accuracy of breeding programs, enabling breeders to develop crop varieties with improved performance, quality, and resilience.

2. **Identification of Chromosomal Rearrangements:**
   » Molecular cytogenetic techniques, such as FISH, enable the detection of chromosomal rearrangements, such as duplications, deletions, inversions, and translocations. These rearrangements can have significant effects on the phenotype of a crop, and their identification is crucial for selecting individuals with improved genetic makeup. The identification of chromosomal rearrangements provides insights into the genetic basis of traits, facilitates the development of strategies to manipulate these traits, and contributes to the development of crop varieties with enhanced characteristics.

3. **Polyploidy and Genome Evolution:**
   » Polyploidy, the condition of having more than two sets of chromosomes, is common in many crop species and plays a significant role in genome evolution. Molecular cytogenetics provides tools to study polyploidy and its impact on genetic diversity and adaptation, contributing to the development of new crop varieties with enhanced characteristics. The study of polyploidy provides insights into the mechanisms of genome evolution, gene expression, and adaptation, offering new opportunities for developing polyploid crops with improved yield, quality, and resilience.

4. **Hybridization and Introgression:**
   » Molecular cytogenetics facilitates the study of hybridization and introgression, where genes from wild relatives are introduced into cultivated varieties. This process can enhance genetic diversity and introduce beneficial traits, such as disease resistance and stress tolerance. The study of hybridization and introgression provides insights into the genetic mechanisms underlying trait inheritance, gene flow, and adaptation, contributing to the development of crop varieties with enhanced resilience and adaptability.

5. **Cytogenetic Analysis of Complex Genomes:**
   » Molecular cytogenetics provides tools to analyze complex genomes, such as those of wheat and maize, which are characterized by large genome sizes and high levels of repetitive sequences. This analysis is essential for understanding the genetic basis of important traits and for developing strategies to manipulate these traits. The analysis of complex genomes

provides insights into the genetic architecture of traits, the dynamics of genome evolution, and the mechanisms of gene regulation, facilitating the development of strategies for crop improvement and genetic research.

Integration with Molecular Biology

The integration of molecular cytogenetics with molecular biology has revolutionized the study of chromosomes and genomes. This integration provides a comprehensive view of the genetic material and its organization, enabling researchers to study the structural and functional organization of chromosomes and to identify genomic variations associated with traits. The integration of molecular cytogenetics with molecular biology enhances our understanding of the genetic basis of phenotypic traits, diseases, and evolutionary processes, offering new opportunities for research and applications in agriculture and medicine.

Integration in Crop Improvement

1. **Molecular Markers and Cytogenetics:**
   » Molecular markers, including single nucleotide polymorphisms (SNPs) and simple sequence repeats (SSRs), are often combined with molecular cytogenetic techniques to investigate the genetic foundation of various traits and to monitor the inheritance of these markers in breeding programs. These markers serve as important tools for identifying genetic variation, allowing breeders to track specific genes or traits through generations. By integrating molecular markers with cytogenetic methods, researchers can gain deeper insights into the genetic mechanisms controlling traits, enhancing the precision and efficiency of breeding efforts aimed at developing improved crop varieties. The integration of molecular markers with cytogenetics enhances the resolution of genetic analyses, enabling researchers to identify and map genes associated with desirable traits. Molecular markers provide a means to monitor the transmission of genetic material in breeding populations, facilitating the selection of individuals with favorable genetic characteristics and improving the efficiency and accuracy of breeding programs.

2. **Functional Genomics:**
   » Functional genomics focuses on uncovering the roles of genes and their interactions within the entire genome. By analyzing gene function and regulatory mechanisms, researchers can devise methods to modify gene expression, leading to the enhancement of desirable crop traits. This field provides valuable insights into the molecular processes driving trait expression, gene networks, and regulatory pathways, which are crucial for both genetic research and crop improvement. When combined with molecular cytogenetics, functional genomics deepens our understanding of the genetic basis behind complex traits, thereby enabling the development of advanced strategies to improve crop performance, resilience, and adaptability to environmental challenges.

3. **Genome Editing Technologies:**
   » Genome editing technologies, such as CRISPR-Cas9, offer new opportunities for manipulating chromosomes and genes. By precisely editing specific genomic regions, researchers can introduce or modify traits, enhancing crop improvement efforts. Genome editing technologies provide tools for studying the effects of genetic modifications on chromosomal structure and function, enabling researchers to develop crop varieties with enhanced characteristics. The integration of genome editing with molecular cytogenetics offers solutions for addressing challenges such as food security, pest control, and environmental sustainability, contributing to the development of sustainable agricultural practices and the improvement of crop performance.

4. **Transcriptomics and Proteomics:**
   » Transcriptomics and proteomics provide insights into gene expression and protein function, contributing to the understanding of the molecular mechanisms underlying agronomic traits. This information is essential for developing strategies to manipulate gene expression and improve crop traits. The integration of transcriptomics and proteomics with molecular cytogenetics provides a comprehensive view of the molecular mechanisms underlying trait expression, gene regulation, and adaptation, offering new opportunities for research and applications in agriculture and medicine.

The Impact of Molecular Cytogenetics on Agriculture

Molecular cytogenetics has had a profound impact on agriculture by providing tools to study and manipulate the genetic makeup of crops. The insights gained from molecular cytogenetic research are essential for developing sustainable agricultural practices and ensuring food security in the face of global challenges. Molecular cytogenetics offers valuable tools for enhancing crop productivity, quality, and resilience, contributing to the development of crop varieties that meet the demands of a growing global population and changing environmental conditions.

Benefits of Molecular Cytogenetics in Agriculture

1. **Increased Productivity and Quality:** Molecular cytogenetics offers powerful tools for the development of new crop varieties with enhanced productivity and quality, playing a crucial role in addressing the increasing global demand for food and supporting food security efforts. By identifying and mapping genes linked to important traits such as yield, quality, and resilience, molecular cytogenetics enables breeders to develop crops with improved performance and adaptability to changing environments. This approach is instrumental in creating crop varieties that not only meet the demands of growing populations but also contribute to sustainable agricultural practices and enhanced food production systems.

2. **Improved Disease Resistance:** Molecular cytogenetics offers valuable insights into the genetic mechanisms of disease resistance, enabling breeders to create crop varieties with enhanced resistance to pathogens and pests. By identifying genes related to disease resistance and understanding the underlying genetic processes, molecular cytogenetics plays a key role

in developing crops with greater resilience. This approach not only helps protect crops from diseases and pests but also contributes to sustainable agricultural practices by reducing the need for chemical interventions and increasing crop productivity under challenging conditions.

3. **Enhanced Stress Tolerance:** Molecular cytogenetics offers essential tools for developing stress-tolerant crop varieties by revealing the genetic foundations of stress tolerance. This knowledge empowers breeders to create new crop varieties with heightened resilience to environmental stresses. By identifying genes linked to stress tolerance and deciphering the genetic mechanisms that enable plants to adapt, molecular cytogenetics plays a crucial role in enhancing crop resilience to abiotic stresses, including drought, heat, and salinity. This approach is vital for improving crop performance in challenging environments, ultimately supporting food security and sustainable agriculture.

4. **Biodiversity Conservation:** Molecular cytogenetics is essential in conserving biodiversity by offering tools to analyze the genetic diversity of crop species and their wild relatives. This information helps in the development of new varieties with enhanced traits while preserving the genetic diversity crucial for future breeding efforts. By exploring the genetic foundations of diversity and adaptation, molecular cytogenetics supports the conservation of genetic resources, ensuring that valuable traits are retained for future generations. This approach plays a pivotal role in maintaining biodiversity and promoting sustainable agricultural practices, safeguarding crop species against environmental challenges and fostering long-term agricultural productivity.

Molecular cytogenetics, through its integration with molecular biology, continues to drive advancements in crop improvement. The insights gained from molecular cytogenetic research are essential for developing sustainable agricultural practices and ensuring food security in the face of global challenges. By leveraging the tools and techniques of molecular cytogenetics, researchers and breeders can develop crop varieties that meet the demands of a growing global population and changing environmental conditions, contributing to the long-term sustainability and resilience of agriculture.

A. Cytogenomics

Cytogenomics is an emerging and interdisciplinary field that integrates cytogenetics with genomics to study the structure, function, and evolution of genomes. This field represents a convergence of cytogenetic and genomic methodologies, providing a comprehensive view of the genome and offering insights that are critical for understanding complex genetic processes and for practical applications, such as crop improvement. By utilizing high-throughput genomic technologies, cytogenomics allows researchers to analyze the organization and variation of genomes at the chromosomal level, facilitating the study of genetic diversity, chromosomal rearrangements, and evolutionary relationships between species. This integration of cytogenetics and genomics enables a more detailed and comprehensive understanding of the genome, contributing to advancements in genetic research, medical diagnostics, and agricultural biotechnology.

Principles of Cytogenomics

Cytogenomics combines the principles of cytogenetics, which focuses on the study of chromosomes, with the principles of genomics, which explores the entire genetic content of an organism. This integration allows for a more detailed and comprehensive understanding of the genome, facilitating the study of the structural and functional organization of chromosomes, the identification of genomic variations, and the exploration of evolutionary relationships between species. The primary goal of cytogenomics is to provide insights into the genetic architecture of complex traits, the mechanisms of genome evolution, and the functional implications of genomic variations.

Key Techniques in Cytogenomics

1. **Next-Generation Sequencing (NGS):**

   » **Principles:** NGS is a high-throughput sequencing technology that allows for the rapid and cost-effective sequencing of entire genomes. It provides a comprehensive view of the genetic material, enabling the identification of genetic variations, gene expression patterns, and regulatory elements. NGS utilizes parallel sequencing of millions of small DNA fragments, allowing researchers to obtain detailed and accurate genetic information on a genome-wide scale.

   » **Applications:** NGS is used in a wide range of applications, including genome sequencing, transcriptome analysis, and epigenomics. It provides insights into the genetic basis of traits and diseases and contributes to the development of personalized medicine. In agriculture, NGS is utilized to study the genetic diversity of crop species, identify genomic regions associated with important agronomic traits, and develop strategies for crop improvement. In medical research, NGS is used to identify genetic mutations associated with diseases, explore the genetic basis of cancer, and develop targeted therapies. In evolutionary biology, NGS provides insights into the mechanisms of speciation, adaptation, and genome evolution.

2. **Comparative Genomic Hybridization (CGH):**

   » **Principles:** CGH is a technique used to detect genomic imbalances, such as duplications and deletions, by comparing the genomic DNA of a test sample with a reference sample. This method provides a comprehensive view of the genome and is useful for identifying copy number variations. CGH involves the hybridization of differentially labeled test and reference DNA samples to a microarray or metaphase chromosomes, allowing researchers to detect genomic regions with altered copy numbers.

   » **Applications:** CGH is used in cancer research, genetic diagnostics, and the study of genomic disorders. It provides insights into the genetic basis of diseases and contributes to the development of personalized medicine. In cancer research, CGH is utilized to identify genomic alterations associated with tumorigenesis, assess tumor heterogeneity, and explore the genetic mechanisms underlying cancer progression. In clinical diagnostics, CGH is used to detect copy number variations associated with genetic disorders, such as developmental delay and congenital anomalies, providing valuable information for patient diagnosis and management. In agricultural research, CGH is used to identify genomic variations associated with agronomic traits, contributing to the development of improved crop varieties.

3. **Chromosome Conformation Capture (3C) Technologies:**

   » **Principles:** 3C technologies, such as Hi-C and Capture-C, are used to study the three-dimensional organization of the genome. These techniques provide insights into the spatial arrangement of chromosomes and the interactions between genomic regions. 3C technologies involve the crosslinking of chromatin, digestion of DNA, and ligation of interacting chromatin fragments, followed by sequencing to identify physical interactions between genomic regions.

   » **Applications:** 3C technologies are used to study gene regulation, chromatin organization, and the role of structural variations in disease. They provide insights into the functional architecture of the genome and contribute to our understanding of genome biology. In medical research, 3C technologies are used to explore the role of chromatin interactions in gene regulation, identify regulatory elements associated with diseases, and investigate the effects of structural variations on genome function. In agricultural research, 3C technologies provide insights into the regulation of gene expression, the effects of chromatin organization on trait expression, and the mechanisms of genome evolution.

4. **Single-Cell Genomics:**

   » **Principles:** Single-cell genomics focuses on analyzing individual cells to explore the genetic and phenotypic variations within a population. This method reveals valuable insights into the diversity among cells and the dynamic nature of cellular processes. Techniques like single-cell RNA sequencing (scRNA-seq) are employed to gather detailed information about gene expression, genetic differences, and cellular functions on a cell-by-cell basis. Single-cell genomics allows researchers to study cellular complexity, helping to uncover the roles of specific cells in various biological processes, diseases, and developmental stages.

   » **Applications:** Single-cell genomics is widely applied in cancer research, developmental biology, and immunology, offering valuable insights into the genetic foundations of diseases and supporting the advancement of personalized medicine. In cancer research, it is utilized to examine tumor heterogeneity, identify cancer stem cells, and investigate the mechanisms behind tumor progression and resistance to treatments. In developmental biology, single-cell genomics enhances our understanding of cell differentiation, lineage specification, and the processes involved in tissue formation. In immunology, it is used to explore the diversity of immune cell populations, unravel the mechanisms governing immune responses, and identify biomarkers for immune-related diseases. This technique is essential for identifying molecular signatures and tailoring therapeutic strategies in various medical fields.

Cytogenomics in Agriculture

In agriculture, cytogenomics offers new opportunities for crop improvement by providing tools to analyze the genetic diversity of crop species and their wild relatives. It enables breeders to identify genomic regions associated with important agronomic traits and to develop new varieties with enhanced characteristics. Cytogenomics provides insights into the genetic architecture of complex traits, the mechanisms of genome evolution, and the functional implications of genomic variations, contributing to the development of strategies for crop improvement and genetic research.

Applications in Crop Improvement

1. **Genetic Diversity and Conservation:** Cytogenomics offers essential tools for evaluating and preserving genetic diversity in crop species and their wild relatives, which serve as a valuable reservoir of genes for improving crop resilience against pests, diseases, and environmental stresses. Assessing this genetic diversity is crucial for conserving genetic resources and promoting sustainable agricultural practices. By exploring the genetic foundations of diversity and adaptation, cytogenomics plays a vital role in safeguarding genetic resources and shaping strategies for crop enhancement. This approach not only preserves biodiversity but also ensures the development of robust crop varieties that can thrive under changing environmental conditions.
2. **Identification of Genomic Variations:** Cytogenomics facilitates the identification of genomic variations, including single nucleotide polymorphisms (SNPs), copy number variations (CNVs), and structural variations, which play a significant role in phenotypic diversity and adaptation. These variations serve as crucial resources for developing marker-assisted breeding strategies and for pinpointing candidate genes for more in-depth research. Understanding these genomic variations provides valuable insights into the genetic architecture of complex traits, the mechanisms underlying adaptation, and the functional impacts of genetic differences. This knowledge is instrumental in advancing strategies for crop improvement and broadening the scope of genetic research, ultimately enhancing agricultural productivity and resilience.
3. **Functional Genomics and Gene Annotation:** Cytogenomics offers valuable insights into the functional roles of genes and their interactions across the genome. By studying gene function and regulation, researchers can devise strategies to manipulate gene expression and enhance crop traits. This integration with functional genomics provides a deeper understanding of the molecular mechanisms involved in trait expression, gene networks, and regulatory pathways, which are essential for improving crops. The combination of functional genomics and cytogenomics enriches our knowledge of the genetic foundations of complex traits, supporting the development of targeted strategies for improving crop performance, resilience, and adaptability in changing environmental conditions.
4. **Genome-Wide Association Studies (GWAS):** Genome-Wide Association Studies (GWAS) is a highly effective method for identifying genetic loci linked to complex traits. By examining genetic variations across a population, GWAS allows researchers to pinpoint genes associated with important traits like yield, disease resistance, and stress tolerance, aiding in the development of enhanced crop varieties. This approach provides valuable insights into the genetic architecture of complex traits, the molecular mechanisms driving trait expression, and the functional consequences of genetic variation. These findings contribute significantly to the development of innovative strategies for crop improvement and genetic research, enabling breeders to create more resilient and productive crop varieties.
5. **Polyploidy and Genome Evolution:** Cytogenomics offers essential tools for investigating polyploidy, a widespread phenomenon in many crop species, and its influence on genetic diversity and adaptation. Understanding polyploidy is critical for uncovering the genetic foundations of traits and for developing methods to manipulate these traits in crop breeding.

Studying polyploidy reveals key insights into genome evolution, gene expression, and the mechanisms of adaptation, opening up opportunities to develop polyploid crops with enhanced yield, quality, and resilience. This research is pivotal for advancing agricultural productivity and creating crops that are better suited to environmental stresses and changing climatic conditions.

Integration with Genomic Technologies

The integration of cytogenomics with genomic technologies has transformed the field, enabling more precise analysis and manipulation of chromosomes and genomes. This integration provides a comprehensive view of the genetic material and its organization, facilitating the study of the structural and functional organization of chromosomes and the identification of genomic variations associated with traits. The integration of cytogenomics with genomic technologies enhances our understanding of the genetic basis of phenotypic traits, diseases, and evolutionary processes, offering new opportunities for research and applications in agriculture and medicine.

Integration in Crop Improvement

1. **Genomic Selection and Prediction:** Genomic selection involves using genome-wide information to predict the breeding value of individuals. This approach allows breeders to select the best individuals for breeding based on their genetic potential, improving the efficiency and accuracy of breeding programs. Genomic selection provides insights into the genetic architecture of complex traits, the mechanisms of trait expression, and the functional implications of genetic variation, contributing to the development of strategies for crop improvement and genetic research.
2. **Transcriptomics and Proteomics:** Transcriptomics and proteomics provide insights into gene expression and protein function, contributing to the understanding of the molecular mechanisms underlying agronomic traits. This information is essential for developing strategies to manipulate gene expression and improve crop traits. The integration of transcriptomics and proteomics with cytogenomics provides a comprehensive view of the molecular mechanisms underlying trait expression, gene regulation, and adaptation, offering new opportunities for research and applications in agriculture and medicine.
3. **Epigenomics and Epigenetic Regulation:** Epigenomics investigates how epigenetic modifications, such as DNA methylation and histone modifications, influence gene expression. This information is vital for understanding the molecular mechanisms that govern development and adaptation, as well as for devising strategies to regulate gene expression. When integrated with cytogenomics, epigenomics offers deeper insights into the regulation of gene expression, the role of epigenetic changes in trait expression, and the processes of adaptation. This combined approach enhances the development of targeted strategies for crop improvement and genetic research, enabling breeders to manipulate gene expression for better yield, quality, and resilience in crops.
4. **Systems Biology and Network Analysis:** Systems biology involves the study of complex biological systems and their interactions. Network analysis provides insights into the functional

interactions between genes, proteins, and other molecules, contributing to our understanding of genome biology and the development of new strategies for crop improvement. The integration of systems biology with cytogenomics provides insights into the regulation of gene expression, the effects of network interactions on trait expression, and the mechanisms of adaptation, contributing to the development of strategies for crop improvement and genetic research.

The Impact of Cytogenomics on Agriculture

Cytogenomics has had a profound impact on agriculture by providing tools to study and manipulate the genetic makeup of crops. The insights gained from cytogenomic research are essential for developing sustainable agricultural practices and ensuring food security in the face of global challenges. Cytogenomics offers valuable tools for enhancing crop productivity, quality, and resilience, contributing to the development of crop varieties that meet the demands of a growing global population and changing environmental conditions.

Benefits of Cytogenomics in Agriculture

1. **Increased Productivity and Quality:** Cytogenomics provides tools to develop new crop varieties with increased productivity and quality, helping to meet the growing demand for food and ensure global food security. By identifying and mapping genes associated with yield, quality, and resilience, cytogenomics facilitates the development of crop varieties with improved performance and adaptability.
2. **Improved Disease Resistance:** Cytogenomics offers valuable insights into the genetic foundations of disease resistance, empowering breeders to create new crop varieties with stronger resistance to pathogens and pests. By identifying key genes linked to disease resistance and unraveling the genetic mechanisms that drive these defenses, cytogenomics plays a crucial role in developing crops with enhanced resilience. This knowledge not only helps in safeguarding crops from diseases and pest infestations but also supports sustainable agricultural practices by reducing the need for chemical interventions, ultimately leading to more robust and productive crop varieties.
3. **Enhanced Stress Tolerance:** Cytogenomics provides tools to develop stress-tolerant crop varieties by providing insights into the genetic basis of stress tolerance, enabling breeders to develop new varieties with enhanced resilience to environmental stresses. By identifying genes associated with stress tolerance and understanding the genetic mechanisms underlying adaptation, cytogenomics contributes to the development of crop varieties with improved resilience to abiotic stresses, such as drought, heat, and salinity.
4. **Biodiversity Conservation:** Cytogenomics plays a pivotal role in conserving biodiversity by offering tools to examine the genetic diversity of crop species and their wild relatives. This information is essential for developing new crop varieties with enhanced traits while preserving the genetic diversity needed for future generations. The genetic foundations of diversity and adaptation, cytogenomics supports the conservation of valuable genetic resources and helps establish sustainable agricultural practices.

Through its integration with advanced genomic technologies, cytogenomics continues to be a driving force in crop improvement. The insights derived from cytogenomic research are crucial for shaping sustainable agricultural strategies, particularly in the context of global challenges such as climate change and food security. By utilizing cytogenomics, researchers and breeders are able to develop crop varieties that address the growing demands of a global population and adapt to shifting environmental conditions, ensuring the long-term sustainability and resilience of agricultural systems.

E. Why Cytogenetics and Cytogenomics?

Cytogenetics and cytogenomics have become indispensable fields in modern biology and agriculture due to their profound impact on understanding and manipulating the genetic architecture of organisms. These fields focus on the study of chromosomes and genomes, providing valuable insights into the structural and functional organization of genetic material. The integration of cytogenetics and cytogenomics has revolutionized genetic research, offering advanced tools and methodologies for analyzing genomes at a chromosomal level. This convergence has significantly contributed to crop improvement by facilitating the identification of genomic regions associated with important agronomic traits, the development of improved crop varieties, and the enhancement of crop resilience to environmental challenges.

Advancements in Understanding Chromosomal Structure and Function

Cytogenetics and cytogenomics have provided significant advancements in understanding the structure and function of chromosomes, which are critical for crop improvement. These fields offer insights into the organization of genetic material within the genome and the mechanisms that control gene expression and regulation. The exploration of chromosomal structure and function has been fundamental in unraveling the complexities of genome organization and the dynamic behavior of genetic material. By analyzing chromosomal architecture, researchers are able to understand the mechanisms that regulate genome stability, gene expression, and the process of inheritance. This knowledge is key to advancing fields such as genetics, molecular biology, and breeding, as it sheds light on how genetic information is organized, maintained, and transmitted across generations, while also contributing to the development of strategies for improving crop traits and managing genetic disorders.

Key Insights from Chromosomal Studies

1. **Chromosome Organization and Structure:** Chromosomes are highly organized structures that contain the genetic material of an organism. They are composed of DNA and proteins, forming a complex structure known as chromatin. The organization of chromosomes is essential for the proper functioning and regulation of genes. Chromosomes are divided into distinct regions, including telomeres and centromeres, which play critical roles in maintaining chromosome stability and ensuring accurate segregation during cell division. Telomeres are repetitive DNA sequences located at the ends of chromosomes, protecting them from degradation and preventing the loss of genetic information during replication. Centromeres are specialized regions that serve as attachment points for spindle fibers during cell division, ensuring the accurate segregation of chromosomes into daughter cells. The study of chromosomal organization has provided insights into the mechanisms that control genome stability and inheritance, contributing to our understanding of genetic disorders

and chromosomal abnormalities.

2. **Gene Expression and Regulation:** The organization of genetic material within chromosomes is closely linked to the regulation of gene expression. Chromosomal regions are organized into euchromatin and heterochromatin, with euchromatin being more accessible and transcriptionally active, while heterochromatin is densely packed and transcriptionally inactive. The regulation of gene expression is controlled by various factors, including chromatin structure, transcription factors, and epigenetic modifications. Cytogenetics and cytogenomics have provided tools to study the spatial and temporal regulation of gene expression, offering insights into the mechanisms that control developmental processes, cellular differentiation, and responses to environmental stimuli. Understanding the regulation of gene expression is essential for identifying key genomic regions that contribute to important agronomic traits and for developing strategies to manipulate these traits in crop improvement.
3. **Chromosomal Rearrangements:** Chromosomal rearrangements, such as duplications, deletions, inversions, and translocations, can have significant effects on the phenotype of an organism. These rearrangements can disrupt gene function, alter gene expression, and create new gene combinations, contributing to genetic diversity and adaptation. The study of chromosomal rearrangements has provided insights into the mechanisms of genome evolution, the genetic basis of phenotypic variation, and the development of genetic disorders. Cytogenetics and cytogenomics offer tools to analyze chromosomal rearrangements and their effects on the phenotype of crops, enabling breeders to identify desirable traits and select individuals with improved genetic makeup. The identification of chromosomal rearrangements is crucial for understanding the genetic architecture of complex traits and for developing strategies to manipulate these traits in crop improvement programs.
4. **Polyploidy and Genome Evolution:** Polyploidy, the condition of having more than two sets of chromosomes, is a common phenomenon in plants and plays a significant role in genome evolution. Polyploidy can lead to increased genetic diversity, enhanced adaptability, and the development of new species. Cytogenetics and cytogenomics provide tools to study polyploidy and its impact on genetic diversity, genome organization, and trait expression. The study of polyploidy has provided insights into the mechanisms of genome duplication, gene dosage effects, and the evolution of genetic pathways. Understanding polyploidy is essential for developing strategies to enhance crop resilience, improve yield, and exploit genetic diversity for crop improvement.

Integration of Molecular Techniques

The integration of molecular techniques, such as fluorescence in situ hybridization (FISH) and next-generation sequencing (NGS), with cytogenetics has revolutionized the study of chromosomes and genomes. These techniques provide tools to analyze the organization and variation of genomes at the chromosomal level, enabling researchers to identify genomic regions associated with important agronomic traits. The combination of molecular and cytogenetic methodologies has facilitated the development of new approaches for studying complex genomes and understanding the genetic basis of phenotypic traits.

Key Molecular Techniques in Cytogenetics and Cytogenomics

1. **Fluorescence In Situ Hybridization (FISH):**

   » **Principles:** FISH is a molecular technique that uses fluorescently labeled DNA probes to hybridize specific DNA sequences on chromosomes. This allows for the direct visualization of chromosomal regions of interest under a fluorescence microscope. FISH provides high-resolution mapping of chromosomal regions, enabling researchers to detect and analyze specific genetic sequences, gene loci, and chromosomal rearrangements.

   » **Applications:** FISH is widely used for gene mapping, identifying chromosomal rearrangements, and studying the organization of the genome. It is applicable in both basic research and clinical diagnostics, providing insights into genetic disorders and chromosomal abnormalities. In crop improvement, FISH is used to identify and map genes associated with desirable traits, such as disease resistance, yield, and stress tolerance. By visualizing specific DNA sequences on chromosomes, FISH provides insights into the organization of the genome and the identification of chromosomal rearrangements that may impact crop performance.

2. **Next-Generation Sequencing (NGS):**

   » **Principles:** NGS is a high-throughput sequencing technology that allows for the rapid and cost-effective sequencing of entire genomes. It provides a comprehensive view of the genetic material, enabling the identification of genetic variations, gene expression patterns, and regulatory elements. NGS utilizes parallel sequencing of millions of small DNA fragments, allowing researchers to obtain detailed and accurate genetic information on a genome-wide scale.

   » **Applications:** NGS is used in a wide range of applications, including genome sequencing, transcriptome analysis, and epigenomics. It provides insights into the genetic basis of traits and diseases and contributes to the development of personalized medicine. In agriculture, NGS is utilized to study the genetic diversity of crop species, identify genomic regions associated with important agronomic traits, and develop strategies for crop improvement. By providing high-resolution sequencing data, NGS enables researchers to analyze the genetic diversity of crop species and their wild relatives, facilitating the development of marker-assisted breeding strategies and the identification of candidate genes for further study.

3. **Comparative Genomic Hybridization (CGH):**

   » **Principles:** CGH is a technique used to detect genomic imbalances, such as duplications and deletions, by comparing the genomic DNA of a test sample with a reference sample. This method provides a comprehensive view of the genome and is useful for identifying copy number variations. CGH involves the hybridization of differentially labeled test and reference DNA samples to a microarray or metaphase chromosomes, allowing researchers to detect genomic regions with altered copy numbers.

   » **Applications:** CGH is used in cancer research, genetic diagnostics, and the study of genomic disorders. It provides insights into the genetic basis of diseases and contributes to the development of personalized medicine. In agricultural research, CGH is used to identify genomic variations associated with agronomic traits, contributing to the development of improved crop varieties. By detecting genomic imbalances, CGH provides insights into the genetic architecture of complex traits and the functional implications of copy number variations in crop improvement.

4. **Chromosome Conformation Capture (3C) Technologies:**
   - **Principles:** 3C technologies, such as Hi-C and Capture-C, are used to study the three-dimensional organization of the genome. These techniques provide insights into the spatial arrangement of chromosomes and the interactions between genomic regions. 3C technologies involve the crosslinking of chromatin, digestion of DNA, and ligation of interacting chromatin fragments, followed by sequencing to identify physical interactions between genomic regions.
   - **Applications:** 3C technologies are used to study gene regulation, chromatin organization, and the role of structural variations in disease. They provide insights into the functional architecture of the genome and contribute to our understanding of genome biology. In agricultural research, 3C technologies provide insights into the regulation of gene expression, the effects of chromatin organization on trait expression, and the mechanisms of genome evolution. By studying the three-dimensional organization of the genome, 3C technologies enable researchers to explore the functional interactions between genes and regulatory elements, facilitating the development of strategies for crop improvement.

Applications in Crop Improvement

Cytogenetics and cytogenomics have numerous applications in crop improvement, offering tools to study and manipulate the genetic makeup of crops. These fields provide valuable insights into the genetic basis of important agronomic traits, such as disease resistance, yield, and stress tolerance. By analyzing the chromosomal structure and composition of plants, breeders can identify key genomic regions associated with desirable traits and select individuals with improved genetic makeup. This knowledge can be used to develop new crop varieties with enhanced characteristics, such as increased productivity, resistance to diseases, and tolerance to environmental stresses.

Key Applications in Crop Improvement

1. **Identification of Desirable Traits:** Cytogenetics and cytogenomics provide tools to identify key genomic regions associated with desirable traits, such as yield, disease resistance, and stress tolerance. By analyzing the chromosomal structure and composition of plants, breeders can identify genomic regions that contribute to important agronomic traits and select individuals with improved genetic makeup. The identification of desirable traits is essential for developing new crop varieties with enhanced characteristics, contributing to increased productivity and resilience to environmental challenges.
2. **Marker-Assisted Breeding:** Marker-assisted breeding is a powerful approach that uses molecular markers to select individuals with favorable genetic makeup. Cytogenetics and cytogenomics provide tools to identify and map genes associated with desirable traits,

facilitating the development of marker-assisted breeding strategies. By using molecular markers to track the inheritance of desirable traits, breeders can accelerate the breeding process and develop crop varieties with improved performance and adaptability. Marker-assisted breeding is widely used in agriculture to enhance traits such as yield, quality, and resilience, contributing to the development of sustainable agricultural practices.

3. **Polyploidy and Genome Evolution:** Polyploidy is a common phenomenon in many crop species and plays a significant role in genome evolution. Cytogenetics and cytogenomics provide tools to study polyploidy and its impact on genetic diversity, genome organization, and trait expression. The study of polyploidy provides insights into the mechanisms of genome duplication, gene dosage effects, and the evolution of genetic pathways. Understanding polyploidy is essential for developing strategies to enhance crop resilience, improve yield, and exploit genetic diversity for crop improvement.
4. **Hybridization and Introgression:** Hybridization and introgression involve the transfer of genes from wild relatives into cultivated varieties, enhancing genetic diversity and introducing beneficial traits. Cytogenetics and cytogenomics facilitate the study of hybridization and introgression, providing insights into the genetic mechanisms underlying trait inheritance, gene flow, and adaptation. The study of hybridization and introgression contributes to the development of crop varieties with enhanced resilience and adaptability, offering new opportunities for improving crop performance and addressing the challenges of food security and environmental sustainability.
5. **Analysis of Complex Genomes:** Many crop species have complex genomes characterized by large genome sizes and high levels of repetitive sequences. Cytogenetics and cytogenomics provide tools to analyze complex genomes, facilitating the study of genetic diversity, chromosomal rearrangements, and evolutionary relationships. The analysis of complex genomes provides insights into the genetic architecture of important traits and the functional implications of genomic variations, contributing to the development of strategies for crop improvement and genetic research.

Cytogenetics and cytogenomics have become essential fields in modern biology and agriculture, providing valuable insights into the genetic architecture of organisms and offering advanced tools for genome analysis and manipulation. These fields have revolutionized genetic research, contributing to advancements in crop improvement, medical diagnostics, and evolutionary biology. By integrating cytogenetic and genomic methodologies, researchers can gain a comprehensive understanding of the genome, facilitating the identification of genomic regions associated with important traits and the development of improved crop varieties. The insights gained from cytogenetic and cytogenomic research are essential for developing sustainable agricultural practices, enhancing crop resilience, and ensuring food security in the face of global challenges.

A. What Are the Needs in the Present Scenario?

In the present scenario, the fields of cytogenetics and cytogenomics play a critical role in addressing some of the most pressing challenges facing agriculture and food security. These challenges include the need to increase food production, combat the impacts of climate change, and enhance the productivity and quality of crops. As the global population continues to grow and environmental conditions become more unpredictable, there is an urgent need for innovative approaches to crop improvement that can meet the demands of a changing world. Cytogenetics and cytogenomics offer powerful tools and insights that are essential for developing sustainable agricultural practices and ensuring global food security.

Addressing Global Food Security

The global population is projected to reach 9.7 billion by 2050, creating a pressing need for increased food production to ensure global food security. This population growth, combined with limited arable land and natural resources, poses significant challenges for the agricultural sector. To meet the growing demand for food, it is essential to develop new crop varieties with enhanced productivity, resilience, and nutritional quality.

The Role of Cytogenetics and Cytogenomics

1. **Enhancing Crop Yield:** Cytogenetics and cytogenomics play a crucial role in understanding the genetic foundations of key agronomic traits, enabling breeders to develop new crop varieties with enhanced yield potential. By examining the chromosomal structure and composition of plants, researchers can identify significant genomic regions associated with traits such as yield, plant height, and grain size. This knowledge is instrumental in creating high-yielding crop varieties, which are essential for boosting food production and ensuring global food security.

    Marker-assisted selection (MAS) and genomic selection (GS) are advanced breeding approaches that leverage molecular markers and genomic data to streamline the process of developing improved crop varieties. These methods allow breeders to rapidly identify and select plants with desirable traits, such as higher yields, reducing the time and resources typically needed in traditional breeding methods. By integrating cytogenetics with MAS and GS, breeders can accelerate the development of high-performing crop varieties that are more adaptable to environmental challenges and meet the demands of a growing population.

2. **Improving Nutritional Quality:** In addition to boosting crop yields, improving the nutritional quality of crops is essential to meet the dietary needs of an expanding global population. Cytogenetics and cytogenomics offer valuable tools for identifying and mapping genes linked to nutritional traits, such as vitamin and mineral content, protein quality, and biofortification. By studying the genetic basis of these traits, breeders can develop crop varieties with enhanced nutritional profiles, which can significantly contribute to better health and nutrition for consumers.

    Genetic engineering and biotechnology further expand the possibilities for improving the nutritional quality of crops. These advanced techniques enable the creation of biofortified varieties with increased levels of essential nutrients, such as iron, zinc, and vitamins, which are vital for addressing malnutrition and improving public health. By integrating cytogenetics with genetic engineering, researchers can develop crops that not only meet the caloric needs of a growing population but also provide enhanced nutritional benefits, supporting overall well-being and reducing nutrient deficiencies globally.

3. **Addressing Pest and Disease Resistance:** Pests and diseases are major threats to crop production and food security. Cytogenetics and cytogenomics offer valuable tools for developing pest- and disease-resistant crop varieties by providing insights into the genetic basis of resistance traits. By identifying and mapping genes associated with resistance to specific pests and pathogens, breeders can develop crop varieties with enhanced resistance, reducing the reliance on chemical pesticides and improving crop productivity. Integrated pest management (IPM) strategies that combine genetic resistance with cultural and biological control methods can further enhance the resilience of crops to pest and disease pressures.

4. **Preserving Genetic Diversity:** Genetic diversity is a key component of food security, as it provides the raw material for crop improvement and adaptation to changing environmental conditions. Cytogenetics and cytogenomics play a crucial role in preserving genetic diversity by providing tools to assess and conserve the genetic resources of crop species and their wild relatives. By understanding the genetic diversity of crop populations, researchers can develop strategies for the conservation and utilization of genetic resources, ensuring the long-term sustainability of agricultural systems.

Combating Climate Change

Climate change poses significant challenges to agriculture, as it leads to increased temperatures, altered precipitation patterns, and the frequency and intensity of extreme weather events. These changes threaten crop productivity and necessitate the development of resilient crop varieties that can withstand environmental stresses. The impacts of climate change on agriculture are multifaceted, affecting crop growth, yield, and quality, as well as the prevalence of pests and diseases.

The Role of Cytogenetics and Cytogenomics

1. **Developing Climate-Resilient Varieties:** Cytogenetics and cytogenomics offer tools to develop climate-resilient crop varieties by providing insights into the genetic basis of stress tolerance. By analyzing the chromosomal structure and composition of plants, researchers can identify genomic regions associated with tolerance to abiotic stresses, such as drought, heat, and salinity. This knowledge facilitates the development of crop varieties with enhanced resilience to environmental stresses, contributing to stable and sustainable agricultural production in the face of climate change. Genomic selection and marker-assisted breeding are valuable approaches for developing climate-resilient varieties, enabling breeders to select individuals with favorable genetic profiles for stress tolerance.
2. **Enhancing Water and Nutrient Use Efficiency:** Water and nutrient use efficiency are critical traits for adapting to the challenges of climate change and resource scarcity. Cytogenetics and cytogenomics provide insights into the genetic basis of these traits, enabling breeders to develop crop varieties that use water and nutrients more efficiently. By improving water and nutrient use efficiency, crops can maintain productivity under suboptimal conditions, reducing the need for irrigation and fertilizers and minimizing environmental impacts. The integration of genomic technologies with precision agriculture practices can further enhance resource use efficiency, optimizing inputs and improving sustainability.

3. **Understanding Adaptation and Evolution:** Climate change drives evolutionary processes and adaptation in natural and agricultural systems. Cytogenetics and cytogenomics offer tools to study the mechanisms of adaptation and evolution, providing insights into how crops respond to changing environmental conditions. By understanding the genetic basis of adaptation, researchers can develop strategies to enhance the adaptability of crops to climate change, ensuring their continued productivity and resilience. The study of polyploidy, hybridization, and introgression contributes to our understanding of adaptation and evolution, offering new opportunities for crop improvement and genetic research.
4. **Predicting and Mitigating Climate Impacts:** Predicting and mitigating the impacts of climate change on agriculture requires a comprehensive understanding of the interactions between genetic, environmental, and management factors. Cytogenetics and cytogenomics provide valuable tools for modeling these interactions, enabling researchers to predict the effects of climate change on crop performance and develop strategies for mitigation. By integrating genetic information with climate models and agronomic practices, researchers can develop adaptive management strategies that enhance the resilience of agricultural systems to climate change.

Enhancing Crop Productivity and Quality

The demand for increased crop productivity and quality is driven by the need to ensure global food security and meet the nutritional needs of the growing population. Crop improvement is essential for achieving these goals, as it provides tools to develop new varieties with enhanced productivity and quality. Enhancing crop productivity and quality is a complex challenge that requires a multidisciplinary approach, integrating genetic, agronomic, and technological innovations.

The Role of Cytogenetics and Cytogenomics

1. **Improving Yield and Biomass:** Cytogenetics and cytogenomics offer valuable tools for enhancing crop productivity by providing insights into the genetic basis of yield-related traits. By identifying and mapping genes associated with yield components, such as grain size, kernel number, and biomass accumulation, breeders can develop crop varieties with improved yield potential. The integration of genomic technologies with phenotypic selection and agronomic management practices can further enhance yield and productivity, contributing to sustainable agricultural production.
2. **Enhancing Quality Traits:** Crop quality is determined by a range of traits, including nutritional content, flavor, texture, and appearance. Cytogenetics and cytogenomics provide tools to identify and map genes associated with quality traits, enabling breeders to develop crop varieties with enhanced quality attributes. By understanding the genetic basis of quality traits, researchers can develop strategies to improve the sensory and nutritional properties of crops, contributing to consumer satisfaction and marketability.
3. **Accelerating Breeding Programs:** The integration of cytogenetics and cytogenomics with modern breeding technologies accelerates the development of new crop varieties by reducing the time and resources required for traditional breeding methods. Marker-assisted selection,

genomic selection, and gene editing are powerful tools that enable breeders to select individuals with favorable genetic profiles, enhancing the efficiency and accuracy of breeding programs. The use of genomic technologies in breeding programs facilitates the rapid development of improved crop varieties that meet the demands of a changing world.

4. **Leveraging Genetic Resources:** The utilization of genetic resources is essential for enhancing crop productivity and quality. Cytogenetics and cytogenomics provide tools to assess and utilize the genetic diversity of crop species and their wild relatives, offering a rich source of genetic variation for crop improvement. By leveraging genetic resources, breeders can develop crop varieties with improved performance, resilience, and adaptability, ensuring the sustainability and resilience of agricultural systems.

G. Its Role and Benefits in Crop Improvement

Cytogenetics and cytogenomics have become fundamental components of crop improvement strategies, providing a deeper understanding of the genetic architecture of plants and enabling the development of improved crop varieties. By studying the structure and function of chromosomes and genomes, these fields offer invaluable insights into the genetic basis of important agronomic traits, such as yield, quality, disease resistance, and stress tolerance. The integration of cytogenetic and genomic methodologies allows researchers and breeders to identify and manipulate key genomic regions associated with desirable traits, leading to the creation of crop varieties that are more productive, resilient, and adaptable to changing environmental conditions.

Role of Cytogenetics and Cytogenomics in Crop Improvement

Cytogenetics and cytogenomics play a crucial role in crop improvement by providing the tools and methodologies necessary to explore and manipulate the genetic makeup of crops. These fields offer comprehensive insights into the organization and function of genetic material, facilitating the identification of genes and genomic regions associated with important agronomic traits. By leveraging the knowledge gained from cytogenetic and genomic studies, researchers can develop new crop varieties with enhanced characteristics, addressing the challenges posed by population growth, climate change, and food security.

Key Roles in Crop Improvement

1. **Understanding Chromosomal Structure and Function:** Cytogenetics and cytogenomics provide detailed insights into the structure and function of chromosomes, revealing the organization of genetic material within the genome. By studying chromosomal architecture, researchers can identify key genomic regions that contribute to important traits, such as yield, quality, and resilience. The understanding of chromosomal structure is essential for elucidating the mechanisms that control gene expression and regulation, enabling the development of strategies to manipulate these processes for crop improvement. Techniques such as karyotyping, fluorescence in situ hybridization (FISH), and chromosome painting are used to visualize and analyze chromosomes, providing a detailed understanding of chromosomal structure and function.

2. **Gene Identification and Mapping:** Cytogenetics and cytogenomics offer tools to identify and map genes associated with desirable traits, facilitating the development of improved crop varieties. By integrating molecular techniques such as FISH and next-generation sequencing (NGS) with cytogenetic analyses, researchers can precisely locate genes on chromosomes and determine their functional roles. Gene mapping is a critical step in understanding the genetic basis of complex traits and developing marker-assisted selection (MAS) strategies that accelerate the breeding process. The identification and mapping of genes provide a foundation for targeted breeding programs, enabling breeders to select individuals with favorable genetic profiles and develop crop varieties with enhanced characteristics.
3. **Studying Chromosomal Rearrangements:** Chromosomal rearrangements, including duplications, deletions, inversions, and translocations, can have significant effects on the phenotype of a crop. Cytogenetics and cytogenomics provide tools to study these rearrangements and their impact on gene function and expression. By analyzing chromosomal composition, researchers can identify rearrangements that contribute to trait variation and adaptation, facilitating the development of crop varieties with improved performance. The study of chromosomal rearrangements is essential for understanding the genetic architecture of complex traits and developing strategies to manipulate these traits in breeding programs.
4. **Exploring Polyploidy and Genome Evolution:** Polyploidy, the presence of multiple sets of chromosomes, is a common phenomenon in plants and plays a significant role in genome evolution. Cytogenetics and cytogenomics provide tools to study polyploidy and its impact on genetic diversity, genome organization, and trait expression. By understanding the mechanisms of polyploidy, researchers can develop strategies to exploit genetic diversity and enhance crop resilience and adaptability. The study of polyploidy contributes to our understanding of genome evolution and the development of crop varieties with improved yield and quality.
5. **Integration with Molecular Techniques:** The integration of molecular techniques, such as FISH, NGS, and comparative genomic hybridization (CGH), with cytogenetics has revolutionized the study of chromosomes and genomes. These techniques provide tools to analyze the organization and variation of genomes at the chromosomal level, enabling researchers to identify genomic regions associated with important traits. The combination of molecular and cytogenetic methodologies offers new opportunities for crop improvement by providing insights into the genetic basis of traits and facilitating the development of improved crop varieties.

Benefits of Cytogenetics and Cytogenomics in Crop Improvement

The application of cytogenetics and cytogenomics in crop improvement offers numerous benefits, enhancing the productivity, quality, and resilience of crop varieties. By providing insights into the genetic basis of important agronomic traits, these fields contribute to the development of sustainable agricultural practices and the achievement of food security goals.

Key Benefits in Crop Improvement

1. **Increased Productivity:** Cytogenetics and cytogenomics provide tools to develop new crop varieties with increased productivity, helping to meet the growing demand for food and

ensure global food security. By analyzing the genetic basis of yield-related traits, breeders can identify key genomic regions associated with productivity and develop high-yielding varieties. Techniques such as genomic selection (GS) and marker-assisted selection (MAS) enable breeders to select individuals with favorable genetic profiles, accelerating the development of productive crop varieties.

2. **Improved Quality:** These fields offer tools to enhance the quality of crops by providing insights into the genetic basis of important traits such as nutritional content, taste, and appearance. By identifying key genomic regions associated with quality traits, breeders can develop new varieties with improved characteristics. The integration of genetic and genomic approaches facilitates the enhancement of sensory and nutritional properties, contributing to consumer satisfaction and marketability.
3. **Disease Resistance:** Cytogenetics and cytogenomics provide tools to develop disease-resistant crop varieties by providing insights into the genetic basis of disease resistance. By identifying and mapping genes associated with resistance to specific pathogens and pests, breeders can develop crop varieties with enhanced resistance, reducing the reliance on chemical pesticides and improving crop productivity. The development of disease-resistant varieties contributes to sustainable agricultural practices and the reduction of crop losses.
4. **Stress Tolerance:** These fields offer tools to develop stress-tolerant crop varieties by providing insights into the genetic basis of stress tolerance. By identifying and mapping genes associated with tolerance to abiotic stresses such as drought, heat, and salinity, breeders can develop crop varieties with enhanced resilience to environmental challenges. The development of stress-tolerant varieties contributes to stable and sustainable agricultural production in the face of climate change and resource scarcity.
5. **Biodiversity Conservation:** Cytogenetics and cytogenomics play a crucial role in conserving biodiversity by providing tools to study the genetic diversity of crop species and their wild relatives. This knowledge can be used to develop new varieties with improved characteristics and to preserve the genetic diversity of crop species for future generations. The conservation of genetic resources is essential for ensuring the sustainability and resilience of agricultural systems and for providing a reservoir of genetic variation for crop improvement. Cytogenetics and cytogenomics are essential fields in crop improvement, offering valuable tools and insights for developing improved crop varieties that meet the demands of a growing global population and changing environmental conditions. By integrating genetic and genomic methodologies, these fields provide a comprehensive understanding of the genetic architecture of crops, facilitating the development of productive, high-quality, and resilient crop varieties. The benefits of cytogenetics and cytogenomics extend beyond crop improvement, contributing to sustainable agricultural practices, biodiversity conservation, and global food security. As the challenges of food production and environmental change continue to evolve, cytogenetics and cytogenomics will play a vital role in shaping the future of agriculture and ensuring the sustainability of global food systems.

Chapter - 2

Organization and Structure of Genome, Genome Size, Organization of Organellar Genomes

A. Gene

A **gene** is a fundamental unit of heredity and a segment of DNA that contains the necessary information to produce a functional product, typically a protein or a functional RNA molecule. Genes are the building blocks of genetic information, encoding the instructions required for the development, functioning, growth, and reproduction of organisms. Each gene is located at a specific position on a chromosome, known as a locus, and can vary in length depending on the complexity of the encoded information. The study of genes is central to the field of genetics, as it provides insights into how traits are passed from one generation to the next and how organisms develop and function.

Key Aspects of Genes

1. **Structure of a Gene:**
   - **Coding Regions (Exons):** Genes are composed of coding regions known as exons, which contain the sequences that are transcribed into RNA and translated into proteins. Exons are the parts of a gene that directly code for the amino acid sequence of a protein.
   - **Non-Coding Regions (Introns):** In addition to exons, genes also contain non-coding regions known as introns. Introns are transcribed into RNA but are typically removed during RNA processing through a process called splicing. The presence of introns can influence gene expression and regulation.
   - **Regulatory Elements:** Genes are associated with various regulatory elements, such as promoters, enhancers, and silencers, which play crucial roles in controlling gene expression. These elements influence the transcription process by determining when, where, and how much a gene is expressed. Promoters are sequences located near the start of a gene that serve as binding sites for transcription factors and RNA polymerase, initiating transcription. Enhancers and silencers are regulatory elements that can be located at a distance from the gene they regulate, and they modulate gene expression by interacting with transcription factors and other proteins.

DOI: 10.1201/9781003717775-2

2. Function of a Gene:

» **Protein and RNA Production:** Genes encode proteins and functional RNAs that perform various biological functions. Proteins are the workhorses of the cell, involved in structural roles, enzymatic activities, signaling, transport, and regulation of cellular processes. Functional RNAs, such as ribosomal RNA (rRNA), transfer RNA (tRNA), and microRNA (miRNA), play roles in protein synthesis and gene regulation.

» **Regulation of Biological Processes:** Genes are essential for the regulation of biological processes, enabling cells to carry out specific functions and respond to internal and external signals. The expression of genes is tightly regulated to ensure that proteins and RNAs are produced at the right time, location, and quantity, allowing cells to maintain homeostasis, differentiate, and adapt to environmental changes.

3. Gene Expression:

» **Transcription and Translation:** Gene expression is the process by which information encoded in a gene is converted into a functional product. This process typically involves two main steps: transcription and translation. During transcription, the DNA sequence of a gene is copied into a complementary RNA molecule by the enzyme RNA polymerase. The resulting RNA molecule, known as messenger RNA (mRNA), serves as a template for protein synthesis. During translation, the mRNA is read by ribosomes, which assemble amino acids into a polypeptide chain based on the sequence encoded in the mRNA.

» **Regulation of Gene Expression:** The regulation of gene expression is critical for controlling the timing and level of protein production. Gene expression can be regulated at multiple levels, including transcriptional, post-transcriptional, translational, and post-translational. Factors such as transcription factors, epigenetic modifications, RNA processing, and protein degradation contribute to the regulation of gene expression, allowing cells to adapt to changing conditions and perform specialized functions.

4. Genetic Variation:

» **Mutations and Polymorphisms:** Genes can undergo mutations, which are changes in the DNA sequence that may affect the gene's function. Mutations can occur spontaneously or be induced by environmental factors, such as radiation or chemicals. While some mutations may lead to genetic disorders or diseases, others contribute to genetic diversity and evolutionary adaptations. Genetic variation also arises from polymorphisms, which are common variations in the DNA sequence that occur in a population. Polymorphisms can influence traits and susceptibility to diseases without causing overt disorders.

» **Role in Evolution:** Genetic variation is a driving force in evolution, as it provides the raw material for natural selection. Variations in genes can lead to differences in traits, allowing organisms to adapt to changing environments and contributing to the diversity of life.

5. Role in Inheritance:

» **Transmission of Genetic Information:** Genes are inherited from parents to offspring, determining inherited traits and contributing to an individual's genetic makeup. Each parent contributes one set of chromosomes to the offspring, resulting in a unique combination of genes that influences physical characteristics, susceptibility to diseases, and other phenotypic traits.

» **Mendelian Inheritance:** The inheritance of genes follows principles described by Gregor Mendel, known as Mendelian inheritance. These principles include the segregation of alleles during gamete formation and the independent assortment of genes, which explain how traits are passed from one generation to the next.

» **Complex Inheritance Patterns:** While Mendelian inheritance describes the inheritance of single-gene traits, many traits are influenced by multiple genes and environmental factors, leading to complex inheritance patterns. Polygenic traits, such as height and skin color, result from the combined effects of multiple genes, while multifactorial traits involve interactions between genetic and environmental factor

B. Genome

A **genome** is the complete set of genetic material present in an organism, encompassing all of its genes and non-coding sequences of DNA or RNA. The genome contains the entirety of an organism's hereditary information, serving as a blueprint for its development, functioning, and reproduction. It encodes all the biological information necessary for life processes, from cellular mechanisms to the overall growth and development of the organism. Genomes can vary significantly in size and complexity across different species, reflecting the diversity of life on Earth. The study of genomes, known as genomics, has become a cornerstone of modern biology, providing insights into the genetic basis of traits, the mechanisms of evolution, and the potential for genetic innovations in medicine and agriculture.

Key Aspects of Genomes

1. **Composition of a Genome:**

» **Coding and Non-Coding DNA:** The genome includes all the DNA sequences present in an organism, comprising both coding regions (genes) and non-coding regions. Coding regions are sequences that contain genes, which are transcribed into RNA and translated into proteins. These regions are directly involved in producing the proteins that perform various functions within the cell. Non-coding DNA, although not translated into proteins, plays essential roles in regulating gene expression, maintaining chromosome structure, and facilitating evolutionary changes. Non-coding regions include intergenic DNA, regulatory sequences, introns, and repetitive elements, such as transposable elements and microsatellites. Despite not encoding proteins, non-coding DNA is involved in crucial biological processes, including the regulation of gene expression, chromatin organization, and the control of genome stability.

» **Regulatory Elements and Non-Coding RNA:** Within the non-coding regions of the genome, various regulatory elements, such as promoters, enhancers, and silencers, are involved in controlling the expression of genes. These elements interact with transcription factors and other proteins to modulate the transcriptional activity of genes, determining when, where, and how much a gene is expressed. In addition to regulatory elements, non-coding RNA molecules, such as microRNAs (miRNAs), long non-coding RNAs (lncRNAs), and small interfering RNAs (siRNAs), play essential roles in post-transcriptional regulation, gene silencing, and the maintenance of genome integrity.

2. Types of Genomes:

» **Prokaryotic Genomes:** Prokaryotic genomes are found in bacteria and archaea, typically consisting of a single circular chromosome. Prokaryotic genomes are generally smaller and contain fewer non-coding regions compared to eukaryotic genomes. The compact nature of prokaryotic genomes reflects their streamlined genetic architecture, which enables efficient replication and adaptation to various environments. Prokaryotic genomes often contain operons, which are clusters of genes that are co-regulated and transcribed together, allowing for coordinated expression of functionally related genes. Additionally, prokaryotic genomes may contain plasmids, which are small, circular DNA molecules that replicate independently and can carry genes for antibiotic resistance, virulence factors, and other adaptive traits.

» **Eukaryotic Genomes:** Eukaryotic genomes are found in plants, animals, fungi, and protists, consisting of multiple linear chromosomes located in the cell nucleus. Eukaryotic genomes are larger and contain a higher proportion of non-coding DNA, reflecting their complexity and diversity. The presence of multiple chromosomes allows for the organization of genetic material into distinct units, each carrying specific genes and regulatory elements. Eukaryotic genomes are characterized by the presence of introns, which are non-coding sequences that are removed during RNA splicing, and a variety of repetitive elements, such as transposable elements and satellite DNA. The size and complexity of eukaryotic genomes are further influenced by processes such as polyploidy, horizontal gene transfer, and genome duplication events.

3. Genome Sequencing:

» **Technological Advances in Sequencing:** Genome sequencing involves determining the complete DNA sequence of an organism's genome, providing insights into its genetic composition and functional potential. Advances in sequencing technologies, such as next-generation sequencing (NGS), have enabled the rapid and cost-effective sequencing of genomes, revolutionizing the fields of genomics, medicine, and agriculture. NGS technologies allow for the parallel sequencing of millions of DNA fragments, generating vast amounts of data that can be assembled into complete genome sequences. The development of third-generation sequencing technologies, such as single-molecule real-time (SMRT) sequencing and nanopore sequencing, offers further improvements in sequencing accuracy, read length, and throughput, facilitating the study of complex genomes and epigenomic modifications.

» **Applications of Genome Sequencing:** Genome sequencing provides a comprehensive view of an organism's genetic material, enabling researchers to explore the genetic basis of traits, study evolutionary relationships, and develop personalized medicine approaches. In agriculture, genome sequencing is used to identify genes associated with important agronomic traits, such as yield, disease resistance, and stress tolerance, contributing to the development of improved crop varieties. In medicine, genome sequencing facilitates the identification of disease-causing mutations, the development of targeted therapies, and the implementation of precision medicine approaches. In conservation biology, genome sequencing is used to assess genetic diversity, monitor endangered species, and inform conservation strategies.

4. **Genomic Variation:**

» **Sources of Genetic Variation:** Genomes exhibit genetic variation, which includes differences in DNA sequences among individuals within a species. This variation can result from mutations, genetic recombination, and horizontal gene transfer. Mutations are changes in the DNA sequence that can occur spontaneously or be induced by environmental factors, leading to genetic diversity and evolutionary adaptations. Genetic recombination, which occurs during meiosis, results in the exchange of genetic material between homologous chromosomes, creating new combinations of alleles and contributing to genetic diversity. Horizontal gene transfer, the movement of genetic material between different species, is a significant source of genetic variation in prokaryotes and can facilitate the acquisition of adaptive traits.

» **Impact on Evolution and Adaptation:** Genetic variation contributes to phenotypic diversity, adaptation to environmental changes, and the evolution of species. Variations in the genome can lead to differences in traits, such as morphology, physiology, and behavior, allowing organisms to adapt to diverse environments and ecological niches. The study of genomic variation provides insights into the mechanisms of evolution, the genetic basis of adaptation, and the processes that drive speciation and diversification.

5. **Applications of Genomics:**

» **Transformative Impact on Medicine:** Genomics has transformative applications in medicine, agriculture, and conservation biology. In medicine, genomics enables the identification of disease-causing genes, the development of targeted therapies, and the implementation of personalized medicine approaches. The study of cancer genomics, for example, provides insights into the genetic alterations that drive tumor development, informing the development of targeted treatments and diagnostic tools. Pharmacogenomics, the study of how genetic variation affects drug response, facilitates the optimization of drug therapy and the reduction of adverse effects, contributing to the advancement of precision medicine.

» **Agricultural Innovations:** In agriculture, genomics is used to improve crop varieties by identifying genes associated with important traits, such as yield, quality, and resilience. Genomic selection and marker-assisted breeding are powerful tools that enable breeders to select individuals with favorable genetic profiles, accelerating the development of improved crop varieties. Genomics also plays a role in the development of genetically modified organisms (GMOs) and gene-edited crops, offering new opportunities for enhancing agricultural productivity and sustainability.

» **Conservation and Biodiversity:** In conservation biology, genomics is used to assess genetic diversity, monitor endangered species, and inform conservation strategies. The study of population genomics provides insights into the genetic structure of populations, the effects of habitat fragmentation, and the impacts of climate change on biodiversity. Genomics also facilitates the identification of cryptic species, the assessment of hybridization events, and the evaluation of genetic health in conservation efforts.

C. Organization and Structure of the Genome

The organization and structure of a genome are crucial for understanding how genetic information is stored, expressed, and regulated within an organism. This organization is complex and varies across different species, reflecting the diversity of life and the specific needs of each organism.

A genome is a dynamic and intricately organized structure comprising all of an organism's genetic material, including genes and non-coding sequences. This organization ensures the proper regulation and expression of genes, maintaining cellular functions and enabling organisms to adapt to their environments. The organization and structure of genomes is essential for unraveling the complexities of genetic processes and the development of novel strategies for genetic research, medicine, and crop improvement.

Key Components of Genome Organization

1. Chromosomal Organization:

Chromosomal organization is a critical aspect of genome structure, serving as the blueprint for genetic storage, expression, and inheritance. Chromosomes, the main units of this organization, are DNA molecules intricately folded and packed with proteins to ensure the integrity and proper function of genetic material. Understanding the components and functions of chromosomal organization helps elucidate how genetic information is managed within cells, allowing for insights into processes such as gene expression, replication, and heredity.

a. Chromosomes

Chromosomes are the fundamental units of genome organization, carrying the genetic information necessary for the development and functioning of all living organisms. In eukaryotes, chromosomes are long, linear molecules of DNA, while in prokaryotes, they are typically circular. This difference in structure reflects the distinct mechanisms of chromosome maintenance and replication between these two types of organisms.

In eukaryotic cells, chromosomes are located within the nucleus, which provides a protective environment for the genetic material. Each chromosome contains a single, continuous DNA molecule that can be millions of base pairs long. The DNA is wound around histone proteins, forming nucleosomes, which are the basic units of chromatin. This organization into chromatin allows the DNA to be compacted efficiently, fitting within the confines of the cell nucleus. In prokaryotes, the circular chromosome is found in the nucleoid region, an area within the cell where the DNA is not enclosed by a membrane. Prokaryotic chromosomes are also associated with proteins, but these proteins differ from histones, reflecting the simpler organization of the prokaryotic genome. The packaging of DNA into chromosomes is essential for several reasons. First, it protects the DNA from damage and degradation, maintaining the integrity of the genetic information. Second, it allows for the efficient segregation of chromosomes during cell division, ensuring that each daughter cell receives a complete set of genetic material. This is particularly important during mitosis and meiosis in eukaryotic cells, where precise chromosome segregation is critical for maintaining genomic stability. Finally, the organization of DNA into chromosomes facilitates the regulation of gene expression by controlling the accessibility of specific regions of the genome to transcription factors and other regulatory proteins.

b. Chromatin Structure

Chromatin is the complex of DNA and proteins that form the structural framework of chromosomes in eukaryotes. It plays a crucial role in regulating gene expression and maintaining genome stability. Chromatin can exist in two main forms: euchromatin and heterochromatin, each with distinct structural and functional characteristics.

Euchromatin is a loosely packed form of chromatin that is transcriptionally active, allowing for gene expression. It is characterized by a more open structure, which permits access to transcription factors and RNA polymerase, facilitating the transcription of genes into RNA. Euchromatin is typically enriched in genes and is associated with regions of the genome that are actively transcribed. This dynamic state of chromatin is essential for cellular processes that require rapid and reversible changes in gene expression, such as development, differentiation, and response to environmental stimuli.

Heterochromatin, on the other hand, is a densely packed form of chromatin that is transcriptionally inactive. It is composed of regions of the genome that are not actively transcribed, including repetitive DNA sequences and gene-poor areas. Heterochromatin plays a critical role in maintaining genome stability by suppressing the activity of transposable elements and protecting the integrity of repetitive sequences. It also contributes to the regulation of gene expression by limiting access to specific regions of the genome, thereby influencing the overall transcriptional landscape of the cell.

The dynamic organization of chromatin is regulated by a variety of factors, including histone modifications, DNA methylation, and chromatin-remodeling complexes. Histone modifications, such as acetylation and methylation, can alter the structure of chromatin and affect its accessibility to transcriptional machinery. DNA methylation, the addition of methyl groups to cytosine residues in DNA, is another important epigenetic modification that influences chromatin structure and gene expression. Chromatin-remodeling complexes are multiprotein assemblies that can reposition, eject, or restructure nucleosomes, thereby modulating the accessibility of chromatin to regulatory proteins.

c. Chromosomal Regions

Chromosomes contain specific regions with distinct functions, each playing a critical role in the maintenance and transmission of genetic information. Three key regions of chromosomes are telomeres, centromeres, and origins of replication.

Telomeres are repetitive DNA sequences located at the ends of linear chromosomes. They serve as protective caps that prevent the degradation of chromosome ends and the loss of genetic information during DNA replication. Telomeres also play a role in maintaining chromosome stability by preventing the ends of chromosomes from being recognized as double-strand breaks, which could lead to chromosomal rearrangements or fusion events. The enzyme telomerase is responsible for adding telomeric repeats to the ends of chromosomes, counteracting the progressive shortening that occurs with each round of DNA replication. Telomere length and maintenance are important factors in cellular aging and senescence, as well as in the prevention of cancer.

Centromeres are regions of chromosomes that serve as attachment points for spindle fibers during cell division. They are essential for the accurate segregation of chromosomes into daughter cells, ensuring that each cell receives the correct complement of genetic material. Centromeres are typically composed of repetitive DNA sequences and specialized proteins that form the kinetochore, a structure that mediates the interaction between chromosomes and the mitotic spindle. The proper function of centromeres is crucial for maintaining genomic stability, as errors in chromosome segregation can lead to aneuploidy, a condition characterized by an abnormal number of chromosomes.

Origins of replication are specific sequences where DNA replication is initiated. These regions are recognized by the replication machinery, which assembles at the origin to begin the process of DNA synthesis. In eukaryotes, multiple origins of replication are distributed along each chromosome, allowing for the simultaneous replication of different regions of the genome and ensuring that the entire genome is duplicated in a timely manner. The regulation of origin activation and the coordination of replication with other cellular processes are essential for maintaining genome integrity and preventing replication stress.

1. Gene Structure and Organization:

Gene structure and organization are fundamental aspects of molecular biology, determining how genetic information is stored, expressed, and regulated within an organism. Genes serve as the basic units of heredity, encoding the instructions necessary for the synthesis of proteins and functional RNAs, which are vital for various biological processes. The organization of genes within the genome varies across different species and reflects the complexity of their genetic and evolutionary histories.

a. Genes

Genes are the fundamental units of heredity and functional elements of the genome, encoding the information necessary for the synthesis of proteins and functional RNAs. In eukaryotic organisms, genes are composed of both coding and non-coding sequences. The coding sequences, known as **exons,** contain the nucleotide sequences that are transcribed into messenger RNA (mRNA) and subsequently translated into proteins. Exons are interspersed with non-coding sequences known as **introns,** which are transcribed into RNA but are removed during RNA processing through a process called splicing.

» **Exons and Introns:** Exons are the expressed sequences within a gene that directly contribute to the final mRNA transcript and determine the amino acid sequence of the corresponding protein. In contrast, introns are intervening sequences that do not code for proteins. Introns are excised from the pre-mRNA during the process of splicing, which occurs in the nucleus before the mRNA is transported to the cytoplasm for translation. The presence of introns allows for alternative splicing, a process by which different combinations of exons can be joined together to produce multiple mRNA variants from a single gene. This increases the diversity of proteins that can be synthesized from a single gene, contributing to the complexity and adaptability of eukaryotic organisms.

» **Regulatory Elements:** In addition to exons and introns, genes are associated with various regulatory elements that control their expression. Promoters are sequences located upstream of the transcription start site that serve as binding sites for RNA polymerase and transcription factors, initiating transcription. Enhancers are distal regulatory elements that can increase the transcriptional activity of specific genes, while silencers repress gene expression. These regulatory elements interact with transcription factors and other proteins to modulate gene expression in response to developmental cues, environmental signals, and cellular conditions.

» **Functional RNAs:** While many genes encode proteins, others produce functional RNAs that are not translated into proteins but play crucial roles in cellular processes. These include ribosomal RNA (rRNA), which forms the structural and functional components of ribosomes; transfer RNA (tRNA), which carries amino acids to the ribosome during protein synthesis; and small nuclear RNA (snRNA) and microRNA (miRNA), which are involved in RNA splicing and gene regulation, respectively. Functional RNAs are essential for maintaining cellular homeostasis and regulating gene expression at various levels.

b. Gene Density

Gene density refers to the number of genes present within a given length of DNA, and it can vary significantly between different species and organisms. The density of genes within a genome is influenced by factors such as genome size, the presence of non-coding DNA, and the organism's evolutionary history.

» **Prokaryotic Genomes:** Prokaryotic genomes, found in bacteria and archaea, are generally compact and characterized by high gene density. Prokaryotic genomes are typically smaller in size, consisting of a single circular chromosome that contains a relatively high proportion of coding DNA. In prokaryotes, genes are often arranged in clusters with little non-coding DNA between them, allowing for efficient utilization of genetic material. This high gene density is a reflection of the streamlined nature of prokaryotic genomes, which have evolved to optimize metabolic efficiency and adapt rapidly to changing environmental conditions.

» **Eukaryotic Genomes:** In contrast to prokaryotes, eukaryotic genomes are larger and contain a higher proportion of non-coding DNA, resulting in lower gene density. Eukaryotic genomes consist of multiple linear chromosomes located within the cell nucleus, and they are characterized by the presence of introns, intergenic regions, and repetitive elements. The non-coding DNA in eukaryotes plays important roles in regulating gene expression, maintaining chromosomal structure, and facilitating evolutionary changes. The distribution and density of genes in eukaryotes are influenced by factors such as genome complexity, the presence of transposable elements, and the evolutionary history of the organism.

» **Evolutionary Considerations:** The variation in gene density between prokaryotes and eukaryotes reflects their distinct evolutionary pressures and life strategies. Prokaryotes have evolved to minimize genome size and maximize genetic efficiency, allowing for rapid growth and adaptation to diverse environments. Eukaryotes, on the other hand, have developed complex regulatory networks and genomic architectures that provide greater flexibility and adaptability, enabling the evolution of multicellularity and diverse biological functions.

c. Operons

Operons are a key feature of prokaryotic gene organization, allowing for the coordinated regulation of functionally related genes. An operon is a cluster of genes that are transcribed together as a single mRNA molecule, enabling the simultaneous expression of multiple genes involved in a common metabolic pathway or cellular process.

- **Structure of an Operon:** An operon typically consists of a promoter, an operator, and one or more structural genes. The promoter is a sequence upstream of the operon that serves as a binding site for RNA polymerase, initiating transcription. The operator is a regulatory sequence located between the promoter and the structural genes, and it acts as a binding site for regulatory proteins that can activate or repress transcription. The structural genes are the coding sequences within the operon that are transcribed into a single polycistronic mRNA, which is subsequently translated into multiple proteins.
- **Regulation of Operons:** The expression of operons is regulated by specific proteins known as repressors and activators. Repressors are proteins that bind to the operator sequence, blocking the binding of RNA polymerase and inhibiting transcription. In contrast, activators are proteins that enhance the binding of RNA polymerase to the promoter, facilitating transcription. The regulation of operons allows prokaryotic cells to respond efficiently to changes in environmental conditions, such as the availability of nutrients or the presence of stress factors. This coordinated regulation of gene expression enables prokaryotes to optimize their metabolic pathways and adapt rapidly to fluctuating environments.
- **Examples of Operons:** One of the most well-known examples of an operon is the lac operon in *Escherichia coli*, which regulates the metabolism of lactose. The lac operon consists of genes that encode enzymes responsible for the transport and breakdown of lactose, and its expression is regulated by the availability of lactose and glucose. Another example is the trp operon, which controls the synthesis of the amino acid tryptophan. The expression of the trp operon is repressed when tryptophan levels are sufficient, preventing the unnecessary production of tryptophan-synthesizing enzymes.
- **Advantages of Operon Organization:** The operon organization offers several advantages to prokaryotic cells, including the ability to regulate the expression of multiple genes simultaneously, conserve energy and resources, and adapt rapidly to environmental changes. By coordinating the expression of functionally related genes, operons enable prokaryotes to efficiently regulate their metabolic pathways and respond to environmental cues, contributing to their evolutionary success and ecological diversity.

2. Non-Coding DNA and Regulatory Elements:

Non-coding DNA constitutes a substantial portion of the genome in many organisms, particularly in eukaryotes. While these DNA segments do not encode proteins, they play vital roles in regulating gene expression, maintaining chromosomal structure, and facilitating evolutionary changes. The understanding of non-coding DNA has evolved significantly, revealing its importance in complex regulatory networks and its contribution to genetic diversity and adaptation. Key components of non-coding DNA include regulatory elements, introns, intergenic DNA, and repetitive sequences, each serving distinct functions in the genome.

a. Non-Coding DNA

Non-coding DNA refers to portions of an organism's genome that do not encode protein sequences. Despite not being translated into proteins, non-coding DNA serves critical functions in regulating biological processes and maintaining genomic integrity. Non-coding DNA can be broadly classified into various categories, including introns, regulatory elements, intergenic regions, and repetitive sequences. The roles of these non-coding regions has transformed our perception of the genome, highlighting its complexity and versatility beyond mere protein-coding capacity.

1. **Introns:**

   » **Role in Gene Expression:** Introns are non-coding sequences found within genes, particularly in eukaryotes. During the transcription of a gene, introns are transcribed into precursor mRNA but are removed during RNA processing through a process called splicing. The removal of introns is essential for generating mature mRNA, which can then be translated into proteins. Introns can influence gene expression through alternative splicing, a mechanism that allows for the production of multiple protein isoforms from a single gene. By enabling different combinations of exons to be joined together, alternative splicing increases the diversity of the proteome and allows cells to adapt to various functional needs.

   » **Regulation of Gene Expression:** Introns can also play regulatory roles by containing sequences that affect transcriptional activity. Some introns harbor enhancer or silencer elements that influence the expression of the gene in which they reside or nearby genes. Additionally, the presence of introns can impact the rate of transcription and the stability of mRNA, further influencing gene expression.

2. **Intergenic DNA:**

   » **Genomic Context:** Intergenic DNA refers to the non-coding regions between genes. These regions, once thought to be "junk DNA," have been found to contain regulatory elements and sequences that contribute to chromatin organization and genome stability. Intergenic regions may also include non-coding RNA genes, such as microRNAs (miRNAs) and long non-coding RNAs (lncRNAs), which play roles in gene regulation and cellular processes.

   » **Regulatory Functions:** Intergenic DNA can contain enhancer elements that regulate gene expression over long distances, influencing genes that are located far from the enhancer itself. These long-range interactions are crucial for the precise regulation of gene expression, especially in complex eukaryotic genomes.

3. **Repetitive Sequences:**

   » **Types and Functions:** Repetitive sequences are abundant in many genomes and can be classified into several types, including transposable elements, microsatellites, and satellite DNA. Each type of repetitive sequence has distinct characteristics and functions, contributing to genetic variation, genome evolution, and chromosomal stability.

   » **Transposable Elements:** Transposable elements, also known as "jumping genes," are mobile genetic elements that can move within the genome. They can replicate and insert themselves into new genomic locations, influencing genome structure and function. Transposable elements can impact gene expression by inserting into or near genes, potentially disrupting their function or altering regulatory regions. They contribute to genetic diversity and have played significant roles in genome evolution, driving the expansion and rearrangement of genomic sequences.

» **Microsatellites:** Microsatellites, also known as short tandem repeats (STRs), are short, repeating sequences of 1 to 6 base pairs. They are highly polymorphic and serve as valuable genetic markers in population genetics and forensic studies. Microsatellites can influence gene expression and genome stability by affecting chromatin structure and gene regulation. Their high mutation rates make them useful for studying genetic variation and evolutionary dynamics.

» **Satellite DNA:** Satellite DNA consists of long, repetitive sequences that are often associated with centromeres and heterochromatin. These sequences play critical roles in maintaining chromosomal structure and stability. Satellite DNA contributes to the formation of centromeres, which are essential for proper chromosome segregation during cell division. The repetitive nature of satellite DNA also influences chromatin organization and can impact gene expression by modulating the accessibility of regulatory elements.

b. Regulatory Elements

Regulatory elements are non-coding DNA sequences that control the expression of genes, playing key roles in the complex regulatory networks that govern cellular processes. These elements include promoters, enhancers, silencers, and insulators, each with specific functions in gene regulation.

1. **Promoters:**

» **Function:** Promoters are DNA sequences located near the transcription start site of a gene, serving as binding sites for RNA polymerase and transcription factors. They initiate transcription by recruiting the transcriptional machinery necessary for RNA synthesis. Promoters are essential for determining the timing, location, and level of gene expression, acting as gatekeepers for transcriptional activity.

» **Core Promoter Elements:** The core promoter contains specific sequences, such as the TATA box, that are recognized by the transcriptional machinery. These elements provide the binding sites for general transcription factors and RNA polymerase, facilitating the assembly of the transcriptional complex.

2. **Enhancers:**

» **Role in Gene Activation:** Enhancers are distal regulatory elements that can increase the transcriptional activity of specific genes. They function by binding transcription factors and other regulatory proteins, which then interact with the promoter region to enhance transcription. Enhancers can act over long distances and can be located upstream, downstream, or within introns of the genes they regulate.

» **Dynamic Regulation:** The activity of enhancers is context-dependent, influenced by factors such as cell type, developmental stage, and environmental conditions. Enhancers can form loops with promoters, bringing them into close proximity and facilitating transcriptional activation.

3. **Silencers:**

   » **Gene Repression:** Silencers are regulatory elements that repress gene expression by binding repressor proteins. These elements inhibit transcription by preventing the binding of transcriptional activators or by recruiting proteins that alter chromatin structure to a repressive state. Silencers are important for maintaining the specificity of gene expression, ensuring that genes are expressed only in appropriate contexts.

   » **Mechanisms of Action:** Silencers can act through various mechanisms, including blocking the binding of activators, recruiting histone deacetylases to promote chromatin compaction, and interfering with enhancer-promoter interactions.

4. **Insulators:**

   » **Boundary Elements:** Insulators are regulatory elements that establish boundaries between different chromatin domains, maintaining the specificity of gene regulation. They prevent the spread of activating or repressing signals between adjacent genomic regions, ensuring that enhancers or silencers do not inappropriately influence neighboring genes.

   » **Function in Chromatin Organization:** Insulators can also influence the three-dimensional organization of the genome by mediating the formation of chromatin loops, which separate active and inactive chromatin domains. This organization is essential for the proper regulation of gene expression and the maintenance of genomic integrity.

4. Genome Size and Complexity:

The size and complexity of genomes have long been subjects of fascination and study in genetics and evolutionary biology. Genome size, measured as the total amount of DNA in an organism's haploid genome, can vary widely among species, leading to intriguing questions about its implications for biological complexity and evolutionary processes. The relationship between genome size and organismal complexity is not straightforward, as highlighted by the C-value paradox and the phenomenon of polyploidy. These concepts offer insights into the factors that influence genome size and the mechanisms that drive genomic and evolutionary change.

a. -Value Paradox

The C-value paradox refers to the observation that there is no direct correlation between the genome size of an organism and its perceived complexity. This paradox arises because organisms with similar levels of biological complexity can have vastly different genome sizes, and conversely, some relatively simple organisms have larger genomes than more complex ones.

1. **Definition and Historical Context:**

   » **C-Value Measurement:** The C-value is the total amount of DNA contained within a haploid genome, typically measured in base pairs or picograms. It represents the genetic content of a single set of chromosomes and provides a metric for comparing genome sizes across species. The term "C-value paradox" was coined to describe the lack of correlation between C-value and organismal complexity.

» **Historical Discovery:** The paradox was first recognized in the mid-20th century when researchers noted that genome sizes did not align with expectations based on organismal complexity. For instance, some amphibians and plants were found to have much larger genomes than humans, despite being less complex in terms of cellular differentiation and behavioral capabilities.

2. **Factors Contributing to the C-Value Paradox:**

» **Non-Coding DNA:** One of the primary contributors to the C-value paradox is the presence of non-coding DNA, which constitutes a significant portion of many genomes, especially in eukaryotes. Non-coding DNA includes introns, regulatory elements, and repetitive sequences that do not directly code for proteins but play crucial roles in gene regulation, chromosomal structure, and genome stability.

» **Repetitive Elements:** Repetitive DNA sequences, such as transposable elements, satellite DNA, and simple sequence repeats, contribute substantially to genome size without necessarily affecting organismal complexity. Transposable elements are particularly abundant in many genomes and can replicate and insert themselves into new locations, expanding the genome over time.

» **Polyploidy:** Whole-genome duplication events, known as polyploidy, can dramatically increase genome size by adding additional sets of chromosomes. Polyploidy is common in plants and some animal lineages and can lead to increased genetic content without a corresponding increase in organismal complexity.

3. **Biological Implications of the C-Value Paradox:**

» **Genome Size and Evolution:** The C-value paradox underscores the complexity of evolutionary processes and the role of genome architecture in shaping organismal traits. Large genomes may offer advantages in terms of genetic diversity and evolutionary potential, but they also impose metabolic costs and challenges in genome replication and maintenance.

» **Functional Significance of Non-Coding DNA:** While much of the non-coding DNA was once considered "junk," it is now recognized as playing important roles in gene regulation, chromatin organization, and evolutionary innovation. The presence of non-coding DNA allows for complex regulatory networks and provides a substrate for genetic variation and adaptation.

» **Research and Technological Advances:** Advances in sequencing technologies and functional genomics have facilitated the exploration of non-coding regions, revealing their contributions to genome function and evolution. Understanding the C-value paradox continues to inform research on genome organization, regulation, and the mechanisms underlying biological complexity.

b. Polyploidy

Polyploidy, the condition of having more than two complete sets of chromosomes, is a significant factor in genome size variation and plays a critical role in the evolution and adaptation of many organisms. Polyploidy is particularly common in plants, where it has been a major driver of speciation and diversity, but it is also found in some animal lineages.

1. **Mechanisms and Types of Polyploidy**

   » **Whole-Genome Duplication:** Polyploidy typically arises from whole-genome duplication events, where an organism inherits additional sets of chromosomes. This can occur through various mechanisms, including errors in meiosis or mitosis, hybridization between species, and endoreduplication (DNA replication without cell division).

   » **Types of Polyploidy:** Polyploidy can be classified into several types based on its origin. Autopolyploidy occurs when an organism duplicates its own genome, resulting in multiple sets of chromosomes from the same species. Allopolyploidy involves the hybridization of two different species, leading to the combination of their genomes in a single organism. Allopolyploids often have enhanced genetic diversity and can exhibit novel traits not found in either parent species.

2. **Biological and Evolutionary Implications of Polyploidy:**

   » **Genetic Diversity and Evolutionary Potential:** Polyploidy increases genetic diversity by providing additional copies of genes, which can undergo mutations and evolve new functions. This genetic redundancy allows for the exploration of new phenotypic traits and can lead to the development of novel adaptations. Polyploid organisms often exhibit increased variability and adaptability, making them more resilient to environmental changes and stresses.

   » **Speciation and Diversity:** Polyploidy plays a significant role in plant speciation, driving the formation of new species that occupy unique ecological niches. The presence of multiple sets of chromosomes in polyploid species enables them to adapt to various environmental conditions, allowing them to exploit new habitats and resources. This adaptability leads to increased biodiversity, as polyploid plants often demonstrate greater resilience and versatility compared to their diploid counterparts. As a result, polyploidy not only contributes to the diversification of plant species but also enhances ecosystem complexity by promoting a wider variety of species interactions and ecological roles.In some cases, polyploidy can facilitate the rapid colonization of new environments and the establishment of stable populations.

   » **Gene Regulation and Expression:** The presence of multiple gene copies in polyploid organisms can influence gene expression and regulation. Polyploidy can result in changes to gene dosage, where the expression levels of genes are altered due to the increased number of copies. This can lead to phenotypic changes and may contribute to the unique characteristics of polyploid species.

3. **Polyploidy in Plants and Animals:**

   » **Plants:** Polyploidy is particularly prevalent in plants, where it has played a significant role in the evolution and diversification of flowering plants, ferns, and other plant groups. Many agriculturally important crops, such as wheat, cotton, and potatoes, are polyploid, benefiting from the increased genetic diversity and adaptability conferred by polyploidy.

» **Animals:** While less common than in plants, polyploidy does occur in some animal lineages, particularly in certain groups of fish, amphibians, and reptiles. In animals, polyploidy can influence traits such as size, growth rate, and reproductive strategies, contributing to their ecological success.

4. Challenges and Considerations of Polyploidy:

» **Genomic Stability:** The increased genomic content in polyploid organisms poses challenges for maintaining genomic stability and accurate chromosome segregation during cell division. Polyploidy can lead to complications in meiosis and an increased risk of chromosomal aberrations, which can affect fertility and viability.

» **Metabolic Costs:** The maintenance of additional genetic material imposes metabolic costs on polyploid organisms, requiring greater resources for DNA replication, transcription, and repair. Balancing these costs with the potential benefits of polyploidy is a key consideration in the evolution and success of polyploid species.

» **Research Opportunities:** Polyploidy presents unique opportunities for research in genetics, evolution, and agriculture. Understanding the mechanisms and consequences of polyploidy can inform breeding programs, conservation efforts, and studies of genome evolution. Advances in genomic technologies and analytical tools continue to enhance our understanding of polyploidy and its impact on biological systems.

5. Genome Evolution and Variation:

Genome evolution and variation are fundamental processes that drive biological diversity and adaptation. These processes are facilitated by mechanisms such as mutation, genetic recombination, and horizontal gene transfer, each contributing to the dynamic nature of genomes and their ability to respond to environmental changes. These mechanisms are critical for understanding the intricate interactions between genetic material and evolutionary forces, which drive the diversity and adaptation of life on Earth. By studying how genetic variations interact with natural selection, mutation, gene flow, and genetic drift, researchers can gain insights into the processes that shape the evolution of species and the complexity of biological systems. This knowledge is fundamental to explaining how organisms adapt to their environments, evolve new traits, and contribute to the vast biodiversity seen across ecosystems.

a. Mutation and Selection

Mutations are changes in the DNA sequence that serve as the primary source of genetic variation. They can occur spontaneously due to errors in DNA replication or be induced by environmental factors such as radiation, chemicals, and viruses. Mutations are essential for evolution as they provide the raw material upon which natural selection acts, enabling organisms to adapt to their environments and evolve over time.

1. **Types of Mutations:**
   - **Point Mutations:** Point mutations involve changes in a single nucleotide base pair in the DNA sequence. These can be classified into several types, including silent mutations (no change in protein function), missense mutations (change in one amino acid), and nonsense mutations (introducing a premature stop codon). Point mutations can have varying effects on gene function, depending on their location and nature.
   - **Insertions and Deletions:** Insertions and deletions (indels) involve the addition or loss of one or more nucleotide base pairs in the DNA sequence. Indels can lead to frameshift mutations, which alter the reading frame of the genetic code, potentially resulting in a nonfunctional protein. These mutations can have significant impacts on gene expression and function, depending on their size and position within the genome.
   - **Copy Number Variations (CNVs):** CNVs are structural variations in the genome that result in the duplication or deletion of large segments of DNA. These variations can affect the number of copies of specific genes, influencing gene dosage and expression. CNVs contribute to genetic diversity and can play important roles in adaptation and evolution.
   - **Chromosomal Mutations:** Chromosomal mutations involve large-scale changes in chromosome structure, including duplications, deletions, inversions, and translocations. These mutations can lead to changes in gene order, dosage, and expression, potentially affecting an organism's phenotype and fitness.

2. **Role of Natural Selection:**
   - **Adaptive Evolution:** Natural selection is the mechanism through which advantageous genetic variations become more prevalent in a population, driving evolutionary change. Organisms with beneficial mutations are more likely to survive, reproduce, and pass these favorable traits on to their offspring. Over generations, this process enables populations to adapt to shifting environmental conditions, enhancing their survival and reproductive success. As natural selection favors these adaptive traits, it can lead to the emergence of new species and contribute to the expansion of biodiversity, shaping the evolutionary trajectory of life on Earth.
   - **Balancing Selection:** In some cases, natural selection can maintain genetic diversity within a population through balancing selection. This occurs when multiple alleles are favored in different environmental contexts or when heterozygous individuals have a selective advantage over homozygous individuals. Balancing selection can help preserve genetic variation and maintain the adaptive potential of populations.
   - **Purifying Selection:** Purifying selection, also known as negative selection, acts to remove deleterious mutations from a population. This process helps maintain the integrity of essential genes and functions by eliminating harmful genetic variants that reduce fitness.

3. **Mutations in Evolutionary Context:**
   - **Neutral Mutations:** Many mutations are neutral, meaning they have no significant effect on

an organism's fitness. These mutations can accumulate in populations over time, providing a reservoir of genetic variation that may become beneficial under changing environmental conditions.

- **Genetic Drift:** Genetic drift is a random, stochastic process that affects the frequency of mutations in a population, especially in smaller populations. It occurs due to the random sampling of alleles during reproduction, causing fluctuations in allele frequencies from one generation to the next. Over time, genetic drift can result in the fixation (permanent presence) or loss of neutral mutations—those that do not provide a selective advantage or disadvantage. While genetic drift acts independently of natural selection, it can interact with selection to influence the genetic diversity of a population, shaping its evolutionary path by altering the frequency of certain traits.

b. Genetic Recombination

Genetic recombination is a critical mechanism that occurs during meiosis, resulting in the exchange of genetic material between homologous chromosomes. This process generates new combinations of alleles, increasing genetic diversity and facilitating adaptation to environmental challenges.

1. **Mechanisms of Recombination:**
   - **Homologous Recombination:** Homologous recombination occurs during meiosis when homologous chromosomes pair and exchange segments of DNA. This process involves the formation of chiasmata, where crossover events occur, allowing the reciprocal exchange of genetic material. Homologous recombination generates new allele combinations, contributing to genetic diversity and the evolutionary potential of populations.
   - **Gene Conversion:** Gene conversion is a non-reciprocal process that can accompany homologous recombination. It involves the unidirectional transfer of genetic material from one homologous chromosome to another, potentially altering allele frequencies and contributing to genetic variation.
2. **Evolutionary Significance of Recombination:**
   - **Increased Genetic Diversity:** Recombination enhances genetic diversity by generating new combinations of alleles during the formation of gametes, providing a mechanism for populations to adapt to changing environments and evolve novel traits. This increased genetic variation is vital for the survival and adaptability of populations, especially when confronted with environmental challenges and selective pressures. By creating a broader genetic pool, recombination allows for greater flexibility in responding to natural selection, ultimately promoting the evolution of traits that improve a population's chances of survival in diverse and fluctuating conditions.
   - **Facilitation of Adaptation:** Recombination allows beneficial mutations to be combined more effectively, facilitating adaptive evolution. By bringing together advantageous alleles from different genetic backgrounds, recombination can accelerate the spread of beneficial traits within populations.

- **Maintenance of Genetic Variation:** Recombination plays a key role in maintaining genetic variation by disrupting linkage disequilibrium, which is the non-random association of alleles at different loci. Through recombination, alleles can assort independently, effectively separating linked alleles and allowing them to recombine in new ways. This process preserves genetic diversity within a population, ensuring a wider range of potential combinations for future generations. By maintaining this genetic variation, recombination increases the population's ability to adapt to environmental changes and selective pressures, contributing to long-term evolutionary potential.

3. **Recombination and Speciation:**

- **Role in Speciation:** Recombination can play a role in speciation by contributing to reproductive isolation and the divergence of populations. In some cases, recombination between divergent populations can lead to the formation of new species by generating novel combinations of alleles that confer reproductive barriers or adaptive advantages.
- **Hybridization and Introgression:** Recombination can occur during hybridization events, where individuals from different species interbreed. This process can result in introgression, the incorporation of genetic material from one species into the gene pool of another, contributing to genetic diversity and evolutionary innovation.

c. Gene Transfer Horizontally

Horizontal gene transfer (HGT) refers to the transfer of genetic material between different species, playing a crucial role in generating genetic variation and enabling the acquisition of new traits. This process is especially important in prokaryotes, where it facilitates the rapid dissemination of advantageous traits, such as antibiotic resistance. HGT allows organisms to quickly adapt to environmental pressures without relying solely on vertical inheritance through reproduction, thereby accelerating evolutionary change and enhancing the adaptability of microbial populations in response to challenges like antibiotic exposure or changing environmental conditions.

4. **Mechanisms of Horizontal Gene Transfer:**

- **Transformation:** Transformation involves the uptake of free DNA from the environment by a recipient cell. This process allows prokaryotes to acquire new genetic material from their surroundings, potentially incorporating beneficial genes into their genomes.
- **Conjugation:** Conjugation is a process in which genetic material is transferred between bacterial cells through direct cell-to-cell contact. This transfer is mediated by a pilus, a specialized structure that facilitates the exchange of plasmids, circular DNA molecules that can carry genes for antibiotic resistance and other adaptive traits.
- **Transduction:** Transduction involves the transfer of genetic material between bacteria via bacteriophages, viruses that infect bacteria. During the infection cycle, phages can accidentally incorporate host DNA into their viral particles, transferring this genetic material to other bacterial cells.

5. **Significance of Horizontal Gene Transfer:**

   » **Acquisition of New Traits:** HGT allows organisms to rapidly acquire new traits, such as antibiotic resistance, metabolic capabilities, and virulence factors. This process enables prokaryotes to adapt quickly to environmental changes and challenges, conferring a selective advantage in diverse ecological niches.

   » **Genetic Diversity and Evolution:** HGT contributes to genetic diversity by introducing novel genes and alleles into populations, facilitating evolutionary change. This process can drive the emergence of new phenotypes and ecological strategies, enhancing the adaptability and resilience of microbial communities.

   » **Impact on Genomic Evolution:** HGT can have significant impacts on genomic evolution by reshaping genome architecture and function. The incorporation of foreign genetic material can lead to genome expansion, the creation of new regulatory networks, and the emergence of novel gene functions.

6. **Horizontal Gene Transfer in Eukaryotes:**

   » **Occurrence in Eukaryotes:** While HGT is more common in prokaryotes, it also occurs in eukaryotes, particularly in unicellular organisms and some plants and animals. In eukaryotes, HGT can contribute to the acquisition of new metabolic pathways, resistance to environmental stressors, and the expansion of gene families.

   » **Evolutionary Implications:** HGT in eukaryotes can influence evolutionary trajectories by introducing new genetic material that enhances fitness and adaptability. Horizontal gene transfer can contribute to the development of complex traits and the diversification of lineages by introducing new genetic material from different species. This exchange of genes enables organisms to acquire novel functions or adapt to new environments more quickly than through traditional evolutionary processes. Over time, the accumulation of horizontally transferred genes can result in the emergence of complex traits, influencing the evolutionary trajectory of a species and contributing to the branching and diversification of lineages. This mechanism is particularly impactful in microbial evolution, where it drives rapid innovation and adaptability.

7. **Functional Genomics:**

Functional genomics is a field of molecular biology that seeks to understand the roles and interactions of genes and other elements within the genome. It moves beyond the mere sequencing of genomes to explore how genes are expressed, regulated, and interact with each other and the environment. The ultimate goal of functional genomics is to elucidate the complex networks that drive biological processes, enabling researchers to link genotypes to phenotypes and to understand the molecular basis of health and disease. This is accomplished through various approaches, including gene expression analysis, epigenomics, and systems biology, each of which provides unique insights into the functioning of the genome.

a. Gene Expression Analysis

Gene expression analysis is a fundamental aspect of functional genomics, focusing on the study of how genes are transcribed into RNA and translated into proteins. Gene expression patterns is crucial for revealing the roles of genes in development, physiology, and disease. Several techniques have been developed to measure gene expression, with RNA sequencing (RNA-seq) and microarray analysis being among the most widely used.

1. **RNA Sequencing (RNA-seq):**
   - **Overview:** RNA-seq is a powerful and versatile technique that uses high-throughput sequencing technology to capture a comprehensive snapshot of the transcriptome. By sequencing the entire population of RNA molecules in a sample, RNA-seq provides detailed information about the presence and quantity of RNA transcripts, including mRNA, non-coding RNA, and small RNAs.
   - **Applications:** RNA-seq is used to study gene expression across different tissues, developmental stages, and environmental conditions. It allows researchers to identify differentially expressed genes, uncover alternative splicing events, and discover novel transcripts. RNA-seq is also instrumental in understanding complex diseases by revealing the molecular pathways and networks involved in disease progression.
   - **Advantages:** RNA-seq offers several advantages over traditional techniques, such as microarrays. It provides high sensitivity, allowing the detection of low-abundance transcripts, and it does not require prior knowledge of the genome sequence, making it suitable for studying non-model organisms. Additionally, RNA-seq can capture the dynamic range of gene expression levels, providing a more accurate representation of the transcriptome.

2. **Microarray Analysis:**
   - **Overview:** Microarray analysis involves the hybridization of labeled RNA samples to a grid of DNA probes on a solid surface, allowing the simultaneous measurement of thousands of gene expression levels. Each probe corresponds to a specific gene, and the intensity of the signal reflects the abundance of the corresponding RNA in the sample.
   - **Applications:** Microarrays have been used extensively to study gene expression profiles in various contexts, including cancer research, drug discovery, and developmental biology. They enable researchers to identify genes involved in specific biological processes and to classify samples based on their expression patterns.
   - **Limitations:** While microarrays provide valuable insights into gene expression, they have limitations compared to RNA-seq. Microarrays are limited by the probes' design, which requires prior knowledge of the genome, and they have a narrower dynamic range of detection. Additionally, microarrays may not detect novel transcripts or low-abundance RNAs as effectively as RNA-seq.

3. **Impact on Research:**

   » **Gene Function and Regulation:** Gene expression analysis provides insights into the function and regulation of genes, revealing how they contribute to the phenotype of an organism. By studying expression patterns, researchers can identify key regulators of biological processes and understand how gene networks respond to internal and external cues.

   » **Disease Mechanisms:** Gene expression studies have been instrumental in uncovering the molecular mechanisms underlying various diseases. By comparing the expression profiles of healthy and diseased tissues, researchers can identify biomarkers for disease diagnosis, prognostic indicators, and potential therapeutic targets.

   » **Developmental Biology:** Understanding the temporal and spatial patterns of gene expression is critical for elucidating the genetic programs that drive development and differentiation. Gene expression analysis provides insights into the complex regulatory networks that control cell fate decisions and tissue morphogenesis.

b. Epigenomics

Epigenomics is the study of epigenetic modifications and their impact on gene expression and genome organization. Unlike genetic changes, which involve alterations in the DNA sequence, epigenetic modifications are reversible chemical changes that can affect gene activity without altering the underlying genetic code. These modifications play critical roles in regulating gene expression, development, differentiation, and adaptation to environmental stimuli.

1. **Epigenetic Modifications:**

   » **DNA Methylation:** DNA methylation involves the addition of a methyl group to the cytosine bases of DNA, typically at CpG dinucleotides. This modification is associated with gene silencing and plays a crucial role in regulating gene expression, maintaining genomic stability, and suppressing transposable elements. DNA methylation patterns are dynamically regulated during development and can be influenced by environmental factors, contributing to phenotypic plasticity and adaptation.

   » **Histone Modifications:** Histone proteins, which package DNA into chromatin, can undergo various post-translational modifications, including acetylation, methylation, phosphorylation, and ubiquitination. These modifications alter chromatin structure and accessibility, influencing gene expression. For example, histone acetylation is generally associated with transcriptional activation, while histone methylation can either activate or repress transcription, depending on the specific residues modified.

   » **Chromatin Remodeling:** Chromatin remodeling refers to the dynamic changes in chromatin structure that regulate access to the DNA by transcription factors and other regulatory proteins. Chromatin remodeling complexes use energy derived from ATP hydrolysis to reposition, eject, or restructure nucleosomes, thereby modulating gene expression and chromatin organization.

2. **Epigenomic Technologies:**

» **ChIP-seq:** Chromatin immunoprecipitation followed by sequencing (ChIP-seq) is a widely used technique for mapping DNA-protein interactions across the entire genome. This method employs antibodies that specifically bind to target proteins, such as transcription factors or histone modifications, allowing researchers to identify the regions of the genome where these proteins are bound. ChIP-seq generates data that reveals the association of these factors with particular genomic regions, offering valuable insights into how gene regulation is controlled and organized, and shedding light on the regulatory mechanisms that influence gene expression and cellular function.

» **Bisulfite Sequencing:** Bisulfite sequencing is a method used to analyze DNA methylation patterns by treating DNA with bisulfite, which converts unmethylated cytosines to uracil while leaving methylated cytosines unchanged. Sequencing the bisulfite-treated DNA reveals the methylation status of individual cytosines, allowing researchers to study methylation patterns at single-base resolution.

» **ATAC-seq:** Assay for transposase-accessible chromatin using sequencing (ATAC-seq) is a technique used to assess chromatin accessibility. By using a transposase enzyme to insert sequencing adapters into open chromatin regions, ATAC-seq identifies accessible regions of the genome, providing insights into regulatory elements and chromatin dynamics.

3. **Epigenomics in Health and Disease:**

» **Developmental Processes:** Epigenetic modifications are crucial for regulating gene expression during development and differentiation. They enable the establishment of cell type-specific gene expression profiles and contribute to the maintenance of cellular identity. Epigenomic studies have revealed the dynamic changes in epigenetic marks that occur during development and how they influence cell fate decisions.

» **Environmental Adaptation:** Epigenetic modifications mediate the effects of environmental factors on gene expression, allowing organisms to adapt to changing conditions. This adaptive potential is critical for responding to stress, pathogens, and other environmental challenges, contributing to the plasticity and resilience of organisms.

» **Disease Mechanisms:** Aberrant epigenetic modifications are implicated in various diseases, including cancer, neurological disorders, and metabolic diseases. Epigenomic studies have identified alterations in DNA methylation and histone modifications that contribute to disease pathogenesis, providing potential targets for therapeutic intervention. Understanding the epigenetic landscape of diseases can inform the development of epigenetic therapies that aim to reverse abnormal epigenetic marks.

c. Systems Biology

Systems biology is an interdisciplinary approach that integrates genomic data with other biological information to model and understand complex biological systems. By analyzing interactions between genes, proteins, and other molecules, systems biology seeks to elucidate the functional architecture of the genome and the dynamic processes that govern cellular function. This holistic perspective enables researchers to study biological systems as integrated networks, providing insights into the emergent properties and behaviors of living organisms.

1. **Integration of Omics Data:**
   - **Multi-Omics Approaches:** Systems biology leverages data from various omics fields, including genomics, transcriptomics, proteomics, metabolomics, and epigenomics, to construct comprehensive models of biological systems. By integrating these diverse data types, researchers can gain a more complete understanding of the interactions and regulatory mechanisms that drive cellular processes.
   - **Network Analysis:** Systems biology employs network analysis to study the interactions between genes, proteins, and metabolites. Biological networks are represented as graphs, with nodes representing molecules and edges representing interactions. Analyzing these networks can reveal key regulators, hubs, and pathways that are critical for cellular function and response to perturbations.

2. **Modeling and Simulation:**
   - **Mathematical Modeling:** Mathematical models are used to simulate and predict the behavior of biological systems. These models can capture the dynamics of gene expression, protein interactions, and metabolic pathways, providing insights into the underlying mechanisms that govern biological processes. Models can be qualitative or quantitative, depending on the complexity of the system and the available data.
   - **Dynamic Simulations:** Dynamic simulations allow researchers to explore the temporal behavior of biological systems under different conditions. By simulating the effects of perturbations, such as genetic mutations or environmental changes, researchers can predict the system's response and identify potential therapeutic targets or intervention strategies.

3. **Applications of Systems Biology:**
   - **Drug Discovery and Development:** Systems biology is integral to drug discovery as it helps identify potential drug targets and predict the effects of drug candidates on complex biological systems. By modeling the interactions between drugs and biological networks, systems biology allows researchers to gain a comprehensive understanding of how drugs influence cellular pathways and molecular processes. This holistic approach can guide the development of more effective therapies by accounting for the broader impacts of a drug, reducing unintended side effects, and enabling the creation of personalized treatment strategies tailored to individual patients' biological profiles.
   - **Understanding Complex Diseases:** Systems biology provides insights into the molecular mechanisms underlying complex diseases, such as cancer, cardiovascular disease, and neurodegenerative disorders. By integrating multi-omics data, researchers can identify dysregulated pathways and interactions that contribute to disease pathogenesis, offering new avenues for diagnosis and treatment.
   - **Synthetic Biology:** Systems biology informs synthetic biology, an emerging field that seeks

to design and engineer new biological systems for various applications. By understanding the principles of biological organization and regulation, researchers can construct synthetic circuits and pathways with desired properties, enabling the development of novel biotechnological solutions

D. Genome size

Genome size, often referred to as the C-value, represents the total amount of DNA contained within the haploid set of chromosomes of an organism. It is a fundamental attribute of the genome, providing a quantitative measure of the genetic material within a cell. Genome size varies widely across the tree of life, ranging from a few hundred thousand base pairs in some prokaryotes to hundreds of billions of base pairs in certain plants and amphibians. This variation reflects the diverse evolutionary histories, ecological adaptations, and genomic architectures that characterize different organisms.

The Concept of Genome Size

The size of a genome is typically measured in base pairs (bp) or its equivalent in mass, such as picograms (pg) of DNA content. This measurement encompasses all the DNA within an organism's genome, including both coding and non-coding regions. The haploid genome, or C-value, represents a single complete set of chromosomes, providing a standard measure for comparing genome sizes across species.

a. Historical Perspective of Genome Size

The historical perspective on genome size reveals the fascinating journey of scientific discovery that has shaped our understanding of genetic material. The exploration of genome size began in the early 20th century and has continued to evolve with advancements in technology and methodology.

Early Discoveries and Theoretical Developments

1. **Initial Observations:**

   » **Pioneering Work:** The concept of genome size emerged with the discovery of nucleic acids in the late 19th century. Friedrich Miescher first identified DNA as a distinct molecule in 1869, but it wasn't until the early 20th century that researchers began to explore its role in heredity. During this period, biologists focused on the physical and chemical properties of chromosomes, recognizing them as carriers of genetic information.

   » **Chromosome Counting:** By the 1920s, cytogeneticists were counting chromosomes in various organisms, noting differences in chromosome numbers and sizes. These observations laid the groundwork for understanding genome size, as researchers began to correlate the physical attributes of chromosomes with the amount of genetic material they contained.

2. **The C-Value Paradox:**

   » **Definition and Emergence:** The term "C-value" was introduced in the mid-20th century to describe the amount of DNA in a haploid genome. Researchers observed that C-values varied widely among species, leading to the realization that genome size did not correlate with organismal complexity. This phenomenon, known as the C-value paradox, challenged prevailing assumptions about the relationship between DNA content and biological sophistication.

- **Early Hypotheses:** Initial attempts to explain the C-value paradox focused on technical limitations and measurement errors. However, as more data accumulated, it became clear that the paradox reflected genuine biological diversity. Scientists proposed various hypotheses, including the presence of non-coding DNA and repetitive elements, to account for the discrepancy between genome size and complexity.

3. **Advancements in Cytogenetics:**

- **Karyotyping and Chromosome Banding:** The development of karyotyping and chromosome banding techniques in the 1950s and 1960s allowed researchers to visualize and analyze chromosomes in greater detail. These methods provided insights into chromosomal structure and organization, revealing the presence of non-coding regions and repetitive sequences that contributed to genome size.
- **Genome Mapping:** Early efforts to map genomes involved the identification of genetic markers and linkage groups, which helped scientists understand the distribution of genes within chromosomes. These studies highlighted the complexity of genomic organization and the potential impact of non-coding DNA on genome size.

Technological Innovations and Breakthroughs

1. **Molecular Biology Revolution:**

- **DNA Sequencing:** The advent of DNA sequencing technologies in the 1970s marked a significant turning point in genomics. The Sanger method, developed by Frederick Sanger and colleagues, enabled the sequencing of entire genomes, providing detailed insights into genetic content and organization. Sequencing efforts revealed the abundance of non-coding DNA and repetitive elements, offering explanations for the C-value paradox.
- **Recombinant DNA Technology:** The development of recombinant DNA technology allowed researchers to manipulate and study specific genes and genomic regions. This innovation facilitated the exploration of gene function and regulation, advancing our understanding of genome complexity and evolution.

2. **Next-Generation Sequencing (NGS):**

- **High-Throughput Sequencing:** The advent of next-generation sequencing (NGS) technologies in the early 2000s transformed the field of genomics by offering rapid, cost-effective sequencing of entire genomes. NGS platforms like Illumina, 454, and Ion Torrent delivered exceptional resolution and scalability, enabling researchers to analyze genome size, structure, and variation across a wide range of organisms. This technological breakthrough allowed for more comprehensive genomic studies, facilitating advancements in fields such as evolutionary biology, medical genetics, and agriculture by making large-scale sequencing accessible and efficient.
- **Applications and Impact:** NGS technologies have been instrumental in advancing comparative

genomics, population genetics, and evolutionary biology. They have facilitated the identification of structural variations, repetitive elements, and regulatory sequences that contribute to genome size variation.

3. **Genomic Databases and Bioinformatics:**
   - » **Data Integration and Analysis:** The proliferation of genomic data has necessitated the development of sophisticated bioinformatics tools and databases. Resources such as GenBank, Ensembl, and the UCSC Genome Browser provide access to genomic sequences, annotations, and comparative data, enabling researchers to explore genome size and complexity across species.
   - » **Comparative Genomics:** Bioinformatics approaches have enabled the comparison of genome size and content across different taxa, revealing patterns of conservation and divergence. These studies have provided insights into the evolutionary forces shaping genome size, such as natural selection, genetic drift, and horizontal gene transfer.

Contemporary Understanding and Research

1. **Non-Coding DNA and Repetitive Elements:**
   - » **Functional Roles:** Recent research has elucidated the functional roles of non-coding DNA and repetitive elements, challenging the notion of "junk DNA." Non-coding regions are now recognized as critical regulators of gene expression, chromatin organization, and genome stability. Repetitive elements, such as transposable elements, contribute to genetic diversity and genome evolution, influencing genome size and complexity.
   - » **Regulatory Networks:** The discovery of regulatory elements, such as enhancers, silencers, and insulators, has highlighted the intricate networks that control gene expression. These elements often reside in non-coding regions, underscoring their importance in shaping genome architecture and function.

2. **Polyploidy and Whole-Genome Duplications:**
   - » **Evolutionary Significance:** Polyploidy, or whole-genome duplication, has emerged as a major driver of genome size variation and evolution. Polyploid organisms possess multiple sets of chromosomes, providing genetic redundancy and the potential for novel adaptations. Polyploidy is particularly prevalent in plants, where it has contributed to speciation, diversification, and ecological success.
   - » **Genomic Consequences:** Whole-genome duplications can lead to increased genome size, as well as structural and functional changes in the genome. Following duplication, redundant gene copies may undergo neofunctionalization, subfunctionalization, or pseudogenization, contributing to the evolution of new traits and functions.

3. **Future Directions:**
   - » **Epigenomics and Chromatin Dynamics:** Advances in epigenomics are shedding light

on the dynamic regulation of genome size and structure. Epigenetic modifications, such as DNA methylation and histone modifications, influence chromatin accessibility and gene expression, playing key roles in genome organization and evolution.

- **Single-Cell Genomics:** The advent of single-cell sequencing technologies is providing insights into the heterogeneity of genome size and function at the cellular level. Single-cell approaches allow researchers to investigate genome dynamics and variability within populations, revealing the impact of genomic architecture on cellular behavior and adaptation.

The historical perspective on genome size reveals the complexities of genomic evolution and the factors that influence genome architecture. From early observations of chromosomal differences to contemporary insights into non-coding DNA and polyploidy, the study of genome size has provided valuable insights into the diversity and adaptability of life. As genomics continues to advance, our understanding of genome size and its implications for biology will continue to deepen, offering new opportunities for research and innovation.

b. Factors Contributing to Genome Size Variation

The variation in genome size across species is a fascinating aspect of evolutionary biology and genomics. While the primary role of the genome is to encode the genetic information necessary for an organism's development and function, genome size does not always correlate with organismal complexity, as illustrated by the C-value paradox. Several factors contribute to the variation in genome size, including non-coding DNA, repetitive elements, and polyploidy.

Non-Coding DNA

Non-coding DNA constitutes a significant portion of many eukaryotic genomes and plays crucial roles in regulating gene expression, maintaining chromosomal structure, and facilitating evolutionary innovation. Although non-coding DNA does not directly code for proteins, its functional importance has become increasingly recognized in recent years.

1. **Types of Non-Coding DNA:**

   - **Introns:** Introns are non-coding sequences found within genes, particularly in eukaryotes. During the transcription of a gene, introns are transcribed into precursor mRNA but are removed during RNA processing through a process called splicing. Introns can influence gene expression through alternative splicing, a mechanism that allows for the production of multiple protein isoforms from a single gene. By enabling different combinations of exons to be joined together, alternative splicing increases the diversity of the proteome and allows cells to adapt to various functional needs.
   - **Regulatory Elements:** Non-coding DNA includes various regulatory elements that control gene expression, such as promoters, enhancers, silencers, and insulators. These elements interact with transcription factors and other regulatory proteins to modulate the transcriptional activity of genes, ensuring precise regulation in response to developmental cues and environmental stimuli.
   - **Intergenic Regions:** Intergenic regions are non-coding sequences located between genes.

These regions, once considered "junk DNA," have been found to contain regulatory elements and sequences that contribute to chromatin organization and genome stability. Intergenic regions may also include non-coding RNA genes, such as microRNAs (miRNAs) and long non-coding RNAs (lncRNAs), which play roles in gene regulation and cellular processes.

2. **Functional Roles of Non-Coding DNA:**

» **Gene Regulation:** Non-coding DNA plays a critical role in regulating gene expression, providing the necessary elements for fine-tuning gene activity. Regulatory elements in non-coding regions interact with transcription factors, coactivators, and repressors to control the timing, location, and level of gene expression, enabling cells to respond to internal and external signals.

» **Chromatin Structure:** Non-coding DNA contributes to the organization of chromatin, the complex of DNA and histone proteins that form the structural framework of chromosomes. Chromatin organization influences the accessibility of DNA to transcription factors and other regulatory proteins, impacting gene expression and cellular function. The dynamic organization of chromatin is essential for maintaining genome stability and regulating gene activity.

» **Evolutionary Innovation:** Non-coding DNA provides a substrate for evolutionary innovation, allowing for the emergence of new regulatory elements and gene functions. Changes in non-coding regions can lead to the evolution of novel gene expression patterns and contribute to the diversity and adaptability of organisms.

Repetitive Elements

Repetitive DNA sequences are abundant in many genomes and can be classified into several types, each with distinct characteristics and functions. Repetitive elements contribute substantially to genome size and play important roles in genome evolution, stability, and regulation.

1. **Types of Repetitive Elements:**

» **Transposable Elements:** Transposable elements, also known as "jumping genes," are mobile genetic elements that can move within the genome. They can replicate and insert themselves into new genomic locations, influencing genome structure and function. Transposable elements can impact gene expression by inserting into or near genes, potentially disrupting their function or altering regulatory regions. They contribute to genetic diversity and have played significant roles in genome evolution, driving the expansion and rearrangement of genomic sequences.

» **Satellite DNA:** Satellite DNA consists of long, repetitive sequences that are often associated with centromeres and heterochromatin. These sequences play critical roles in maintaining chromosomal structure and stability. Satellite DNA contributes to the formation of centromeres, which are essential for proper chromosome segregation during cell division. The repetitive nature of satellite DNA also influences chromatin organization and can impact gene expression by modulating the accessibility of regulatory elements.

» **Microsatellites:** Microsatellites, also known as short tandem repeats (STRs), are short, repeating sequences of 1 to 6 base pairs. They are highly polymorphic and serve as valuable genetic markers in population genetics and forensic studies. Microsatellites can influence gene expression and genome stability by affecting chromatin structure and gene regulation. Their high mutation rates make them useful for studying genetic variation and evolutionary dynamics.

2. **Roles and Impacts of Repetitive Elements:**

» **Genomic Expansion:** Repetitive elements, particularly transposable elements, are major contributors to genome expansion. Their ability to replicate and insert into new locations can lead to increases in genome size over evolutionary time scales. The accumulation of repetitive elements is a key factor in the variation of genome sizes observed across species.

» **Regulation of Gene Expression:** Repetitive elements can influence gene expression by modulating chromatin structure and accessibility. Some transposable elements contain regulatory sequences that can act as enhancers, silencers, or promoters, affecting the transcriptional activity of nearby genes. The regulatory potential of repetitive elements highlights their role in shaping gene expression networks and cellular functions.

» **Genomic Stability and Evolution:** Repetitive elements contribute to genomic instability by promoting recombination and chromosomal rearrangements. These events can lead to genetic diversity and the evolution of new traits, but they also pose challenges for maintaining genome integrity. Mechanisms that suppress transposable element activity, such as DNA methylation and RNA interference, are crucial for preserving genome stability.

Polyploidy

Polyploidy, the condition of having more than two complete sets of chromosomes, is a significant factor in genome size variation and plays a critical role in the evolution and adaptation of many organisms. Polyploidy is particularly common in plants, where it has been a major driver of speciation and diversity, but it is also found in some animal lineages.

1. **Mechanisms and Types of Polyploidy:**

» **Whole-Genome Duplication:** Polyploidy typically arises from whole-genome duplication events, where an organism inherits additional sets of chromosomes. This can occur through various mechanisms, including errors in meiosis or mitosis, hybridization between species, and endoreduplication (DNA replication without cell division).

» **Types of Polyploidy:** Polyploidy can be classified into several types based on its origin. Autopolyploidy occurs when an organism duplicates its own genome, resulting in multiple sets of chromosomes from the same species. Allopolyploidy involves the hybridization of two different species, leading to the combination of their genomes in a single organism. Allopolyploids often have enhanced genetic diversity and can exhibit novel traits not found in either parent species.

2. **Biological and Evolutionary Implications of Polyploidy:**

» **Genetic Diversity and Evolutionary Potential:** Polyploidy increases genetic diversity by providing additional copies of genes, which can undergo mutations and evolve new functions. This genetic redundancy allows for the exploration of new phenotypic traits and can lead to the development of novel adaptations. Polyploid organisms often exhibit increased variability and adaptability, making them more resilient to environmental changes and stresses.

» **Speciation and Diversity:** Polyploidy is a major driver of speciation in plants, contributing to the emergence of new species with distinct ecological niches. Polyploid species can exploit new habitats and resources, leading to increased biodiversity and ecosystem complexity. In some cases, polyploidy can facilitate the rapid colonization of new environments and the establishment of stable populations.

» **Gene Regulation and Expression:** The presence of multiple gene copies in polyploid organisms can influence gene expression and regulation. Polyploidy can result in changes to gene dosage, where the expression levels of genes are altered due to the increased number of copies. This can lead to phenotypic changes and may contribute to the unique characteristics of polyploid species.

3. **Polyploidy in Plants and Animals:**

» **Plants:** Polyploidy is particularly prevalent in plants, where it has played a significant role in the evolution and diversification of flowering plants, ferns, and other plant groups. Many agriculturally important crops, such as wheat, cotton, and potatoes, are polyploid, benefiting from the increased genetic diversity and adaptability conferred by polyploidy.

» **Animals:** While less common than in plants, polyploidy does occur in some animal lineages, particularly in certain groups of fish, amphibians, and reptiles. In animals, polyploidy can influence traits such as size, growth rate, and reproductive strategies, contributing to their ecological success.

4. **Challenges and Considerations of Polyploidy:**

» **Genomic Stability:** The increased genomic content in polyploid organisms poses challenges for maintaining genomic stability and accurate chromosome segregation during cell division. Polyploidy can lead to complications in meiosis and an increased risk of chromosomal aberrations, which can affect fertility and viability.

» **Metabolic Costs:** The maintenance of additional genetic material imposes metabolic costs on polyploid organisms, requiring greater resources for DNA replication, transcription, and repair. Balancing these costs with the potential benefits of polyploidy is a key consideration in the evolution and success of polyploid species.

» **Research Opportunities:** Polyploidy presents unique opportunities for research in genetics, evolution, and agriculture. Understanding the mechanisms and consequences of polyploidy can inform breeding programs, conservation efforts, and studies of genome evolution. Advances in genomic technologies and analytical tools continue to enhance our understanding of polyploidy and its impact on biological systems.

Genome size variation is shaped by a multitude of factors, including non-coding DNA, repetitive elements, and polyploidy. These components influence the evolutionary trajectories of species, contributing to the diversity and adaptability of life on Earth. As our understanding of genome architecture deepens, we gain valuable insights into the fundamental principles that govern genetic organization, expression, and evolution.

c. Importance of Genome Size

Genome size is a crucial aspect of genomic biology that has profound implications for an organism's biology, evolution, and adaptability. Genome size is essential for insights into genomic function, organismal complexity, and the evolutionary processes that shape the diversity of life. The study of genome size also informs various practical applications in research, agriculture, and medicine.

Genomic Architecture and Function

The size of a genome plays a significant role in determining its structure, organization, and function. Genome size influences the arrangement of genetic material within cells and impacts the regulation of gene expression and chromatin dynamics.

1. **Chromatin Organization:**
   - **Structure of Chromatin:** Chromatin is the complex of DNA and histone proteins that forms the structural framework of chromosomes. The organization of chromatin is influenced by genome size, with larger genomes often having more intricate chromatin structures. This complexity arises from the need to compact and organize extensive genetic material within the limited space of the cell nucleus.
   - **Gene Regulation:** The dynamic organization of chromatin influences the accessibility of DNA to transcription factors and other regulatory proteins. In larger genomes, the presence of non-coding regions and repetitive elements can modulate chromatin structure, affecting gene expression patterns. Chromatin remodeling and epigenetic modifications play key roles in regulating gene activity, allowing for precise control of cellular processes.
2. **Gene Density and Expression:**
   - **Impact on Gene Density:** Genome size affects gene density, with larger genomes typically exhibiting lower gene density due to the abundance of non-coding DNA. In prokaryotes, where genomes are generally smaller and more compact, gene density is high, and non-coding regions are minimal. In contrast, eukaryotic genomes often contain extensive non-coding regions, leading to lower gene density and more complex regulatory networks.
   - **Regulation of Gene Expression:** The distribution and density of genes within a genome can influence the regulation of gene expression. In larger genomes, non-coding regions provide additional regulatory elements, such as enhancers, silencers, and insulators, that modulate gene activity. These elements contribute to the fine-tuning of gene expression, enabling cells to respond to developmental cues and environmental changes.

Evolutionary Implications

Genome size is a key factor in evolutionary biology, influencing genetic diversity, adaptability, and speciation. The study of genome size provides insights into the mechanisms of evolution and the forces that shape the diversity of life.

1. **Genetic Diversity and Adaptability:**
   - **Role in Genetic Diversity:** Larger genomes often harbor greater genetic diversity, providing a reservoir of genetic material that can be harnessed for adaptation and evolution. This diversity arises from the presence of multiple gene copies, repetitive elements, and non-coding regions that serve as substrates for genetic variation.
   - **Adaptive Potential:** Genome size influences an organism's adaptive potential by providing the genetic flexibility to respond to environmental challenges. The presence of redundant gene copies and regulatory elements in larger genomes allows for the evolution of novel traits and the exploration of new ecological niches.

2. **Speciation and Diversification:**
   - **Influence on Speciation:** Genome size can play a role in speciation by contributing to reproductive isolation and the divergence of populations. Polyploidy, for example, is a significant driver of speciation in plants, enabling the emergence of new species with distinct ecological adaptations.
   - **Diversity and Evolution:** Genome size variation contributes to the diversity and evolutionary dynamics of populations. Differences in genome size can lead to changes in gene expression, phenotypic traits, and reproductive strategies, influencing the evolutionary trajectories of species.

Applications in Research and Industry

The study of genome size has important applications in research, agriculture, and medicine. Understanding genome size informs various practical endeavors, from breeding programs to biotechnological innovations.

1. **Breeding and Conservation:**
   - **Crop Improvement:** Knowledge of genome size is essential for plant breeding programs aimed at developing new crop varieties with desirable traits, such as increased yield, disease resistance, and environmental tolerance. Genome size data can guide the selection of breeding strategies and the identification of genetic markers for trait improvement.
   - **Conservation Biology:** Genome size information is valuable for conservation efforts, particularly in preserving genetic diversity and protecting endangered species. Understanding genome size can inform the management of genetic resources and the development of conservation strategies that maintain biodiversity.

2. **Biotechnology and Medicine:**

   » **Genomic Technologies:** Advances in genomic technologies, such as genome editing and synthetic biology, rely on a deep understanding of genome size and architecture. Genome size data informs the design and implementation of biotechnological applications, enabling precise manipulation of genetic material for research and industrial purposes.

   » **Medical Applications:** Insights into genome size and complexity are crucial for understanding the genetic basis of diseases and developing targeted therapies. Genome size data can inform the diagnosis and treatment of genetic disorders, contributing to the advancement of personalized medicine and genomics-based healthcare.

d. Challenges and Future Directions

Despite significant advancements in understanding genome size, several challenges remain. Ongoing research continues to explore the complexities of genome architecture, the role of non-coding DNA, and the mechanisms underlying genome evolution.

1. **Complexity of Non-Coding DNA:**

   » **Functional Understanding:** The functional roles of non-coding DNA remain an active area of investigation. While progress has been made in elucidating the regulatory functions of non-coding elements, many aspects of their biology are still not fully understood. Continued research into non-coding regions is essential for uncovering their contributions to genome function and evolution.

   » **Epigenetic Regulation:** Advances in epigenomics are providing insights into the dynamic regulation of genome size and structure. Epigenetic modifications, such as DNA methylation and histone modifications, influence chromatin accessibility and gene expression, playing key roles in genome organization and evolution. Understanding the interplay between epigenetic regulation and genome size is a critical area of research.

2. **Mechanisms of Genome Expansion:**

   » **Transposable Elements:** Transposable elements are major contributors to genome expansion and variability. Understanding the mechanisms by which these elements replicate and integrate into genomes is crucial for elucidating the dynamics of genome evolution. Research into transposable elements can provide insights into their impact on genetic diversity and genomic stability.

   » **Polyploidy and Whole-Genome Duplications:** Polyploidy is a significant factor in genome size variation, particularly in plants. Whole-genome duplications can lead to increased genome size and the evolution of new traits. Continued exploration of polyploidy and its consequences for genomic architecture and adaptation is essential for understanding the role of genome size in evolution.

3. Integration of Genomic Data:

- **Multi-Omics Approaches:** Integrating genome size data with other genomic information, such as transcriptomics, proteomics, and metabolomics, provides a more comprehensive understanding of genome function and evolution. Multi-omics approaches hold promise for unraveling the complexity of genomic systems and their interactions with the environment.
- **Comparative Genomics:** Comparative genomics allows for the analysis of genome size and content across different species, revealing patterns of conservation and divergence. These studies provide insights into the evolutionary forces shaping genome size and inform our understanding of genetic diversity and adaptation.

e. Measuring Genome Size

The measurement of genome size is a fundamental aspect of genomics, providing essential insights into the amount of genetic material within an organism. Genome size is crucial for exploring genomic architecture, evolutionary biology, and genetic diversity. Genome size can be measured using various units and methodologies, each offering unique insights and levels of precision.

1. Units of Measurement

Genome size requires precise units of measurement to quantify and compare the amount of genetic material across different species. The two primary units used are base pairs and picograms, each providing unique insights into the structure and content of genomes.

Base Pairs (bp)

Definition and Significance:

Base pairs (bp) are the fundamental units of DNA, consisting of pairs of nucleotides that form the double helix structure of the DNA molecule. In this configuration, adenine (A) pairs with thymine (T), and cytosine (C) pairs with guanine (G), creating a stable structure through hydrogen bonding. The measurement of genome size in base pairs represents the linear sequence of these nucleotide pairs, offering a direct indication of an organism's genetic information. The use of base pairs as a unit of measurement is standard in molecular biology and provides a clear understanding of the genome's coding potential.

Expression in Larger Units:

1. **Kilobases (kb):** One kilobase equals 1,000 base pairs. This unit is often used to describe small genomic features, such as individual genes or regulatory elements.
2. **Megabases (Mb):** One megabase equals 1,000,000 base pairs. Megabases are commonly used to express the size of entire chromosomes or large genomic regions, making it easier to discuss the genomic content of more complex organisms.
3. **Gigabases (Gb):** One gigabase equals 1,000,000,000 base pairs. This unit is particularly useful for describing the size of large genomes, such as those of mammals and plants. For example, the human genome is approximately 3.2 Gb, illustrating the vast amount of genetic information contained within a single set of chromosomes.

Applications and Importance:

Measuring genome size in base pairs is crucial for understanding the organization of genetic material within an organism. It facilitates the comparison of genomes across species, providing insights into evolutionary relationships and genetic diversity. For instance, comparing the 135 Mb genome of *Arabidopsis thaliana* with the 3.2 Gb human genome highlights the differences in complexity and gene content between plants and mammals.

Moreover, base pair measurement aids in identifying genomic regions responsible for specific traits or diseases. By analyzing the base pair sequences, researchers can pinpoint mutations, gene duplications, or deletions that contribute to phenotypic variation or pathological conditions.

Interpretation and Implications:

The size of a genome, measured in base pairs, can offer insights into an organism's evolutionary history and genetic architecture. Larger genomes may contain more genes, regulatory elements, and non-coding regions, contributing to greater biological complexity and adaptability. However, the C-value paradox—where genome size does not always correlate with organismal complexity—highlights the role of non-coding DNA and repetitive sequences in shaping genome architecture. This paradox underscores the need for a deeper understanding of how non-coding regions contribute to the regulation of gene expression and the evolution of complex traits.

Picograms (pg)

Definition and Conversion:

Picograms (pg) are a unit of mass used to quantify the total DNA content in a haploid genome. One picogram is equivalent to approximately 978 megabases of DNA, providing a measure of the physical quantity of genetic material within a cell. This unit is particularly useful in cytogenetics, where the focus is on the mass and structure of DNA rather than its linear sequence.

Advantages and Applications:

1. **Comparative Analysis:** Measuring genome size in picograms allows for the comparison of DNA content across different species and cell types. This approach is valuable in evolutionary biology, where researchers can assess changes in genome size associated with adaptation and speciation.
2. **Chromosomal Studies:** Picogram measurements are crucial for studying chromosomal structure and organization. They provide insights into the amount of genetic material contained within individual chromosomes, aiding in the identification of chromosomal rearrangements and polyploidy.
3. **Practical Applications:** In practical applications, picogram measurements are used to determine the DNA content of various organisms, which is essential for breeding programs, conservation efforts, and medical research. For example, understanding the DNA content of crop species can guide the development of new varieties with improved traits.

Role in Cytogenetics:

In cytogenetics, measuring genome size in picograms is valuable for studying chromosomal aberrations and genome stability. Changes in DNA content, such as duplications or deletions, can lead to genomic instability and disease. Picogram measurements provide a reliable method for detecting these changes and understanding their impact on cellular function and development.

Implications for Evolution and Adaptation:

The measurement of genome size in picograms offers insights into the evolutionary processes that shape genome architecture. Variations in DNA content can reflect the evolutionary pressures faced by different species, such as environmental challenges or ecological niches. By studying genome size, researchers can gain a deeper understanding of how organisms adapt to their environments and the genetic mechanisms underlying these adaptations.

2. Methods of Measurement

The accurate measurement of genome size is essential for understanding genetic diversity, evolution, and the functional organization of genomes. Various methods have been developed to quantify genome size, each offering distinct advantages and limitations.

Flow Cytometry

Principle and Technique:

Flow cytometry is a powerful technique used to measure genome size by analyzing the fluorescence intensity of stained nuclei. This method involves staining cells with a DNA-specific fluorescent dye, such as propidium iodide or DAPI, which binds to the DNA and emits fluorescence when excited by a laser. As the cells pass through the flow cytometer, the fluorescence emitted by each cell is measured, providing an estimate of DNA content based on fluorescence intensity.

Advantages and Applications:

1. **Rapid Analysis:** Flow cytometry offers rapid analysis, allowing for the measurement of genome size in a large number of cells in a short period. This makes it an ideal technique for high-throughput studies and large-scale population analyses.
2. **High Throughput:** The ability to analyze thousands of cells per second enables researchers to obtain statistically significant data and assess variations in genome size within populations or across species.
3. **Precision and Accuracy:** Flow cytometry provides precise and accurate estimates of DNA content, allowing for comparisons of genome sizes across different species, populations, and developmental stages. This precision is particularly valuable in studies of polyploidy, where the accurate measurement of DNA content is crucial for determining ploidy levels.
4. **Applications in Various Fields:** Flow cytometry is widely used in plant and animal breeding,

conservation biology, and cancer research, where accurate measurement of genome size is critical for understanding genetic diversity, cellular processes, and disease mechanisms.

Limitations and Considerations:

Despite its advantages, flow cytometry has limitations that researchers must consider. The technique requires viable cells, and the accuracy of the measurements depends on the quality of the fluorescent dye and the calibration of the instrument. Sample preparation can also influence results, as the staining process must be carefully controlled to ensure consistent and reliable data. Additionally, flow cytometry may not provide detailed information about the specific sequences or regions within the genome, making it less suitable for studies that require sequence-level resolution.

Innovations and Improvements:

Recent innovations in flow cytometry have improved its sensitivity and resolution, allowing for more precise measurements of genome size. Advances in fluorescent dyes and instrumentation have enhanced the ability to detect subtle differences in DNA content, making flow cytometry a valuable tool for genomic research. Additionally, the integration of flow cytometry with other techniques, such as fluorescence-activated cell sorting (FACS), has expanded its applications, enabling researchers to isolate specific cell populations based on genome size or other characteristics.

Feulgen Staining

Principle and Procedure:

The Feulgen reaction is a classical cytochemical method for quantifying DNA content based on the binding of a specific stain to DNA molecules. The procedure involves treating cells with hydrochloric acid to hydrolyze the DNA, followed by staining with Schiff's reagent, which binds to the aldehyde groups generated by hydrolysis. The intensity of the resulting color is proportional to the DNA content, allowing for quantitative measurement.

Historical Significance:

Feulgen staining has been historically used for estimating genome size in various organisms, particularly before the advent of modern sequencing technologies. It provided early insights into the variation of genome sizes across species and contributed to the development of cytogenetic techniques. The method was instrumental in the early studies of chromosomal structure and function, laying the groundwork for our understanding of genomic organization.

Applications and Limitations:

1. **Chromosomal Studies:** Feulgen staining is valuable for studying chromosomal structure and organization, as well as for identifying chromosomal abnormalities. It allows researchers to visualize chromosomes and assess their size, number, and morphology, providing insights into the genetic architecture of cells.
2. **Comparative Genomics:** By comparing DNA content across different species, researchers can gain insights into the evolutionary dynamics of genome size. Feulgen staining has

been used to investigate the relationship between genome size and organismal complexity, contributing to the understanding of the C-value paradox.

3. **Limitations:** Despite its historical significance, Feulgen staining has limitations. The method requires meticulous sample preparation, and the accuracy of the measurements can be affected by factors such as staining intensity, tissue fixation, and section thickness. Feulgen staining is less precise than modern techniques, such as flow cytometry and DNA sequencing, and may not provide detailed information about specific genomic regions or sequences.

Innovations and Developments:

While Feulgen staining is considered a traditional method, recent advancements have improved its accuracy and applicability. Innovations in staining protocols and imaging technologies have enhanced the resolution and reproducibility of Feulgen staining, making it a valuable complement to modern genomic techniques. Additionally, the integration of Feulgen staining with digital image analysis has enabled more precise quantification of DNA content, expanding its applications in genomic research.

DNA Sequencing

Advances in Sequencing Technologies:

DNA sequencing has revolutionized genomics by providing direct and comprehensive measurements of genome size and content. Advances in sequencing technologies, such as next-generation sequencing (NGS), have enabled the rapid and cost-effective sequencing of entire genomes, offering unparalleled insights into genetic architecture.

1. **Next-Generation Sequencing (NGS):** NGS platforms, such as Illumina, PacBio, and Oxford Nanopore, provide high-throughput and high-resolution data, allowing researchers to explore genome size and content in unprecedented detail. NGS technologies can sequence millions of DNA fragments simultaneously, providing a comprehensive view of the genome's linear sequence.
2. **Third-Generation Sequencing:** Third-generation sequencing technologies, such as single-molecule real-time (SMRT) sequencing and nanopore sequencing, offer long-read capabilities, enabling the assembly of complex genomes with high accuracy. These technologies are particularly useful for resolving repetitive regions and structural variations that contribute to genome size variation.

Comprehensive Genomic Analysis:

DNA sequencing provides a comprehensive view of the genome, including both coding and non-coding regions, repetitive elements, and structural variations. By sequencing the entire genome, researchers can accurately determine genome size, identify gene content, and explore the functional roles of genetic elements.

1. **Functional Genomics:** Sequencing-based approaches facilitate the discovery of novel genes, regulatory elements, and genomic variations that contribute to biological complexity. By analyzing the complete genome sequence, researchers can gain insights into the genetic basis of traits, diseases, and evolutionary adaptations.

2. **Comparative Genomics:** DNA sequencing allows for the comparison of genome sizes and contents across different species, revealing patterns of conservation and divergence. Comparative genomics provides insights into the evolutionary forces shaping genome architecture and informs our understanding of genetic diversity and adaptation.

Applications in Genomics:

1. **Population Genetics:** DNA sequencing is used to study genetic variation within and between populations, providing insights into evolutionary dynamics, gene flow, and adaptation. By analyzing genome-wide data, researchers can identify genetic markers associated with phenotypic traits and explore the genetic basis of adaptation to environmental changes.
2. **Evolutionary Biology:** Sequencing-based approaches facilitate the study of genome evolution, including the identification of gene duplications, deletions, and rearrangements that contribute to genomic diversity. By comparing genome sequences across species, researchers can reconstruct evolutionary relationships and trace the origins of specific traits or adaptations.
3. **Medical Research:** DNA sequencing has transformative applications in medicine, enabling the identification of disease-causing genes, the development of targeted therapies, and the advancement of personalized medicine. Sequencing-based approaches provide insights into the genetic basis of complex diseases, offering new avenues for diagnosis and treatment.

Limitations and Challenges:

Despite its transformative impact, DNA sequencing has limitations. The accuracy of genome size estimation depends on the completeness and quality of the assembled genome sequence, which can be affected by repetitive regions, sequencing errors, and gaps in coverage. Additionally, the analysis and interpretation of large-scale sequencing data require substantial computational resources and expertise, posing challenges for researchers in terms of data management and analysis.

Future Directions and Innovations:

The field of DNA sequencing continues to evolve, with ongoing advancements aimed at improving accuracy, efficiency, and affordability. Innovations in sequencing technologies and bioinformatics tools are enhancing our ability to measure genome size and explore genomic complexity. The integration of sequencing data with other omics information, such as transcriptomics and epigenomics, provides a more comprehensive understanding of genome function and evolution, paving the way for new discoveries and applications in genomics.

Implications and Future Directions

The measurement of genome size has significant implications for understanding genomic architecture, evolution, and diversity. As genomic technologies continue to advance, researchers can explore genome size with greater precision and depth, uncovering new insights into the complexities of genetic material.

Comparative Genomics and Evolutionary Biology

Patterns of Genome Size Variation:

1. **Evolutionary Insights:** Measuring genome size across diverse species provides insights into the evolutionary forces shaping genome architecture. Comparative studies reveal patterns of genome size variation, highlighting the roles of non-coding DNA, repetitive elements, and polyploidy in genomic evolution. These patterns offer valuable information about the adaptive strategies and evolutionary pressures faced by different organisms.
2. **Adaptive Evolution:** Genome size variation can reflect an organism's adaptation to specific ecological niches or environmental conditions. For example, polyploidy in plants is often associated with increased adaptability to changing environments, while the expansion of repetitive elements may contribute to genomic flexibility and resilience. By studying genome size variation, researchers can gain insights into the genetic mechanisms underlying adaptation and speciation.
3. **Evolutionary Paradoxes:** The study of genome size also highlights evolutionary paradoxes, such as the C-value paradox, where genome size does not correlate with organismal complexity. By exploring the functional roles of non-coding DNA and repetitive sequences, researchers can elucidate the evolutionary significance of genome size variation and the factors that contribute to the diversity of life.

The C-Value Paradox:

1. **Non-Coding DNA:** The C-value paradox underscores the importance of non-coding DNA in shaping genome architecture and complexity. Non-coding regions, once considered "junk DNA," are now recognized as critical components of the genome, playing roles in gene regulation, chromatin organization, and evolutionary innovation. By investigating the functions of non-coding DNA, researchers can gain insights into the mechanisms that drive genome evolution and complexity.
2. **Repetitive Elements:** Repetitive DNA sequences, such as transposable elements and satellite DNA, contribute significantly to genome size variation. These elements can influence genome stability, gene expression, and evolutionary dynamics, offering explanations for the discrepancies between genome size and complexity. The study of repetitive elements provides insights into the evolutionary processes that shape genome size and the adaptive potential of different species.

Applications in Agriculture and Medicine

Crop Improvement and Breeding:

1. **Genetic Diversity:** Understanding genome size is essential for plant breeding programs aimed at developing new crop varieties with desirable traits, such as increased yield, disease resistance, and environmental tolerance. Genome size data can guide the selection of breeding strategies, inform the identification of genetic markers, and facilitate the discovery of beneficial traits.

2. **Polyploidy in Crops:** Many agriculturally important crops, such as wheat, cotton, and potatoes, are polyploid, benefiting from the increased genetic diversity and adaptability conferred by polyploidy. By studying genome size and polyploidy, researchers can develop strategies for crop improvement and optimize breeding programs to enhance crop productivity and resilience.
3. **Conservation of Genetic Resources:** Genome size information is valuable for the conservation of genetic resources and the protection of endangered species. Understanding genome size can inform the management of genetic diversity in crop populations, guide the conservation of wild relatives, and contribute to the preservation of biodiversity.

Medical Research and Diagnostics:

1. **Genetic Disorders:** In medicine, measuring genome size provides insights into the genetic basis of diseases and informs the development of targeted therapies. Genome size data can aid in the diagnosis and treatment of genetic disorders, contributing to the advancement of personalized medicine and genomics-based healthcare.
2. **Cancer Research:** Genome size measurement is particularly relevant in cancer research, where changes in DNA content can indicate chromosomal abnormalities and genomic instability. By analyzing genome size, researchers can identify genetic alterations associated with cancer development and progression, informing the development of diagnostic tools and therapeutic strategies.
3. **Therapeutic Applications:** Insights into genome size and complexity can guide the development of gene therapies and other targeted treatments. Understanding the genetic architecture of diseases allows researchers to design interventions that address the underlying causes of pathological conditions, offering new avenues for medical innovation and patient care.

Challenges and Future Research

Technical Limitations and Innovations:

1. **Measurement Challenges:** Despite advancements in measurement techniques, challenges remain in accurately estimating genome size, particularly in organisms with large or complex genomes. Continued innovation in sequencing technologies and bioinformatics tools will enhance the precision and efficiency of genome size measurement, addressing the limitations of current methodologies.
2. **Data Integration:** Integrating genome size data with other omics information, such as transcriptomics, proteomics, and epigenomics, provides a holistic understanding of genome function and evolution. Multi-omics approaches hold promise for unraveling the complexities of genomic systems and their interactions with the environment, offering new insights into the dynamic nature of genomes.

Integration of Multi-Omics Data:

1. **Systems Biology:** The integration of genome size data with systems biology approaches allows researchers to model and understand complex biological systems. By analyzing interactions

between genes, proteins, and other molecules, systems biology provides insights into the functional architecture of the genome and the dynamic processes that govern cellular function.

2. **Personalized Medicine:** Multi-omics approaches enable the development of personalized medicine, where genomic data is integrated with clinical information to tailor treatments to individual patients. By considering genome size and other genetic factors, researchers can design personalized interventions that optimize therapeutic outcomes and improve patient care.
3. **Biotechnological Innovations:** Advances in genomics and biotechnology offer new opportunities for innovation in various fields, from agriculture to medicine. By leveraging genome size data and multi-omics information, researchers can develop novel biotechnological solutions that address global challenges, such as food security, environmental sustainability, and human health.

f. Variation in Genome Size

Genome size, which refers to the total amount of DNA contained within one copy of a single genome, can vary dramatically among different species. This variation reflects the diverse evolutionary histories, ecological niches, and genomic architectures of organisms. The differences in genome size are influenced by several factors, including non-coding DNA, repetitive elements, and whole-genome duplications.

1. Differences Among Species

The diversity in genome size is particularly evident when comparing different biological domains and kingdoms, such as prokaryotes and eukaryotes, and within various groups of plants and animals. Each of these groups exhibits unique genomic characteristics that contribute to their genome size variation.

Prokaryotes

Genomic Characteristics:

- » **Small and Compact Genomes:** Prokaryotic genomes, which are found in bacteria and archaea, are typically small and compact. They range from about 160,000 base pairs in *Mycoplasma genitalium* to over 12 million base pairs in some cyanobacteria. This compactness reflects a streamlined genome that contains primarily essential genes necessary for survival and reproduction.
- » **High Gene Density:** Prokaryotic genomes are characterized by high gene density, meaning that a large proportion of the genome consists of coding sequences. This high gene density is due to the relative scarcity of non-coding DNA and the efficient organization of genetic material. Many prokaryotic genomes lack introns, which are non-coding regions within genes, further contributing to their compact size.

Genomic Streamlining:

- » **Streamlined Genomes:** The streamlined nature of prokaryotic genomes is an adaptation to their often harsh and competitive environments. Prokaryotes tend to minimize unnecessary genetic material, maintaining only the genes required for their specific ecological niches.

This genetic economy allows for rapid replication and adaptation, enabling prokaryotes to thrive in diverse environments.

- **Minimal Genomes:** Some prokaryotes possess minimal genomes, which represent the smallest set of genes required for cellular life. For instance, *Mycoplasma genitalium* has one of the smallest known genomes among free-living organisms, reflecting its specialized lifestyle as a parasitic bacterium.

Eukaryotes

Genomic Complexity:

- **Larger and More Complex Genomes:** Eukaryotic genomes are generally larger and more complex than prokaryotic genomes. They range from around 10 million base pairs in some fungi to over 100 billion base pairs in certain plant species. Eukaryotic genomes are characterized by the presence of multiple linear chromosomes, extensive non-coding regions, and a variety of regulatory elements.
- **Non-Coding DNA:** A significant proportion of eukaryotic genomes consists of non-coding DNA, which includes introns, intergenic regions, and regulatory elements. Non-coding DNA plays essential roles in regulating gene expression, maintaining chromosomal structure, and facilitating evolutionary innovation. This complexity contributes to the larger size of eukaryotic genomes compared to prokaryotes.

Diversity Among Eukaryotes:

- **Fungi and Protists:** Fungi and protists exhibit a wide range of genome sizes, reflecting their diverse lifestyles and ecological roles. Some fungi have compact genomes with minimal non-coding DNA, while others possess larger genomes with extensive repetitive elements and regulatory sequences.
- **Plants and Animals:** Among eukaryotes, plants and animals exhibit significant variation in genome size. For example, the genome of the model plant *Arabidopsis thaliana* is approximately 135 million base pairs, while the genome of the lungfish *Protopterus aethiopicus* is over 130 billion base pairs. Similarly, the human genome is about 3.2 billion base pairs, while the genome of the marbled lungfish is one of the largest known.

Plants and Animals

Plant Genomes:

- **Large and Variable Genomes:** Plant genomes are known for their large size and variability. This variation is often attributed to the presence of polyploidy, transposable elements, and repetitive sequences. Polyploidy, or whole-genome duplication, is common in plants and can result in significant increases in genome size.
- **Repetitive Elements:** Plant genomes often contain a high proportion of repetitive DNA, including transposable elements and satellite DNA. These repetitive sequences contribute to genome expansion and provide opportunities for genetic innovation and adaptation.

Animal Genomes:

- **Diverse Genome Sizes:** Animal genomes exhibit a wide range of sizes, reflecting the diversity of life forms and ecological niches. Some animals, such as insects and birds, have relatively small genomes, while others, such as amphibians and lungfish, possess some of the largest genomes known.
- **Genome Size and Complexity:** In animals, genome size does not always correlate with organismal complexity. For example, the genome of the marbled lungfish is much larger than that of humans, despite being less complex. This discrepancy highlights the role of non-coding DNA and repetitive elements in shaping genome architecture.

2. Factors Influencing Genome Size

The variation in genome size among different species is influenced by several key factors, including non-coding DNA, repetitive elements, and polyploidy. These factors contribute to the complexity and adaptability of genomes, shaping the evolutionary trajectories of organisms.

Non-Coding DNA

Role and Importance:

- **Regulation of Gene Expression:** Non-coding DNA plays a crucial role in regulating gene expression. It contains regulatory elements such as promoters, enhancers, silencers, and insulators that control the transcription of genes. These elements interact with transcription factors and other regulatory proteins to modulate gene activity in response to developmental cues and environmental signals.
- **Chromosomal Structure and Stability:** Non-coding DNA contributes to the organization and stability of chromosomes. It includes sequences that maintain chromosomal integrity, such as telomeres and centromeres, which are essential for proper chromosome segregation during cell division. Non-coding regions also play roles in chromatin organization, influencing the accessibility of DNA to regulatory proteins.

Facilitation of Evolutionary Changes:

- **Evolutionary Innovation:** Non-coding DNA provides a substrate for evolutionary innovation. Changes in non-coding regions can lead to the emergence of new regulatory elements and gene functions, contributing to the diversity and adaptability of organisms. Non-coding DNA acts as a reservoir of genetic material that can be co-opted for new functions, driving evolutionary change.
- **Genetic Variation and Adaptation:** Non-coding DNA contributes to genetic variation and adaptation by providing opportunities for mutations and recombination. Changes in regulatory sequences can alter gene expression patterns, leading to phenotypic variation and the evolution of new traits. Non-coding regions can also influence the evolution of complex traits and behaviors, contributing to the diversity of life.

Repetitive Elements

Types and Functions:

- **Transposable Elements:** Transposable elements, also known as "jumping genes," are mobile genetic elements that can move within the genome. They can replicate and insert themselves into new genomic locations, leading to genome expansion over time. Transposable elements can influence gene expression by inserting into or near genes, potentially disrupting their function or altering regulatory regions.
- **Satellite DNA and Simple Sequence Repeats:** Satellite DNA consists of long, repetitive sequences that are often associated with centromeres and heterochromatin. These sequences play critical roles in maintaining chromosomal structure and stability. Simple sequence repeats (SSRs), or microsatellites, are short, repeating sequences of 1 to 6 base pairs. They are highly polymorphic and serve as valuable genetic markers in population genetics and evolutionary studies.

Impact on Genome Architecture and Evolution:

- **Genome Expansion:** Repetitive elements are major contributors to genome expansion. Their ability to replicate and insert into new locations can lead to increases in genome size over evolutionary timescales. The accumulation of repetitive elements is a key factor in the variation of genome sizes observed across species.
- **Regulation of Gene Expression:** Repetitive elements can influence gene expression by modulating chromatin structure and accessibility. Some transposable elements contain regulatory sequences that can act as enhancers, silencers, or promoters, affecting the transcriptional activity of nearby genes. The regulatory potential of repetitive elements highlights their role in shaping gene expression networks and cellular functions.
- **Genomic Stability and Evolution:** Repetitive elements contribute to genomic instability by promoting recombination and chromosomal rearrangements. These events can lead to genetic diversity and the evolution of new traits, but they also pose challenges for maintaining genome integrity. Mechanisms that suppress transposable element activity, such as DNA methylation and RNA interference, are crucial for preserving genome stability.

Polyploidy

Definition and Occurrence:

- **Polyploidy:** Polyploidy is the condition of having more than two complete sets of chromosomes. It is a significant factor in genome size variation and plays a critical role in the evolution and adaptation of many organisms. Polyploidy is particularly common in plants, where it has been a major driver of speciation and diversity, but it is also found in some animal lineages.
- **Whole-Genome Duplications:** Polyploidy typically arises from whole-genome duplication events, where an organism inherits additional sets of chromosomes. This can occur through various mechanisms, including errors in meiosis or mitosis, hybridization between species, and endoreduplication (DNA replication without cell division).

Biological and Evolutionary Implications:

- **Genetic Diversity and Evolutionary Potential:** Polyploidy increases genetic diversity by providing additional copies of genes, which can undergo mutations and evolve new functions. This genetic redundancy allows for the exploration of new phenotypic traits and can lead to the development of novel adaptations. Polyploid organisms often exhibit increased variability and adaptability, making them more resilient to environmental changes and stresses.
- **Speciation and Diversity:** Polyploidy is a major driver of speciation in plants, contributing to the emergence of new species with distinct ecological niches. Polyploid species can exploit new habitats and resources, leading to increased biodiversity and ecosystem complexity. In some cases, polyploidy can facilitate the rapid colonization of new environments and the establishment of stable populations.
- **Gene Regulation and Expression:** The presence of multiple gene copies in polyploid organisms can influence gene expression and regulation. Polyploidy can result in changes to gene dosage, where the expression levels of genes are altered due to the increased number of copies. This can lead to phenotypic changes and may contribute to the unique characteristics of polyploid species.

3. The C-Value Paradox

The C-value paradox refers to the observation that genome size does not correlate with an organism's perceived complexity. This paradox arises from the presence of non-coding DNA and repetitive elements, which contribute to genome size without necessarily increasing organismal complexity.

Lack of Correlation with Complexity

Discrepancies in Genome Size and Complexity:

- **Complexity and Genome Size:** The C-value paradox highlights the lack of correlation between genome size and organismal complexity. Some relatively simple organisms, such as certain amphibians and plants, have larger genomes than more complex organisms like humans. This discrepancy challenges the assumption that larger genomes correspond to greater biological complexity and functional capacity.
- **Examples of the Paradox:** The marbled lungfish, *Protopterus aethiopicus*, has one of the largest known genomes, yet it is considered less complex than mammals. Similarly, some amphibians and plants possess genomes that far exceed those of more complex organisms, such as birds and mammals. These examples illustrate the disconnect between genome size and the perceived complexity of an organism.

Factors Contributing to the Paradox:

- **Non-Coding DNA and Repetitive Elements:** The presence of extensive non-coding DNA and repetitive elements contributes to the C-value paradox. These genomic components can occupy significant portions of the genome without directly influencing organismal complexity. Instead, they play roles in regulating gene expression, maintaining chromosomal stability, and facilitating evolutionary innovation.

» **Genomic Architecture and Evolution:** The C-value paradox underscores the complexity of genomic architecture and the evolutionary processes that shape genome size. Large genomes may offer advantages in terms of genetic diversity and adaptability, but they also impose metabolic costs and challenges in genome replication and maintenance. Understanding the factors that contribute to genome size variation provides insights into the balance between genetic complexity and functional efficiency.

Evolutionary Implications

Adaptation and Evolutionary Potential:

» **Genetic Diversity and Adaptation:** Large genomes may harbor greater genetic diversity, providing a reservoir of genetic material that can be harnessed for adaptation and evolution. This diversity arises from the presence of multiple gene copies, regulatory elements, and non-coding regions that serve as substrates for genetic variation. The evolutionary potential of large genomes is enhanced by their ability to explore new phenotypic traits and ecological niches.

» **Adaptive Strategies:** The C-value paradox highlights the adaptive strategies employed by different organisms to navigate their environments. While some species have evolved streamlined genomes with minimal non-coding DNA, others have expanded their genomes to accommodate complex regulatory networks and adaptive traits. These strategies reflect the diverse evolutionary pressures faced by organisms and the dynamic nature of genomic evolution.

Challenges and Considerations:

» **Metabolic Costs:** Large genomes impose metabolic costs on organisms, requiring greater resources for DNA replication, transcription, and repair. Balancing these costs with the potential benefits of genetic diversity is a key consideration in the evolution of genome size. Understanding the trade-offs between genome size and metabolic efficiency provides insights into the constraints and opportunities of genomic evolution.

» **Genomic Stability:** The increased genomic content in large genomes poses challenges for maintaining genomic stability and accurate chromosome segregation during cell division. Polyploidy and repetitive elements can lead to complications in meiosis and an increased risk of chromosomal aberrations, which can affect fertility and viability. Mechanisms that preserve genome stability, such as DNA repair pathways and epigenetic regulation, are crucial for the success of large-genome organisms.

g. Implications of Genome Size

The size of an organism's genome has profound implications for its biology, evolution, and adaptability. Genome size influences various biological processes, from gene expression regulation to evolutionary potential, and provides insights into the interactions between genetic material and environmental factors. Genome size helps elucidate the mechanisms of genome evolution, the complexity of genetic networks, and the adaptive strategies employed by organisms.

1. Evolutionary Potential

Genome size plays a significant role in shaping an organism's evolutionary potential by influencing genetic diversity and adaptability. Larger genomes often harbor more genetic material, which serves as a reservoir for evolutionary innovation and adaptation.

Genetic Diversity

Reservoir of Diversity:

» **Genetic Material:** Larger genomes typically contain more genetic material, providing a broader base for genetic variation and evolutionary change. This genetic diversity arises from the presence of multiple gene copies, repetitive elements, and extensive non-coding regions that can undergo mutations and recombination. These genomic components serve as substrates for genetic variation, enabling organisms to explore new phenotypic traits and adapt to changing environments.

» **Functional Redundancy:** The presence of multiple copies of genes in larger genomes can lead to functional redundancy, where duplicate genes may acquire new functions or regulatory patterns over time. This redundancy allows for the diversification of gene function and contributes to the emergence of novel traits, enhancing an organism's evolutionary potential.

Evolutionary Innovation:

» **Adaptation to Environments:** Genome size influences an organism's ability to adapt to different environmental conditions. Larger genomes provide the genetic flexibility needed to respond to diverse ecological challenges, such as climate change, habitat alteration, and biotic interactions. The genetic variation inherent in larger genomes allows populations to adapt to selective pressures, promoting survival and evolutionary success.

» **Exploration of New Niches:** Genetic diversity within large genomes facilitates the exploration of new ecological niches. The presence of a diverse array of genes and regulatory elements enables organisms to occupy a wide range of habitats and adapt to various environmental conditions. This adaptability is crucial for species that inhabit dynamic ecosystems or face rapid environmental changes.

Adaptability and Speciation

Polyploidy and Speciation:

» **Role of Polyploidy:** Polyploidy, the condition of having more than two complete sets of chromosomes, is a significant factor in the evolutionary potential of species. Polyploid organisms often exhibit increased genetic diversity and variability, which can lead to the emergence of new species. The additional genetic material provided by polyploidy serves as a substrate for natural selection and diversification, facilitating speciation and the evolution of novel traits.

» **Adaptive Radiation:** Polyploidy can promote adaptive radiation, where a single ancestral species rapidly diversifies into multiple descendant species that exploit different ecological niches. This process is particularly common in plants, where polyploidy has played a key role in the diversification of angiosperms. The increased genetic variation and flexibility associated with polyploidy enable species to colonize new environments and adapt to various ecological challenges.

Speciation Mechanisms:

- **Reproductive Isolation:** Genome size can influence the mechanisms of reproductive isolation, a key factor in speciation. Differences in genome size and content can lead to reproductive barriers between populations, preventing gene flow and facilitating the divergence of species. For example, changes in chromosomal structure or number can result in hybrid sterility or inviability, promoting speciation.
- **Genomic Divergence:** The presence of large genomes with extensive non-coding regions provides opportunities for genomic divergence between populations. Non-coding DNA can accumulate mutations that alter gene regulation and expression, leading to phenotypic differences and the evolution of new species. This genomic divergence contributes to the diversity and complexity of life on Earth.

2. Regulation of Gene Expression

Genome size affects the regulation of gene expression through its influence on chromatin organization, gene density, and the complexity of regulatory networks. These factors play a crucial role in determining the functional architecture of the genome and the dynamic processes that govern cellular function.

Chromatin Organization

Structural Complexity:

- **Chromatin Structure:** Genome size impacts chromatin organization, influencing how DNA is packaged and accessed within the cell nucleus. Larger genomes often exhibit more complex chromatin structures, with extensive non-coding regions and repetitive elements that contribute to chromatin compaction and organization. The dynamic organization of chromatin affects the accessibility of DNA to transcription factors and other regulatory proteins, modulating gene expression.
- **Euchromatin and Heterochromatin:** Chromatin exists in two main forms: euchromatin and heterochromatin. Euchromatin is loosely packed and transcriptionally active, allowing for gene expression, while heterochromatin is densely packed and transcriptionally inactive, playing roles in maintaining genome stability and regulating gene expression. The balance between euchromatin and heterochromatin is influenced by genome size and the distribution of non-coding DNA.

Epigenetic Regulation:

- **Role of Epigenetics:** Epigenetic modifications, such as DNA methylation and histone modifications, play critical roles in regulating gene expression and chromatin organization.

These modifications can alter the accessibility of chromatin and the activity of genes, influencing cellular processes and development. The complexity of epigenetic regulation is often correlated with genome size, as larger genomes possess more regulatory elements and epigenetic marks that modulate gene activity.

» **Environmental Responses:** Epigenetic regulation allows organisms to respond to environmental cues by altering gene expression patterns. Larger genomes, with their extensive regulatory networks, provide greater flexibility in epigenetic responses, enabling organisms to adapt to changing environments and stresses.

Gene Density and Expression

Impact on Gene Density:

» **Distribution of Genes:** Genome size influences gene density, with larger genomes typically having lower gene density due to the presence of extensive non-coding DNA. This distribution affects the overall architecture of the genome and the spatial organization of genes within chromosomes. The arrangement of genes and non-coding regions can impact the regulation of gene expression and the interactions between genetic elements.

» **Functional Implications:** Lower gene density in larger genomes can result in more complex regulatory networks, where non-coding regions play critical roles in controlling gene expression. These networks involve interactions between promoters, enhancers, silencers, and other regulatory elements that modulate the transcriptional activity of genes. The complexity of these networks contributes to the diversity and adaptability of organisms with large genomes.

Gene Expression and Regulation:

» **Transcriptional Complexity:** Genome size affects the complexity of transcriptional regulation, with larger genomes often possessing intricate regulatory pathways that govern gene expression. The presence of extensive non-coding DNA and repetitive elements provides opportunities for the evolution of novel regulatory mechanisms that fine-tune gene activity in response to developmental and environmental signals.

» **Regulatory Interactions:** The interactions between genes and regulatory elements in larger genomes contribute to the diversity of gene expression patterns. These interactions involve the coordination of multiple regulatory factors, such as transcription factors, RNA molecules, and chromatin remodelers, that influence the spatial and temporal expression of genes. The complexity of these interactions is a hallmark of larger genomes, enabling precise control over cellular processes and functions.

3. Challenges and Considerations

The maintenance of large genomes presents several challenges and considerations, including metabolic costs and genomic stability. Balancing these factors with the potential benefits of genome size is crucial for the evolution and success of species.

Metabolic Costs

Resource Allocation:

- **DNA Replication and Repair:** Maintaining a large genome imposes significant metabolic costs on an organism, requiring substantial resources for DNA replication, transcription, and repair. The synthesis and maintenance of extensive genetic material demand energy and cellular resources, influencing an organism's metabolic efficiency and fitness.
- **Transcriptional Burden:** The transcription of large genomes involves the production of numerous RNA molecules, contributing to the metabolic demands of the cell. The regulation and processing of these transcripts require additional resources, impacting the overall metabolic balance of the organism. Efficient resource allocation is essential for maintaining genome integrity and supporting cellular functions.

Balancing Costs and Benefits:

- **Trade-Offs:** The evolutionary success of species with large genomes depends on balancing the metabolic costs of maintaining extensive genetic material with the potential benefits of increased genetic diversity and adaptability. Organisms must navigate the trade-offs between genome size and metabolic efficiency, optimizing their genetic architecture to meet ecological and evolutionary challenges.
- **Adaptive Strategies:** Some organisms employ adaptive strategies to mitigate the metabolic costs of large genomes. For example, polyploid plants may exhibit increased photosynthetic efficiency or enhanced nutrient uptake, offsetting the energetic demands of maintaining additional genetic material. These strategies highlight the dynamic interplay between genome size and metabolic adaptation.

Genomic Stability

Challenges to Stability:

- **Chromosomal Rearrangements:** Large genomes are more susceptible to genomic instability, such as chromosomal rearrangements, duplications, deletions, and translocations. These structural changes can disrupt gene function, alter regulatory networks, and compromise genome integrity. The increased genomic content in large genomes poses challenges for accurate chromosome segregation during cell division, affecting fertility and viability.
- **Mutational Load:** The accumulation of mutations in large genomes can impact genomic stability and evolutionary fitness. Mutations may arise from errors in DNA replication, exposure to environmental mutagens, or transposable element activity. The management of mutational load is critical for preserving genome integrity and ensuring the successful transmission of genetic material across generations.

Mechanisms of Stability:

- **DNA Repair Pathways:** Mechanisms that maintain genomic stability, such as DNA repair pathways, are essential for preserving genome integrity. These pathways involve the detection and repair of DNA damage, preventing the accumulation of mutations and maintaining chromosomal structure. Efficient DNA repair is particularly important for organisms with large genomes, where the potential for genomic instability is higher.

- **Epigenetic Regulation:** Epigenetic mechanisms play a crucial role in maintaining genomic stability by regulating chromatin organization and gene expression. Epigenetic marks, such as DNA methylation and histone modifications, contribute to the suppression of transposable elements and the stabilization of chromosomal architecture. These regulatory processes are vital for protecting large genomes from structural and functional disruptions.

4. Research and Technological Advances

Advances in sequencing technologies and genomic research have facilitated the exploration of genome size and content, providing insights into the genetic architecture of diverse organisms. These technologies have informed research in genetics, evolution, and conservation biology.

Sequencing Technologies

Advancements in Sequencing:

- **Next-Generation Sequencing (NGS):** Advances in sequencing technologies, such as next-generation sequencing (NGS), have revolutionized the study of genomes by enabling the rapid and cost-effective sequencing of entire genomes. NGS platforms, such as Illumina, PacBio, and Oxford Nanopore, provide high-throughput and high-resolution data, allowing researchers to explore genome size and content in unprecedented detail.
- **Third-Generation Sequencing:** Third-generation sequencing technologies, such as single-molecule real-time (SMRT) sequencing and nanopore sequencing, offer long-read capabilities, enabling the assembly of complex genomes with high accuracy. These technologies are particularly useful for resolving repetitive regions and structural variations that contribute to genome size variation.

Insights into Genome Architecture:

- **Comprehensive Genomic Analysis:** Sequencing-based approaches provide a comprehensive view of the genome, including both coding and non-coding regions, repetitive elements, and structural variations. By sequencing the entire genome, researchers can accurately determine genome size, identify gene content, and explore the functional roles of genetic elements.
- **Comparative Genomics:** DNA sequencing allows for the comparison of genome sizes and contents across different species, revealing patterns of conservation and divergence. Comparative genomics provides insights into the evolutionary forces shaping genome architecture and informs our understanding of genetic diversity and adaptation.

Applications in Breeding and Conservation

Crop Improvement and Breeding:

- **Genetic Diversity and Selection:** Understanding genome size is essential for plant breeding programs aimed at developing new crop varieties with desirable traits, such as increased yield, disease resistance, and environmental tolerance. Genome size data can guide the selection of breeding strategies, inform the identification of genetic markers, and facilitate the discovery of beneficial traits.
- **Polyploidy in Crops:** Many agriculturally important crops, such as wheat, cotton, and potatoes, are polyploid, benefiting from the increased genetic diversity and adaptability conferred by polyploidy. By studying genome size and polyploidy, researchers can develop strategies for crop improvement and optimize breeding programs to enhance crop productivity and resilience.

Conservation of Genetic Resources:

- **Conservation Biology:** Genome size information is valuable for the conservation of genetic resources and the protection of endangered species. Understanding genome size can inform the management of genetic diversity in crop populations, guide the conservation of wild relatives, and contribute to the preservation of biodiversity.
- **Assessment of Genetic Diversity:** Genome size data can be used to assess genetic diversity within populations, informing conservation efforts and management strategies. By analyzing genome size variation, researchers can identify populations at risk of genetic erosion and develop strategies to maintain genetic diversity and resilience.

D. Organization of organellar genomes

The organization of organellar genomes, specifically those of mitochondria and chloroplasts, is a fascinating area of study in genetics and molecular biology. These organelles contain their genetic material, separate from the nuclear genome, and exhibit unique structural and functional characteristics. The organellar genomes have been instrumental in understanding evolutionary processes, cellular energy production, and metabolic functions.

a. Mitochondrial Genomes

The mitochondrial genome plays a critical role in the energy production, metabolism, and cell signaling processes that occur within eukaryotic cells. The structure, organization, and function of mitochondrial genomes provide insight into their evolutionary origins and their importance in cellular activities.

Structure and Organization

Mitochondria possess their distinct genomes, which are separate from the nuclear genome. These genomes exhibit unique structural characteristics that reflect their evolutionary origins and functional roles within the cell.

1. Circular DNA

Circular Genome Structure:

» **Comparative Structure:** The mitochondrial genome is typically circular, resembling the genomes of bacteria. This circular structure is compact and highly efficient, allowing mitochondria to maintain essential genetic information within a limited space. The compact nature of mitochondrial DNA (mtDNA) is crucial for its efficient replication and transcription within the organelle.

» **Size Variation:** The size of the mitochondrial genome varies among different species, reflecting differences in their evolutionary history and metabolic requirements. For instance, human mtDNA is approximately 16,569 base pairs long, encoding 37 genes essential for mitochondrial function. In contrast, some plant and protist mitochondrial genomes can be much larger, containing additional genes and regulatory sequences that contribute to their complexity.

» **Genome Organization:** The circular organization of mtDNA facilitates its replication and maintenance. Unlike linear chromosomes, circular DNA can be easily replicated without the need for specialized structures, such as telomeres, that protect linear chromosome ends. This feature is advantageous for the rapid and continuous replication of mtDNA in actively dividing cells.

Multiple Copies:

» **Redundancy and Stability:** Mitochondria contain multiple copies of their genome, with each mitochondrion housing several circular DNA molecules. This redundancy is crucial for maintaining mitochondrial function, as it allows for the compensation of potential DNA damage or mutations. Multiple copies of mtDNA ensure that sufficient genetic material is available for the production of essential proteins and RNAs, even if some copies are damaged.

» **Dynamic Copy Number:** The number of mtDNA copies per mitochondrion can vary depending on the cell type, developmental stage, and environmental conditions. Cells with high energy demands, such as muscle cells and neurons, often contain more mitochondria and mtDNA copies to support their increased metabolic activity. The ability to adjust mtDNA copy number provides a mechanism for cells to regulate mitochondrial function in response to changing physiological needs.

» **Implications for Disease:** The presence of multiple mtDNA copies also has implications for mitochondrial diseases. Mutations in mtDNA can lead to a heteroplasmic state, where both normal and mutant copies coexist within a cell. The proportion of mutant mtDNA can influence the severity of mitochondrial diseases, highlighting the importance of mtDNA copy number and integrity in maintaining cellular health.

2. Gene Content

Protein-Coding Genes:

» **Essential Proteins:** The mitochondrial genome encodes essential proteins that are integral components of the electron transport chain (ETC) and ATP synthesis. In humans, mtDNA encodes 13 proteins, including NADH dehydrogenase, cytochrome c oxidase, and ATP synthase subunits. These proteins are critical for the process of oxidative phosphorylation, where electrons are transferred through a series of complexes in the inner mitochondrial membrane to generate ATP.

» **Role in Energy Metabolism:** The proteins encoded by mtDNA play a central role in cellular energy metabolism. They facilitate the transfer of electrons and protons across the inner mitochondrial membrane, creating a proton gradient that drives ATP synthesis. This process is the primary means of energy production in eukaryotic cells, highlighting the importance of mtDNA-encoded proteins in sustaining life.

» **Conservation Across Species:** The genes encoding these proteins are highly conserved across different species, reflecting their fundamental role in cellular energy production. The conservation of mitochondrial protein-coding genes underscores the evolutionary importance of mitochondria as energy-generating organelles.

Ribosomal and Transfer RNAs:

» **Mitochondrial Ribosomes:** Mitochondrial genomes also encode ribosomal RNAs (rRNAs), including the 12S and 16S rRNAs, which are components of the mitochondrial ribosomes. Mitochondrial ribosomes are responsible for translating mtDNA-encoded mRNAs into proteins within the organelle. The distinct structure of mitochondrial ribosomes reflects their adaptation to the unique environment of the mitochondrion.

» **Transfer RNAs (tRNAs):** Human mtDNA includes 22 tRNA genes, which are necessary for mitochondrial protein synthesis. These tRNAs are responsible for delivering amino acids to the mitochondrial ribosomes during translation, facilitating the assembly of proteins encoded by mtDNA. The set of mitochondrial tRNAs is sufficient to decode the mitochondrial genetic code, which has some variations compared to the universal genetic code used in nuclear genomes.

» **Semiautonomous Nature:** The presence of rRNA and tRNA genes within the mitochondrial genome highlights the organelle's semiautonomous nature. While mitochondria rely on nuclear-encoded proteins for many of their functions, they possess the genetic machinery to synthesize some of their proteins independently. This autonomy is a remnant of their evolutionary origins as free-living bacteria.

3. Non-Coding Regions

D-Loop:

» **Regulatory Region:** The mitochondrial genome contains non-coding regions, such as the displacement loop (D-loop), which serves as the origin of replication and transcription. The D-loop is a critical regulatory region that influences the replication and expression of mitochondrial DNA. It contains essential sequences for the initiation of mtDNA replication and transcription, as well as binding sites for regulatory proteins.

- **Replication Control:** The D-loop plays a key role in controlling the replication of mtDNA. It contains the origin of replication for the heavy (H) strand and is involved in the initiation of replication for both the H and light (L) strands. The regulation of mtDNA replication is tightly linked to the cell cycle and energy demands, ensuring that mitochondrial function is maintained under various physiological conditions.
- **Transcriptional Regulation:** The D-loop also regulates the transcription of mtDNA. It contains promoters for both the H and L strands, which drive the expression of mitochondrial genes. The regulation of transcription within the D-loop is essential for coordinating the production of mitochondrial proteins and RNAs, ensuring efficient energy production and metabolic function.
- **Genetic Variation:** The D-loop is a hotspot for genetic variation and mutations due to its non-coding nature and regulatory significance. Variations in the D-loop can impact mtDNA replication and transcription, leading to changes in mitochondrial function. The high mutation rate in the D-loop makes it a useful marker for studying population genetics and evolutionary relationships.

Evolutionary Significance

The evolutionary history of mitochondria provides insights into the origin and diversification of eukaryotic cells. The structure and function of mitochondrial genomes reflect their evolutionary past and their integration into host cells.

1. Endosymbiotic Theory

Origins of Mitochondria:

- **Endosymbiotic Event:** The endosymbiotic theory posits that mitochondria originated from free-living proteobacteria that were engulfed by ancestral eukaryotic cells. This endosymbiotic event is believed to have occurred over a billion years ago, leading to the integration of mitochondria into the host cell. The engulfed bacteria formed a symbiotic relationship with the host, providing it with a source of ATP in exchange for a stable environment and nutrients.
- **Evidence Supporting the Theory:** Several lines of evidence support the endosymbiotic theory. The circular nature of mitochondrial DNA and its resemblance to bacterial genomes are key indicators of this evolutionary origin. Additionally, the presence of bacterial-like ribosomes and the similarity of mitochondrial proteins to those found in proteobacteria further support the theory.
- **Evolutionary Integration:** Over time, mitochondria became fully integrated into the eukaryotic cell, losing many of their original genes to the host nucleus. This integration allowed for the specialization of mitochondria as energy-producing organelles, essential for the complexity and diversity of eukaryotic life. The endosymbiotic theory highlights the importance of symbiosis in the evolution of life, demonstrating how cooperative interactions between different organisms can lead to new biological innovations.

2. Genome Reduction

Gene Transfer to Nucleus:

- **Loss of Genes:** During evolution, many mitochondrial genes were transferred to the nuclear genome, resulting in a reduced mitochondrial genome. This gene transfer was facilitated by the exchange of genetic material between the mitochondrion and the host nucleus, a process that likely occurred over millions of years. The loss of genes from the mitochondrial genome reflects the integration of mitochondrial function with the host cell's regulatory networks.
- **Functional Integration:** The transfer of mitochondrial genes to the nucleus allowed for more efficient coordination of cellular processes. Nuclear-encoded mitochondrial proteins are imported into the mitochondria, where they contribute to various functions, including oxidative phosphorylation, metabolic regulation, and organelle maintenance. This integration ensures that mitochondrial activity is tightly linked to the overall metabolic state of the cell.
- **Evolutionary Implications:** The reduction of the mitochondrial genome has significant evolutionary implications. It highlights the dynamic nature of genome evolution, where gene transfer and loss can lead to the specialization and adaptation of organelles. The integration of mitochondrial function with the nuclear genome also underscores the importance of cellular compartmentalization in the evolution of complex life forms.

Functional Roles

Mitochondria are essential organelles with diverse functional roles in cellular processes. Their unique genetic and biochemical characteristics enable them to perform critical functions related to energy production, metabolism, and cell signaling.

1. Energy Production

Oxidative Phosphorylation:

- **ATP Synthesis:** Mitochondria are often referred to as the powerhouse of the cell due to their role in generating ATP through oxidative phosphorylation. This process occurs in the inner mitochondrial membrane, where a series of protein complexes, known as the electron transport chain (ETC), transfer electrons from NADH and FADH2 to oxygen. The energy released during electron transfer is used to pump protons across the inner membrane, creating a proton gradient.
- **Chemiosmotic Coupling:** The proton gradient generated by the ETC drives ATP synthesis through chemiosmotic coupling. Protons flow back into the mitochondrial matrix through ATP synthase, a protein complex that catalyzes the conversion of ADP and inorganic phosphate into ATP. This process is the primary means of energy production in eukaryotic cells, providing the ATP required for various cellular functions.
- **Efficiency and Regulation:** The efficiency of oxidative phosphorylation is influenced by various factors, including the availability of substrates, oxygen levels, and the integrity of mitochondrial membranes. Mitochondrial genomes encode key components of the ETC,

ensuring the proper assembly and function of the oxidative phosphorylation machinery. The regulation of mitochondrial energy production is critical for maintaining cellular energy homeostasis and responding to changes in metabolic demand.

2. Metabolic Pathways

Biosynthesis and Metabolism:

» **Citric Acid Cycle:** Mitochondria are involved in various metabolic pathways, including the citric acid cycle (Krebs cycle), which occurs in the mitochondrial matrix. This cycle oxidizes acetyl-CoA to produce NADH and FADH2, which are used in the ETC to generate ATP. The citric acid cycle also produces metabolic intermediates used in the synthesis of amino acids, nucleotides, and other biomolecules.

» **Fatty Acid Oxidation:** Mitochondria are the site of fatty acid oxidation, a process that breaks down fatty acids into acetyl-CoA units. This process provides an important source of energy, especially during periods of fasting or increased energy demand. Fatty acid oxidation is tightly regulated to ensure the efficient utilization of lipid stores and the prevention of lipid accumulation.

» **Amino Acid Metabolism:** Mitochondria play a role in amino acid metabolism, including the catabolism of branched-chain amino acids and the interconversion of amino acids and their derivatives. These processes are essential for maintaining amino acid homeostasis and supporting protein synthesis and other metabolic functions.

3. Apoptosis and Cell Signaling

Role in Apoptosis:

» **Programmed Cell Death:** Mitochondria play a pivotal role in programmed cell death, or apoptosis, a process that eliminates damaged or unwanted cells. During apoptosis, mitochondria release pro-apoptotic factors, such as cytochrome c, into the cytosol. Cytochrome c activates a cascade of proteolytic enzymes called caspases, leading to the dismantling of cellular components and cell death.

» **Regulation of Apoptosis:** The regulation of mitochondrial-mediated apoptosis involves complex interactions between pro-apoptotic and anti-apoptotic proteins. Members of the Bcl-2 family of proteins regulate the permeability of the mitochondrial outer membrane, controlling the release of apoptotic factors. The balance between these proteins determines the cell's susceptibility to apoptosis, influencing tissue homeostasis and immune responses.

Cell Signaling:

» **Reactive Oxygen Species (ROS):** Mitochondria are a major source of reactive oxygen species (ROS), which are byproducts of oxidative phosphorylation. While high levels of ROS can cause cellular damage, low levels serve as signaling molecules that regulate various cellular processes, including cell proliferation, differentiation, and stress responses.

- **Calcium Signaling:** Mitochondria play a role in calcium signaling by sequestering and releasing calcium ions in response to cellular signals. Calcium uptake by mitochondria influences ATP production and the regulation of metabolic pathways, while calcium release modulates cellular processes such as muscle contraction and neurotransmitter release.
- **Integration with Cellular Networks:** Mitochondria are integrated into cellular signaling networks, coordinating metabolic responses and stress adaptations. Their ability to sense and respond to changes in cellular energy status and environmental conditions highlights their role as central regulators of cellular homeostasis and function.

b. Chloroplast Genomes

Chloroplast genomes play a crucial role in the life processes of plants and algae. As the site of photosynthesis and various biosynthetic pathways, chloroplasts contain genetic information essential for their function. These genomes provide insights into the evolutionary history, structural organization, and functional roles of chloroplasts within cells.

Structure and Organization

Chloroplast genomes exhibit unique structural characteristics that are integral to their function in photosynthesis and other cellular processes. Understanding the organization of these genomes reveals the complexity and efficiency of chloroplasts as semi-autonomous organelles.

1. Circular DNA

Circular Genome Structure:

- **Characteristics and Size:** Chloroplast genomes are typically circular DNA molecules, ranging in size from 120,000 to 160,000 base pairs. They are generally larger than mitochondrial genomes, reflecting their greater gene content and complexity. For example, the chloroplast genome of *Arabidopsis thaliana* is approximately 154,478 base pairs long and encodes around 120 genes, including those involved in photosynthesis and other essential metabolic processes.
- **Compact and Efficient:** The circular structure of chloroplast DNA (cpDNA) allows for the compact and efficient organization of genetic information. This configuration facilitates the replication, transcription, and repair of chloroplast genomes, ensuring the proper functioning of the organelle. The compactness of cpDNA also minimizes the occurrence of errors during these processes, contributing to the stability of the chloroplast genome.
- **Comparison with Bacterial Genomes:** The circular nature of cpDNA is reminiscent of bacterial genomes, supporting the endosymbiotic theory of chloroplast origin. This structural similarity underscores the evolutionary relationship between chloroplasts and their cyanobacterial ancestors, highlighting the retention of prokaryotic features in chloroplast genomes.

Multiple Copies:

- **Redundancy and Stability:** Chloroplasts contain multiple copies of their genome, with each chloroplast housing several circular DNA molecules. This redundancy ensures the maintenance of chloroplast function and compensates for potential DNA damage or mutations. The presence of multiple cpDNA copies allows chloroplasts to continue functioning even if some copies are damaged, providing a buffer against genetic disruptions.

» **Dynamic Regulation:** The number of cpDNA copies can vary depending on the cell type, developmental stage, and environmental conditions. Cells with high photosynthetic activity, such as mesophyll cells in leaves, often contain more chloroplasts and cpDNA copies to support increased metabolic demands. This dynamic regulation of cpDNA copy number enables plants to adapt to changing environmental conditions and optimize their photosynthetic efficiency.

» **Implications for Genetic Engineering:** The presence of multiple cpDNA copies also has implications for genetic engineering. Chloroplast transformation, where foreign genes are inserted into the chloroplast genome, can achieve high levels of gene expression due to the abundance of cpDNA copies. This feature makes chloroplasts attractive targets for producing recombinant proteins and other biotechnological applications.

2. Gene Content

Photosynthetic Genes:

» **Components of Photosystems:** Chloroplast genomes encode proteins involved in photosynthesis, including components of photosystems I and II, the cytochrome b6f complex, ATP synthase, and the Calvin cycle. These genes are essential for capturing light energy and converting it into chemical energy. Photosystem I and II proteins, such as P700 and P680, play crucial roles in light absorption and electron transfer, driving the synthesis of ATP and NADPH.

» **Electron Transport Chain:** The chloroplast genome also encodes proteins that form the electron transport chain within the thylakoid membranes. This chain facilitates the transfer of electrons from water to NADP+, generating a proton gradient that drives ATP synthesis. Key components of this chain include plastocyanin, plastoquinone, and the cytochrome b6f complex, all of which are encoded by cpDNA.

» **Calvin Cycle Enzymes:** The Calvin cycle, responsible for carbon fixation, relies on enzymes encoded by the chloroplast genome, such as ribulose-1,5-bisphosphate carboxylase/oxygenase (RuBisCO) and phosphoribulokinase. These enzymes catalyze the conversion of carbon dioxide into organic compounds, such as glucose, which serve as energy sources for plant growth and development.

Ribosomal and Transfer RNAs:

» **Chloroplast Ribosomes:** Chloroplast genomes encode ribosomal RNAs (rRNAs), which are components of the chloroplast ribosomes. Chloroplast ribosomes resemble bacterial ribosomes in structure and function, reflecting the evolutionary origin of chloroplasts from cyanobacteria. The chloroplast genome of *Arabidopsis thaliana*, for example, includes 4 rRNA genes that contribute to the assembly of functional ribosomes within the organelle.

- **Transfer RNAs (tRNAs):** Chloroplast genomes also encode transfer RNAs (tRNAs) required for protein synthesis. The cpDNA of *Arabidopsis thaliana* contains 30 tRNA genes, each corresponding to specific amino acids needed for protein assembly. These tRNAs facilitate the translation of chloroplast-encoded mRNAs into proteins, ensuring the production of essential photosynthetic and metabolic proteins within the organelle.
- **Autonomous Protein Synthesis:** The presence of rRNA and tRNA genes within the chloroplast genome highlights the organelle's capacity for autonomous protein synthesis. While chloroplasts rely on nuclear-encoded proteins for many functions, they possess the genetic machinery to synthesize some proteins independently, reflecting their semi-autonomous nature.

3. Inverted Repeats

Structural Features:

- **Definition and Role:** Chloroplast genomes often contain inverted repeats (IRs), which are sequences that are identical but oriented in opposite directions. These repeats play a critical role in maintaining genome stability and facilitating the repair of DNA damage. The presence of IRs contributes to the structural organization of the chloroplast genome, ensuring its integrity and proper function.
- **Genome Stabilization:** Inverted repeats help stabilize the chloroplast genome by promoting homologous recombination, a process that repairs DNA breaks and maintains genetic integrity. This stabilization is crucial for preserving the functionality of cpDNA, particularly in the face of environmental stresses that can induce DNA damage.
- **Implications for Evolution:** The presence of IRs in chloroplast genomes has evolutionary implications. The expansion and contraction of IRs can lead to changes in genome size and structure, influencing the evolution of chloroplast genomes across different plant lineages. The conservation of IRs among diverse plant species underscores their importance in chloroplast genome evolution and stability.

Evolutionary Significance

The evolutionary history of chloroplasts provides insights into the origin and diversification of photosynthetic eukaryotes. The structure and function of chloroplast genomes reflect their evolutionary past and their integration into host cells.

1. Endosymbiotic Theory

Origins of Chloroplasts:

- **Endosymbiotic Event:** The endosymbiotic theory posits that chloroplasts originated from free-living cyanobacteria that were engulfed by ancestral eukaryotic cells. This endosymbiotic event is believed to have occurred over a billion years ago, leading to the evolution of chloroplasts and the establishment of photosynthetic eukaryotes. The engulfed cyanobacteria formed a symbiotic relationship with the host cell, providing it with the ability to capture and convert light energy into chemical energy.

- **Evidence Supporting the Theory:** Several lines of evidence support the endosymbiotic theory. The similarities between chloroplast DNA and bacterial genomes, such as the circular structure and gene content, are key indicators of this evolutionary origin. Additionally, the presence of bacterial-like ribosomes and the similarity of chloroplast proteins to those found in cyanobacteria further support the theory.
- **Evolutionary Integration:** Over time, chloroplasts became fully integrated into the eukaryotic cell, losing many of their original genes to the host nucleus. This integration allowed for the specialization of chloroplasts as photosynthetic organelles, essential for the diversity and success of photosynthetic eukaryotes. The endosymbiotic theory highlights the importance of symbiosis in the evolution of life, demonstrating how cooperative interactions between different organisms can lead to new biological innovations.

2. Genome Reduction

Gene Transfer to Nucleus:

- **Loss of Genes:** Like mitochondria, chloroplasts have undergone genome reduction through the transfer of genes to the nuclear genome. This gene transfer was facilitated by the exchange of genetic material between the chloroplast and the host nucleus, a process that likely occurred over millions of years. The loss of genes from the chloroplast genome reflects the integration of chloroplast function with the host cell's regulatory networks.
- **Functional Integration:** The transfer of chloroplast genes to the nucleus allowed for more efficient coordination of cellular processes. Nuclear-encoded chloroplast proteins are imported into the chloroplast, where they contribute to various functions, including photosynthesis, metabolic regulation, and organelle maintenance. This integration ensures that chloroplast activity is tightly linked to the overall metabolic state of the cell.
- **Evolutionary Implications:** The reduction of the chloroplast genome has significant evolutionary implications. It highlights the dynamic nature of genome evolution, where gene transfer and loss can lead to the specialization and adaptation of organelles. The integration of chloroplast function with the nuclear genome also underscores the importance of cellular compartmentalization in the evolution of complex life forms.

Functional Roles

Chloroplasts are essential organelles with diverse functional roles in cellular processes. Their unique genetic and biochemical characteristics enable them to perform critical functions related to photosynthesis, metabolism, and cellular regulation.

1. Photosynthesis

Energy Conversion:

- **Light Capture and Conversion:** Chloroplasts are responsible for photosynthesis, the process by which light energy is captured and converted into chemical energy in the form of ATP and NADPH. Photosynthesis occurs in two stages: the light-dependent reactions and the

Calvin cycle. During the light-dependent reactions, chlorophyll and other pigments in the thylakoid membranes absorb light energy, exciting electrons that are transferred through a series of electron carriers.

» **Production of ATP and NADPH:** The energy captured during the light-dependent reactions is used to generate ATP and NADPH. These energy-rich molecules are produced through photophosphorylation, where the proton gradient generated by the electron transport chain drives ATP synthesis. NADPH is produced through the reduction of NADP+ by electrons from photosystem I, providing reducing power for subsequent biosynthetic processes.

» **Calvin Cycle:** The ATP and NADPH generated in the light-dependent reactions are used in the Calvin cycle, which occurs in the chloroplast stroma. The Calvin cycle is responsible for carbon fixation, where carbon dioxide is converted into organic compounds, such as glucose, through a series of enzyme-catalyzed reactions. Key enzymes involved in the Calvin cycle include RuBisCO, which catalyzes the initial step of carbon fixation.

2. Biosynthesis

Production of Metabolites:

» **Synthesis of Amino Acids:** Chloroplasts are involved in the biosynthesis of various amino acids, which are the building blocks of proteins. The chloroplasts house pathways for the synthesis of essential amino acids, such as glutamine, serine, and glycine, which play critical roles in plant metabolism and development.

» **Fatty Acid Biosynthesis:** Chloroplasts are also the site of fatty acid biosynthesis, where acetyl-CoA is converted into long-chain fatty acids. These fatty acids are important components of cellular membranes and serve as precursors for the synthesis of lipids and other complex molecules. The regulation of fatty acid biosynthesis in chloroplasts is crucial for maintaining membrane fluidity and integrity.

» **Secondary Metabolites:** In addition to primary metabolites, chloroplasts contribute to the production of secondary metabolites, which have roles in plant defense and adaptation. These compounds include terpenoids, flavonoids, and alkaloids, which are involved in protecting plants from herbivores, pathogens, and environmental stresses. The biosynthesis of secondary metabolites in chloroplasts is tightly regulated and influenced by environmental cues.

3. Regulation of Cellular Processes

Redox Signaling:

» **Role of Reactive Oxygen Species (ROS):** Chloroplasts are involved in redox signaling, a process that regulates cellular responses to environmental changes. The production of reactive oxygen species (ROS) during photosynthesis serves as a signal for activating stress response pathways and modulating gene expression. ROS are byproducts of electron transport in chloroplasts and play roles in signaling pathways that regulate plant growth and defense.

» **Signal Transduction:** Redox signaling in chloroplasts involves the modulation of redox-sensitive proteins and transcription factors, which influence the expression of genes involved in stress responses, metabolism, and development. The balance between ROS production and scavenging is critical for maintaining cellular homeostasis and preventing oxidative damage.

» **Environmental Adaptation:** Chloroplasts play a role in the plant's adaptation to changing environmental conditions, such as light intensity, temperature, and water availability. Redox signaling pathways in chloroplasts allow plants to adjust their photosynthetic activity, metabolic processes, and stress responses in response to environmental cues, ensuring survival and growth under diverse conditions.

c. Comparison of Mitochondrial and Chloroplast Genomes

Mitochondria and chloroplasts are essential organelles found in eukaryotic cells, each playing a unique role in cellular energy metabolism and biosynthesis. Despite sharing some similarities due to their endosymbiotic origins, these organelles differ significantly in terms of genome structure, functional specialization, and evolutionary history.

1. Structural Differences

Genome Size

Chloroplast Genomes:

» **Larger Size:** Chloroplast genomes are typically larger than mitochondrial genomes, ranging from 120,000 to 160,000 base pairs in length. This larger size reflects the greater complexity and gene content of chloroplasts, which are involved in a wide range of metabolic processes beyond energy production. For instance, the chloroplast genome of *Arabidopsis thaliana* is approximately 154,478 base pairs long and encodes around 120 genes, including those essential for photosynthesis and other biosynthetic pathways.

» **Diverse Gene Content:** The chloroplast genome encodes numerous proteins that are integral to photosynthesis, such as components of photosystems I and II, the cytochrome b6f complex, and ATP synthase. In addition, chloroplast genomes encode proteins involved in the Calvin cycle, responsible for carbon fixation, as well as enzymes required for the biosynthesis of amino acids, lipids, and secondary metabolites. This diverse gene content reflects the multifaceted roles of chloroplasts in plant metabolism and growth.

» **Structural Features:** Chloroplast genomes often contain inverted repeats (IRs), which are sequences that are identical but oriented in opposite directions. These repeats play a crucial role in maintaining genome stability and facilitating DNA repair. The presence of IRs contributes to the structural organization of the chloroplast genome, ensuring its integrity and proper function.

Mitochondrial Genomes:

» **Smaller Size:** Mitochondrial genomes are generally smaller than chloroplast genomes, with sizes ranging from 15,000 to 100,000 base pairs, depending on the organism. For example, the human mitochondrial genome is approximately 16,569 base pairs long and encodes 37 genes. This smaller size is due to the streamlined nature of mitochondrial genomes, which have undergone extensive gene loss and transfer to the nuclear genome over evolutionary time.

- **Limited Gene Content:** The mitochondrial genome encodes a limited number of proteins, primarily those involved in the electron transport chain and ATP synthesis. In humans, mtDNA encodes 13 proteins essential for oxidative phosphorylation, as well as 22 tRNAs and 2 rRNAs required for mitochondrial protein synthesis. The focus on energy production reflects the specialized role of mitochondria in cellular respiration and energy metabolism.
- **Structural Simplicity:** Mitochondrial genomes are typically circular and do not contain inverted repeats like chloroplast genomes. Instead, they feature non-coding regions, such as the displacement loop (D-loop), which serve as regulatory elements for replication and transcription. The structural simplicity of mitochondrial genomes allows for efficient replication and maintenance, despite their smaller size and gene content.

2. Functional Specialization

Energy Production vs. Photosynthesis

Mitochondria:

- **ATP Production:** Mitochondria are primarily responsible for ATP production through oxidative phosphorylation, a process that occurs in the inner mitochondrial membrane. The electron transport chain, composed of protein complexes encoded by both mitochondrial and nuclear genomes, transfers electrons from NADH and FADH2 to oxygen, creating a proton gradient that drives ATP synthesis. This process is the main source of energy for eukaryotic cells, highlighting the critical role of mitochondria in cellular metabolism.
- **Metabolic Pathways:** In addition to ATP production, mitochondria are involved in various metabolic pathways, including the citric acid cycle, fatty acid oxidation, and amino acid metabolism. These pathways generate metabolic intermediates used in biosynthesis and provide energy for cellular processes. Mitochondria also play a role in the regulation of apoptosis and cell signaling, further underscoring their importance in maintaining cellular homeostasis.

Chloroplasts:

- **Photosynthesis:** Chloroplasts are specialized for photosynthesis, the process by which light energy is captured and converted into chemical energy in the form of ATP and NADPH. Photosynthesis occurs in two stages: the light-dependent reactions and the Calvin cycle. During the light-dependent reactions, chlorophyll and other pigments in the thylakoid membranes absorb light energy, exciting electrons that are transferred through a series of electron carriers. This process generates ATP and NADPH, which are used in the Calvin cycle to fix carbon dioxide into organic compounds, such as glucose.
- **Biosynthesis:** Beyond photosynthesis, chloroplasts are involved in the biosynthesis of various metabolites, including amino acids, fatty acids, and secondary metabolites. These compounds play essential roles in plant growth, development, and defense against environmental stresses. Chloroplasts also contribute to the regulation of cellular processes through redox signaling, which involves the production of reactive oxygen species (ROS) as signaling molecules that modulate gene expression and stress responses.

» **Integration with Plant Metabolism:** Chloroplasts are integrated into the overall metabolic network of plant cells, interacting with other organelles and cellular components to coordinate metabolic processes. This integration allows plants to adapt to changing environmental conditions, optimize resource use, and maintain energy balance.

3. Evolutionary Origins

Endosymbiotic Events

Mitochondria:

» **Proteobacterial Origin:** Mitochondria are believed to have originated from free-living proteobacteria that were engulfed by ancestral eukaryotic cells. This endosymbiotic event, estimated to have occurred over a billion years ago, led to the integration of mitochondria into the host cell as energy-producing organelles. The engulfed bacteria formed a symbiotic relationship with the host, providing it with a source of ATP in exchange for a stable environment and nutrients.

» **Evidence Supporting the Theory:** Several lines of evidence support the endosymbiotic theory of mitochondrial origin. The circular nature of mitochondrial DNA and its resemblance to bacterial genomes are key indicators of this evolutionary origin. Additionally, the presence of bacterial-like ribosomes and the similarity of mitochondrial proteins to those found in proteobacteria further support the theory. The endosymbiotic origin of mitochondria is also supported by the presence of double membranes, which are reminiscent of the engulfing process that led to their integration into eukaryotic cells.

» **Genome Reduction:** Over time, mitochondria underwent genome reduction, with many of their original genes being transferred to the nuclear genome. This gene transfer facilitated the integration of mitochondrial function with the host cell's regulatory networks, allowing for more efficient coordination of cellular processes. The reduction of the mitochondrial genome highlights the dynamic nature of genome evolution and the role of endosymbiosis in shaping eukaryotic cells.

Chloroplasts:

» **Cyanobacterial Origin:** Chloroplasts are believed to have originated from free-living cyanobacteria that were engulfed by ancestral eukaryotic cells. This endosymbiotic event is thought to have occurred after the origin of mitochondria, leading to the evolution of photosynthetic eukaryotes. The engulfed cyanobacteria formed a symbiotic relationship with the host cell, providing it with the ability to capture and convert light energy into chemical energy through photosynthesis.

» **Evidence Supporting the Theory:** The endosymbiotic theory of chloroplast origin is supported by the similarities between chloroplast DNA and bacterial genomes, such as the circular structure and gene content. The presence of bacterial-like ribosomes and the similarity of chloroplast proteins to those found in cyanobacteria further support this evolutionary origin. Chloroplasts also possess double membranes, indicative of their endosymbiotic origin.

- **Genome Reduction:** Like mitochondria, chloroplasts have undergone genome reduction through the transfer of genes to the nuclear genome. This gene transfer has allowed for the integration of chloroplast function with the host cell's regulatory networks, enhancing cellular coordination and efficiency. The reduction of the chloroplast genome underscores the importance of endosymbiosis in the evolution of complex life forms and the specialization of organelles for distinct functions.

d. Significance in Evolution and Biotechnology

The significance of organellar genomes, specifically those of mitochondria and chloroplasts, extends beyond their essential roles in cellular processes. These genomes provide invaluable insights into evolutionary biology and have significant applications in biotechnology and conservation genetics.

1. Evolutionary Insights

Tracing Ancestry

Understanding Evolutionary History:

- **Molecular Phylogenetics:** The study of organellar genomes, such as mitochondrial DNA (mtDNA) and chloroplast DNA (cpDNA), plays a crucial role in molecular phylogenetics, the field of biology that reconstructs evolutionary relationships among organisms. By comparing DNA sequences from different species, researchers can infer phylogenetic trees that depict the evolutionary pathways and common ancestry of species. These phylogenetic analyses provide a comprehensive understanding of the tree of life and the genetic connections between diverse organisms.
- **Mitochondrial Eve and Common Ancestors:** One of the landmark findings from the study of mitochondrial genomes is the concept of "Mitochondrial Eve," the most recent common matrilineal ancestor of all living humans. By analyzing mtDNA sequences from different human populations, researchers have traced human evolutionary history and migration patterns, revealing insights into the origins and spread of Homo sapiens. Similarly, chloroplast genomes help trace the ancestry of plant species, shedding light on the evolutionary history of photosynthetic eukaryotes.
- **Divergence and Speciation:** The analysis of organellar genomes also provides insights into the processes of divergence and speciation. By comparing organellar DNA sequences among closely related species, researchers can identify genetic differences that contribute to reproductive isolation and speciation. This information is vital for understanding how new species arise and adapt to different ecological niches, contributing to the biodiversity of life on Earth.

Comparative Genomics:

- **Genomic Conservation and Variation:** Comparative genomics focuses on analyzing and comparing the genomes of different species to uncover both conserved and variable genomic regions. Organellar genomes, such as those of mitochondria and chloroplasts, are particularly suited for comparative studies due to their relatively small size and high mutation rates. By examining conserved sequences, researchers can identify essential genes and regulatory elements that play critical roles in organellar function and have been maintained through evolutionary processes. In contrast, variable regions offer valuable insights into how organelles have adapted to environmental pressures, shedding light on the evolutionary mechanisms that drive diversity in organellar function and structure across species.

» **Horizontal Gene Transfer:** The study of organellar genomes has uncovered occurrences of horizontal gene transfer (HGT), where genetic material is exchanged between different species. HGT significantly influences the evolution of organellar genomes, with genes being transferred not only between organelles and the nuclear genome but also across species. These gene transfer events contribute to the complexity of organellar evolution, highlighting the fluid nature of genetic exchange. Understanding HGT sheds light on the mechanisms driving genome evolution, offering insights into how genetic material can be shared across diverse lineages, shaping the functionality and adaptability of both organelles and entire organisms.

» **Adaptive Evolution:** Organellar genomes provide evidence of adaptive evolution, where genetic changes confer a selective advantage in specific environments. For example, variations in mitochondrial genomes are linked to adaptations to different metabolic demands, such as high-altitude environments or extreme temperatures. Similarly, chloroplast genomes exhibit adaptations related to photosynthetic efficiency and tolerance to environmental stresses. These adaptive changes highlight the role of organellar genomes in shaping the evolutionary trajectory of species.

2. Biotechnological Applications

Genetic Engineering

Transforming Organelles:

» **Chloroplast Transformation:** Chloroplast transformation is a genetic engineering technique that involves the insertion of foreign genes into the chloroplast genome. This approach offers several advantages over nuclear transformation, including high levels of gene expression, maternal inheritance (preventing gene flow through pollen), and the capacity to express multiple genes from operons. Chloroplast transformation has been successfully used to produce recombinant proteins, vaccines, and biopharmaceuticals, offering potential applications in agriculture, medicine, and biotechnology.

» **Mitochondrial Genome Editing:** Advances in genetic engineering have enabled the development of techniques for editing mitochondrial genomes. Mitochondrial genome editing holds promise for the treatment of mitochondrial diseases, which result from mutations in mtDNA. By precisely altering mtDNA, researchers aim to correct pathogenic mutations and restore mitochondrial function, offering potential therapeutic applications for patients with mitochondrial disorders.

Applications in Agriculture:

» **Improving Crop Traits:** Genetic engineering of chloroplast genomes can enhance crop traits, such as resistance to pests and diseases, tolerance to environmental stresses, and increased nutritional content. For example, the expression of insecticidal proteins in chloroplasts can provide plants with resistance to herbivorous insects, reducing the need for chemical pesticides. Similarly, engineering chloroplasts to produce essential nutrients can improve the nutritional quality of crops, addressing deficiencies in human diets.

» **Biofuel Production:** Chloroplast transformation can be utilized to optimize the production of biofuels by engineering plants and algae to produce high levels of lipid-rich biomass. These biofuels offer a sustainable alternative to fossil fuels, contributing to renewable energy solutions. By enhancing photosynthetic efficiency and lipid biosynthesis pathways in chloroplasts, researchers aim to increase the yield and quality of biofuels derived from photosynthetic organisms.

Applications in Medicine:

» **Pharmaceutical Production:** The production of pharmaceutical compounds in chloroplasts offers a cost-effective and scalable platform for producing vaccines, antibodies, and therapeutic proteins. Chloroplasts can be engineered to express foreign proteins at high levels, facilitating the large-scale production of pharmaceuticals for various medical applications. This approach has the potential to reduce production costs and increase access to life-saving medications.

» **Gene Therapy:** The ability to edit organellar genomes presents new opportunities for gene therapy, particularly for mitochondrial diseases. By targeting specific mutations in mtDNA, researchers can develop therapies that correct genetic defects and restore normal cellular function. Gene therapy approaches targeting organellar genomes hold promise for treating a range of genetic disorders and improving patient outcomes.

3. Conservation Genetics

Biodiversity and Conservation

Assessing Genetic Diversity:

» **Mitochondrial DNA as a Marker:** Mitochondrial DNA is a valuable marker for assessing genetic diversity and population structure in endangered species. Due to its maternal inheritance and high mutation rate, mtDNA provides a reliable means of tracing lineage and studying population dynamics. Researchers use mtDNA to assess the genetic health of populations, identify distinct lineages, and monitor changes in genetic diversity over time.

» **Chloroplast DNA in Plants:** In plants, chloroplast DNA serves as a valuable tool for studying genetic diversity and determining phylogenetic relationships among species. Chloroplast DNA markers offer insights into the evolutionary history of plant species, aiding in the identification of genetically distinct populations and informing conservation priorities. By analyzing chloroplast DNA, researchers can track lineage divergence and understand population structure, providing crucial information for conserving biodiversity. This knowledge helps conservationists develop strategies to protect genetic resources, ensuring the preservation of plant diversity and maintaining the resilience of ecosystems in the face of environmental challenges.

Conservation Strategies:

» **Identifying Conservation Units:** Organellar genomes are used to identify conservation units, such as evolutionarily significant units (ESUs) and management units (MUs), within species. These units represent populations that are genetically distinct and should be prioritized for conservation efforts. By delineating conservation units based on organellar DNA, researchers can develop targeted strategies to protect biodiversity and prevent the loss of genetic diversity.

» **Restoration and Reintroduction:** The genetic information obtained from organellar genomes informs restoration and reintroduction efforts for endangered species. By understanding the genetic makeup of source and recipient populations, conservationists can select individuals that are genetically compatible and have a high likelihood of successful establishment. This approach enhances the success of restoration and reintroduction programs, contributing to the recovery of threatened species.

Implications for Ecosystem Health:

» **Maintaining Genetic Diversity:** Genetic diversity is essential for the adaptability and resilience of species in changing environments. Organellar genomes provide insights into the genetic diversity of populations, helping to identify areas of concern and guide conservation efforts. Maintaining genetic diversity ensures that species can adapt to environmental changes, resist diseases, and continue to provide ecosystem services.

» **Conserving Evolutionary Potential:** The study of organellar genomes contributes to the conservation of evolutionary potential by preserving the genetic resources needed for future adaptation and evolution. By protecting genetic diversity, conservationists can ensure that species have the capacity to evolve in response to environmental challenges, safeguarding biodiversity for future generations.

Chapter - 3

Nuclear DNA Organization

A. Nuclear DNA organization

The organization of nuclear DNA is a complex and highly structured process that allows for the efficient storage, replication, and expression of genetic information within the nucleus of eukaryotic cells. This intricate organization is crucial for maintaining genomic integrity and facilitating the regulation of gene expression.

1. Chromatin Structure

The structure and organization of chromatin within the nucleus of eukaryotic cells are essential for the efficient packaging, accessibility, and regulation of the genetic material. Chromatin is a complex of DNA and proteins that serve as the fundamental unit of DNA organization, playing a crucial role in genome management. Understanding the components and levels of chromatin organization provides insights into how cells control gene expression, maintain genomic integrity, and adapt to various physiological conditions.

Components of Chromatin

Chromatin is composed of DNA and histone proteins, forming a dynamic and intricate structure that facilitates the regulation of genetic information. The chromatin components are essential for understanding how DNA is packaged and accessed within the nucleus.

DNA and Histone Proteins

Nucleosome Structure:

» **Nucleosome Core Particle:** The basic repeating unit of chromatin is the nucleosome, which consists of DNA wrapped around an octamer of histone proteins. This octamer is composed of two copies each of the core histones: H2A, H2B, H3, and H4. Approximately 147 base pairs of DNA are wrapped around the histone octamer, forming the nucleosome core particle. This structure is the first level of DNA packaging, reducing the linear length of DNA and facilitating its compaction within the nucleus.

DOI: 10.1201/9781003717775-3

» **Histone Tails and Modifications:** The histone proteins have N-terminal tails that extend beyond the nucleosome core. These tails are subject to various post-translational modifications, including methylation, acetylation, phosphorylation, and ubiquitination. These modifications play crucial roles in regulating chromatin structure and function by altering the interaction between histones and DNA or recruiting other chromatin-associated proteins. For example, acetylation of histone tails is associated with transcriptional activation by loosening chromatin structure and promoting access to DNA.

Histone-DNA Interaction:

» **Electrostatic Interactions:** The interaction between DNA and histones is primarily mediated by electrostatic forces. DNA is negatively charged due to its phosphate backbone, while histones are positively charged, allowing for strong ionic interactions. These interactions facilitate the stable association of DNA with histones, enabling the formation of nucleosomes and higher-order chromatin structures.

» **Role in Gene Regulation:** The histone-DNA interaction is dynamic and can be modulated by chromatin remodeling complexes and histone modifications. This dynamic nature is essential for regulating gene expression, as it determines the accessibility of DNA to transcription factors and other regulatory proteins. The ability to alter chromatin structure allows cells to respond to environmental cues and developmental signals, ensuring appropriate gene expression patterns.

Linker DNA and Histone H1

Linker DNA:

» **Connecting Nucleosomes:** Nucleosomes are connected by stretches of linker DNA, which vary in length among different organisms. In humans, linker DNA typically spans 20 to 90 base pairs. This linker DNA serves as a spacer between nucleosomes, providing flexibility to the chromatin fiber and enabling higher-order folding and compaction.

» **Regulatory Regions:** Linker DNA often contains regulatory elements, such as enhancers and promoters, that are critical for gene expression. The accessibility of these regions is influenced by the positioning of nucleosomes and the presence of histone modifications, highlighting the role of linker DNA in the regulation of transcription.

Histone H1:

» **Linker Histone Function:** Histone H1, known as the linker histone, binds to the linker DNA and the entry/exit sites of DNA on the nucleosome. This binding stabilizes the nucleosome and facilitates the formation of higher-order chromatin structures, such as the 30-nanometer fiber. Histone H1 plays a critical role in compacting chromatin and regulating its accessibility, influencing processes such as DNA replication, repair, and transcription.

» **Role in Chromatin Compaction:** Histone H1 is involved in the transition from the "beads-on-a-string" structure of chromatin to more condensed forms. By promoting chromatin folding and compaction, histone H1 helps to organize the genome within the limited space of the nucleus, allowing for efficient management of genetic information.

Levels of Chromatin Organization

Chromatin exists in multiple levels of organization, each contributing to the overall structure and function of the genome. The organization of chromatin influences gene expression, DNA replication, and genome stability.

Euchromatin and Heterochromatin

Euchromatin:

» **Transcriptionally Active Regions:** Euchromatin is less densely packed chromatin associated with actively transcribed genome regions. It is characterized by an open and accessible chromatin conformation, allowing transcription factors and RNA polymerase to bind to DNA and initiate gene expression. Euchromatin is enriched with histone modifications that promote transcription, such as histone acetylation.

» **Role in Gene Expression:** Euchromatin contains most of the genes that are expressed during a cell's life cycle, reflecting its importance in regulating cellular processes and responding to environmental signals. The dynamic nature of euchromatin allows cells to rapidly activate or repress genes in response to changing conditions, ensuring proper cellular function and adaptation.

Heterochromatin:

» **Transcriptionally Inactive Regions:** Heterochromatin is densely packed chromatin that is often transcriptionally inactive. It is divided into two types: constitutive heterochromatin and facultative heterochromatin. Constitutive heterochromatin is permanently compacted and consists of repetitive sequences, such as satellite DNA and transposable elements, playing roles in maintaining genome stability. Facultative heterochromatin is reversible and can transition between active and inactive states, allowing for the regulation of gene expression during development and differentiation.

» **Genome Stability and Silencing:** Heterochromatin is involved in silencing repetitive DNA sequences and transposable elements, preventing their mobilization and maintaining genome integrity. It also plays a role in chromosome organization and segregation during cell division, contributing to the structural stability of the genome. The formation of heterochromatin is regulated by specific histone modifications, such as histone H3 lysine 9 methylation, and the recruitment of chromatin-binding proteins.

Chromatin Remodeling

Chromatin Remodeling Complexes:

» **ATP-Dependent Complexes:** Chromatin remodeling complexes are ATP-dependent molecular machines that alter chromatin structure by repositioning, ejecting, or restructuring nucleosomes. These complexes, such as SWI/SNF, ISWI, and CHD, use the energy from ATP hydrolysis to modify nucleosome positioning, allowing for changes in chromatin accessibility and gene expression.

- **Gene Regulation:** Chromatin remodeling complexes play a crucial role in regulating gene expression by modulating the accessibility of DNA to transcriptional machinery. They are involved in various cellular processes, including transcription, DNA replication, repair, and recombination. By dynamically altering chromatin structure, these complexes enable cells to respond to developmental cues and environmental signals, ensuring proper regulation of gene expression.

Histone Modifications:

- **Post-Translational Modifications:** Histone proteins undergo a wide range of post-translational modifications, including methylation, acetylation, phosphorylation, ubiquitination, and sumoylation. These modifications occur on specific amino acid residues within the histone tails and serve as molecular signals that influence chromatin structure and function.
- **Regulation of Chromatin Dynamics:** Histone modifications regulate chromatin dynamics by altering the interaction between histones and DNA or by recruiting specific chromatin-associated proteins. For example, histone acetylation is associated with transcriptional activation, as it reduces the positive charge on histones and decreases their affinity for DNA, resulting in a more open chromatin structure. In contrast, histone methylation can be associated with either activation or repression, depending on the specific residues modified.

Role in Epigenetic Regulation:

- **Epigenetic Memory:** Histone modifications contribute to epigenetic memory, the heritable transmission of gene expression patterns without changes in the underlying DNA sequence. These modifications establish and maintain cell-type-specific gene expression profiles, allowing for the differentiation and development of multicellular organisms.
- **Interaction with Non-Coding RNAs:** Non-coding RNAs, such as long non-coding RNAs (lncRNAs) and small interfering RNAs (siRNAs), interact with chromatin and histone modifiers to regulate gene expression. These interactions play roles in the formation of heterochromatin, X-chromosome inactivation, and genomic imprinting, highlighting the complexity of epigenetic regulation.

2. Chromosomal Organization

The organization of DNA into chromosomes is a fundamental aspect of nuclear DNA structure in eukaryotic cells. Chromosomes are vital for ensuring the proper segregation of genetic material during cell division and play a crucial role in the regulation of gene expression. The arrangement of DNA into distinct chromosomal structures allows for the efficient management of large genomes and the maintenance of genomic integrity. Understanding the organization and dynamics of chromosomes provides insights into the mechanisms that control cellular processes, development, and adaptation.

Chromosome Structure

The structure of chromosomes is a complex and highly organized system that enables the storage and transmission of genetic information. Eukaryotic chromosomes are linear DNA molecules, and their organization is essential for the replication, segregation, and expression of genetic material. The linear nature of chromosomes, along with specific structural regions like centromeres and telomeres, is key to understanding their function and dynamics.

Linear DNA Molecules

Eukaryotic Chromosome Structure:

» **Double-Stranded DNA Helix:** At the core of each eukaryotic chromosome is a single, continuous double-stranded DNA helix. This DNA helix is a polymer of nucleotides, each consisting of a sugar, phosphate group, and nitrogenous base. The linear arrangement of DNA in chromosomes is critical for the accurate replication and segregation of genetic material, as it allows for the precise alignment and pairing of homologous chromosomes during meiosis and mitosis.

» **Chromatin Fiber:** In eukaryotic cells, DNA does not exist as a naked helix but is complexed with histone proteins to form chromatin. This chromatin fiber is organized into repeating units called nucleosomes, where DNA is wrapped around an octamer of histone proteins. This arrangement not only compacts the DNA but also regulates access to genetic information, influencing gene expression and chromosomal behavior.

» **Higher-Order Structures:** Beyond the nucleosome level, chromatin further compacts into higher-order structures. These include the 30-nanometer fiber, looped domains, and ultimately the fully condensed chromosome during cell division. The hierarchical organization of chromatin into these structures is facilitated by histone H1, non-histone proteins, and the scaffolding provided by the nuclear matrix.

» **Gene Organization:** Eukaryotic chromosomes are organized into distinct regions that contain genes, regulatory elements, and non-coding sequences. Genes are arranged linearly along the DNA molecule, each with specific sequences that regulate their expression, including promoters, enhancers, and silencers. The arrangement of these elements within the chromosome influences gene expression patterns and the overall function of the genome.

Implications for Genome Function:

» **Replication Efficiency:** The linear nature of eukaryotic chromosomes allows for the replication of large genomes within a limited time frame. DNA replication begins at multiple origins along each chromosome, ensuring that the entire genome is copied efficiently and accurately during the S phase of the cell cycle. The replication machinery progresses bidirectionally from these origins, meeting at termination sites to complete the replication process.

» **Segregation Accuracy:** Linear chromosomes enable the precise segregation of genetic material during cell division. During mitosis, sister chromatids are separated and distributed equally to daughter cells, ensuring that each cell receives an identical set of chromosomes.

In meiosis, homologous chromosomes undergo recombination and segregation, leading to the generation of haploid gametes with a diverse combination of alleles.

» **Genetic Recombination:** The linear arrangement of DNA in chromosomes facilitates genetic recombination, a process that increases genetic diversity by exchanging genetic material between homologous chromosomes. This exchange occurs during meiosis and contributes to the variation observed within populations, providing a substrate for evolution and adaptation.

Centromeres and Telomeres

Centromeres:

» **Role in Chromosome Segregation:** Centromeres are specialized regions of the chromosome that play a critical role in ensuring accurate segregation of chromosomes during cell division. They are the sites where the kinetochore, a protein complex, assembles and attaches to spindle fibers during mitosis and meiosis. The interaction between the kinetochore and spindle fibers facilitates the movement of chromosomes to opposite poles of the cell, ensuring that each daughter cell receives an equal complement of chromosomes.

» **Centromeric DNA Sequences:** Centromeres are characterized by specific DNA sequences that vary among species but often consist of repetitive elements. In humans, centromeres are composed of alpha satellite DNA, a highly repetitive sequence that forms a large heterochromatic region. These sequences provide the structural basis for kinetochore formation and the binding of centromere-associated proteins, which are essential for centromere function and stability.

» **Epigenetic Regulation:** The identity and function of centromeres are not solely determined by DNA sequence but are also regulated by epigenetic modifications. Centromeric chromatin is enriched with the histone variant CENP-A, which replaces the standard histone H3 in nucleosomes. CENP-A nucleosomes are crucial for the recruitment of kinetochore proteins and the maintenance of centromere identity, ensuring proper chromosome segregation during cell division.

» **Centromere Evolution and Diversity:** Centromeres exhibit considerable diversity in size, sequence, and organization among different species. This diversity reflects the evolutionary pressures and adaptations that centromeres have undergone. Despite their variability, centromeres share common structural and functional features that highlight their essential role in genome stability and inheritance.

Telomeres:

» **Protection of Chromosome Ends:** Telomeres are repetitive DNA sequences located at the ends of linear chromosomes. They serve as protective caps that prevent the degradation of chromosome ends and the loss of genetic information during DNA replication. Telomeres also prevent the fusion of chromosome ends, maintaining chromosomal integrity and stability.

» **Telomeric DNA Sequences:** In humans, telomeres consist of tandem repeats of the sequence TTAGGG, which can extend for several kilobases. These repeats are bound by a specialized protein complex known as shelterin, which protects telomeres from being recognized as DNA damage and regulates telomere length and maintenance.

- **Telomere Replication and Maintenance:** The replication of telomeres poses a unique challenge due to the end-replication problem, where conventional DNA polymerases cannot fully replicate the ends of linear DNA molecules. To address this, telomeres are maintained by the enzyme telomerase, which extends telomeric repeats using an RNA template. Telomerase activity is essential for maintaining telomere length and preventing telomere shortening, which is associated with cellular aging and senescence.
- **Telomeres and Aging:** Telomere length is a key factor in cellular aging and has been linked to the aging process in organisms. As cells divide, telomeres gradually shorten, leading to replicative senescence and the cessation of cell division. This process acts as a tumor suppressive mechanism, limiting the proliferative capacity of cells with critically short telomeres. However, excessive telomere shortening can contribute to age-related diseases and cellular dysfunction.

Chromosome Territories

Chromosome territories refer to the spatial arrangement of chromosomes within the nucleus, where each chromosome occupies a distinct, non-overlapping region. This organization is crucial for the regulation of gene expression and the maintenance of nuclear architecture. The concept of chromosome territories provides insights into how the three-dimensional organization of the genome influences cellular processes and gene regulation.

Nuclear Organization

Spatial Arrangement of Chromosomes:

- **Chromosome Positioning:** Within the nucleus, chromosomes are organized into discrete territories that are spatially separated from one another. This non-random positioning of chromosomes is influenced by factors such as gene density, transcriptional activity, and chromosomal size. Gene-rich chromosomes tend to be located towards the nuclear interior, while gene-poor chromosomes are often positioned at the nuclear periphery.
- **Functional Implications:** The spatial arrangement of chromosomes within the nucleus has important implications for gene expression and cellular function. The positioning of genes relative to nuclear landmarks, such as the nuclear envelope and nucleolus, can influence their transcriptional activity. For example, genes located near the nuclear periphery are often associated with transcriptional repression, while those in the nuclear interior are more likely to be actively transcribed.
- **Dynamic Nature of Territories:** Chromosome territories are not static structures but are highly dynamic and can change in response to developmental cues, environmental signals, and cellular stress. During differentiation, chromosomes undergo reorganization to establish cell-type-specific gene expression patterns. Additionally, changes in nuclear architecture, such as chromatin compaction and decompaction, influence the spatial organization of chromosomes and their accessibility to transcriptional machinery.

Regulation of Gene Expression:

- **Proximity to Regulatory Elements:** The arrangement of chromosome territories influences the proximity of genes to regulatory elements, such as enhancers and silencers. This spatial organization facilitates the interaction of genes with distant regulatory sequences, enabling the coordinated regulation of gene expression. The formation of chromatin loops and transcriptional hubs allows for the clustering of co-regulated genes, promoting efficient transcriptional responses to external signals.
- **Nuclear Compartments and Domains:** The nucleus contains distinct compartments and domains that contribute to the regulation of gene expression. For example, the nucleolus is a nuclear subcompartment involved in ribosomal RNA synthesis and ribosome biogenesis. The presence of nucleolus-associated domains (NADs) and lamina-associated domains (LADs) further highlights the role of nuclear organization in modulating gene expression. These domains are associated with specific chromatin states and transcriptional activities, influencing the spatial distribution of chromosomes within the nucleus.

Interchromosomal Interactions

Interactions Between Chromosomes:

- **Transcriptional Hubs:** Chromosome territories facilitate interchromosomal interactions, where genes located on different chromosomes can come into physical proximity. These interactions play roles in the regulation of gene expression and the formation of transcriptional hubs, where multiple genes are co-regulated. Transcriptional hubs are dynamic structures that assemble and disassemble in response to cellular signals, allowing for coordinated gene expression across the genome.
- **Long-Range Enhancer-Promoter Interactions:** Interchromosomal interactions enable long-range enhancer-promoter contacts, where enhancers located on one chromosome can regulate the expression of genes on another chromosome. These interactions are mediated by chromatin looping and are facilitated by architectural proteins, such as CTCF and cohesin. Long-range enhancer-promoter interactions are essential for the precise regulation of gene expression during development and differentiation.

Role in Genome Organization:

- **Functional Clustering of Genes:** Interchromosomal interactions contribute to the functional clustering of genes within the nucleus. Genes with related functions or shared regulatory pathways can be brought together through interchromosomal contacts, enabling their co-regulation and coordinated expression. This functional clustering is important for the efficient execution of cellular programs and the integration of complex signaling networks.
- **Impact on Nuclear Architecture:** The dynamic nature of interchromosomal interactions influences the overall architecture of the nucleus. Changes in the spatial organization of chromosomes and the formation of interchromosomal contacts can alter the distribution of chromatin domains and nuclear compartments. This architectural flexibility allows cells to adapt to changing conditions and maintain genomic integrity.

3. Nuclear Matrix and Scaffold

The nuclear matrix and scaffold are essential structural components within the nucleus of eukaryotic cells, contributing significantly to the organization, regulation, and functionality of nuclear DNA. These structures provide a framework that not only supports chromatin organization but also plays a critical role in maintaining the structural integrity of the nucleus, regulating gene expression, and facilitating DNA replication and repair processes.

Nuclear Matrix

The nuclear matrix is a complex and dynamic network of fibrous proteins that provides structural support to the nucleus. It serves as an architectural scaffold, organizing chromatin and influencing the spatial arrangement of chromosomes and nuclear bodies. The nuclear matrix plays a crucial role in maintaining nuclear shape and integrity, especially during cell division.

Structural Support

Fibrous Network of Proteins:

- **Composition:** The nuclear matrix is composed of a variety of proteins, including structural proteins, RNA-binding proteins, and components of the nuclear pore complexes. This fibrous network forms a three-dimensional framework that spans the entire nucleus, providing mechanical support and maintaining its structural integrity. The matrix is believed to include proteins such as lamins, which form a mesh-like structure underlying the inner nuclear membrane, and other nuclear matrix proteins like matrin-3 and SAF-A (scaffold attachment factor A).
- **Role in Nuclear Architecture:** The nuclear matrix is integral to the architecture of the nucleus, as it anchors chromatin and nuclear bodies, such as the nucleolus, speckles, and Cajal bodies. By organizing these components, the nuclear matrix ensures the proper distribution and functioning of nuclear processes. It acts as a scaffold that helps position and maintain the spatial organization of these structures, thereby playing a role in the compartmentalization and regulation of nuclear functions.
- **Stability During Cell Division:** During cell division, the nuclear matrix plays a vital role in maintaining nuclear stability and shape. It provides a scaffold that supports chromatin condensation and segregation, ensuring the accurate distribution of genetic material to daughter cells. The nuclear matrix helps to re-establish nuclear organization after mitosis, facilitating the reformation of nuclear compartments and chromatin territories. The integrity of the nuclear matrix is crucial for mitotic spindle assembly and the proper alignment and segregation of chromosomes.

Influence on Chromatin Organization:

- **Chromatin Anchorage:** The nuclear matrix serves as an anchoring site for chromatin loops, which are formed by the attachment of chromatin fibers to specific matrix-associated regions (MARs) or scaffold/matrix attachment regions (SARs). These loops play a critical role in organizing the genome into functional domains and regulating gene expression by modulating the accessibility of genes to transcriptional machinery. MARs are DNA sequences that bind to the nuclear matrix, facilitating the formation of chromatin loops that enhance or suppress transcriptional activity depending on their spatial arrangement.

» **Chromatin Compaction and Dynamics:** The nuclear matrix influences chromatin compaction and dynamics by regulating the positioning and movement of chromatin fibers. It facilitates the formation of higher-order chromatin structures, such as topologically associating domains (TADs), which are involved in the spatial organization of the genome and the regulation of long-range chromatin interactions. The compaction state of chromatin is closely linked to gene expression, with more compact chromatin generally being less accessible to transcriptional machinery.

Regulation of Gene Expression

The nuclear matrix is actively involved in the regulation of gene expression by providing a platform for the interaction of transcription factors, chromatin remodelers, and other regulatory proteins with specific genomic regions. This interaction influences the transcriptional activity of genes and modulates the cellular response to internal and external signals.

Anchoring Chromatin Loops:

Gene Expression Regulation:

» **Spatial Organization of Genes:** The anchoring of chromatin loops to the nuclear matrix allows for the spatial organization of genes and regulatory elements within the nucleus. This organization facilitates the interaction of promoters with enhancers and other regulatory sequences, enhancing the transcriptional activation of genes. The formation of chromatin loops enables the clustering of co-regulated genes, promoting coordinated gene expression in response to developmental cues and environmental stimuli.

» **Enhancer-Promoter Interactions:** Enhancers are regulatory elements that can be located far from their target genes. The nuclear matrix helps bring enhancers into close proximity with their target promoters, facilitating the formation of enhancer-promoter loops that enhance transcriptional activation. This spatial organization is crucial for the precise regulation of gene expression during development and differentiation.

Transcriptional Complex Assembly:

» **Platform for Complex Assembly:** The nuclear matrix serves as a platform for the assembly of transcriptional complexes, which are multi-protein structures that mediate the initiation and regulation of transcription. These complexes include RNA polymerase II, transcription factors, co-activators, and chromatin remodelers. By providing a scaffold for the assembly of these complexes, the nuclear matrix ensures the precise regulation of gene expression.

» **Regulation of Transcriptional Machinery:** The assembly of transcriptional complexes at the nuclear matrix facilitates the coordination of transcriptional initiation and elongation. The matrix-associated proteins help recruit and stabilize transcriptional machinery at specific genomic loci, enhancing the efficiency and specificity of transcriptional responses.

Interaction with Regulatory Proteins:

Transcription Factors and Co-Regulators:

- **Binding and Modulation:** The nuclear matrix interacts with transcription factors and co-regulators, influencing their binding to target genes and modulating transcriptional activity. This interaction is regulated by post-translational modifications, such as phosphorylation and acetylation, which affect the stability and activity of transcription factors. The nuclear matrix also recruits chromatin remodelers and histone modifiers, which alter chromatin structure and regulate gene accessibility.
- **Role in Gene Silencing and Activation:** The nuclear matrix can facilitate both gene activation and silencing by recruiting different sets of regulatory proteins. For example, matrix-associated proteins can recruit repressive complexes to silence genes by promoting chromatin compaction, or they can recruit activators to enhance transcription by facilitating chromatin opening.

Non-Coding RNAs:

- **Regulation of Gene Expression:** Non-coding RNAs, such as long non-coding RNAs (lncRNAs) and microRNAs (miRNAs), play a role in the regulation of gene expression by interacting with the nuclear matrix. These RNAs can guide chromatin-modifying complexes to specific genomic regions, modulating chromatin structure and transcriptional activity. The nuclear matrix provides a structural basis for the interaction of non-coding RNAs with chromatin and regulatory proteins, influencing the regulation of gene expression.
- **Epigenetic Regulation:** Non-coding RNAs associated with the nuclear matrix can influence epigenetic regulation by guiding histone modifiers and DNA methyltransferases to specific loci. This targeting can lead to changes in chromatin accessibility and gene expression, contributing to the establishment and maintenance of epigenetic states.

Nuclear Scaffold

The nuclear scaffold is a proteinaceous structure within the nucleus that contributes to the higher-order organization of chromatin. It serves as a framework for the attachment of chromatin loops and plays a crucial role in DNA replication, repair, and the maintenance of genome integrity.

Higher-Order Chromatin Structure

Formation of Chromatin Loops:

- **Looped Domains:** The nuclear scaffold facilitates the formation of chromatin loops, which are fundamental units of higher-order chromatin structure. These loops are anchored to the scaffold at specific attachment sites, creating looped domains that organize the genome into functional regions. The formation of chromatin loops influences gene expression by bringing distant regulatory elements into proximity with target genes, enabling long-range enhancer-promoter interactions.
- **Role of Scaffold Attachment Regions (SARs):** SARs are specific DNA sequences that

interact with the nuclear scaffold, providing anchor points for chromatin loops. These regions play a critical role in establishing the structural framework of the genome, influencing the spatial arrangement of chromatin and regulating access to genetic information.

Topologically Associating Domains (TADs):

» **Genome Compartmentalization:** TADs are regions of the genome that interact more frequently with themselves than with other regions, forming distinct chromatin domains. The nuclear scaffold plays a role in the establishment and maintenance of TADs, which contribute to the spatial organization of the genome and the regulation of gene expression. TADs are characterized by specific boundaries, often marked by the binding of architectural proteins such as CTCF and cohesin, which mediate the formation of chromatin loops and influence the three-dimensional organization of the genome.

» **Impact on Gene Regulation:** TADs compartmentalize the genome into distinct functional units, influencing the regulation of gene expression by segregating active and inactive chromatin regions. The organization of genes within TADs can affect their accessibility to transcriptional machinery and regulatory elements, contributing to the fine-tuning of gene expression in response to cellular signals.

Regulation of Gene Accessibility

Modulation of Chromatin Dynamics:

» **Chromatin Remodeling:** The nuclear scaffold regulates gene accessibility by modulating chromatin dynamics and compaction. It influences the positioning and movement of chromatin fibers, affecting the accessibility of genes to transcriptional machinery. The scaffold provides a structural basis for the dynamic reorganization of chromatin, allowing cells to respond to developmental cues and environmental signals by altering gene expression patterns.

» **Regulatory Protein Interaction:** The scaffold interacts with chromatin remodelers and histone modifiers, facilitating the recruitment of these proteins to specific genomic loci. These interactions play a critical role in regulating chromatin accessibility by modifying histone tails and altering nucleosome positioning.

Influence on Epigenetic Regulation:

» **Establishment of Epigenetic Marks:** The nuclear scaffold interacts with epigenetic regulators, such as chromatin remodelers and histone modifiers, which modify chromatin structure and regulate gene expression. These interactions influence the establishment and maintenance of epigenetic marks, such as DNA methylation and histone modifications, which play a critical role in regulating chromatin accessibility and gene expression. The nuclear scaffold provides a framework for the integration of epigenetic signals with chromatin structure, influencing the regulation of gene expression and cellular function.

» **Epigenetic Memory and Stability:** The nuclear scaffold contributes to the stability of epigenetic states by maintaining the spatial organization of chromatin and facilitating the transmission of epigenetic information through cell divisions. This stability is essential for preserving cell identity and maintaining the integrity of epigenetic landscapes.

Role in DNA Replication and Repair

The nuclear scaffold is involved in DNA replication and repair by providing a structural basis for the organization of replication origins and the recruitment of repair factors. It ensures the efficient duplication of genetic material and the maintenance of genome integrity in response to DNA damage.

DNA Replication:

Organization of Replication Origins:

» **Spatial Arrangement:** The nuclear scaffold plays a role in the organization of replication origins, which are specific sequences where DNA replication is initiated. The scaffold provides a structural basis for the spatial arrangement of replication origins within the nucleus, ensuring the efficient and coordinated initiation of DNA replication. This organization allows for the replication of large genomes within a limited time frame, ensuring that the entire genome is copied accurately during the S phase of the cell cycle.

» **Regulation of Origin Activity:** The positioning of replication origins within the nuclear scaffold influences their activation and timing. Origins located in different nuclear environments may have varying levels of activity, contributing to the regulation of replication timing and the coordination of DNA synthesis with other cellular processes.

Replication Fork Stability:

» **Support for Replication Machinery:** The nuclear scaffold contributes to the stability of replication forks, which are structures that form at the replication origins during DNA replication. The scaffold provides a framework for the recruitment of replication factors, such as DNA polymerases and helicases, which are essential for the progression of the replication fork. By stabilizing replication forks, the nuclear scaffold ensures the efficient duplication of genetic material and prevents the collapse of replication forks, which can lead to genome instability and DNA damage.

» **Response to Replication Stress:** The nuclear scaffold plays a role in responding to replication stress, which can arise from obstacles encountered by the replication machinery. It facilitates the recruitment of repair factors and checkpoint proteins to stalled replication forks, helping to resolve replication issues and maintain genome integrity.

DNA Repair:

Recruitment of Repair Factors:

» **Assembly of Repair Complexes:** The nuclear scaffold plays a critical role in the maintenance of genome integrity by facilitating the recruitment of DNA repair factors to sites of DNA damage. It provides a structural basis for the assembly of repair complexes, which mediate the recognition and repair of DNA lesions. These complexes include proteins involved in various repair pathways, such as base excision repair, nucleotide excision repair, and double-strand break repair.

- **Regulation of Repair Pathway Choice:** The scaffold influences the choice of repair pathway by modulating the accessibility of repair factors to damaged sites. It helps direct the repair process based on the type and severity of DNA damage, ensuring that the appropriate repair mechanisms are activated to restore genome integrity.

Coordination of Repair Pathways:

- **Integration with Chromatin Structure:** The nuclear scaffold coordinates the activation of DNA repair pathways, ensuring the efficient and accurate repair of DNA damage. It regulates the access of repair factors to damaged sites and modulates the interaction of repair proteins with chromatin and other nuclear structures. This coordination is essential for preserving genome integrity and preventing the accumulation of mutations that can lead to genetic diseases and cancer.
- **Maintenance of Genome Stability:** The scaffold plays a crucial role in maintaining genome stability by facilitating the repair of DNA double-strand breaks and other critical lesions. It helps orchestrate the repair process, ensuring that DNA is accurately restored and that chromatin structure is re-established following repair.

The nuclear matrix and scaffold are vital components of the nuclear architecture, providing structural support and regulatory functions that are essential for the organization and maintenance of nuclear DNA. Through their roles in chromatin organization, gene expression regulation, and DNA replication and repair, these structures contribute to the overall stability and functionality of the genome.

4. Epigenetic Regulation

Epigenetic regulation is a critical aspect of nuclear DNA organization that influences gene expression without altering the underlying DNA sequence. Epigenetic mechanisms enable cells to respond to environmental cues and maintain stable gene expression patterns across cell divisions. These processes are essential for development, differentiation, and the adaptability of organisms to changing environments.

DNA Methylation

DNA methylation is one of the most studied and well-characterized epigenetic modifications. It involves the addition of methyl groups to the DNA molecule, typically at cytosine bases within CpG dinucleotides. This modification is crucial for regulating gene expression and maintaining genomic integrity.

Addition of Methyl Groups

Mechanism and Process:

- **CpG Dinucleotides:** DNA methylation primarily occurs at cytosine residues in CpG dinucleotides, where a cytosine is followed by a guanine nucleotide connected by a phosphate. CpG sites are not uniformly distributed throughout the genome and are often clustered in regions known as CpG islands, which are frequently found near the promoter regions of genes.
- **DNA Methyltransferases (DNMTs):** The enzymes responsible for adding methyl groups

to cytosine residues are DNA methyltransferases. Three major types of DNMTs are involved in mammalian DNA methylation: DNMT1, DNMT3A, and DNMT3B. DNMT1 is the maintenance methyltransferase that copies methylation patterns during DNA replication, ensuring that the epigenetic mark is preserved in daughter cells. DNMT3A and DNMT3B are involved in de novo methylation, establishing new methylation patterns during early development and in response to environmental signals.

- **Stable and Heritable Mark:** DNA methylation is considered a stable epigenetic mark, capable of being inherited through cell divisions. Once established, methylation marks can persist over many generations, providing a mechanism for long-term gene silencing and regulation of genomic elements.

Functional Roles:

- **Gene Silencing:** DNA methylation is commonly associated with transcriptional repression. When CpG islands in the promoter region of a gene are methylated, the access of transcription factors and RNA polymerase to the DNA is hindered, leading to reduced gene expression. This silencing mechanism is crucial for regulating tissue-specific genes and maintaining cellular identity.
- **Transposable Element Suppression:** Methylation of transposable elements prevents their expression and transposition, protecting the genome from instability caused by these mobile genetic elements. This suppression is vital for maintaining genomic integrity and preventing mutations that can arise from transposon activity.
- **X-Chromosome Inactivation:** In female mammals, one of the two X chromosomes is inactivated to achieve dosage compensation. This process, known as X-chromosome inactivation (XCI), is mediated by DNA methylation, which silences genes on the inactive X chromosome, ensuring equal gene expression levels between males and females.
- **Genomic Imprinting:** DNA methylation plays a crucial role in genomic imprinting, where specific genes are expressed in a parent-of-origin-specific manner. Imprinted genes are marked by differential methylation patterns established during gamete formation, leading to monoallelic expression and contributing to developmental regulation.

Role in Development and Disease

Developmental Functions:

- **Cell Differentiation:** DNA methylation is essential for normal development and cellular differentiation. It regulates the expression of lineage-specific genes, enabling the formation of specialized cell types. During development, methylation patterns are dynamically altered to activate or silence genes in a tissue-specific manner, guiding the differentiation process.
- **Embryonic Development:** In early embryogenesis, global DNA demethylation occurs, followed by the establishment of new methylation patterns that guide cell fate decisions. This reprogramming is critical for the activation of pluripotency genes and the subsequent differentiation of embryonic stem cells into diverse cell lineages.

Disease Associations:

- **Cancer:** Aberrant DNA methylation patterns are commonly observed in cancer. Hypermethylation of tumor suppressor gene promoters can lead to their silencing, contributing to uncontrolled cell proliferation and tumor development. Conversely, hypomethylation of oncogenes and repetitive sequences can promote genomic instability and tumorigenesis. These epigenetic changes are often considered hallmarks of cancer and are being explored as potential biomarkers for diagnosis and therapy.
- **Neurological Disorders:** DNA methylation abnormalities are implicated in various neurological disorders, such as Rett syndrome, Fragile X syndrome, and schizophrenia. Mutations in genes involved in methylation or its interpretation can lead to altered gene expression patterns and neurological deficits, highlighting the importance of proper methylation regulation in brain development and function.
- **Aging and Environmental Factors:** Changes in DNA methylation are associated with aging and can be influenced by environmental factors such as diet, stress, and exposure to toxins. These epigenetic alterations may contribute to age-related diseases and underscore the impact of the environment on epigenetic regulation. Understanding these changes is crucial for developing strategies to mitigate the effects of aging and environmental stressors on health.

Histone Modifications

Histone modifications are chemical changes to the histone proteins around which DNA is wrapped, affecting chromatin structure and gene expression. These modifications occur on specific amino acid residues within the histone tails and can either activate or repress transcription, depending on the type and location of the modification.

Chemical Modifications

Types of Modifications:

- **Acetylation:** Histone acetylation involves the addition of acetyl groups to lysine residues on histone tails, generally associated with transcriptional activation. This modification is mediated by histone acetyltransferases (HATs) and reversed by histone deacetylases (HDACs). Acetylation reduces the positive charge on histones, decreasing their affinity for the negatively charged DNA and leading to a more open chromatin structure.
- **Methylation:** Histone methylation involves the addition of methyl groups to lysine or arginine residues and can be associated with either transcriptional activation or repression, depending on the specific residue and the number of methyl groups added. For example, methylation of histone H3 at lysine 4 (H3K4) is associated with active transcription, while methylation at lysine 9 (H3K9) and lysine 27 (H3K27) is linked to gene repression and heterochromatin formation. This modification is carried out by histone methyltransferases and removed by histone demethylases.
- **Phosphorylation:** Histone phosphorylation occurs on serine or threonine residues and is often associated with chromatin remodeling and DNA repair. Phosphorylation can

influence chromatin dynamics and gene expression by altering histone-DNA interactions and recruiting specific binding proteins. This modification is particularly important during cell cycle transitions and in response to DNA damage.

» **Ubiquitination:** Histone ubiquitination involves the attachment of ubiquitin molecules to lysine residues, influencing chromatin structure and transcriptional regulation. This modification can target histones for degradation or alter their interaction with other chromatin-associated proteins, playing roles in gene activation and silencing.

Mechanisms of Regulation:

» **Chromatin Accessibility:** Histone modifications influence chromatin accessibility by altering the compaction state of chromatin and modulating the binding of transcriptional machinery. Acetylation generally promotes an open chromatin conformation, facilitating transcription, while certain methylation marks can lead to a more closed and repressive chromatin state. The dynamic regulation of histone modifications allows cells to respond to developmental cues and environmental signals by modulating gene expression patterns.

» **Recruitment of Effector Proteins:** Histone modifications serve as docking sites for effector proteins that mediate downstream effects on chromatin structure and gene expression. For example, acetylated histones can recruit bromodomain-containing proteins that activate transcription, while methylated histones can attract proteins with chromodomains that promote chromatin compaction and gene silencing. These interactions play a crucial role in coordinating the assembly of transcriptional complexes and regulating gene activity.

» **Epigenetic Memory:** Histone modifications contribute to the establishment and maintenance of epigenetic memory, allowing cells to retain specific gene expression patterns across cell divisions. This memory is crucial for maintaining cellular identity and ensuring the faithful transmission of epigenetic states. The interplay between different histone modifications and their regulatory proteins ensures that epigenetic marks are faithfully propagated and interpreted during development and differentiation.

Chromatin Accessibility

Transcriptional Activation and Repression:

» **Role of Acetylation:** Histone acetylation is closely associated with transcriptional activation. By reducing the electrostatic interactions between histones and DNA, acetylation facilitates the binding of transcription factors and RNA polymerase to gene promoters, enhancing transcriptional activity. The dynamic regulation of acetylation levels allows cells to respond rapidly to environmental cues and developmental signals.

» **Dual Role of Methylation:** Histone methylation can have both activating and repressive effects on gene expression, depending on the context. Methylation of H3K4 is associated with active transcription and is enriched at gene promoters and enhancers, while methylation of H3K27 and H3K9 is linked to transcriptional repression and the formation of heterochromatin, a densely packed form of chromatin that is transcriptionally inactive. The specific combination of histone modifications, known as the "histone code," provides a complex regulatory mechanism for fine-tuning gene expression.

Chromatin Remodeling Complexes:

» **ATP-Dependent Remodeling:** Chromatin remodeling complexes utilize ATP hydrolysis to reposition, eject, or restructure nucleosomes, thereby altering chromatin accessibility and regulating gene expression. These complexes play a critical role in enabling transcriptional machinery access to DNA and facilitating the establishment of specific chromatin states. They are involved in various cellular processes, including transcription, DNA repair, and replication, by dynamically modifying chromatin structure.

» **Impact on Development and Differentiation:** Chromatin remodeling is essential for developmental processes and cellular differentiation, as it allows for the dynamic regulation of gene expression in response to developmental cues and environmental stimuli. The coordinated action of remodeling complexes and histone modifications ensures that genes are expressed in a temporally and spatially regulated manner. This regulation is vital for the proper development of tissues and organs, as well as the maintenance of stem cell pluripotency.

Non-Coding RNAs

Non-coding RNAs (ncRNAs) are RNA molecules that do not code for proteins but play essential roles in regulating gene expression and chromatin organization. These RNAs modulate transcriptional and post-transcriptional processes, contributing to the complexity of gene regulation.

Regulatory RNAs

Types of Non-Coding RNAs:

» **MicroRNAs (miRNAs):** miRNAs are short, approximately 22-nucleotide-long RNAs that regulate gene expression post-transcriptionally by binding to complementary sequences in target mRNAs. This binding leads to mRNA degradation or translational repression, fine-tuning protein levels in the cell. miRNAs are involved in a wide range of biological processes, including development, differentiation, and immune responses.

» **Long Non-Coding RNAs (lncRNAs):** lncRNAs are longer RNA molecules that can act as scaffolds for chromatin modifiers and transcriptional regulators. They participate in a wide range of regulatory functions, including chromatin remodeling, transcriptional activation or repression, and RNA processing. lncRNAs can influence gene expression at both the transcriptional and post-transcriptional levels, acting as scaffolds for chromatin modifiers and transcriptional regulators.

» **Small Interfering RNAs (siRNAs):** siRNAs are similar in size to miRNAs and are involved in RNA interference (RNAi), a process that leads to the degradation of specific mRNA targets. siRNAs play roles in defense against viral infections and the regulation of transposable elements, contributing to genome stability.

Mechanisms of Action:

- **Transcriptional Regulation:** ncRNAs can modulate transcriptional activity by interacting with chromatin, transcription factors, and RNA polymerase complexes. lncRNAs, for example, can recruit chromatin-modifying enzymes to specific genomic loci, influencing histone modifications and chromatin accessibility. This regulation allows ncRNAs to modulate gene expression in response to cellular signals and environmental changes.
- **Scaffolding Functions:** lncRNAs can serve as molecular scaffolds, bringing together multiple protein complexes to coordinate gene regulation. They can facilitate the assembly of transcriptional complexes at gene promoters or enhancers, thereby modulating transcriptional output. This scaffolding function enables lncRNAs to integrate multiple regulatory signals and coordinate complex gene expression programs.
- **Epigenetic Regulation:** ncRNAs are involved in the establishment and maintenance of epigenetic marks. For instance, lncRNAs can recruit DNA methyltransferases or histone-modifying enzymes to specific loci, contributing to the regulation of gene expression and the formation of repressive chromatin states. This epigenetic regulation plays a crucial role in maintaining cellular identity and ensuring the proper transmission of epigenetic information during development and differentiation.

Gene Silencing and Activation

Role of miRNAs and siRNAs:

- **Post-Transcriptional Gene Silencing:** miRNAs and siRNAs mediate post-transcriptional gene silencing by guiding RNA-induced silencing complexes (RISCs) to complementary mRNA targets. This targeting leads to mRNA degradation or translational repression, reducing the levels of specific proteins in the cell. This regulatory mechanism is crucial for fine-tuning gene expression and maintaining cellular homeostasis.
- **Regulation of Developmental Pathways:** miRNAs play critical roles in regulating developmental pathways by modulating the expression of key transcription factors and signaling molecules. They contribute to cell fate decisions, tissue differentiation, and the maintenance of stem cell pluripotency. By fine-tuning the expression of target genes, miRNAs ensure the precise regulation of developmental processes.

Function of lncRNAs:

- **Transcriptional Activation:** lncRNAs can activate gene expression by recruiting transcriptional activators or chromatin remodelers to specific loci. They can also interact with transcription factors to enhance their binding to target promoters, facilitating the initiation of transcription. This regulation is essential for coordinating complex gene expression programs in response to developmental and environmental signals.
- **Regulation of Chromatin Structure:** lncRNAs influence chromatin structure and accessibility by interacting with chromatin-modifying complexes. They can promote the formation of open chromatin states at active genes or contribute to the establishment of repressive chromatin at

silenced loci. This regulation plays a crucial role in maintaining cellular identity and ensuring the proper transmission of epigenetic information during development and differentiation.

5. Dynamic Nature of Nuclear Organization

The organization of nuclear DNA is not static; it is highly dynamic and responsive to cellular signals and environmental changes. This dynamic nature is crucial for regulating gene expression, ensuring proper cellular function, and allowing cells to adapt to changing conditions. The ability of the nucleus to reorganize its structure and modify chromatin dynamics is essential for development, differentiation, and response to environmental stimuli.

Nuclear Reorganization

Nuclear reorganization refers to the rearrangement of chromatin and nuclear architecture in response to various signals, including developmental cues and environmental stimuli. This reorganization is a fundamental aspect of cellular adaptation and plays a critical role in determining cell fate and function.

Developmental Cues

Influence on Chromatin and Chromosomes:

- **Cell Differentiation:** During development, cells undergo differentiation, transforming from pluripotent stem cells into specialized cell types. This process involves extensive nuclear reorganization, where chromatin structure is remodeled to activate or repress lineage-specific genes. For example, in stem cells, chromatin is generally more open and accessible, allowing for the expression of pluripotency genes. As cells differentiate, chromatin becomes more compact in regions that need to be silenced, while becoming more open in regions that promote the expression of differentiation-specific genes.
- **Lineage Commitment:** Nuclear organization is intricately linked to lineage commitment, where cells become committed to a specific developmental pathway. This commitment is marked by changes in the spatial arrangement of chromatin and chromosomes, leading to the establishment of a specific gene expression program. For instance, during hematopoiesis, where blood stem cells differentiate into various blood cell types, nuclear reorganization ensures that genes required for each lineage are appropriately activated or repressed.
- **Transcriptional Regulation:** Developmental cues such as transcription factors and signaling pathways play crucial roles in nuclear reorganization. They guide the recruitment of chromatin remodelers and histone modifiers to specific genomic loci, altering chromatin architecture and facilitating the transcription of target genes. The interaction of these cues with nuclear architecture ensures the precise regulation of gene expression necessary for development and differentiation.

Role in Establishing Cell Identity:

- **Epigenetic Memory:** Nuclear reorganization contributes to the establishment of epigenetic memory, where specific patterns of gene expression are maintained across cell divisions. This memory is critical for preserving cellular identity, ensuring that differentiated cells retain their specialized functions. Epigenetic modifications, such as DNA methylation and histone modifications, work in concert with nuclear organization to stabilize gene expression patterns.

» **Nuclear Architecture:** The architecture of the nucleus, including the spatial arrangement of chromatin territories and nuclear bodies, influences gene expression and cell identity. Nuclear compartments, such as the nucleolus and nuclear speckles, are dynamically reorganized during differentiation, reflecting changes in gene expression demands. These compartments serve as hubs for transcriptional regulation, RNA processing, and ribosome biogenesis, contributing to cellular specialization.

Response to Environmental Stimuli

Impact on Chromatin Structure:

» **Stress Response:** Cells constantly encounter environmental stresses, such as heat shock, oxidative stress, and DNA damage. These stresses trigger rapid changes in nuclear organization, leading to the activation of stress-responsive genes. For example, in response to heat shock, heat shock proteins (HSPs) are rapidly upregulated, requiring chromatin remodeling to allow transcriptional access. The dynamic repositioning of chromatin enables cells to mount an effective stress response, promoting survival and adaptation.

» **Nutrient Availability:** The availability of nutrients can influence nuclear organization and gene expression. For instance, under nutrient-rich conditions, growth-promoting genes are activated, while nutrient scarcity leads to the repression of these genes and the activation of stress-response pathways. Nuclear reorganization facilitates these changes by modulating chromatin accessibility and the spatial arrangement of regulatory elements.

» **Signal Transduction:** Cellular signaling pathways play a pivotal role in nuclear reorganization by transmitting external signals to the nucleus, leading to changes in gene expression. Signaling molecules, such as hormones and growth factors, can induce the recruitment of transcription factors and chromatin modifiers to specific genomic loci, altering nuclear architecture and gene activity. For example, the hormone estrogen can activate estrogen receptor (ER) signaling, leading to chromatin remodeling and the upregulation of target genes involved in cell proliferation.

Cellular Responses and Adaptation:

» **Gene Expression Regulation:** Environmental stimuli trigger changes in gene expression patterns through nuclear reorganization. This regulation is crucial for cellular adaptation, enabling cells to respond to changes in their environment by activating or repressing specific genes. The dynamic nature of nuclear organization ensures that cells can rapidly adjust their gene expression profiles to meet changing demands.

» **Chromatin Remodeling and Dynamics:** Chromatin remodeling complexes, such as the SWI/SNF and NuRD complexes, play critical roles in mediating nuclear responses to environmental stimuli. These complexes alter nucleosome positioning and composition, allowing for the rapid and reversible changes in chromatin structure required for gene regulation. The dynamic interplay between chromatin remodelers and nuclear architecture enables cells to fine-tune gene expression in response to external cues.

Chromatin Dynamics

Chromatin dynamics refer to the ability of chromatin to undergo structural changes, allowing for the regulation of gene expression, DNA repair, and replication. These dynamics are essential for maintaining genomic stability and facilitating cellular responses to internal and external signals.

Nucleosome Remodeling

Nucleosome remodeling plays a crucial role in regulating chromatin structure and accessibility, enabling dynamic changes that are essential for gene expression, DNA repair, and replication. This process is mediated by ATP-dependent chromatin remodeling complexes, which reposition, eject, or restructure nucleosomes.

Mechanisms of Nucleosome Positioning:

» **ATP-Dependent Remodeling Complexes:** Nucleosome remodeling is facilitated by ATP-dependent chromatin remodeling complexes, such as the SWI/SNF, ISWI, CHD, and INO80 complexes. These complexes utilize the energy from ATP hydrolysis to reposition, evict, or restructure nucleosomes, thereby modulating chromatin accessibility. Each complex has distinct mechanisms and functions, contributing to the dynamic regulation of chromatin.

» **Nucleosome Sliding and Ejection:** Remodeling complexes can slide nucleosomes along the DNA to expose or occlude specific regions, facilitating access to transcription factors and other DNA-binding proteins. In some cases, nucleosomes can be entirely ejected from the DNA, allowing for the assembly of large protein complexes, such as the transcriptional machinery. This repositioning is essential for the activation and repression of genes in response to cellular signals.

» **Histone Variant Exchange:** Some remodeling complexes, such as the SWR1 complex, are involved in the exchange of histone variants within nucleosomes. Histone variants can confer distinct structural and functional properties to nucleosomes, influencing chromatin dynamics and gene expression. For example, the incorporation of the histone variant H2A.Z is associated with transcriptional activation and the regulation of chromatin accessibility.

Role in Gene Expression:

» **Transcriptional Activation:** Nucleosome remodeling is a key step in the activation of transcription. By repositioning nucleosomes, remodeling complexes create an open chromatin environment that allows transcription factors and RNA polymerase II to access promoter regions and initiate transcription. This dynamic regulation is crucial for the timely and coordinated expression of genes in response to developmental cues and environmental stimuli.

» **Gene Repression:** Remodeling complexes can also contribute to gene repression by promoting the formation of closed chromatin structures that inhibit transcription. This repression can be achieved through the sliding of nucleosomes to occlude transcription factor binding sites or the recruitment of repressive histone variants that stabilize compact chromatin configurations. The balance between activation and repression is critical for maintaining proper gene expression patterns and cellular homeostasis.

Role in DNA Repair and Replication:

» **DNA Repair:** Nucleosome remodeling is essential for DNA repair processes, as it facilitates the access of repair factors to damaged DNA sites. Remodeling complexes can remove or reposition nucleosomes to expose DNA lesions, allowing repair proteins to recognize and rectify the damage. This dynamic response is crucial for maintaining genomic integrity and preventing the accumulation of mutations.

» **DNA Replication:** During DNA replication, nucleosome remodeling is required to accommodate the passage of the replication machinery along the DNA template. Remodeling complexes assist in the disassembly and reassembly of nucleosomes at replication forks, ensuring the accurate duplication of genetic material. This process is vital for preserving chromatin structure and epigenetic marks during cell division.

Chromatin Looping

Chromatin looping is a fundamental aspect of chromatin organization, enabling long-range interactions between distant genomic regions. These loops facilitate the regulation of gene expression by bringing enhancers, promoters, and other regulatory elements into close proximity, forming functional transcriptional hubs.

Mechanisms of Loop Formation:

» **CTCF and Cohesin Complexes:** Chromatin loops are primarily mediated by architectural proteins such as CTCF (CCCTC-binding factor) and cohesin. CTCF is a DNA-binding protein that serves as an insulator, demarcating loop boundaries and preventing inappropriate interactions between neighboring chromatin domains. Cohesin is a ring-shaped protein complex that encircles DNA and stabilizes loop structures, facilitating the physical interaction of genomic regions.

» **Dynamic Looping:** Chromatin loops are dynamic structures that can form and dissolve in response to cellular signals. This dynamic nature allows cells to modulate gene expression by altering the spatial arrangement of regulatory elements. The regulation of loop dynamics is critical for coordinating gene expression during development, differentiation, and cellular responses to environmental changes.

Role in Gene Regulation:

» **Enhancer-Promoter Interactions:** Chromatin loops enable enhancers, which are distal regulatory elements, to interact with target gene promoters, enhancing transcriptional activation. These interactions are crucial for the precise regulation of gene expression, as enhancers can modulate the activity of multiple genes within a chromatin domain. The specificity and strength of enhancer-promoter interactions are influenced by the three-dimensional organization of the genome.

» **Transcriptional Hubs:** Chromatin looping facilitates the formation of transcriptional hubs, where multiple genes and regulatory elements are co-localized within a confined nuclear space. These hubs serve as sites of coordinated gene regulation, allowing for the efficient recruitment of transcriptional machinery and the synchronized activation of gene networks. The organization of transcriptional hubs is dynamic and can change in response to developmental cues and environmental stimuli.

Impact on Development and Differentiation:

» **Lineage-Specific Gene Expression:** Chromatin loops play a crucial role in establishing lineage-specific gene expression patterns during development. By modulating the interactions between enhancers and promoters, chromatin loops enable the selective activation of genes required for cell fate decisions and tissue differentiation. The spatial organization of chromatin loops is a key determinant of cellular identity and function.

» **Cellular Adaptation:** The dynamic regulation of chromatin loops allows cells to adapt to changing environmental conditions by modulating gene expression. In response to external signals, such as hormones, growth factors, and stressors, chromatin loops can reorganize to facilitate the activation of adaptive gene programs. This adaptability is essential for cellular resilience and survival in fluctuating environments.

» **Chromatin Domain Insulation:** Chromatin loops also contribute to the insulation of chromatin domains, preventing inappropriate interactions between neighboring domains. This insulation is crucial for maintaining the integrity of gene expression programs and preventing cross-talk between regulatory elements. The organization of chromatin loops ensures that genes are expressed in a context-specific manner, supporting the proper functioning of cellular processes.

6. Nucleolar Organization

The nucleolus is a distinct and prominent nuclear compartment within eukaryotic cells that plays a crucial role in ribosome biogenesis and the regulation of various cellular processes. It is not membrane-bound but is a highly organized and dynamic structure formed around specific chromosomal regions known as nucleolar organizer regions (NORs). The nucleolus is essential for maintaining cellular homeostasis, responding to stress, and regulating the cell cycle.

Ribosome Biogenesis

Ribosome biogenesis is the primary function of the nucleolus, involving the synthesis and processing of ribosomal RNA (rRNA) and the assembly of ribosomal subunits. This complex process is critical for protein synthesis and overall cellular function.

rRNA Synthesis and Processing

rRNA Synthesis:

- **Transcription of rDNA:** Ribosomal RNA is synthesized from ribosomal DNA (rDNA) genes, which are organized in tandem repeats on certain chromosomes. In humans, these rDNA repeats are found on the short arms of acrocentric chromosomes (13, 14, 15, 21, and 22). RNA polymerase I is the enzyme responsible for transcribing rDNA into a large precursor rRNA molecule (45S pre-rRNA).
- **Pre-rRNA Processing:** The 45S pre-rRNA undergoes extensive processing, including cleavage, chemical modification, and folding. This processing results in the production of mature rRNA molecules, specifically 18S, 5.8S, and 28S rRNAs, which are integral components of ribosomal subunits. The processing involves small nucleolar RNAs (snoRNAs) and various ribonucleoprotein complexes that facilitate cleavage and modification.

Assembly of Ribosomal Subunits:

- **Association with Ribosomal Proteins:** Once processed, the mature rRNAs are assembled with ribosomal proteins imported from the cytoplasm. The small subunit (40S) is formed by the association of 18S rRNA with ribosomal proteins, while the large subunit (60S) is composed of 5.8S, 28S, and 5S rRNAs along with additional proteins.
- **Export to Cytoplasm:** The assembled ribosomal subunits are exported from the nucleolus to the cytoplasm through nuclear pores. This export is mediated by specific transport receptors and is crucial for the initiation of protein synthesis in the cytoplasm, where ribosomes translate mRNA into polypeptides.

Nucleolar Organizer Regions (NORs)

Chromosomal Foundation:

- **Location of rDNA Repeats:** Nucleolar organizer regions (NORs) are chromosomal loci containing multiple copies of rDNA repeats. These regions serve as the foundation for nucleolar formation, as they are the sites where rRNA synthesis is initiated. The presence and activity of NORs are directly correlated with the size and number of nucleoli within a cell.
- **NOR Activity:** The activity of NORs determines the cell's capacity for ribosome production. Highly active NORs are associated with larger and more numerous nucleoli, reflecting the cell's increased demand for protein synthesis. This activity can vary between different cell types and physiological conditions, highlighting the dynamic nature of nucleolar organization.

Formation and Maintenance:

- **Nucleolar Assembly:** The nucleolus assembles around active NORs, driven by the transcription of rDNA and the processing of rRNA. Proteins involved in ribosome biogenesis, such as fibrillarin and nucleolin, localize to the nucleolus, contributing to its structure and function. The assembly process is tightly regulated and responsive to cellular signals, ensuring the nucleolus adapts to the cell's needs.

» **Nucleolar Disassembly:** During cell division, particularly mitosis, the nucleolus disassembles as transcription is temporarily halted. Following mitosis, the nucleolus reassembles in daughter cells, a process dependent on the reactivation of rDNA transcription. This cycle of disassembly and reassembly underscores the nucleolus's dynamic nature and its role in coordinating cell division and ribosome production.

Nucleolar Functions

Beyond ribosome biogenesis, the nucleolus plays multiple roles in regulating cellular processes, including stress response, cell cycle progression, and the sequestration of regulatory factors. These functions highlight the nucleolus as a central hub for cellular regulation.

Stress Response and Cell Cycle Regulation

Cellular Stress Sensing:

» **DNA Damage Response:** The nucleolus acts as a sensor for cellular stress, such as DNA damage, oxidative stress, and nutrient deprivation. In response to stress, nucleolar activity is modulated, impacting ribosome production and cellular metabolism. For instance, DNA damage can lead to nucleolar disruption and the release of stress response proteins, such as p53, which is stabilized in the nucleoplasm to initiate DNA repair and cell cycle arrest.

» **Nutrient Sensing:** The nucleolus is also involved in sensing nutrient availability. Under nutrient-rich conditions, the nucleolus supports robust ribosome production to meet the demands of protein synthesis. Conversely, nutrient scarcity triggers a reduction in nucleolar activity, conserving resources and modulating metabolic pathways to adapt to the stress.

Regulation of Cell Cycle:

» **Cell Cycle Checkpoints:** The nucleolus plays a role in regulating cell cycle progression by interacting with cell cycle regulators. It is involved in the sequestration and release of key factors that influence cell cycle checkpoints, ensuring that cells do not proceed through the cycle under unfavorable conditions.

» **Mitotic Regulation:** During mitosis, the nucleolus disassembles, and its components are redistributed. The reformation of the nucleolus in daughter cells is a tightly regulated process that coordinates with cell cycle progression. This regulation ensures that ribosome biogenesis resumes promptly after division, supporting cellular growth and proliferation.

Sequestration of Regulatory Factors

Nucleolar Sequestration:

» **Transcription Factors:** The nucleolus serves as a hub for the sequestration of various transcription factors, modulating their activity and availability. This sequestration influences gene expression patterns and cellular responses to environmental changes. For example, under stress conditions, certain transcription factors are sequestered in the nucleolus to prevent their interaction with target genes, thus modulating stress-responsive pathways.

- **Cell Cycle Regulators:** Key cell cycle regulators, such as cyclins and cyclin-dependent kinases, can be sequestered within the nucleolus. This sequestration provides a mechanism for controlling cell cycle progression and ensures that cells only divide when appropriate signals are present.

Impact on Gene Expression and Cellular Function:

- **Regulation of mRNA Processing:** The nucleolus influences mRNA processing by sequestering RNA-binding proteins and splicing factors. This regulation affects the maturation and stability of mRNA transcripts, impacting protein synthesis and cellular function. By modulating the availability of these factors, the nucleolus can fine-tune gene expression in response to changing cellular conditions.
- **Role in Senescence and Aging:** The nucleolus is implicated in cellular senescence and aging, processes characterized by the irreversible arrest of cell proliferation. Changes in nucleolar morphology and activity are associated with aging-related decline in cellular function. The nucleolus's involvement in stress responses and metabolic regulation contributes to its role in aging and longevity.

Significance of Nucleolar Organization

The nucleolus is a dynamic and multifunctional nuclear compartment that plays a central role in cellular homeostasis. Its organization and functions are intricately linked to the regulation of gene expression, cellular stress responses, and ribosome production. Understanding the complexities of nucleolar organization provides insights into the fundamental mechanisms that govern cellular processes and the interplay between nuclear structure and function. The nucleolus's ability to adapt to changing cellular environments underscores its importance as a key regulator of cellular physiology and adaptation.

B. Nuclear and Cytoplasmic genome interactions and signal transduction

1. Nuclear and Cytoplasmic Genome Interactions

The interactions between nuclear and cytoplasmic genomes are central to the understanding of cellular biology, reflecting a complex and intricate network of communication that underpins cellular function and integrity. This interaction is essential for the regulation of various cellular processes, from energy production to metabolic regulation, and plays a crucial role in cellular adaptation, development, and evolution.

Evolutionary Origins and Integration of Organelle Genomes

The origins of mitochondria and chloroplasts can be traced back to a pivotal evolutionary event known as endosymbiosis. This theory posits that these organelles originated from free-living prokaryotes that were engulfed by ancestral eukaryotic cells. Over time, a symbiotic relationship developed, leading to the permanent integration of these prokaryotic cells as organelles within the host cell. Mitochondria are thought to have evolved from an ancestral alpha-proteobacterium, which entered into a symbiotic relationship with a primitive eukaryotic cell. This relationship allowed the host cell to benefit from the bacterium's ability to generate ATP through aerobic respiration, providing a significant evolutionary advantage. Similarly, chloroplasts are believed to have originated from a cyanobacterium, which conferred the ability to perform photosynthesis to its eukaryotic host, revolutionizing the energy dynamics of the cell.

As a result of this endosymbiotic event, a substantial portion of the genetic material from the original prokaryotic cells was transferred to the host cell's nuclear genome, leading to the highly integrated and interdependent relationship observed today. Despite this transfer of genetic material, mitochondria and chloroplasts retained a small, specialized set of genes essential for their function, primarily related to energy production and photosynthesis. The integration of these organelle genomes into the host cell's regulatory network required the evolution of intricate communication systems that ensure proper coordination between the nuclear and organelle genomes. This integration was driven by evolutionary pressures to maintain and optimize the core functions of these organelles, such as ATP production in mitochondria and carbon fixation in chloroplasts.

Mitochondrial Genome and Function:

The mitochondrial genome (mtDNA) is a circular DNA molecule, typically inherited maternally, that encodes essential components of the mitochondrial electron transport chain, tRNAs, and rRNAs. The retention of this genetic material is crucial for mitochondrial functions, such as ATP production via oxidative phosphorylation. The mtDNA encodes key proteins involved in the respiratory complexes, which are necessary for the transfer of electrons and the generation of an electrochemical gradient across the inner mitochondrial membrane. This gradient drives the synthesis of ATP, the primary energy currency of the cell.

Mitochondria play a vital role in regulating cellular metabolism and energy homeostasis. They are involved in various metabolic pathways, including the tricarboxylic acid (TCA) cycle, fatty acid oxidation, and amino acid metabolism. Mitochondria also serve as signaling hubs that integrate metabolic and stress signals to regulate cellular processes such as apoptosis, autophagy, and inflammation. In addition to their role in energy metabolism, mitochondria are involved in the regulation of cellular redox balance, calcium homeostasis, and the synthesis of essential biomolecules, such as heme and steroids. The dynamic nature of mitochondria, including their ability to undergo fission, fusion, and mitophagy, is crucial for maintaining mitochondrial quality and function, particularly in response to cellular stress or damage.

Mutations in mtDNA can lead to mitochondrial diseases, which often affect tissues with high energy demands, such as muscle and nerve tissues. These mutations can disrupt the normal function of the electron transport chain, leading to impaired energy production, increased oxidative stress, and mitochondrial dysfunction. Mitochondrial diseases can manifest as a wide range of clinical symptoms, including muscle weakness, neurodegenerative disorders, and metabolic abnormalities. Understanding the genetic and biochemical basis of these diseases is critical for developing targeted therapies and interventions.

Chloroplast Genome and Function:

The chloroplast genome (cpDNA) is also a circular DNA molecule that encodes genes essential for photosynthesis and other metabolic processes. Chloroplasts are responsible for converting light energy into chemical energy through photosynthesis, a process that is fundamental to the survival of plants and the ecosystems that depend on them. The cpDNA encodes proteins involved in the photosynthetic machinery, including the light-harvesting complexes, electron transport chain, and ATP synthase. Chloroplasts play a crucial role in the synthesis of organic compounds, such as glucose, through the Calvin cycle. They are also involved in the synthesis of essential biomolecules, such as fatty acids, amino acids, and secondary metabolites. Chloroplasts contribute to the regulation of plant growth, development, and stress responses, providing a dynamic platform for integrating environmental signals and metabolic pathways.

The expression of chloroplast genes is tightly regulated in response to environmental conditions, particularly light, which is the primary driver of photosynthesis. Chloroplasts can adjust their genetic output to optimize photosynthetic efficiency and respond to changes in light intensity, temperature, and nutrient availability. This dynamic regulation allows plants to adapt to changing environmental conditions and maintain optimal growth and productivity. Chloroplasts also engage in a variety of signaling interactions with the nucleus, allowing the cell to coordinate the expression of nuclear and chloroplast genes in response to environmental and developmental cues. This coordination is essential for optimizing photosynthetic performance and ensuring that the energy needs of the cell are met. Understanding the molecular mechanisms underlying chloroplast gene expression and signaling is essential for improving crop yields and developing sustainable agricultural practices.

Coordination and Communication Between Nuclear and Cytoplasmic Genomes

The coordination between nuclear and cytoplasmic genomes is achieved through a sophisticated network of signals and molecular interactions. This coordination ensures that cellular activities are harmonized and optimized for the cell's physiological needs and environmental conditions.

Nuclear-Encoded Proteins and Organelle Biogenesis:

A significant portion of proteins required for mitochondrial and chloroplast function is encoded by nuclear genes. These proteins are synthesized in the cytosol and then imported into the respective organelles through specialized transport mechanisms. The import of nuclear-encoded proteins into organelles is a highly regulated process, influenced by factors such as the cell cycle, developmental stage, and environmental conditions. For mitochondria, the import machinery involves several complexes, including the translocase of the outer membrane (TOM) and the translocase of the inner membrane (TIM). These complexes recognize mitochondrial targeting sequences on precursor proteins and facilitate their translocation into the mitochondrial matrix or inner membrane. Similarly, the import of chloroplast proteins involves the translocon of the outer chloroplast membrane (TOC) and the translocon of the inner chloroplast membrane (TIC) complexes, which guide proteins into the chloroplast stroma or thylakoid membranes.

The import of nuclear-encoded proteins into organelles is essential for maintaining organelle function and composition. This process allows the cell to modulate organelle activity in response to changing needs and challenges, thus maintaining cellular homeostasis and adaptability. The coordination of organelle biogenesis and protein import is crucial for ensuring that organelles can meet the metabolic and energy demands of the cell.

Retrograde Signaling:

Retrograde signaling refers to the communication from organelles back to the nucleus, conveying information about the organelle's functional status and metabolic demands. This signaling is crucial for adjusting nuclear gene expression to optimize organelle function and ensure cellular health. For example, mitochondrial retrograde signaling can be activated in response to mitochondrial stress or dysfunction, leading to changes in nuclear gene expression that promote mitochondrial biogenesis, repair, and stress response pathways.

Similarly, chloroplasts can communicate with the nucleus through retrograde signals that inform the cell about the photosynthetic efficiency and the need for adjustments in light-harvesting complexes or metabolic pathways. These signals can include reactive oxygen species (ROS), metabolites, or changes in the redox state, which influence nuclear gene expression to enhance photosynthetic performance and protect against photodamage. The coordination of retrograde signaling involves multiple pathways and signaling molecules, such as calcium ions, ROS, and transcription factors. These signals can activate specific nuclear genes that regulate organelle function and cellular adaptation to stress. Understanding the molecular mechanisms underlying retrograde signaling is essential for elucidating how cells respond to changes in organelle function and environmental conditions.

Cross-Talk Between Signaling Pathways:

The integration of signals from different cellular compartments involves cross-talk between various signaling pathways. This cross-talk allows the cell to integrate multiple signals and produce a coordinated response. For instance, calcium signaling, often triggered by environmental stimuli, can interact with mitochondrial signaling pathways to influence cellular metabolism and stress responses. Similarly, hormonal signaling pathways, such as those involving abscisic acid or salicylic acid in plants, can modulate both nuclear and organelle functions to regulate growth, development, and defense mechanisms.

Cross-talk between signaling pathways provides a mechanism for fine-tuning cellular responses and ensuring that the cell can adapt to complex and dynamic environments. This integration of signals enables the cell to prioritize specific pathways and allocate resources based on the immediate needs and challenges it faces. Understanding the interactions between signaling pathways is critical for uncovering the complex regulatory networks that govern cellular behavior and adaptation.

Epigenetic Regulation and Gene Expression:

Epigenetic modifications play a critical role in the coordination of nuclear and cytoplasmic genomes. These modifications, including DNA methylation and histone modifications, can alter chromatin structure and gene accessibility, thereby influencing gene expression. Signal transduction pathways can modulate epigenetic marks, leading to changes in gene expression that reflect the cell's adaptive responses to internal and external cues. This epigenetic regulation provides a mechanism for the cell to fine-tune gene expression and maintain long-term changes in cellular behavior in response to environmental changes or developmental signals.

Epigenetic regulation is essential for establishing and maintaining cellular identity and function, allowing cells to respond to environmental cues while preserving their specialized roles. In plants, epigenetic modifications can influence processes such as flowering time, stress responses, and disease resistance. In animals, epigenetic regulation plays a role in development, differentiation, and disease pathogenesis. Understanding the interplay between epigenetic modifications and nuclear-cytoplasmic interactions is crucial for elucidating the mechanisms of cellular adaptation and evolution. The regulation of gene expression through epigenetic mechanisms allows cells to maintain a stable yet flexible gene expression profile that can adapt to environmental and developmental changes. This regulation is critical for ensuring that cellular processes are finely tuned to meet the demands of a changing environment, supporting growth, development, and survival.

Organellar Biogenesis and Dynamics:

The biogenesis and dynamics of organelles such as mitochondria and chloroplasts are tightly regulated processes that are essential for cellular function and adaptation. Organelle biogenesis involves the coordinated expression of nuclear and organellar genes to produce the necessary components for organelle formation and maintenance. This coordination is particularly important during cell division, differentiation, and stress responses, where changes in organelle number and function are required to meet cellular demands.

Mitochondrial dynamics, including fission and fusion, are regulated by both nuclear and mitochondrial factors, allowing the cell to adapt its energy production and metabolic capacity. Similarly, chloroplast biogenesis and dynamics are influenced by environmental conditions, such as light and nutrient availability, ensuring that photosynthetic capacity is matched to the plant's growth and developmental needs. The regulation of organelle biogenesis and dynamics involves complex signaling networks that integrate information from the nucleus, cytoplasm, and environment. These networks coordinate the expression of genes involved in organelle replication, division, and function, ensuring that cellular processes are optimized for growth and adaptation. Understanding the molecular mechanisms underlying organelle biogenesis and dynamics is essential for elucidating the processes that govern cellular development and evolution.

Interplay with Cellular Metabolism:

The interaction between nuclear and cytoplasmic genomes is closely linked to cellular metabolism. Mitochondria and chloroplasts are central to metabolic pathways such as oxidative phosphorylation and photosynthesis, which provide the energy and metabolic intermediates necessary for cellular function. The regulation of these pathways involves the integration of signals from the nucleus, cytoplasm, and environment, allowing the cell to optimize energy production and metabolic efficiency.

For instance, in response to nutrient availability or metabolic demands, the nuclear genome can modulate the expression of genes involved in mitochondrial or chloroplast function, thereby adjusting the organelle's metabolic output. This interplay between nuclear and cytoplasmic genomes ensures that cellular metabolism is flexible and responsive to changing conditions, supporting cellular growth, development, and survival. The integration of metabolic signals with nuclear and cytoplasmic interactions allows cells to coordinate energy production and utilization with cellular processes such as growth, division, and differentiation. This coordination is essential for maintaining cellular homeostasis and supporting the diverse functions of multicellular organisms. Understanding the interplay between cellular metabolism and nuclear-cytoplasmic interactions is crucial for elucidating the mechanisms that underlie cellular function and adaptation.

Impact on Cellular Health and Disease:

The coordination and communication between nuclear and cytoplasmic genomes have significant implications for cellular health and disease. Disruptions in these interactions can lead to a variety of diseases, particularly those affecting tissues with high energy demands, such as the brain, heart, and muscles. Mitochondrial diseases, for example, often result from mutations in either the mitochondrial or nuclear genomes that impair mitochondrial function, leading to energy deficits and cellular dysfunction.

Similarly, defects in chloroplast function can affect plant growth and development, reducing photosynthetic efficiency and stress tolerance. The mechanisms underlying nuclear-cytoplasmic interactions is therefore crucial for developing strategies to diagnose, treat, and prevent diseases associated with organelle dysfunction. In addition to genetic mutations, environmental factors such as oxidative stress, toxins, and pathogens can disrupt nuclear-cytoplasmic interactions, leading to cellular damage and disease. The study of these interactions provides insights into the cellular processes that protect against damage and maintain cellular integrity. Identifying the molecular pathways involved in these protective mechanisms is essential for developing therapeutic strategies to combat diseases associated with organelle dysfunction.

Biotechnological Applications:

The study of nuclear and cytoplasmic genome interactions has important biotechnological applications, particularly in fields such as agriculture, medicine, and synthetic biology. In agriculture, manipulating the expression of nuclear and organellar genes can enhance crop traits such as yield, stress tolerance, and nutritional content. Techniques such as genetic engineering and CRISPR-Cas9 gene editing offer powerful tools for modifying these interactions to improve crop performance and sustainability.

In medicine, understanding the role of nuclear-cytoplasmic interactions in disease pathogenesis can inform the development of targeted therapies and interventions. Strategies that enhance or restore proper communication between nuclear and organellar genomes may provide new avenues for treating mitochondrial diseases and other conditions associated with organelle dysfunction. The ability to manipulate nuclear-cytoplasmic interactions also holds promise for synthetic biology, where researchers aim to design and construct novel biological systems with tailored functions. By engineering interactions between nuclear and cytoplasmic genomes, scientists can create organisms with enhanced capabilities for bioproduction, bioremediation, and bioenergy generation. Understanding the principles governing these interactions is crucial for advancing the field of synthetic biology and developing innovative solutions to global challenges.

Future Directions and Research:

Ongoing research into nuclear and cytoplasmic genome interactions continues to uncover new insights into the complexity of cellular regulation and adaptation. Advances in genomic and proteomic technologies, along with improved understanding of signaling networks and epigenetic regulation, are driving progress in this field. Future studies are likely to focus on elucidating the precise molecular mechanisms underlying these interactions, exploring their role in cellular plasticity and evolution, and harnessing their potential for biotechnological innovation.

The interactions between nuclear and cytoplasmic genomes provide a foundation for exploring the diversity and adaptability of life on Earth. These interactions are central to the evolution of complex multicellular organisms, enabling them to adapt to changing environments and develop diverse functional capabilities. As research continues to advance, the potential applications of this knowledge in fields such as biotechnology, agriculture, and medicine are vast and hold promise for addressing some of the world's most pressing challenges.

Coordination and Communication:

The coordination between nuclear and cytoplasmic genomes is achieved through a sophisticated network of signals. Nuclear genes encode proteins that are imported into mitochondria and chloroplasts to assist in their functions. Conversely, these organelles send signals to the nucleus to modulate gene expression based on metabolic needs and environmental conditions. This bidirectional communication ensures the cell can adapt to changes and maintain efficient energy production.

Gene Expression Regulation:

Gene expression in the nucleus can be influenced by signals originating from the cytoplasm. For instance, retrograde signaling from mitochondria to the nucleus can alter the expression of nuclear genes in response to mitochondrial dysfunction or stress. This interaction is crucial for maintaining cellular energy balance and responding to oxidative stress.

2. Signal Transduction Pathways

Signal transduction pathways are intricate networks of molecular interactions that allow cells to sense and respond to various external and internal signals. These pathways enable cells to adapt to changing environments, communicate with each other, and coordinate complex physiological processes. Signal transduction is fundamental to cellular functions such as growth, differentiation, apoptosis, immune responses, and homeostasis.

Signal transduction begins with the recognition of a signal molecule, or ligand, by a specific receptor, typically located on the cell surface or within the cell. Upon ligand binding, the receptor undergoes a conformational change that activates intracellular signaling cascades. These cascades often involve a series of phosphorylation events mediated by protein kinases, leading to the activation of various effector proteins. Ultimately, these signaling events culminate in a cellular response, such as a change in gene expression, metabolic activity, or cytoskeletal organization.

The complexity and specificity of signal transduction pathways arise from the diverse array of receptors, second messengers, and downstream effectors involved. Each component of a signaling pathway contributes to the precise regulation of cellular responses, allowing cells to integrate multiple signals and generate appropriate outcomes. Additionally, cross-talk between different signaling pathways adds another layer of regulation, enabling cells to fine-tune their responses based on the context and nature of the signals received.

Components of Signal Transduction

Signal transduction pathways consist of several key components that work together to transmit signals from the extracellular environment to the cell's interior, ultimately leading to a physiological response.

These components include receptors, second messengers, protein kinases, and transcription factors. Each plays a unique role in the signaling process, contributing to the specificity and regulation of cellular responses.

Receptors

Receptors are proteins located on the cell surface or within the cell that bind to specific signaling molecules, initiating the signal transduction process. They serve as the primary sensors of extracellular signals and play a crucial role in determining the specificity of a signaling pathway. There are several major types of receptors, each with distinct structures and mechanisms of action:

G Protein-Coupled Receptors (GPCRs):

G Protein-Coupled Receptors (GPCRs) represent one of the largest and most diverse families of cell surface receptors. They play a critical role in numerous physiological processes, including sensory perception, neurotransmission, and immune responses. GPCRs are characterized by their seven-transmembrane domain structure and their ability to activate heterotrimeric G proteins upon ligand binding.

When a ligand binds to a GPCR, it induces a conformational change in the receptor, leading to the activation of an associated G protein. G proteins are composed of three subunits: alpha (α), beta (β), and gamma (γ). In the inactive state, the alpha subunit is bound to GDP and is associated with the beta and gamma subunits. Ligand binding to the GPCR promotes the exchange of GDP for GTP on the alpha subunit, resulting in its activation and dissociation from the beta and gamma subunits.

Activated G proteins can then interact with various downstream effectors, such as adenylyl cyclase, phospholipase C, and ion channels, to propagate the signal. The specific downstream pathways activated depend on the type of G protein involved (e.g., Gs, Gi, Gq), allowing GPCRs to mediate a wide range of cellular responses. The diversity of GPCRs and their ligands enables cells to respond to a vast array of stimuli, including hormones, neurotransmitters, lipids, and peptides. GPCR signaling is tightly regulated through mechanisms such as receptor desensitization, internalization, and degradation, ensuring that cellular responses are appropriately controlled.

Receptor Tyrosine Kinases (RTKs):

Receptor Tyrosine Kinases (RTKs) are a class of cell surface receptors with intrinsic tyrosine kinase activity. They play a pivotal role in regulating cell growth, differentiation, metabolism, and survival. RTKs are activated by ligands such as growth factors, cytokines, and hormones. Upon ligand binding, RTKs undergo dimerization, a process that brings two receptor molecules together. This dimerization activates the intrinsic kinase activity of the receptors, leading to the autophosphorylation of specific tyrosine residues within their cytoplasmic domains. These phosphorylated tyrosine residues serve as docking sites for various signaling proteins, initiating a cascade of downstream signaling events. RTK signaling pathways involve a multitude of downstream effectors, including small GTPases (e.g., Ras), lipid kinases (e.g., PI3K), and MAP kinases (e.g., ERK). These pathways regulate critical cellular processes, such as proliferation, survival, migration, and differentiation. Dysregulation of RTK signaling is associated with various diseases, particularly cancer. Mutations or overexpression of RTKs can lead to constitutive activation of signaling pathways, driving uncontrolled cell proliferation and survival. Targeting RTKs with specific inhibitors is a therapeutic strategy for treating certain cancers.

Nuclear Receptors:

Nuclear receptors are a class of intracellular receptors that respond to lipid-soluble ligands, such as steroid hormones, thyroid hormones, and retinoic acids. Unlike cell surface receptors, nuclear receptors function as transcription factors that directly regulate gene expression. Upon binding to their ligands, nuclear receptors undergo conformational changes that enable them to interact with specific DNA sequences known as hormone response elements (HREs) in the promoter regions of target genes. This interaction recruits coactivators or corepressors, modulating the transcriptional activity of the target genes. Nuclear receptors play essential roles in regulating a wide range of physiological processes, including metabolism, development, reproduction, and immune responses. They act as key regulators of gene expression in response to hormonal and environmental signals, allowing cells to adapt to changing conditions. The ability of nuclear receptors to modulate gene expression makes them attractive targets for therapeutic interventions. Agonists and antagonists of nuclear receptors are used to treat various conditions, such as inflammation, cancer, and metabolic disorders.

Ion Channel Receptors:

Ion channel receptors are ligand-gated channels that open in response to the binding of specific ligands, allowing ions to flow across the cell membrane. These receptors play a critical role in processes such as synaptic transmission, muscle contraction, and sensory perception. Ion channel receptors are classified based on the ions they conduct (e.g., cation channels, anion channels) and the ligands that activate them (e.g., neurotransmitters, hormones). Ligand binding induces conformational changes in the receptor, leading to the opening of the ion channel and the influx or efflux of ions, which can depolarize or hyperpolarize the cell membrane. The rapid and transient nature of ion channel signaling allows for precise temporal control of cellular responses. Ion channels are subject to various regulatory mechanisms, including desensitization, modulation by second messengers, and phosphorylation, ensuring that ion fluxes are tightly controlled.

Ion channel receptors are important targets for pharmacological interventions in conditions such as epilepsy, pain, and cardiovascular diseases. Modulators of ion channels can alter neuronal excitability, muscle contractility, and other physiological processes, offering therapeutic benefits in diverse clinical contexts.

Second Messengers

Second messengers are small molecules that propagate and amplify signals within the cell, serving as intermediaries that link receptor activation to downstream cellular responses. They play a crucial role in modulating the intensity and duration of signaling events, allowing cells to respond to stimuli with precision and specificity.

Cyclic AMP (cAMP):

Cyclic AMP (cAMP) is a second messenger derived from ATP by the action of the enzyme adenylyl cyclase. It is one of the most well-studied second messengers and is involved in regulating a wide range of physiological processes, including metabolism, gene expression, and cell growth. The production of cAMP is typically triggered by the activation of G protein-coupled receptors (GPCRs) that stimulate adenylyl cyclase through the Gs protein. Once generated, cAMP activates protein kinase A (PKA), a serine/threonine kinase that phosphorylates specific target proteins, leading to changes in their activity and function. cAMP signaling pathways are involved in processes such as glycogen metabolism, lipid metabolism, and cell proliferation. For example, in response to the hormone epinephrine, cAMP is produced in liver and muscle cells, leading to the activation of PKA and the phosphorylation of enzymes involved in glycogen breakdown, thereby increasing glucose availability for energy production. The cAMP signaling pathway is tightly regulated by phosphodiesterases (PDEs), which degrade cAMP into AMP, terminating the signal. This regulation ensures that cAMP levels are precisely controlled, allowing for rapid and transient signaling events. Dysregulation of cAMP signaling is associated with various diseases, including heart failure, asthma, and certain cancers. Pharmacological modulation of cAMP levels, either by targeting adenylyl cyclase or PDEs, is a therapeutic strategy for managing these conditions.

Calcium Ions (Ca^{2+}):

Calcium ions (Ca^{2+}) serve as a universal second messenger in many signaling pathways, mediating a wide range of cellular processes, including muscle contraction, neurotransmitter release, and gene expression. The versatility of calcium signaling is attributed to its ability to interact with a diverse array of calcium-binding proteins and enzymes. Intracellular calcium levels are tightly regulated by calcium channels, pumps, and exchangers located in the plasma membrane and intracellular organelles, such as the endoplasmic reticulum (ER) and mitochondria. In response to various stimuli, calcium channels open, allowing Ca^{2+} to flow into the cytoplasm, where it can activate calcium-dependent proteins and signaling cascades. Calcium signaling is often initiated by the activation of GPCRs or receptor tyrosine kinases (RTKs), which can stimulate the production of inositol triphosphate (IP_3) and diacylglycerol (DAG). IP_3 binds to its receptor on the ER membrane, triggering the release of calcium from the ER into the cytoplasm. The resulting increase in cytosolic calcium levels activates various calcium-dependent proteins, such as calmodulin and protein kinase C (PKC), leading to diverse cellular responses. Calcium signaling is highly dynamic and spatially regulated, allowing for the generation of complex calcium oscillations and waves that encode specific information. These calcium signals can influence processes such as cell proliferation, differentiation, apoptosis, and synaptic plasticity. Disruptions in calcium signaling are associated with numerous diseases, including neurodegenerative disorders, cardiovascular diseases, and cancer. Therapeutic strategies that target calcium channels and pumps are being explored for their potential to modulate calcium signaling in disease contexts.

Inositol Triphosphate (IP_3) and Diacylglycerol (DAG):

Inositol triphosphate (IP_3) and diacylglycerol (DAG) are second messengers generated by the hydrolysis of phosphatidylinositol 4,5-bisphosphate (PIP_2) by the enzyme phospholipase C (PLC). They play a critical role in mediating signaling pathways activated by GPCRs and RTKs. IP_3 is a soluble molecule that diffuses through the cytoplasm and binds to its receptor on the ER membrane, leading to the release of calcium ions from the ER into the cytoplasm. This increase in intracellular calcium levels triggers calcium-dependent signaling pathways, contributing to diverse cellular responses. DAG, on the other hand, remains in the plasma membrane and acts as a cofactor for the activation of protein kinase C (PKC). PKC phosphorylates various target proteins involved in cell growth, differentiation, and immune responses, thereby modulating their activity and function. The IP_3/DAG signaling pathway is involved in a wide range of physiological processes, including hormone secretion, neurotransmitter release, and immune cell activation. It is also implicated in the regulation of cell proliferation, migration, and apoptosis. Dysregulation of IP_3/DAG signaling is associated with several diseases, including cancer, cardiovascular diseases, and neurodegenerative disorders. Targeting components of this pathway, such as PLC, PKC, and IP_3 receptors, is a potential therapeutic strategy for modulating aberrant signaling in disease states. The coordinated regulation of IP_3 and DAG signaling ensures that cellular responses are finely tuned and context-dependent. The interplay between these second messengers and other signaling pathways allows for the integration of multiple signals and the generation of precise cellular outcomes.

Protein Kinases

Protein kinases are enzymes that catalyze the transfer of phosphate groups from ATP to specific amino acid residues on target proteins, typically serine, threonine, or tyrosine residues. Phosphorylation by protein kinases is a common mechanism for regulating protein activity, localization, and interactions, thereby modulating various cellular processes.

Mitogen-Activated Protein Kinases (MAPKs):

Mitogen-Activated Protein Kinases (MAPKs) are a family of serine/threonine kinases that play a central role in transducing extracellular signals into intracellular responses. They are involved in regulating cell growth, differentiation, stress responses, and apoptosis. The MAPK signaling cascade is a highly conserved pathway that is activated by a wide range of stimuli, including growth factors, cytokines, and environmental stressors.

The MAPK signaling cascade consists of three tiers of kinases: MAP kinase kinase kinases (MAPKKKs), MAP kinase kinases (MAPKKs), and MAP kinases (MAPKs). Upon activation by upstream signals, MAPKKKs phosphorylate and activate MAPKKs, which in turn phosphorylate and activate MAPKs. Activated MAPKs translocate to the nucleus, where they phosphorylate and regulate various transcription factors and other target proteins.

There are several MAPK pathways, including the extracellular signal-regulated kinase (ERK) pathway, the c-Jun N-terminal kinase (JNK) pathway, and the p38 MAPK pathway. Each pathway is activated by specific stimuli and regulates distinct cellular responses. For example, the ERK pathway is primarily activated by growth factors and is involved in promoting cell proliferation and differentiation, while the JNK and p38 pathways are activated by stress signals and are involved in regulating apoptosis and inflammatory responses. The specificity and regulation of MAPK signaling are achieved through the assembly of multi-protein complexes known as signaling modules or signalosomes. These complexes include scaffold proteins that bring together the components of the MAPK cascade, ensuring efficient signal transmission and preventing cross-talk with other pathways.

Dysregulation of MAPK signaling is associated with various diseases, particularly cancer, where constitutive activation of the ERK pathway can drive uncontrolled cell proliferation and survival.

Targeting MAPK pathways with specific inhibitors is a therapeutic strategy for treating certain cancers and inflammatory diseases.

Protein Kinase A (PKA):

Protein Kinase A (PKA) is a serine/threonine kinase that is activated by cyclic AMP (cAMP), a second messenger generated in response to the activation of G protein-coupled receptors (GPCRs). PKA is a key mediator of cAMP-dependent signaling pathways and is involved in regulating a wide range of physiological processes, including metabolism, gene expression, and cell growth.

PKA is composed of two regulatory and two catalytic subunits. In the absence of cAMP, the regulatory subunits inhibit the catalytic activity of PKA by binding to the catalytic subunits. Upon the production of cAMP, cAMP molecules bind to the regulatory subunits, causing a conformational change that releases and activates the catalytic subunits. Activated PKA phosphorylates specific target proteins on serine or threonine residues, modulating their activity and function. PKA signaling is involved in processes such as glycogen metabolism, lipid metabolism, and cell proliferation. For example, in response to the hormone epinephrine, PKA is activated in liver and muscle cells, leading to the phosphorylation of enzymes involved in glycogen breakdown, thereby increasing glucose availability for energy production. The specificity and regulation of PKA signaling are achieved through the compartmentalization of PKA and its substrates within specific cellular microdomains. A-kinase anchoring proteins (AKAPs) play a crucial role in targeting PKA to specific subcellular locations, ensuring that PKA signaling is spatially and temporally regulated. Dysregulation of PKA signaling is associated with various diseases, including heart failure, diabetes, and certain cancers. Pharmacological modulation of PKA activity, either by targeting cAMP levels or PKA itself, is a therapeutic strategy for managing these conditions.

Protein Kinase C (PKC):

Protein Kinase C (PKC) is a family of serine/threonine kinases that are activated by diacylglycerol (DAG) and calcium ions, second messengers generated in response to the activation of phospholipase C (PLC) by G protein-coupled receptors (GPCRs) or receptor tyrosine kinases (RTKs). PKC plays a critical role in regulating various cellular processes, including cell growth, differentiation, apoptosis, and immune responses. The PKC family consists of multiple isoforms, each with distinct regulatory and catalytic domains. PKC isoforms are classified into three categories based on their activation requirements: conventional PKCs (cPKCs), which require both DAG and calcium for activation; novel PKCs (nPKCs), which require DAG but not calcium; and atypical PKCs (aPKCs), which are activated by other mechanisms. Upon activation, PKC phosphorylates a wide range of substrates, including transcription factors, enzymes, and cytoskeletal proteins, modulating their activity and function. PKC signaling is involved in processes such as hormone secretion, neurotransmitter release, and immune cell activation. The specificity and regulation of PKC signaling are achieved through the compartmentalization of PKC isoforms within specific cellular microdomains. Scaffold proteins and lipid rafts play a crucial role in targeting PKC to specific subcellular locations, ensuring that PKC signaling is spatially and temporally regulated. Dysregulation of PKC signaling is associated with several diseases, including cancer, cardiovascular diseases, and neurodegenerative disorders. Targeting specific PKC isoforms with inhibitors or activators is a potential therapeutic strategy for modulating aberrant PKC signaling in disease states. The coordinated regulation of PKC signaling ensures that cellular responses are finely tuned and context-dependent. The interplay between PKC and other signaling pathways allows for the integration of multiple signals and the generation of precise cellular outcomes.

Akt (Protein Kinase B):

Akt, also known as Protein Kinase B (PKB), is a serine/threonine kinase that plays a central role in transducing signals from growth factors, cytokines, and hormones. It is a key regulator of cell survival, growth, proliferation, and metabolism. Akt is a major component of the phosphoinositide 3-kinase (PI3K)/Akt signaling pathway, which is activated by receptor tyrosine kinases (RTKs) and other receptors. The activation of Akt is initiated by the recruitment of PI3K to activated receptors, leading to the production of phosphatidylinositol 3,4,5-trisphosphate (PIP_3) in the plasma membrane. PIP_3 serves as a docking site for Akt and phosphoinositide-dependent kinase-1 (PDK1), facilitating the phosphorylation and activation of Akt by PDK1 and the mammalian target of rapamycin complex 2 (mTORC2). Once activated, Akt phosphorylates a wide range of substrates involved in regulating apoptosis, cell cycle progression, glucose metabolism, and protein synthesis. For example, Akt phosphorylates and inhibits pro-apoptotic proteins such as BAD and caspase-9, promoting cell survival. Akt also activates the mTOR pathway, enhancing protein synthesis and cell growth. The specificity and regulation of Akt signaling are achieved through the compartmentalization of Akt and its substrates within specific cellular microdomains. Scaffold proteins and lipid rafts play a crucial role in targeting Akt to specific subcellular locations, ensuring that Akt signaling is spatially and temporally regulated. Dysregulation of Akt signaling is associated with various diseases, particularly cancer, where constitutive activation of the PI3K/Akt pathway can drive uncontrolled cell proliferation and survival. Targeting Akt and its upstream regulators with specific inhibitors is a therapeutic strategy for treating certain cancers and metabolic disorders. The coordinated regulation of Akt signaling ensures that cellular responses are finely tuned and context-dependent. The interplay between Akt and other signaling pathways allows for the integration of multiple signals and the generation of precise cellular outcomes.

Transcription Factors

Transcription factors are proteins that regulate gene expression by binding to specific DNA sequences in the promoter regions of target genes. They play a critical role in translating extracellular signals into changes in gene expression, allowing cells to respond to various physiological and environmental cues. Transcription factors are activated or modulated by various signaling pathways, including those involving protein kinases and second messengers.

Nuclear Factor-kappa B (NF-κB):

Nuclear Factor-kappa B (NF-κB) is a family of transcription factors that play a central role in regulating immune and inflammatory responses, cell survival, and apoptosis. NF-κB is activated by a wide range of stimuli, including cytokines, pathogens, stress signals, and DNA damage. In the resting state, NF-κB is sequestered in the cytoplasm by inhibitor proteins known as IκBs. Upon activation by upstream signaling pathways, such as those involving receptor-interacting protein kinases (RIPKs) and

IκB kinases (IKKs), IκBs are phosphorylated and degraded, releasing NF-κB. The released NF-κB translocates to the nucleus, where it binds to specific DNA sequences and regulates the expression of target genes involved in immune responses, inflammation, cell proliferation, and survival. NF-κB signaling is tightly regulated through various mechanisms, including feedback inhibition, post-translational modifications, and interactions with coactivators and corepressors. The specificity of NF-κB signaling is determined by the composition of NF-κB dimers, the nature of the activating signals, and the cellular context. Dysregulation of NF-κB signaling is associated with several diseases, including inflammatory disorders, autoimmune diseases, and cancer. Aberrant activation of NF-κB can lead to chronic inflammation, uncontrolled cell proliferation, and resistance to apoptosis. Targeting NF-κB signaling with specific inhibitors is a therapeutic strategy for managing inflammatory diseases and certain cancers The coordinated regulation of NF-κB signaling ensures that cellular responses are finely tuned and context-dependent. The interplay between NF-κB and other signaling pathways allows for the integration of multiple signals and the generation of precise cellular outcomes.

Signal Transducers and Activators of Transcription (STATs):

Signal Transducers and Activators of Transcription (STATs) are a family of transcription factors that mediate signaling pathways activated by cytokines and growth factors. STATs play a crucial role in regulating immune responses, cell growth, differentiation, and survival.

STATs are activated by receptor-associated kinases, such as Janus kinases (JAKs), in response to cytokine receptor engagement. Upon activation, STATs are phosphorylated on specific tyrosine residues, leading to their dimerization and translocation to the nucleus. In the nucleus, STATs bind to specific DNA sequences and regulate the expression of target genes involved in various physiological processes. The JAK/STAT signaling pathway is highly conserved and is involved in mediating the effects of a wide range of cytokines and growth factors, including interferons, interleukins, and colony-stimulating factors. STATs can form homo- or heterodimers, allowing for diverse transcriptional responses depending on the nature of the activating signals. Dysregulation of STAT signaling is associated with several diseases, including immune disorders, cancer, and metabolic diseases. Aberrant activation of STATs can lead to uncontrolled cell proliferation, immune dysfunction, and resistance to apoptosis. Targeting STAT signaling with specific inhibitors is a therapeutic strategy for managing conditions associated with dysregulated cytokine signaling. The specificity and regulation of STAT signaling are achieved through various mechanisms, including the modulation of STAT phosphorylation, interactions with coactivators and corepressors, and feedback inhibition. The coordinated regulation of STAT signaling ensures that cellular responses are finely tuned and context-dependent.

Activator Protein 1 (AP-1):

Activator Protein 1 (AP-1) is a family of transcription factors composed of heterodimers formed by members of the JUN, FOS, ATF, and MAF protein families. AP-1 plays a central role in regulating gene expression in response to a wide range of stimuli, including growth factors, cytokines, stress signals, and pathogens. AP-1 transcription factors are activated by various signaling pathways, including the MAPK pathway, which regulates the phosphorylation and activation of JUN and FOS proteins. Once activated, AP-1 dimers translocate to the nucleus, where they bind to specific DNA sequences known as AP-1 binding sites and regulate the expression of target genes involved in cell proliferation, differentiation, apoptosis, and inflammation. The specificity and regulation of AP-1 signaling are achieved through the composition of AP-1 dimers, the nature of the activating signals, and the cellular context. Different combinations of JUN, FOS, ATF, and MAF proteins can form distinct AP-1 complexes with unique transcriptional activities. Dysregulation of AP-1 signaling is associated with several diseases, including cancer, inflammatory disorders, and neurodegenerative diseases. Aberrant activation of AP-1 can lead to uncontrolled cell proliferation, chronic inflammation, and resistance to apoptosis. Targeting AP-1 signaling with specific inhibitors is a potential therapeutic strategy for managing conditions associated with dysregulated gene expression. The coordinated regulation of AP-1 signaling ensures that cellular responses are finely tuned and context-dependent. The interplay between AP-1 and other signaling pathways allows for the integration of multiple signals and the generation of precise cellular outcomes.

Hypoxia-Inducible Factor (HIF):

Hypoxia-Inducible Factor (HIF) is a transcription factor that plays a key role in regulating gene expression in response to low oxygen levels (hypoxia). HIF is a heterodimer composed of an oxygen-sensitive alpha (α) subunit and a constitutively expressed beta (β) subunit. Under normoxic conditions, the HIF-α subunit is rapidly degraded by the ubiquitin-proteasome system, preventing HIF activation. In response to hypoxia, the degradation of the HIF-α subunit is inhibited, allowing it to accumulate and dimerize with the HIF-β subunit. The HIF heterodimer translocates to the nucleus, where it binds to specific DNA sequences known as hypoxia-responsive elements (HREs) and regulates the expression of target genes involved in angiogenesis, metabolism, and cell survival.

HIF signaling is involved in various physiological processes, including the adaptation to hypoxia, regulation of erythropoiesis, and modulation of metabolism. It plays a crucial role in tissue development, wound healing, and the response to ischemic conditions. Dysregulation of HIF signaling is associated with several diseases, including cancer, cardiovascular diseases, and chronic kidney disease. Aberrant activation of HIF can lead to increased angiogenesis, altered metabolism, and resistance to apoptosis. Targeting HIF signaling with specific inhibitors is a therapeutic strategy for managing conditions associated with dysregulated oxygen sensing. The specificity and regulation of HIF signaling are achieved through various mechanisms, including the modulation of HIF stability, interactions with coactivators and corepressors, and feedback inhibition. The coordinated regulation of HIF signaling ensures that cellular responses are finely tuned and context-dependent.

Key Signal Transduction Pathways

Signal transduction pathways are the conduits through which cells translate extracellular signals into precise cellular responses. These pathways are complex networks of interactions involving various signaling molecules, second messengers, and effector proteins. Key signal transduction pathways play critical roles in regulating diverse cellular processes, including cell growth, differentiation, survival, metabolism, and immune responses.

MAPK/ERK Pathway

The Mitogen-Activated Protein Kinase (MAPK) pathway, specifically the Extracellular Signal-Regulated Kinase (ERK) pathway, is a key signaling cascade involved in regulating cell growth, differentiation, and survival. This pathway is activated by a variety of extracellular signals, including growth factors, cytokines, and stress stimuli, and plays a crucial role in cellular responses to these signals. The MAPK/ERK pathway is initiated by the activation of receptor tyrosine kinases (RTKs) or other receptors, such as G protein-coupled receptors (GPCRs), in response to ligand binding. Upon activation, these receptors trigger the recruitment and activation of the small GTPase Ras, which serves as a molecular switch to initiate the signaling cascade. Activated Ras recruits and activates the serine/threonine kinase Raf, the first kinase in the MAPK/ERK cascade. Raf phosphorylates and activates MEK (MAPK/ERK kinase), a dual-specificity kinase that phosphorylates and activates ERK (Extracellular Signal-Regulated Kinase). Activated ERK translocates to the nucleus, where it phosphorylates various transcription factors, leading to changes in gene expression that drive cellular responses such as proliferation, differentiation, and survival. The MAPK/ERK pathway is tightly regulated by feedback mechanisms and cross-talk with other signaling pathways to ensure appropriate cellular responses. Negative feedback loops involving phosphatases, such as dual-specificity phosphatases (DUSPs), help to terminate ERK signaling and prevent excessive activation. Dysregulation of the MAPK/ERK pathway is implicated in various diseases, particularly cancer. Aberrant activation of the pathway can lead to uncontrolled cell proliferation, survival, and resistance to apoptosis, contributing to tumor development and progression. Targeting components of the MAPK/ERK pathway, such as Raf and MEK inhibitors, is a therapeutic strategy for treating cancers with dysregulated ERK signaling.

The specificity of the MAPK/ERK pathway is achieved through the assembly of multi-protein complexes known as signaling modules or signalosomes. These complexes include scaffold proteins that bring together the components of the MAPK cascade, ensuring efficient signal transmission and preventing cross-talk with other pathways. This compartmentalization allows for the spatial and temporal regulation of ERK signaling, enabling precise control of cellular responses.

PI3K/Akt Pathway

The Phosphoinositide 3-Kinase (PI3K)/Akt pathway is a key signaling pathway that regulates cell survival, growth, proliferation, and metabolism. This pathway is activated by various extracellular signals, including growth factors, cytokines, and hormones, and plays a critical role in cellular responses to these signals. The PI3K/Akt pathway is initiated by the activation of receptor tyrosine kinases (RTKs) or other receptors, such as G protein-coupled receptors (GPCRs), in response to ligand binding. Upon activation, these receptors recruit and activate PI3K, a lipid kinase that phosphorylates phosphatidylinositol 4,5-bisphosphate (PIP2) to generate phosphatidylinositol 3,4,5-trisphosphate (PIP3), a second messenger that recruits and activates downstream signaling proteins. One of the key downstream effectors of PIP3 is the serine/threonine kinase Akt (Protein Kinase B). Akt is recruited to the plasma membrane by PIP3, where it is phosphorylated and activated by phosphoinositide-dependent kinase 1 (PDK1) and the mammalian target of rapamycin complex 2 (mTORC2). Once activated, Akt phosphorylates a wide range of substrates involved in regulating apoptosis, cell cycle progression, glucose metabolism, and protein synthesis. The PI3K/Akt pathway plays a critical role in promoting cell survival by inhibiting pro-apoptotic proteins and activating anti-apoptotic proteins. For example, Akt phosphorylates and inhibits the pro-apoptotic protein BAD, promoting cell survival. Akt also activates the mTOR pathway, enhancing protein synthesis and cell growth. The specificity and regulation of the PI3K/Akt pathway are achieved through the compartmentalization of signaling components within specific cellular microdomains. Scaffold proteins and lipid rafts play a crucial role in targeting PI3K, Akt, and their substrates to specific subcellular locations, ensuring that signaling is spatially and temporally regulated. Dysregulation of the PI3K/Akt pathway is associated with various diseases, particularly cancer. Aberrant activation of the pathway can drive uncontrolled cell proliferation, survival, and resistance to apoptosis, contributing to tumor development and progression. Targeting components of the PI3K/Akt pathway, such as PI3K and Akt inhibitors, is a therapeutic strategy for treating cancers with dysregulated Akt signaling. The coordinated regulation of the PI3K/Akt pathway ensures that cellular responses are finely tuned and context-dependent. The interplay between the PI3K/Akt pathway and other signaling pathways allows for the integration of multiple signals and the generation of precise cellular outcomes.

JAK/STAT Pathway

The Janus Kinase (JAK)/Signal Transducer and Activator of Transcription (STAT) pathway is a key signaling pathway that mediates the effects of cytokines and growth factors on immune responses, cell growth, differentiation, and survival. This pathway is highly conserved and plays a critical role in regulating various physiological processes. The JAK/STAT pathway is initiated by the binding of cytokines or growth factors to their respective receptors, which are typically associated with receptor-associated kinases known as Janus kinases (JAKs). Upon ligand binding, the receptors undergo conformational changes that activate the associated JAKs through autophosphorylation. Activated JAKs phosphorylate specific tyrosine residues on the receptor, creating docking sites for STAT proteins. STATs are then recruited to the receptor and phosphorylated by JAKs on specific tyrosine residues. Phosphorylated STATs dimerize and translocate to the nucleus, where they bind to specific DNA sequences known as gamma-interferon activation sites (GAS) and regulate the expression of target genes involved in immune responses, cell growth, and differentiation. The JAK/STAT pathway is tightly regulated by various mechanisms, including negative feedback inhibition by suppressor of cytokine signaling (SOCS) proteins, which inhibit JAK activity and promote receptor degradation. This regulation ensures that cytokine signaling is precisely controlled, allowing for appropriate cellular responses. Dysregulation of the JAK/STAT pathway is associated with several diseases, including immune disorders, cancer, and metabolic diseases. Aberrant activation of STATs can lead to uncontrolled cell proliferation, immune dysfunction, and resistance to apoptosis. Targeting JAK/STAT signaling with specific inhibitors is a therapeutic strategy for managing conditions associated with dysregulated cytokine signaling. The specificity and regulation of the JAK/STAT pathway are achieved through the composition of STAT dimers, the nature of the activating cytokines, and the cellular context. The coordinated regulation of JAK/STAT signaling ensures that cellular responses are finely tuned and context-dependent.

The interplay between the JAK/STAT pathway and other signaling pathways allows for the integration of multiple signals and the generation of precise cellular outcomes. The JAK/STAT pathway is involved in mediating the effects of a wide range of cytokines and growth factors, including interferons, interleukins, and colony-stimulating factors.

Wnt Signaling Pathway

The Wnt signaling pathway is a highly conserved signaling cascade that plays a critical role in regulating embryonic development, cell proliferation, differentiation, and stem cell maintenance. This pathway is activated by a family of secreted glycoproteins known as Wnt proteins, which bind to specific cell surface receptors to initiate the signaling cascade. The Wnt signaling pathway can be divided into two main branches: the canonical (β-catenin-dependent) pathway and the non-canonical (β-catenin-independent) pathway. The canonical Wnt pathway is the most well-studied and is involved in regulating cell fate decisions, proliferation, and differentiation. In the absence of Wnt signaling, the destruction complex, composed of proteins such as Axin, adenomatous polyposis coli (APC), and glycogen synthase kinase 3 (GSK3), targets β-catenin for degradation. This prevents β-catenin from accumulating in the cytoplasm and translocating to the nucleus, where it would otherwise regulate gene expression. Upon Wnt binding to its receptor, Frizzled, and co-receptor, LRP5/6, the destruction complex is inhibited, allowing β-catenin to accumulate in the cytoplasm and translocate to the nucleus. In the nucleus, β-catenin forms a complex with transcription factors of the TCF/LEF family, regulating the expression of target genes involved in cell proliferation, differentiation, and stem cell maintenance. The non-canonical Wnt pathways, including the planar cell polarity (PCP) pathway and the Wnt/Ca^{2+} pathway, regulate processes such as cell polarity, migration, and calcium signaling. These pathways do not involve β-catenin and are activated by distinct sets of Wnt ligands and receptors. The Wnt signaling pathway is tightly regulated by various mechanisms, including feedback inhibition, cross-talk with other signaling pathways, and the involvement of secreted Wnt antagonists, such as Dickkopf (DKK) and secreted frizzled-related proteins (sFRPs). This regulation ensures that Wnt signaling is precisely controlled, allowing for appropriate cellular responses. Dysregulation of Wnt signaling is associated with several diseases, particularly cancer. Aberrant activation of the Wnt/β-catenin pathway can lead to uncontrolled cell proliferation and survival, contributing to tumor development and progression. Targeting components of the Wnt signaling pathway, such as β-catenin and its downstream effectors, is a therapeutic strategy for treating cancers with dysregulated Wnt signaling. The coordinated regulation of Wnt signaling ensures that cellular responses are finely tuned and context-dependent. The interplay between the Wnt pathway and other signaling pathways allows for the integration of multiple signals and the generation of precise cellular outcomes.

The Wnt signaling pathway is also involved in regulating the self-renewal and differentiation of stem cells, making it a key player in tissue homeostasis and regeneration. Understanding the molecular mechanisms underlying Wnt signaling is essential for developing strategies to manipulate stem cell behavior and promote tissue regeneration.

Calcium Signaling

Calcium signaling is a fundamental signaling mechanism that plays a central role in regulating a wide range of cellular processes, including muscle contraction, neurotransmitter release, gene expression, and apoptosis. Calcium ions (Ca^{2+}) serve as a universal second messenger, mediating diverse physiological responses by interacting with calcium-binding proteins and enzymes. Intracellular calcium levels are tightly regulated by calcium channels, pumps, and exchangers located in the plasma membrane and intracellular organelles, such as the endoplasmic reticulum (ER) and mitochondria. In response to various stimuli, calcium channels open, allowing Ca^{2+} to flow into the cytoplasm, where it can activate calcium-dependent proteins and signaling cascades.

Calcium signaling is often initiated by the activation of G protein-coupled receptors (GPCRs) or receptor tyrosine kinases (RTKs), which can stimulate the production of inositol triphosphate (IP_3) and diacylglycerol (DAG). IP_3 binds to its receptor on the ER membrane, triggering the release of calcium from the ER into the cytoplasm. The resulting increase in cytosolic calcium levels activates various calcium-dependent proteins, such as calmodulin and protein kinase C (PKC), leading to diverse cellular responses.Calcium signaling is highly dynamic and spatially regulated, allowing for the generation of complex calcium oscillations and waves that encode specific information. These calcium signals can influence processes such as cell proliferation, differentiation, apoptosis, and synaptic plasticity. The specificity and regulation of calcium signaling are achieved through the compartmentalization of calcium channels, pumps, and sensors within specific cellular microdomains. The interplay between calcium signaling and other signaling pathways allows for the integration of multiple signals and the generation of precise cellular outcomes. Dysregulation of calcium signaling is associated with numerous diseases, including neurodegenerative disorders, cardiovascular diseases, and cancer. Therapeutic strategies that target calcium channels, pumps, and sensors are being explored for their potential to modulate calcium signaling in disease contexts. The coordinated regulation of calcium signaling ensures that cellular responses are finely tuned and context-dependent. The versatility of calcium signaling is attributed to its ability to interact with a diverse array of calcium-binding proteins and enzymes, allowing it to mediate a wide range of physiological processes. Calcium signaling also plays a critical role in the regulation of apoptosis, a programmed cell death process essential for maintaining tissue homeostasis and eliminating damaged or unwanted cells. Calcium ions can modulate the activity of key apoptotic proteins, such as caspases and Bcl-2 family proteins, influencing cell survival and death decisions.

3. Integration of Nuclear and Cytoplasmic Signaling

The integration of nuclear and cytoplasmic signaling is a fundamental aspect of cellular biology, essential for coordinating complex cellular responses to internal and external stimuli. This integration allows cells to process diverse signals from their environment and cytoplasm, ensuring that appropriate and cohesive responses are produced. The nucleus, acting as the command center of the cell, plays a central role in interpreting signals and orchestrating cellular functions such as gene expression, metabolism, growth, differentiation, and survival. Cellular signaling pathways are intricate networks of interactions that enable cells to respond to changes in their environment and maintain homeostasis. These pathways involve the transmission of signals from receptors on the cell surface or within the cell to various intracellular targets, ultimately leading to changes in cellular behavior. The integration of these pathways within the cell requires precise coordination between nuclear and cytoplasmic signaling components. The nucleus and cytoplasm are distinct cellular compartments with specialized functions. The nucleus contains the genetic material and serves as the site of transcription, where DNA is transcribed into RNA. The cytoplasm, on the other hand, is the site of translation and various metabolic processes, housing numerous organelles that perform specialized functions. The integration of signaling between these compartments is achieved through a complex interplay of molecular interactions and regulatory mechanisms. The integration of nuclear and cytoplasmic signaling is mediated by several key mechanisms, including retrograde signaling, cross-talk between pathways, and epigenetic regulation. These mechanisms ensure that signals are transmitted accurately and efficiently, allowing cells to adapt to changing conditions and respond to various stimuli.

Retrograde Signaling

Retrograde signaling is a vital communication process that allows organelles, particularly mitochondria and chloroplasts, to convey information back to the nucleus. This type of signaling is essential for regulating nuclear gene expression in response to changes in organelle function, enabling cells to adapt to metabolic shifts and environmental stressors.

Mitochondrial Retrograde Signaling

Mitochondria are dynamic organelles responsible for energy production through oxidative phosphorylation, regulation of metabolic pathways, and maintenance of cellular redox balance. They play a critical role in sensing and responding to cellular stress and damage. Mitochondrial retrograde signaling refers to the communication from mitochondria to the nucleus, which modulates nuclear gene expression in response to changes in mitochondrial function.

Mitochondrial dysfunction can result from various factors, including oxidative stress, nutrient deprivation, or genetic mutations. When mitochondria experience stress or damage, they initiate retrograde signaling pathways that activate nuclear genes involved in stress responses, metabolic adaptation, and mitochondrial biogenesis. This signaling ensures that the cell can adapt to changes in energy demands and maintain cellular homeostasis. One of the key players in mitochondrial retrograde signaling is the transcription factor nuclear factor erythroid 2-related factor 2 (NRF2). In response to oxidative stress, NRF2 translocates to the nucleus and induces the expression of antioxidant genes, promoting cellular defense mechanisms against oxidative damage. Similarly, the transcription factor activating transcription factor 4 (ATF4) is activated in response to mitochondrial stress and regulates the expression of genes involved in amino acid metabolism, redox balance, and apoptosis. Mitochondrial retrograde signaling also involves the release of signaling molecules such as reactive oxygen species (ROS), calcium ions, and metabolites, which can influence nuclear gene expression. For example, ROS produced by mitochondria can activate signaling pathways that induce the expression of genes involved in detoxification and repair processes. The integration of mitochondrial retrograde signaling with other cellular signaling pathways, such as the MAPK and PI3K/Akt pathways, allows for coordinated cellular responses to stress and metabolic changes. This cross-talk ensures that cells can respond to complex stimuli and maintain cellular homeostasis. Dysregulation of mitochondrial retrograde signaling is associated with various diseases, including neurodegenerative disorders, cancer, and metabolic syndromes.

Chloroplast Retrograde Signaling

Chloroplasts are the sites of photosynthesis in plants and certain algae, playing a crucial role in energy conversion and carbon fixation. They are also involved in various metabolic processes, such as the synthesis of fatty acids, amino acids, and secondary metabolites. Chloroplast retrograde signaling refers to the communication from chloroplasts to the nucleus, which regulates nuclear gene expression in response to changes in chloroplast function and environmental conditions.Chloroplasts can sense environmental changes, such as light intensity, temperature, and nutrient availability, and relay this information to the nucleus through retrograde signaling pathways. This signaling ensures that nuclear gene expression is adjusted to optimize photosynthetic efficiency and metabolic function.

One of the key pathways involved in chloroplast retrograde signaling is the generation of ROS, which can act as signaling molecules to modulate nuclear gene expression. Under stress conditions, such as high light intensity or drought, chloroplasts produce ROS that can activate transcription factors involved in stress responses and acclimation processes. Another important component of chloroplast retrograde signaling is the accumulation of specific metabolites, such as plastidial isoprenoids and tetrapyrroles, which can influence nuclear gene expression. For example, the accumulation of the tetrapyrrole intermediate Mg-protoporphyrin IX (Mg-ProtoIX) can trigger the expression of nuclear genes involved in chlorophyll biosynthesis and stress responses. Chloroplast retrograde signaling also involves the release of signaling proteins, such as WHIRLY1 and PTM (plastid-to-nucleus retrograde signaling molecule), which translocate to the nucleus and regulate gene expression. These proteins act as mediators of the communication between chloroplasts and the nucleus, ensuring that nuclear gene expression is coordinated with chloroplast function. The integration of chloroplast retrograde signaling with other signaling pathways, such as the abscisic acid (ABA) signaling pathway, allows for coordinated cellular responses to environmental stressors. This cross-talk ensures that plants can adapt to changing environmental conditions and maintain growth and productivity. Chloroplast retrograde signaling is essential for improving crop yields and developing strategies to enhance plant resilience to environmental stress. Manipulating these signaling pathways through genetic engineering or biotechnological approaches can enhance plant stress tolerance and improve agricultural productivity.

Cross-Talk Between Pathways

Cross-talk between signaling pathways refers to the interactions and interconnections between different signaling cascades within a cell. This cross-talk allows cells to integrate multiple signals and produce coordinated responses, ensuring that cellular functions are finely tuned and context-dependent. Cross-talk between pathways adds complexity and flexibility to cellular signaling networks, enabling cells to adapt to diverse stimuli and environmental changes.

MAPK and PI3K/Akt Pathways

The MAPK (Mitogen-Activated Protein Kinase) and PI3K/Akt (Phosphoinositide 3-Kinase/Akt) pathways are two major signaling pathways involved in regulating cell growth, proliferation, survival, and metabolism. These pathways often exhibit cross-talk, where components of one pathway influence or modulate the activity of the other, allowing for the integration of signals and the coordination of cellular responses. The MAPK/ERK pathway is typically activated by growth factors and mitogens and plays a crucial role in promoting cell proliferation and differentiation. The pathway involves a cascade of phosphorylation events, leading to the activation of ERK (Extracellular Signal-Regulated Kinase), which translocates to the nucleus and regulates the expression of genes involved in cell cycle progression and survival. The PI3K/Akt pathway is activated by various extracellular signals, including growth factors and hormones, and regulates cell survival, growth, and metabolism. Upon activation, PI3K generates the second messenger PIP3, which recruits and activates Akt, a serine/threonine kinase that phosphorylates various substrates involved in apoptosis inhibition, glucose metabolism, and protein synthesis. Cross-talk between the MAPK and PI3K/Akt pathways occurs at multiple levels, allowing for the integration of signals and the fine-tuning of cellular responses. For example, both pathways can converge on shared downstream targets, such as mTOR (mammalian target of rapamycin), a key regulator of protein synthesis and cell growth. The integration of these pathways ensures that cells can respond to growth signals with coordinated changes in proliferation, survival, and metabolism. Additionally, cross-talk between the MAPK and PI3K/Akt pathways can involve feedback loops and regulatory proteins that modulate the activity of key signaling components. For instance, Akt can phosphorylate and inhibit Raf, a MAPK pathway component, thereby providing negative feedback regulation. Conversely, ERK can phosphorylate and inhibit the activity of certain PI3K components, modulating the strength and duration of Akt signaling. The cross-talk between these pathways also allows for the integration of additional signals, such as stress and inflammation, which can influence cell fate decisions. For example, the activation of stress-activated MAPK pathways, such as JNK and p38, can modulate the activity of the PI3K/Akt pathway, influencing apoptosis and survival responses. Dysregulation of cross-talk between the MAPK and PI3K/Akt pathways is associated with various diseases, particularly cancer. Aberrant signaling can lead to uncontrolled cell proliferation, survival, and resistance to apoptosis, contributing to tumor development and progression. Targeting the cross-talk between these pathways with specific inhibitors or combination therapies is a potential strategy for treating cancers with dysregulated signaling.

Calcium and cAMP Signaling

Calcium and cyclic AMP (cAMP) are two major second messengers involved in cellular signaling, mediating a wide range of physiological processes. These signaling pathways often exhibit cross-talk, where changes in calcium levels can influence cAMP signaling, and vice versa, allowing for the integration of signals and the coordination of cellular responses.

Calcium signaling is initiated by the release of calcium ions (Ca^{2+}) from intracellular stores, such as the endoplasmic reticulum (ER), or by the influx of Ca^{2+} through plasma membrane channels. Calcium ions act as a universal second messenger, activating various calcium-dependent proteins and enzymes, such as calmodulin, protein kinase C (PKC), and calcium/calmodulin-dependent protein kinases (CaMKs). cAMP signaling is initiated by the activation of adenylyl cyclase, an enzyme that converts ATP into cAMP in response to the activation of G protein-coupled receptors (GPCRs). cAMP acts as a second messenger that activates protein kinase A (PKA), a serine/threonine kinase that phosphorylates various target proteins involved in metabolism, gene expression, and cell growth.

Cross-talk between calcium and cAMP signaling can occur through several mechanisms, allowing for the integration of signals and the modulation of cellular responses. For example, calcium can modulate the activity of adenylyl cyclase, influencing cAMP production and PKA activation. Calcium can also activate phosphodiesterases (PDEs), enzymes that degrade cAMP, thereby modulating the strength and duration of cAMP signaling. Conversely, cAMP can influence calcium signaling by modulating the activity of calcium channels and pumps. For instance, PKA can phosphorylate and modulate the activity of L-type calcium channels, influencing calcium influx and cellular responses. cAMP can also regulate the activity of calcium pumps, such as the sarcoplasmic/endoplasmic reticulum Ca^{2+} ATPase (SERCA), modulating calcium homeostasis and signaling. The cross-talk between calcium and cAMP signaling is involved in various physiological processes, including neurotransmitter release, muscle contraction, and hormone secretion. It allows cells to integrate multiple signals and produce coordinated responses, ensuring that cellular functions are finely tuned and context-dependent. Dysregulation of cross-talk between calcium and cAMP signaling is associated with several diseases, including cardiovascular diseases, neurodegenerative disorders, and metabolic syndromes. The molecular mechanisms underlying this cross-talk is essential for developing therapeutic strategies to target these diseases. The coordinated regulation of calcium and cAMP signaling ensures that cellular responses are finely tuned and context-dependent. The interplay between these signaling pathways allows for the integration of multiple signals and the generation of precise cellular outcomes.

JAK/STAT and TGF-β Pathways

The Janus Kinase (JAK)/Signal Transducer and Activator of Transcription (STAT) pathway and the Transforming Growth Factor-beta (TGF-β) pathway are two major signaling pathways involved in regulating immune responses, cell growth, differentiation, and tissue homeostasis. These pathways often exhibit cross-talk, where components of one pathway influence or modulate the activity of the other, allowing for the integration of signals and the coordination of cellular responses.

The JAK/STAT pathway is activated by cytokines and growth factors, leading to the phosphorylation and activation of STAT transcription factors. Activated STATs translocate to the nucleus, where they regulate the expression of target genes involved in immune responses, cell proliferation, and differentiation. The TGF-β pathway is activated by the binding of TGF-β ligands to their receptors, leading to the phosphorylation and activation of Smad transcription factors. Activated Smads translocate to the nucleus, where they regulate the expression of target genes involved in cell proliferation, differentiation, and tissue remodeling. Cross-talk between the JAK/STAT and TGF-β pathways occurs at multiple levels, allowing for the integration of signals and the fine-tuning of cellular responses. For example, STATs and Smads can physically interact and form complexes, modulating each other's transcriptional activity and influencing the expression of shared target genes. The cross-talk between these pathways also allows for the integration of additional signals, such as inflammatory cytokines and growth factors, which can influence cell fate decisions. For instance, the activation of the JAK/STAT pathway by pro-inflammatory cytokines can modulate the activity of the TGF-β pathway, influencing immune responses and tissue remodeling. The interplay between the JAK/STAT and TGF-β pathways is involved in various physiological processes, including immune regulation, wound healing, and tissue fibrosis. It allows cells to integrate multiple signals and produce coordinated responses, ensuring that cellular functions are finely tuned and context-dependent. Dysregulation of cross-talk between the JAK/STAT and TGF-β pathways is associated with several diseases, including cancer, autoimmune disorders, and fibrotic diseases. Understanding the molecular mechanisms underlying this cross-talk is essential for developing therapeutic strategies to target these diseases. The coordinated regulation of the JAK/STAT and TGF-β pathways ensures that cellular responses are finely tuned and context-dependent. The interplay between these signaling pathways allows for the integration of multiple signals and the generation of precise cellular outcomes.

Epigenetic Regulation

Epigenetic regulation is a critical mechanism by which cells integrate nuclear and cytoplasmic signaling to modulate gene expression and cellular behavior. Epigenetic modifications, such as DNA methylation and histone modifications, alter chromatin structure and gene accessibility, thereby influencing the transcriptional activity of genes. These modifications provide a means for cells to maintain long-lasting changes in gene expression in response to various signals, allowing for cellular memory and adaptation to environmental changes.

DNA Methylation

DNA methylation is an epigenetic modification that involves the addition of a methyl group to the 5-carbon of cytosine residues within CpG dinucleotides, leading to the formation of 5-methylcytosine. This modification is catalyzed by DNA methyltransferases (DNMTs) and is typically associated with gene silencing and transcriptional repression.

DNA methylation plays a crucial role in regulating gene expression by altering the accessibility of DNA to transcription factors and the transcriptional machinery. Methylation of promoter regions can inhibit the binding of transcription factors, leading to gene silencing. Conversely, demethylation of promoter regions can activate gene expression by facilitating the binding of transcription factors. Signal transduction pathways can modulate DNA methylation patterns, leading to changes in gene expression in response to various stimuli. For example, signaling pathways activated by growth factors, hormones, or stress signals can influence the expression and activity of DNMTs, leading to alterations in DNA methylation patterns and gene expression. DNA methylation is involved in various physiological processes, including development, differentiation, and imprinting. It plays a critical role in establishing and maintaining cellular identity by regulating the expression of lineage-specific genes and preventing the expression of inappropriate genes. Dysregulation of DNA methylation is associated with several diseases, including cancer, neurodegenerative disorders, and developmental disorders. Aberrant DNA methylation patterns can lead to the silencing of tumor suppressor genes or the activation of oncogenes, contributing to tumor development and progression. Understanding the mechanisms underlying DNA methylation and its regulation by signaling pathways is essential for developing therapeutic strategies to target these diseases. The coordinated regulation of DNA methylation ensures that cellular responses are finely tuned and context-dependent. The interplay between DNA methylation and other epigenetic modifications, such as histone modifications, allows for the integration of multiple signals and the generation of precise cellular outcomes.

Histone Modifications

Histone modifications are epigenetic modifications that involve the covalent addition or removal of chemical groups to histone proteins, which are the core components of nucleosomes. These modifications include acetylation, methylation, phosphorylation, and ubiquitination, and they play a critical role in regulating chromatin structure and gene expression. Histone acetylation involves the addition of acetyl groups to lysine residues on histone tails, catalyzed by histone acetyltransferases (HATs). This modification is typically associated with transcriptional activation, as it reduces the positive charge on histones, leading to a more open chromatin structure and increased accessibility of DNA to transcription factors. Histone methylation involves the addition of methyl groups to lysine or arginine residues on histone tails, catalyzed by histone methyltransferases (HMTs). This modification can be associated with either transcriptional activation or repression, depending on the specific lysine residue that is methylated. For example, methylation of histone H3 lysine 4 (H3K4) is associated with active gene expression, while methylation of histone H3 lysine 9 (H3K9) is associated with gene repression.

Signal transduction pathways can modulate histone modifications, leading to changes in gene expression in response to various stimuli. For example, signaling pathways activated by growth factors, stress signals, or metabolic changes can influence the activity of HATs and HMTs, leading to alterations in histone modification patterns and gene expression. Histone modifications play a crucial role in regulating gene expression during development, differentiation, and response to environmental changes. They provide a means for cells to maintain cellular memory and adapt to changing conditions by modulating the transcriptional activity of specific genes. Dysregulation of histone modifications is associated with several diseases, including cancer, cardiovascular diseases, and neurodegenerative disorders. Aberrant histone modification patterns can lead to inappropriate gene expression, contributing to disease development and progression. Understanding the mechanisms underlying histone modifications and their regulation by signaling pathways is essential for developing therapeutic strategies to target these diseases. The coordinated regulation of histone modifications ensures that cellular responses are finely tuned and context-dependent. The interplay between histone modifications and other epigenetic modifications, such as DNA methylation, allows for the integration of multiple signals and the generation of precise cellular outcomes.

Chromatin Remodeling

Chromatin remodeling is a dynamic process that involves the alteration of chromatin structure to regulate gene accessibility and transcriptional activity. It is mediated by chromatin remodeling complexes, which use ATP-dependent mechanisms to reposition, eject, or restructure nucleosomes, allowing for changes in chromatin organization. Chromatin remodeling complexes are composed of multiple subunits, including ATPases that provide the energy for remodeling and other subunits that confer specificity and regulatory functions. These complexes play a critical role in regulating gene expression, DNA replication, repair, and recombination. Signal transduction pathways can influence chromatin remodeling by modulating the activity of chromatin remodeling complexes, leading to changes in gene expression in response to various stimuli. For example, signaling pathways activated by growth factors, stress signals, or metabolic changes can influence the recruitment and activity of chromatin remodeling complexes, leading to alterations in chromatin structure and gene accessibility. Chromatin remodeling is involved in various physiological processes, including development, differentiation, and response to environmental changes. It plays a critical role in establishing and maintaining cellular identity by regulating the expression of lineage-specific genes and facilitating the activation or repression of specific gene networks. Dysregulation of chromatin remodeling is associated with several diseases, including cancer, developmental disorders, and neurodegenerative disorders. Aberrant chromatin remodeling can lead to inappropriate gene expression and chromatin organization, contributing to disease development and progression. Understanding the mechanisms underlying chromatin remodeling and its regulation by signaling pathways is essential for developing therapeutic strategies to target these diseases. The coordinated regulation of chromatin remodeling ensures that cellular responses are finely tuned and context-dependent. The interplay between chromatin remodeling and other epigenetic modifications, such as DNA methylation and histone modifications, allows for the integration of multiple signals and the generation of precise cellular outcomes.

Implications and Applications

The integration of nuclear and cytoplasmic signaling and the understanding of signal transduction pathways have profound implications for various fields, including medicine, agriculture, and biotechnology. These insights provide valuable knowledge for developing targeted therapies, improving crop traits, and advancing biotechnological applications.

Disease Mechanisms

Aberrations in signal transduction pathways and nuclear-cytoplasmic interactions are often implicated in various diseases, including cancer, neurodegenerative disorders, cardiovascular diseases, and metabolic syndromes. Understanding the molecular mechanisms underlying these aberrations is crucial for developing targeted therapies and interventions.

Cancer:

Cancer is a complex disease characterized by uncontrolled cell proliferation, survival, and metastasis. Dysregulation of signaling pathways, such as the MAPK, PI3K/Akt, and JAK/STAT pathways, is a hallmark of cancer. Mutations or alterations in components of these pathways can lead to constitutive activation of signaling, driving tumor development and progression. For example, mutations in the Ras gene, a key component of the MAPK pathway, are frequently observed in various cancers, leading to persistent activation of the pathway and uncontrolled cell proliferation. Similarly, mutations or overexpression of receptor tyrosine kinases (RTKs), such as EGFR and HER2, can result in aberrant activation of the PI3K/Akt pathway, promoting cell survival and resistance to apoptosis.

Neurodegenerative Disorders:

Neurodegenerative disorders, such as Alzheimer's disease, Parkinson's disease, and Huntington's disease, are characterized by the progressive degeneration of neurons and the accumulation of toxic protein aggregates. Dysregulation of signaling pathways involved in neuronal survival, synaptic plasticity, and protein homeostasis is implicated in the pathogenesis of these disorders. For example, dysregulation of calcium signaling and oxidative stress can contribute to neuronal damage and apoptosis in neurodegenerative disorders. The accumulation of misfolded proteins, such as amyloid-beta and tau in Alzheimer's disease, can activate stress-responsive signaling pathways, leading to neuroinflammation and cell death.

Cardiovascular Diseases:

Cardiovascular diseases, such as atherosclerosis, hypertension, and heart failure, are major causes of morbidity and mortality worldwide. Dysregulation of signaling pathways involved in vascular homeostasis, inflammation, and cardiac remodeling is implicated in the development and progression of cardiovascular diseases.

For example, aberrant activation of the renin-angiotensin-aldosterone system (RAAS) and inflammatory signaling pathways can contribute to vascular dysfunction, inflammation, and fibrosis in cardiovascular diseases. The activation of the NF-κB pathway, a key regulator of inflammation, is associated with atherosclerosis and cardiac remodeling. The molecular mechanisms underlying cardiovascular diseases is crucial for developing therapeutic strategies to target disease-associated pathways. Approaches such as targeting RAAS components, modulating inflammatory signaling, and promoting vascular repair are being explored for their potential to improve cardiovascular health and reduce disease burden.

Biotechnological Applications

The integration of nuclear and cytoplasmic signaling and the understanding of signal transduction pathways have significant implications for biotechnological applications, particularly in agriculture and biotechnology. These insights provide valuable knowledge for developing strategies to enhance crop traits, improve agricultural productivity, and advance biotechnological innovations.

Agriculture:

In agriculture, manipulating nuclear-cytoplasmic interactions and signal transduction pathways can enhance crop traits such as stress tolerance, yield, and disease resistance. Understanding the molecular mechanisms underlying plant responses to environmental stressors, such as drought, salinity, and pathogens, is essential for developing strategies to improve crop resilience and productivity. For example, manipulating signaling pathways involved in stress responses, such as the abscisic acid (ABA) and salicylic acid (SA) pathways, can enhance plant tolerance to abiotic and biotic stressors. Genetic engineering and CRISPR-Cas9 technology can be employed to modify key signaling components and enhance desirable traits in crops. Biotechnological approaches, such as transgenic technology and gene editing, are being used to introduce specific genes or modify existing genes in crops to improve traits such as yield, nutritional content, and resistance to pests and diseases. These approaches offer the potential to develop crops with enhanced performance and sustainability.

Biotechnology:

In biotechnology, understanding signal transduction pathways provides valuable insights for developing novel therapeutic strategies, bioproduction systems, and synthetic biology applications. These insights can inform the design and engineering of biological systems with tailored functions and capabilities. For example, manipulating signaling pathways involved in cell growth and metabolism can enhance the production of valuable biomolecules, such as proteins, enzymes, and metabolites, in microbial and mammalian cell systems. Synthetic biology approaches, such as the construction of synthetic gene circuits and metabolic pathways, offer the potential to engineer organisms with enhanced bioproduction capabilities. In therapeutic applications, targeting specific components of signal transduction pathways offers potential avenues for treating various diseases. Small molecule inhibitors, monoclonal antibodies, and RNA-based therapies are being developed to modulate signaling pathways for therapeutic benefit. The molecular mechanisms underlying signal transduction pathways is essential for advancing biotechnological innovations and developing strategies to address global challenges in health, agriculture, and the environment.

Therapeutic Interventions

Targeting specific components of signal transduction pathways offers potential therapeutic avenues for treating various diseases. Understanding the molecular mechanisms underlying signaling pathways provides valuable insights for developing targeted therapies and interventions that modulate these pathways for therapeutic benefit.

Small Molecule Inhibitors:

Small molecule inhibitors are a class of therapeutic agents that target specific components of signaling pathways, modulating their activity and function. These inhibitors are designed to bind to key proteins involved in signaling cascades, such as kinases, phosphatases, and receptors, thereby blocking or altering their activity. For example, kinase inhibitors target specific kinases involved in signaling pathways, such as the MAPK, PI3K/Akt, and JAK/STAT pathways, to modulate their activity and influence cellular responses. These inhibitors are used to treat various cancers, inflammatory diseases, and metabolic disorders. Small molecule inhibitors offer several advantages as therapeutic agents, including oral bioavailability, ease of synthesis, and the ability to target intracellular proteins. However, challenges such as off-target effects and resistance mechanisms need to be addressed to optimize their therapeutic efficacy.

Monoclonal Antibodies:

Monoclonal antibodies are a class of therapeutic agents that target specific proteins involved in signaling pathways, modulating their activity and function. These antibodies are designed to bind to extracellular or cell surface proteins, such as receptors and ligands, thereby blocking or altering their activity. For example, monoclonal antibodies targeting receptor tyrosine kinases (RTKs), such as HER2 and EGFR, are used to treat various cancers by blocking receptor activation and downstream signaling. Monoclonal antibodies targeting cytokines, such as TNF-α and IL-6, are used to treat inflammatory diseases by modulating immune responses. Monoclonal antibodies offer several advantages as therapeutic agents, including high specificity, long half-life, and the ability to target extracellular proteins. However, challenges such as immunogenicity and high production costs need to be addressed to optimize their therapeutic efficacy.

RNA-Based Therapies:

RNA-based therapies are a class of therapeutic agents that target specific components of signaling pathways by modulating gene expression. These therapies are designed to deliver RNA molecules, such as small interfering RNA (siRNA), antisense oligonucleotides, or messenger RNA (mRNA), to influence the expression of target genes. For example, siRNA and antisense oligonucleotides can be used to silence or degrade specific mRNA transcripts, reducing the expression of disease-associated genes. mRNA-based therapies can be used to deliver therapeutic proteins, such as vaccines or enzymes, to treat various diseases. RNA-based therapies offer several advantages as therapeutic agents, including the ability to target virtually any gene, rapid development timelines, and the potential for personalized medicine. However, challenges such as delivery efficiency and stability need to be addressed to optimize their therapeutic efficacy. The integration of nuclear and cytoplasmic signaling and the understanding of signal transduction pathways provide valuable insights for developing therapeutic strategies that target specific signaling components for therapeutic benefit. These insights offer the potential to address various diseases and improve health outcomes.

C. Inheritance and expression of organellar DNA

Organellar DNA, found in mitochondria and chloroplasts, represents a unique aspect of eukaryotic cellular biology. Unlike the nuclear genome, which is inherited biparentally and subject to Mendelian inheritance, organellar DNA follows distinct inheritance patterns and has specialized mechanisms of expression. The study of organellar genomes offers insights into evolutionary biology, cellular function, and genetic diversity. Mitochondria and chloroplasts are endosymbiotic organelles that originated from ancient prokaryotic cells. Over millions of years, these organelles have retained a portion of their ancestral genomes, which encode essential proteins and RNAs required for their function. Despite their reduced size compared to the nuclear genome, organellar genomes play critical roles in cellular energy production, metabolism, and homeostasis. The inheritance and expression of organellar DNA are subject to unique regulatory mechanisms that ensure the proper functioning of these organelles within the cell. Understanding these processes is essential for elucidating the complex interactions between nuclear and organellar genomes and their contributions to cellular physiology and evolution.

Mitochondrial DNA Inheritance

Mitochondrial DNA (mtDNA) is a small, circular DNA molecule located within the mitochondria, the energy-producing organelles of eukaryotic cells. The inheritance of mtDNA follows distinct patterns that differ from the Mendelian inheritance of nuclear DNA. Understanding the mechanisms and implications of mtDNA inheritance is crucial for studying mitochondrial diseases, population genetics, and evolutionary biology.

Maternal Inheritance of mtDNA

One of the most distinctive features of mitochondrial DNA inheritance is its maternal mode of transmission. In most eukaryotic organisms, mtDNA is inherited exclusively from the mother. This pattern of inheritance is observed across a wide range of species, including animals, plants, and fungi, and is a consequence of the mechanisms involved in fertilization and early embryonic development. During fertilization, the sperm contributes its nuclear DNA to the zygote, while the egg provides both nuclear and mitochondrial DNA. However, the sperm's mitochondria are typically degraded or excluded from the developing embryo, resulting in the exclusive transmission of maternal mtDNA. Several mechanisms have been proposed to explain the maternal inheritance of mtDNA, including the active degradation of paternal mitochondria and the selective replication of maternal mtDNA.

The maternal inheritance of mtDNA has significant implications for genetic studies and evolutionary biology. Because mtDNA is inherited as a single haplotype from the mother, it can be used to trace maternal lineages and study the evolutionary relationships between populations. This property of mtDNA has been exploited in fields such as phylogenetics and population genetics, where it is used to construct maternal ancestry trees and assess genetic diversity. The study of mtDNA inheritance has also revealed the presence of heteroplasmy, a condition in which multiple mtDNA variants coexist within a single cell or organism. Heteroplasmy can result from mutations in mtDNA or the coexistence of mtDNA from different maternal lineages. The proportion of different mtDNA variants can vary between tissues and over time, leading to phenotypic diversity and implications for mitochondrial diseases. Mitochondrial diseases are a group of genetic disorders caused by mutations in mtDNA or nuclear genes that affect mitochondrial function. These diseases often present with a wide range of symptoms, reflecting the essential roles of mitochondria in energy production and metabolism. The maternal inheritance of mtDNA means that mitochondrial diseases can be passed from mother to offspring, with varying degrees of severity depending on the level of heteroplasmy and tissue distribution of the mutant mtDNA.

Mechanisms of mtDNA Replication and Segregation

The replication and segregation of mitochondrial DNA are tightly regulated processes that ensure the maintenance of mtDNA copy number and distribution within cells. These processes are critical for mitochondrial biogenesis, cellular homeostasis, and the inheritance of mtDNA during cell division.

Mitochondrial DNA replication is initiated at specific regions within the mitochondrial genome, known as the origins of replication. Unlike the nuclear genome, which undergoes tightly regulated cell cycle-dependent replication, mtDNA replication occurs continuously throughout the cell cycle. This continuous replication allows for the maintenance of mtDNA copy number and the rapid adaptation of mitochondria to changing cellular energy demands. The replication of mtDNA is carried out by a dedicated set of proteins, including the mitochondrial DNA polymerase (POLG), the mitochondrial DNA helicase (TWINKLE), and the mitochondrial single-stranded DNA-binding protein (mtSSB). These proteins work in concert to unwind the mtDNA, synthesize new strands, and ensure the fidelity of mtDNA replication. Segregation of mtDNA occurs during cell division, where replicated mtDNA is distributed to daughter mitochondria and subsequently to daughter cells. Unlike nuclear DNA, which is precisely partitioned during mitosis, mtDNA segregation is a stochastic process, with mtDNA molecules randomly distributed between daughter mitochondria. This random segregation can lead to variations in mtDNA copy number and heteroplasmy levels between cells and tissues.

The regulation of mtDNA replication and segregation is influenced by various factors, including cellular energy status, mitochondrial dynamics, and the availability of replication proteins. Mitochondrial dynamics, including fission and fusion, play a crucial role in maintaining mtDNA integrity and distribution. Fission allows for the removal of damaged mitochondria through a process known as mitophagy, while fusion facilitates the mixing of mtDNA and mitochondrial contents, maintaining mitochondrial function and homogeneity. Dysregulation of mtDNA replication and segregation can lead to mitochondrial dysfunction and the development of mitochondrial diseases. Mutations in genes encoding mtDNA replication proteins, such as POLG, are associated with a range of mitochondrial disorders, including progressive external ophthalmoplegia and mitochondrial DNA depletion syndromes. Understanding the molecular mechanisms of mtDNA replication and segregation is essential for developing therapeutic strategies to target these disorders.

Genetic Bottlenecks and mtDNA Evolution

The inheritance and evolution of mitochondrial DNA are influenced by genetic bottlenecks, which are periods of reduced mtDNA diversity that occur during early development. Genetic bottlenecks can result in rapid shifts in mtDNA haplotype frequencies and play a crucial role in shaping mtDNA diversity and evolution. During early embryonic development, a small number of mtDNA molecules are selected to populate the developing germline. This selection results in a genetic bottleneck, where only a subset of mtDNA variants is transmitted to the next generation. The genetic bottleneck can lead to significant changes in mtDNA haplotype frequencies, resulting in the rapid fixation or loss of mtDNA variants.

The genetic bottleneck has important implications for the inheritance of mtDNA and the expression of mitochondrial diseases. The random nature of the bottleneck means that offspring can inherit varying proportions of mutant and wild-type mtDNA, leading to differences in disease severity and phenotypic expression. Understanding the factors influencing the genetic bottleneck is essential for predicting the transmission of mitochondrial diseases and developing therapeutic interventions. The evolution of mtDNA is shaped by various factors, including mutation, selection, and genetic drift. mtDNA has a high mutation rate compared to nuclear DNA, leading to the accumulation of genetic variation over time. The high mutation rate is attributed to the oxidative environment within mitochondria and the lack of protective histones and robust DNA repair mechanisms. Natural selection acts on mtDNA variants, influencing their frequency within populations. mtDNA variants that confer a selective advantage, such as increased energy efficiency or stress resistance, can increase in frequency over time. Conversely, deleterious mtDNA variants may be purged from populations through purifying selection. Genetic drift, the random fluctuation of allele frequencies, also plays a significant role in mtDNA evolution, particularly in small populations. The effects of genetic drift are amplified by the genetic bottleneck, leading to rapid changes in mtDNA diversity and the potential fixation of neutral or mildly deleterious variants. The study of mtDNA evolution provides valuable insights into the evolutionary history and adaptation of species. mtDNA is commonly used in phylogenetic studies to reconstruct evolutionary relationships and trace population histories. Understanding the evolutionary dynamics of mtDNA is essential for interpreting phylogenetic data and assessing the impact of genetic variation on mitochondrial function and disease.

Chloroplast DNA Inheritance

Chloroplast DNA (cpDNA) is the genetic material found in chloroplasts, the organelles responsible for photosynthesis and other essential metabolic processes in plants and algae. Like mitochondrial DNA, cpDNA exhibits unique inheritance patterns that differ from the Mendelian inheritance of nuclear DNA. The study of cpDNA inheritance provides insights into plant genetics, evolution, and breeding.

Maternal and Paternal Inheritance of cpDNA

Chloroplast DNA inheritance is typically maternal in most angiosperms (flowering plants), meaning that cpDNA is inherited exclusively from the mother. However, in some plant species, chloroplast inheritance can be biparental or even paternal. Understanding the mechanisms and implications of cpDNA inheritance is crucial for plant breeding and evolutionary studies.

Maternal Inheritance of cpDNA:

In most angiosperms, cpDNA is inherited maternally, similar to the inheritance of mitochondrial DNA. During fertilization, the chloroplasts from the pollen (male gamete) are usually excluded or degraded, resulting in the exclusive transmission of maternal cpDNA from the ovule (female gamete) to the offspring. This pattern of inheritance ensures the maternal transmission of chloroplasts and their genetic material. Maternal inheritance of cpDNA is advantageous for plants because it maintains the integrity of the chloroplast genome and prevents the introduction of incompatible chloroplast types from the paternal parent. This exclusivity ensures that chloroplasts, which play a crucial role in photosynthesis and energy production, are transmitted stably across generations. Maternal inheritance also has implications for plant breeding and genetics. Because cpDNA is maternally inherited, it can be used to trace maternal lineages and study the evolutionary relationships between plant species. Additionally, understanding cpDNA inheritance can inform breeding strategies aimed at improving traits such as photosynthetic efficiency and stress tolerance.

Paternal and Biparental Inheritance of cpDNA:

While maternal inheritance is predominant in most angiosperms, some plant species exhibit paternal or biparental inheritance of cpDNA. Paternal inheritance occurs when cpDNA is transmitted exclusively from the paternal parent, while biparental inheritance involves the transmission of cpDNA from both parents. Paternal inheritance of cpDNA is observed in some gymnosperms (e.g., pines) and certain angiosperms. In these species, chloroplasts from the pollen contribute to the chloroplast population in the offspring, resulting in the paternal transmission of cpDNA. Paternal inheritance can lead to greater genetic diversity in chloroplast genomes and may influence traits such as photosynthetic efficiency and adaptation to environmental conditions. Biparental inheritance of cpDNA is rare and occurs in some plant species where chloroplasts from both parents are transmitted to the offspring. This pattern of inheritance can result in heteroplasmy, where multiple cpDNA haplotypes coexist within a single individual. Biparental inheritance can increase genetic diversity and provide opportunities for recombination and adaptation. The mechanisms underlying paternal and biparental inheritance of cpDNA are not fully understood but may involve factors such as the retention of chloroplasts during fertilization, the regulation of chloroplast division, and the selective replication of specific cpDNA haplotypes.

Mechanisms of cpDNA Replication and Segregation

Chloroplast DNA replication and segregation are tightly regulated processes that ensure the maintenance of cpDNA copy number and distribution within cells. These processes are essential for chloroplast biogenesis, cellular homeostasis, and the inheritance of cpDNA during cell division.

Replication of cpDNA:

Chloroplast DNA replication is initiated at specific regions within the chloroplast genome, known as the origins of replication. Unlike the nuclear genome, cpDNA replication is not tightly coupled to the cell cycle and can occur continuously throughout the cell cycle. This continuous replication allows for the maintenance of cpDNA copy number and the rapid adaptation of chloroplasts to changing environmental conditions. The replication of cpDNA is carried out by a dedicated set of proteins, including the chloroplast DNA polymerase, helicase, and single-stranded DNA-binding proteins. These proteins work in concert to unwind the cpDNA, synthesize new strands, and ensure the fidelity of cpDNA replication. Chloroplast DNA replication is regulated by various factors, including light, developmental signals, and environmental conditions. Light is a key regulator of cpDNA replication, as it influences the expression of nuclear-encoded proteins involved in chloroplast biogenesis and function. Additionally, developmental signals and environmental cues, such as nutrient availability and temperature, can modulate cpDNA replication to optimize chloroplast function.

Segregation of cpDNA:

Segregation of cpDNA occurs during cell division, where replicated cpDNA is distributed to daughter chloroplasts and subsequently to daughter cells. Unlike nuclear DNA, which is precisely partitioned during mitosis, cpDNA segregation is a stochastic process, with cpDNA molecules randomly distributed between daughter chloroplasts. This random segregation can lead to variations in cpDNA copy number and heteroplasmy levels between cells and tissues. Chloroplast division is a tightly regulated process that involves the coordination of nuclear and chloroplast-encoded proteins.

The division of chloroplasts is mediated by a complex machinery that includes the FtsZ protein, a tubulin-like protein that forms a contractile ring at the chloroplast division site. The regulation of chloroplast division is influenced by various factors, including light, developmental signals, and the availability of replication proteins. The regulation of cpDNA replication and segregation is essential for maintaining chloroplast function and integrity. Chloroplasts are dynamic organelles that can undergo fusion and fission, processes that allow for the mixing of cpDNA and chloroplast contents, maintaining chloroplast function and homogeneity. Fusion facilitates the sharing of cpDNA and other chloroplast components, while fission allows for the removal of damaged chloroplasts through a process known as chlorophagy. Dysregulation of cpDNA replication and segregation can lead to chloroplast dysfunction and the development of chloroplast-associated disorders. Mutations in genes encoding cpDNA replication proteins are associated with various chloroplast diseases, including variegation and chlorosis. Understanding the molecular mechanisms of cpDNA replication and segregation is essential for developing therapeutic strategies to target these disorders.

Genetic Bottlenecks and cpDNA Evolution

The inheritance and evolution of chloroplast DNA are influenced by genetic bottlenecks, which are periods of reduced cpDNA diversity that occur during early development. Genetic bottlenecks can result in rapid shifts in cpDNA haplotype frequencies and play a crucial role in shaping cpDNA diversity and evolution.

Genetic Bottlenecks in cpDNA Inheritance:

During early embryonic development, a small number of cpDNA molecules are selected to populate the developing germline. This selection results in a genetic bottleneck, where only a subset of cpDNA variants is transmitted to the next generation. The genetic bottleneck can lead to significant changes in cpDNA haplotype frequencies, resulting in the rapid fixation or loss of cpDNA variants. The genetic bottleneck has important implications for the inheritance of cpDNA and the expression of chloroplast-associated traits. The random nature of the bottleneck means that offspring can inherit varying proportions of mutant and wild-type cpDNA, leading to differences in phenotypic expression and chloroplast function. Understanding the factors influencing the genetic bottleneck is essential for predicting the transmission of chloroplast-associated traits and developing breeding strategies.

Evolution of cpDNA:

The evolution of cpDNA is shaped by various factors, including mutation, selection, and genetic drift. cpDNA has a relatively low mutation rate compared to nuclear DNA, but the accumulation of genetic variation over time can lead to the development of novel cpDNA haplotypes. The low mutation rate is attributed to the protective environment within chloroplasts and the presence of robust DNA repair mechanisms. Natural selection acts on cpDNA variants, influencing their frequency within populations. cpDNA variants that confer a selective advantage, such as increased photosynthetic efficiency or stress resistance, can increase in frequency over time. Conversely, deleterious cpDNA variants may be purged from populations through purifying selection. Genetic drift, the random fluctuation of allele frequencies, also plays a significant role in cpDNA evolution, particularly in small populations. The effects of genetic drift are amplified by the genetic bottleneck, leading to rapid changes in cpDNA diversity and the potential fixation of neutral or mildly deleterious variants. The study of cpDNA evolution provides valuable insights into the evolutionary history and adaptation of plant species.

cpDNA is commonly used in phylogenetic studies to reconstruct evolutionary relationships and trace population histories. Understanding the evolutionary dynamics of cpDNA is essential for interpreting phylogenetic data and assessing the impact of genetic variation on chloroplast function and plant traits.

Expression of Organellar DNA

The expression of organellar DNA, including mitochondrial DNA (mtDNA) and chloroplast DNA (cpDNA), involves unique regulatory mechanisms that differ from those of nuclear DNA. The expression of organellar genomes is essential for the proper functioning of mitochondria and chloroplasts, which play critical roles in cellular energy production, metabolism, and homeostasis.

Transcription and Translation of mtDNA

Mitochondrial DNA (mtDNA) encodes essential proteins and RNAs required for mitochondrial function, including components of the electron transport chain and ATP synthase. The expression of mtDNA involves distinct transcription and translation processes that are adapted to the unique environment of mitochondria.

Transcription of mtDNA:

The transcription of mtDNA is initiated at specific promoter regions within the mitochondrial genome, known as the heavy strand promoter (HSP) and light strand promoter (LSP). These promoters direct the transcription of the heavy and light strands of mtDNA, respectively, leading to the production of polycistronic RNA transcripts that are processed into individual mRNAs, rRNAs, and tRNAs. Mitochondrial transcription is carried out by a dedicated set of transcription machinery, including the mitochondrial RNA polymerase (POLRMT), transcription factor A (TFAM), and transcription factor B (TFB). These proteins work in concert to initiate and regulate mtDNA transcription, ensuring the production of RNA molecules required for mitochondrial protein synthesis. Mitochondrial transcription is regulated by various factors, including nuclear-encoded proteins, metabolic signals, and environmental conditions. The expression of nuclear-encoded transcription factors, such as NRF1 and NRF2, influences mtDNA transcription by modulating the expression of mitochondrial transcription machinery. Additionally, metabolic signals, such as changes in energy status or redox balance, can influence mtDNA transcription to optimize mitochondrial function.

Translation of mtDNA:

The translation of mtDNA-encoded mRNAs occurs within the mitochondrial matrix, where mitochondrial ribosomes (mitoribosomes) carry out protein synthesis. Mitoribosomes are composed of mtDNA-encoded rRNAs and nuclear-encoded ribosomal proteins, reflecting the cooperative interaction between nuclear and mitochondrial genomes in mitochondrial protein synthesis. Mitochondrial translation involves the use of mitochondrial tRNAs, which are encoded by mtDNA and recognize specific codons in mitochondrial mRNAs. The translation process is regulated by various factors, including translation initiation factors, elongation factors, and mitochondrial chaperones that facilitate protein folding and assembly. The expression of mtDNA is subject to unique regulatory mechanisms that ensure the coordination of mitochondrial and nuclear gene expression. The coordination of mtDNA expression with nuclear-encoded mitochondrial proteins is essential for the assembly and function of the electron transport chain and other mitochondrial complexes. Dysregulation of mtDNA expression is associated with various mitochondrial diseases, characterized by defects in mitochondrial function and energy production. Mutations in mtDNA-encoded genes or nuclear-encoded factors involved in mtDNA expression can lead to impaired mitochondrial function and the development of disorders such as mitochondrial myopathy, Leber's hereditary optic neuropathy, and MELAS syndrome.

Transcription and Translation of cpDNA

Chloroplast DNA (cpDNA) encodes essential proteins and RNAs required for chloroplast function, including components of the photosynthetic machinery and chloroplast ribosomes. The expression of cpDNA involves distinct transcription and translation processes that are adapted to the unique environment of chloroplasts.

Transcription of cpDNA:

The transcription of cpDNA is initiated at specific promoter regions within the chloroplast genome, known as the plastid-encoded polymerase (PEP) promoters and nuclear-encoded polymerase (NEP) promoters. These promoters direct the transcription of cpDNA, leading to the production of polycistronic RNA transcripts that are processed into individual mRNAs, rRNAs, and tRNAs. Chloroplast transcription is carried out by two distinct RNA polymerases: the plastid-encoded polymerase (PEP) and the nuclear-encoded polymerase (NEP). PEP is a multi-subunit RNA polymerase similar to bacterial RNA polymerase, reflecting the endosymbiotic origin of chloroplasts. NEP is a single-subunit RNA polymerase similar to mitochondrial RNA polymerase, providing an additional layer of transcriptional regulation. The regulation of cpDNA transcription is influenced by various factors, including light, developmental signals, and environmental conditions. Light is a key regulator of chloroplast transcription, as it influences the expression of nuclear-encoded proteins involved in chloroplast biogenesis and function. Additionally, developmental signals and environmental cues, such as nutrient availability and temperature, can modulate cpDNA transcription to optimize chloroplast function.

Translation of cpDNA:

The translation of cpDNA-encoded mRNAs occurs within the chloroplast stroma, where chloroplast ribosomes carry out protein synthesis. Chloroplast ribosomes are composed of cpDNA-encoded rRNAs and nuclear-encoded ribosomal proteins, reflecting the cooperative interaction between nuclear and chloroplast genomes in chloroplast protein synthesis.

Chloroplast translation involves the use of chloroplast tRNAs, which are encoded by cpDNA and recognize specific codons in chloroplast mRNAs. The translation process is regulated by various factors, including translation initiation factors, elongation factors, and chloroplast chaperones that facilitate protein folding and assembly. The expression of cpDNA is subject to unique regulatory mechanisms that ensure the coordination of chloroplast and nuclear gene expression. The coordination of cpDNA expression with nuclear-encoded chloroplast proteins is essential for the assembly and function of the photosynthetic apparatus and other chloroplast complexes. Dysregulation of cpDNA expression is associated with various chloroplast disorders, characterized by defects in photosynthesis and chloroplast function. Mutations in cpDNA-encoded genes or nuclear-encoded factors involved in cpDNA expression can lead to impaired chloroplast function and the development of disorders such as variegation and chlorosis. The molecular mechanisms underlying cpDNA expression are essential for developing strategies to enhance crop traits, improve photosynthetic efficiency, and increase agricultural productivity. Approaches such as genetic engineering, CRISPR-Cas9 technology, and biotechnological interventions are being explored for their potential to manipulate chloroplast gene expression and enhance plant performance.

Coordination of Organellar and Nuclear Gene Expression

The coordination of organellar and nuclear gene expression is essential for the proper functioning of mitochondria and chloroplasts, which rely on both organellar-encoded and nuclear-encoded proteins to perform their functions. The integration of organellar and nuclear gene expression involves complex regulatory mechanisms that ensure the synchronized production of proteins required for organellar biogenesis and function.

Nuclear Control of Organellar Gene Expression:

The expression of organellar genomes is regulated by nuclear-encoded proteins that control transcription, translation, and post-translational processes within mitochondria and chloroplasts. These nuclear-encoded factors include transcription factors, RNA-binding proteins, ribosomal proteins, and chaperones that facilitate protein folding and assembly. The regulation of organellar gene expression by nuclear factors allows for the integration of organellar function with cellular and environmental cues. For example, the expression of nuclear-encoded mitochondrial and chloroplast proteins is modulated by metabolic signals, stress responses, and developmental signals, ensuring that organellar function is optimized for cellular needs. Nuclear control of organellar gene expression also involves retrograde signaling pathways, where organelles communicate with the nucleus to modulate nuclear gene expression in response to changes in organellar function. Retrograde signaling ensures that the expression of nuclear-encoded organellar proteins is adjusted based on the metabolic status and stress conditions of mitochondria and chloroplasts.

Coordination of Protein Import and Assembly:

The coordination of organellar and nuclear gene expression is essential for the import and assembly of proteins required for organellar function. Mitochondria and chloroplasts rely on nuclear-encoded proteins that are synthesized in the cytosol and imported into the organelles, where they are assembled into functional complexes. The import of nuclear-encoded proteins into mitochondria and chloroplasts is mediated by specific targeting signals and translocation machinery. Mitochondrial proteins typically contain mitochondrial targeting sequences that direct them to the mitochondrial matrix, while chloroplast proteins contain transit peptides that guide them to the chloroplast stroma. The assembly of organellar protein complexes involves the coordinated expression and interaction of organellar-encoded and nuclear-encoded proteins. This coordination ensures the proper assembly and function of key organellar complexes, such as the electron transport chain in mitochondria and the photosynthetic apparatus in chloroplasts. The coordination of organellar and nuclear gene expression is critical for cellular energy production, metabolism, and homeostasis. Dysregulation of this coordination can lead to organellar dysfunction and the development of various diseases and disorders.

D. Variation in DNA content - C value paradox

The **"C value paradox"** refers to the seemingly contradictory relationship between the amount of DNA in an organism's genome (its **"C value"**) and its biological complexity or evolutionary status.

While one might expect more complex organisms to possess larger amounts of genetic material, this is often not the case. Some simple organisms have much larger genomes than more complex ones.

The C value paradox highlights the complexity of genome evolution and the role of non-coding DNA in shaping the genetic architecture of organisms. It challenges the assumption that genome size directly correlates with organismal complexity and prompts scientists to explore the functional significance of non-coding and repetitive DNA sequences. The resolution of the C value paradox involves examining the diverse elements within genomes, such as transposable elements, gene duplications, and non-coding RNA, which contribute to genome size without necessarily adding to organismal complexity.

The Concept of the C Value

The term "C value" refers to the amount of DNA contained within a haploid nucleus (such as a gamete) of an organism, typically measured in picograms or base pairs. It provides an estimate of the genome size, which includes both coding and non-coding regions of DNA. Understanding the concept of the C value is crucial for studying genome evolution, diversity, and organization.

Definition and Measurement

The C value is a fundamental concept in genomics and molecular biology, providing insights into the genetic material that constitutes an organism's genome. It represents the total amount of DNA within a haploid cell, typically measured in picograms (pg) or base pairs (bp). The C value serves as a quantitative measure of genome size, encompassing both coding and non-coding regions of DNA.

Definition of C Value:

The C value is defined as the total DNA content within the nucleus of a haploid cell, such as a sperm or egg cell. It reflects the entire genome's size, including all the genetic material inherited from one parent. The C value encompasses the complete set of chromosomes, which carry the genes and regulatory elements that govern an organism's development, function, and reproduction. In diploid organisms, such as humans, the C value represents half of the total DNA content within somatic cells, which contain two sets of chromosomes—one from each parent. In haploid organisms, such as bacteria and some algae, the C value represents the total DNA content within a single cell, as these organisms contain only one set of chromosomes.

Measurement of C Value:

The measurement of C value involves quantifying the total DNA content within a cell's nucleus, providing insights into genome size and organization. Several techniques have been developed to measure C value, each with its advantages and limitations:

1. **Flow Cytometry:** Flow cytometry is a widely used technique for measuring C value by analyzing the fluorescence intensity of DNA-binding dyes. Cells are stained with fluorescent dyes that bind specifically to DNA, and the fluorescence intensity is measured as the cells pass through a flow cytometer. The intensity of fluorescence correlates with the DNA content, allowing for the estimation of C value.

2. **Feulgen Staining:** Feulgen staining is a classical cytochemical technique that involves staining DNA with Schiff's reagent after hydrolysis with hydrochloric acid. The intensity

of the staining is proportional to the DNA content, allowing for the estimation of C value. Feulgen staining is particularly useful for visualizing chromosomes and assessing DNA content in fixed cells.

3. **DNA Sequencing:** Advances in DNA sequencing technologies have enabled the direct measurement of C value by sequencing and assembling the entire genome. High-throughput sequencing platforms provide a comprehensive and accurate estimate of genome size by quantifying the total number of base pairs within the genome. This approach is particularly useful for organisms with large and complex genomes.

4. **Spectrophotometry:** Spectrophotometry involves measuring the absorbance of ultraviolet (UV) light by DNA solutions. DNA absorbs UV light at a wavelength of 260 nm, and the absorbance is proportional to the DNA concentration. By measuring the absorbance of DNA samples, researchers can estimate the C value based on known standards.

The measurement of C value provides valuable insights into the genetic diversity and complexity of organisms. It allows researchers to compare genome sizes across different species, investigate the evolutionary relationships between organisms, and explore the functional significance of non-coding and repetitive DNA sequences.

Genome Size and Biological Complexity

The relationship between genome size and biological complexity is a central theme in genomics and evolutionary biology. Genome size, measured by the C value, varies widely among organisms and does not always correlate with an organism's perceived complexity. This lack of correlation, known as the C value paradox, challenges the notion that larger genomes are associated with more complex organisms.

Genome Size Variation:

Genome size varies dramatically across different taxa, from the smallest viral genomes consisting of a few thousand base pairs to the largest plant and animal genomes containing hundreds of billions of base pairs. This variation is influenced by several factors, including the presence of non-coding DNA, repetitive sequences, and transposable elements. For example, among eukaryotes, genome sizes can range from approximately 10 million base pairs in some fungi to over 150 billion base pairs in certain plant species. Within animals, genome sizes can vary from around 50 million base pairs in small insects to over 100 billion base pairs in amphibians and lungfish.

Complexity and Genome Size:

Biological complexity is often characterized by the diversity of cell types, tissue organization, and the complexity of physiological processes and behaviors. One might intuitively expect that more complex organisms, such as mammals, would have larger genomes to accommodate the additional genetic information required for their complexity. However, this is not always the case.

The lack of a consistent correlation between genome size and complexity is exemplified by the comparison of genome sizes among vertebrates. For instance, some amphibians have significantly larger genomes than mammals, despite their simpler body plans and life cycles. Similarly, certain plants with large genomes, such as lilies and onions, do not exhibit a corresponding increase in complexity compared to other plants with smaller genomes. The C value paradox highlights the decoupling of genome size from organismal complexity, prompting researchers to explore the factors contributing to genome size variation. It underscores the importance of understanding the role of non-coding DNA and repetitive elements in shaping genome architecture and evolution.

Implications of the C Value Paradox:

The C value paradox has significant implications for our understanding of genome evolution and function. It challenges the assumption that genome size directly reflects the number of protein-coding genes or the complexity of an organism's phenotype. Instead, it suggests that non-coding and repetitive DNA play crucial roles in genome organization, regulation, and evolution. The C value paradox also highlights the importance of non-coding RNA, regulatory elements, and epigenetic modifications in shaping gene expression and organismal complexity. It underscores the need to study the functional significance of non-coding regions and their contributions to genome evolution and adaptation. The resolution of the C value paradox involves examining the diverse elements within genomes, such as transposable elements, gene duplications, and non-coding RNA, which contribute to genome size without necessarily adding to organismal complexity. By exploring the molecular and evolutionary processes underlying genome size variation, researchers can gain insights into the dynamic nature of genomes and their role in shaping biodiversity.

Causes of the C Value Paradox

The C value paradox arises from the discrepancy between genome size and organismal complexity, challenging the assumption that larger genomes contain more genes or functional elements. Several factors contribute to the C value paradox, including the presence of non-coding DNA, repetitive sequences, and transposable elements. Understanding these factors provides insights into the mechanisms underlying genome size variation and the evolutionary forces shaping genome architecture.

Non-Coding DNA

Non-coding DNA refers to the portions of an organism's genome that do not encode proteins. Despite not being translated into proteins, non-coding DNA plays essential roles in regulating gene expression, maintaining genome stability, and contributing to evolutionary processes. The presence of non-coding DNA is a significant factor contributing to the C value paradox, as it accounts for much of the genome size variation observed among organisms.

Types of Non-Coding DNA:

Non-coding DNA encompasses a diverse array of sequences with distinct functions and evolutionary origins. Some of the main types of non-coding DNA include:

1. **Introns:** Introns are non-coding sequences found within genes that are transcribed into RNA but removed during RNA processing. While introns do not encode proteins, they play essential roles in regulating gene expression, alternative splicing, and mRNA stability. The presence of introns contributes to the complexity of gene regulation and increases the overall genome size.
2. **Regulatory Elements:** Regulatory elements are non-coding sequences that control the

transcriptional activity of genes. These elements include promoters, enhancers, silencers, and insulators, which interact with transcription factors and other regulatory proteins to modulate gene expression. The complexity and diversity of regulatory elements contribute to the fine-tuning of gene expression and the development of complex traits.

3. **Intergenic Regions:** Intergenic regions are non-coding sequences located between genes. These regions may contain regulatory elements, non-coding RNA genes, and repetitive sequences. Intergenic regions play a role in maintaining genome structure and organization, and their size and composition can influence genome size variation.

4. **Non-Coding RNAs:** Non-coding RNAs (ncRNAs) are RNA molecules transcribed from non-coding DNA regions that do not encode proteins. ncRNAs include microRNAs (miRNAs), long non-coding RNAs (lncRNAs), and small interfering RNAs (siRNAs), which are involved in gene regulation, RNA processing, and chromatin remodeling. The presence of ncRNAs adds to the functional complexity of non-coding DNA.

Role of Non-Coding DNA in the C Value Paradox:

Non-coding DNA accounts for a substantial portion of genome size variation and contributes to the C value paradox. While protein-coding genes represent a small fraction of the genome, non-coding DNA can constitute the majority of an organism's genetic material. For example, in the human genome, only about 1-2% of the DNA encodes proteins, while the remaining 98-99% consists of non-coding sequences.

The presence of extensive non-coding DNA challenges the assumption that genome size directly correlates with the number of protein-coding genes or organismal complexity. Instead, non-coding DNA plays critical roles in regulating gene expression, maintaining genome stability, and facilitating evolutionary innovation. Non-coding DNA contributes to the functional complexity of genomes by providing regulatory elements that control gene expression in response to developmental cues and environmental signals. It also serves as a reservoir of genetic variation that can drive evolutionary adaptation and diversification. The resolution of the C value paradox involves recognizing the functional significance of non-coding DNA and its contributions to genome evolution and organismal complexity. By exploring the roles of non-coding DNA in gene regulation, genome architecture, and evolutionary processes, researchers can gain insights into the dynamic nature of genomes and their role in shaping biodiversity.

Repetitive DNA

Repetitive DNA refers to sequences that are repeated multiple times within the genome. These sequences can vary in size, organization, and function, and they contribute significantly to genome size variation and the C value paradox. Repetitive DNA includes both tandem repeats and interspersed repeats, which can influence genome structure, stability, and evolution.

Types of Repetitive DNA:

Repetitive DNA encompasses a wide range of sequences with distinct characteristics and functions. Some of the main types of repetitive DNA include:

1. **Tandem Repeats:** Tandem repeats are sequences that are repeated in a head-to-tail fashion. They include microsatellites (short tandem repeats) and minisatellites, which consist of short repeat units ranging from a few base pairs to several hundred base pairs. Tandem repeats are often located in non-coding regions and can influence genome structure and function.
2. **Interspersed Repeats:** Interspersed repeats are sequences that are scattered throughout the genome. They include transposable elements (TEs), such as long interspersed nuclear elements (LINEs) and short interspersed nuclear elements (SINEs), which can move within the genome and create new copies. Interspersed repeats can influence gene expression, genome stability, and evolution.
3. **Satellite DNA:** Satellite DNA consists of large arrays of tandemly repeated sequences, often found in centromeric and telomeric regions of chromosomes. Satellite DNA plays a role in chromosome structure and segregation during cell division, and its amplification can contribute to genome size variation.
4. **Ribosomal DNA (rDNA):** Ribosomal DNA consists of repeated copies of genes encoding ribosomal RNA (rRNA), which are essential for ribosome biogenesis and protein synthesis. rDNA repeats are organized into clusters and contribute to the structural and functional organization of the genome.

Role of Repetitive DNA in the C Value Paradox:

Repetitive DNA is a major contributor to genome size variation and the C value paradox. While protein-coding genes represent a small fraction of the genome, repetitive DNA can constitute a significant portion of an organism's genetic material. For example, in some plant genomes, repetitive DNA can account for over 80% of the total genome size. The presence of extensive repetitive DNA challenges the assumption that genome size directly correlates with the number of protein-coding genes or organismal complexity. Instead, repetitive DNA plays critical roles in shaping genome architecture, stability, and evolution. Repetitive DNA contributes to the structural organization of the genome by providing centromeric and telomeric sequences that are essential for chromosome segregation and stability. It also serves as a substrate for homologous recombination, facilitating genome rearrangements and adaptation.

The expansion and contraction of repetitive DNA can lead to rapid changes in genome size, contributing to the C value paradox. Transposable elements, in particular, are dynamic components of the genome that can increase genome size through replication and insertion events. The activity of transposable elements can also influence gene expression, genome stability, and evolutionary innovation. The resolution of the C value paradox involves recognizing the functional significance of repetitive DNA and its contributions to genome evolution and organismal complexity. By exploring the roles of repetitive DNA in genome organization, stability, and adaptation, researchers can gain insights into the dynamic nature of genomes and their role in shaping biodiversity.

Transposable Elements

Transposable elements (TEs), also known as mobile genetic elements or "jumping genes," are sequences of DNA that can change their position within the genome. They are a major component of repetitive DNA and contribute significantly to genome size variation and the C value paradox. Transposable elements play essential roles in genome evolution, diversity, and regulation.

Types of Transposable Elements:

Transposable elements are classified into two main types based on their mechanism of transposition: retrotransposons and DNA transposons.

1. **Retrotransposons:** Retrotransposons are transposable elements that move within the genome via an RNA intermediate. They are transcribed into RNA, which is then reverse-transcribed into DNA and inserted into a new genomic location. Retrotransposons can be further divided into long terminal repeat (LTR) retrotransposons and non-LTR retrotransposons, such as LINEs and SINEs.
   - **LTR Retrotransposons:** LTR retrotransposons are characterized by the presence of long terminal repeats at both ends of the element. They are similar to retroviruses in structure and replication and can contribute to genome expansion through replication and insertion events.
   - **Non-LTR Retrotransposons:** Non-LTR retrotransposons, including LINEs and SINEs, lack long terminal repeats and replicate via a mechanism known as target-primed reverse transcription. LINEs are autonomous elements that encode the proteins required for their replication, while SINEs are non-autonomous elements that rely on the replication machinery of LINEs.
2. **DNA Transposons:** DNA transposons are transposable elements that move within the genome via a DNA intermediate. They are excised from their original location and inserted into a new site within the genome. DNA transposons encode a transposase enzyme that catalyzes their excision and insertion, allowing them to move within the genome.

Role of Transposable Elements in the C Value Paradox:

Transposable elements are a major contributor to genome size variation and the C value paradox. They can account for a significant portion of an organism's genome, with their abundance varying widely among different species. For example, in the maize genome, transposable elements constitute over 80% of the total genome size. The presence of transposable elements challenges the assumption that genome size directly correlates with the number of protein-coding genes or organismal complexity. Instead, transposable elements play critical roles in shaping genome architecture, stability, and evolution.

Transposable elements contribute to genome expansion through their ability to replicate and insert into new genomic locations. Their activity can lead to the creation of new copies, increasing in genome size. Additionally, transposable elements can facilitate genome rearrangements, gene duplications, and the generation of genetic diversity.

The activity of transposable elements can influence gene expression and regulation by inserting into or near genes, regulatory elements, or other functional sequences. They can create new regulatory elements, such as enhancers and silencers, or disrupt existing regulatory networks, leading to changes in gene expression patterns. Transposable elements also play a role in genome defense mechanisms, such as the RNA interference (RNAi) pathway, which targets and silences transposable elements to maintain genome stability. This interaction between transposable elements and genome defense mechanisms can drive the co-evolution of genomes and transposable elements. The resolution of the C value paradox involves recognizing the functional significance of transposable elements and their contributions to genome evolution and organismal complexity. By exploring the roles of transposable elements in genome organization, regulation, and adaptation, researchers can gain insights into the dynamic nature of genomes and their role in shaping biodiversity.

Implications and Significance

The C value paradox has profound implications for our understanding of genome evolution, regulation, and functionality. It challenges traditional notions of genome size and complexity, prompting researchers to explore the diverse elements within genomes that contribute to their size and organization. Understanding the C value paradox is essential for elucidating the mechanisms underlying genome evolution and the role of non-coding DNA in shaping genetic diversity.

Genome Evolution and Diversity

The C value paradox provides valuable insights into genome evolution and diversity, highlighting the complexity and dynamism of genetic material. It challenges the notion that genome size directly correlates with organismal complexity, prompting researchers to explore the factors contributing to genome size variation and the evolutionary forces shaping genome architecture.

Genome Evolution:

The evolution of genomes is a complex process influenced by various factors, including mutation, selection, genetic drift, and recombination. The C value paradox underscores the importance of non-coding DNA, repetitive sequences, and transposable elements in shaping genome evolution and diversity.

1. **Mutation:** Mutations are changes in the DNA sequence that can lead to genetic variation and diversity. The high mutation rate of non-coding regions and repetitive DNA contributes to the accumulation of genetic variation over time, driving genome evolution and adaptation. Mutations in non-coding DNA can influence gene regulation, expression, and the development of complex traits.
2. **Selection:** Natural selection acts on genetic variation, influencing the frequency of alleles within populations. Non-coding DNA and repetitive sequences can be subject to selective pressures, leading to the retention or elimination of specific sequences based on their functional significance. The C value paradox highlights the role of non-coding DNA in providing a reservoir of genetic variation that can drive evolutionary innovation.
3. **Genetic Drift:** Genetic drift is the random fluctuation of allele frequencies within populations, particularly in small populations. The effects of genetic drift are amplified by the presence of non-coding DNA and repetitive sequences, which can lead to rapid changes in genome size

and composition. The C value paradox underscores the impact of genetic drift on genome evolution and diversity.

4. **Recombination:** Recombination is the exchange of genetic material between homologous chromosomes during meiosis, leading to the generation of genetic diversity. Repetitive DNA and transposable elements can facilitate recombination events, contributing to genome rearrangements and the generation of novel genetic combinations. The C value paradox highlights the role of recombination in shaping genome architecture and evolution.

Genome Diversity:

The C value paradox emphasizes the diversity of genome sizes and compositions across different taxa, reflecting the complex interplay of evolutionary forces shaping genetic material. Genome diversity is influenced by the presence of non-coding DNA, repetitive sequences, and transposable elements, which contribute to the variation in genome size and organization.

1. **Non-Coding DNA:** Non-coding DNA accounts for a significant portion of genome diversity, with its abundance and composition varying widely among different species. The presence of extensive non-coding DNA challenges the assumption that genome size directly reflects the number of protein-coding genes or organismal complexity. Instead, non-coding DNA plays critical roles in regulating gene expression, maintaining genome stability, and facilitating evolutionary innovation.

2. **Repetitive DNA:** Repetitive DNA is a major contributor to genome diversity, with its abundance and distribution varying among different taxa. The expansion and contraction of repetitive DNA can lead to rapid changes in genome size, contributing to the C value paradox. Repetitive DNA plays essential roles in shaping genome structure, stability, and evolution.

3. **Transposable Elements:** Transposable elements are dynamic components of the genome that contribute to genome diversity through their ability to replicate and insert into new genomic locations. Their activity can influence gene expression, genome stability, and evolutionary innovation. The presence of transposable elements challenges the assumption that genome size directly correlates with organismal complexity.

The resolution of the C value paradox involves recognizing the functional significance of non-coding DNA, repetitive sequences, and transposable elements in shaping genome evolution and diversity. By exploring the molecular and evolutionary processes underlying genome size variation, researchers can gain insights into the dynamic nature of genomes and their role in shaping biodiversity.

Non-Coding DNA and Gene Regulation

The C value paradox underscores the importance of non-coding DNA in gene regulation and the development of complex traits. Non-coding DNA, once considered "junk DNA," plays essential roles in regulating gene expression, maintaining genome stability, and facilitating evolutionary processes. Understanding the functional significance of non-coding DNA is crucial for elucidating the mechanisms underlying gene regulation and organismal complexity.

Gene Regulation:

Gene regulation is the process by which cells control the expression of genes in response to developmental cues and environmental signals. Non-coding DNA plays a critical role in gene regulation by providing regulatory elements that control the transcriptional activity of genes.

1. **Regulatory Elements:** Regulatory elements are non-coding sequences that modulate the transcriptional activity of genes. These elements include promoters, enhancers, silencers, and insulators, which interact with transcription factors and other regulatory proteins to control gene expression. The complexity and diversity of regulatory elements contribute to the fine-tuning of gene expression and the development of complex traits.
2. **Non-Coding RNAs:** Non-coding RNAs (ncRNAs) are RNA molecules transcribed from non-coding DNA regions that do not encode proteins. ncRNAs, including microRNAs (miRNAs), long non-coding RNAs (lncRNAs), and small interfering RNAs (siRNAs), play essential roles in gene regulation, RNA processing, and chromatin remodeling. ncRNAs can influence gene expression by modulating mRNA stability, translation, and chromatin structure.
3. **Epigenetic Modifications:** Epigenetic modifications, such as DNA methylation and histone modifications, are chemical modifications that alter chromatin structure and gene accessibility. These modifications are influenced by non-coding DNA and play a crucial role in regulating gene expression, maintaining cellular identity, and facilitating developmental processes.

Functional Significance of Non-Coding DNA:

The C value paradox highlights the functional significance of non-coding DNA in shaping gene regulation and organismal complexity. Non-coding DNA provides a reservoir of genetic elements that can influence gene expression, genome architecture, and evolutionary adaptation.

1. **Regulatory Networks:** Non-coding DNA contributes to the complexity of regulatory networks that control gene expression in response to developmental cues and environmental signals. The presence of diverse regulatory elements allows for the fine-tuning of gene expression and the development of complex traits, such as tissue differentiation, organ development, and physiological responses.
2. **Genome Stability:** Non-coding DNA plays a role in maintaining genome stability by providing centromeric and telomeric sequences essential for chromosome segregation and integrity. It also serves as a substrate for homologous recombination, facilitating genome rearrangements and adaptation.
3. **Evolutionary Innovation:** Non-coding DNA provides a reservoir of genetic variation that can drive evolutionary innovation and diversification. The presence of non-coding DNA allows for the exploration of novel regulatory elements, gene expression patterns, and developmental pathways, contributing to the evolution of complex traits and biodiversity.

The resolution of the C value paradox involves recognizing the functional significance of non-coding DNA in gene regulation and organismal complexity. By exploring the roles of non-coding DNA in genome organization, regulation, and evolution, researchers can gain insights into the dynamic nature of genomes and their role in shaping biodiversity.

Transposable Elements and Genome Innovation

The C value paradox highlights the role of transposable elements (TEs) in genome innovation and evolution. Transposable elements are dynamic components of the genome that contribute to genome size variation, regulation, and diversity. Understanding the role of TEs in shaping genome architecture and function is essential for elucidating the mechanisms underlying genome evolution and adaptation.

Transposable Elements and Genome Architecture:

Transposable elements play a crucial role in shaping genome architecture by influencing the organization, stability, and function of genetic material. Their ability to move within the genome and create new copies contributes to genome expansion and rearrangement.

1. **Genome Expansion:** Transposable elements can increase genome size through replication and insertion events. The activity of TEs can lead to the creation of new copies, resulting in an increase in genome size and complexity. TEs account for a significant portion of genome size variation and contribute to the C value paradox.
2. **Genome Rearrangement:** Transposable elements facilitate genome rearrangements by promoting homologous recombination and genome shuffling. Their presence can influence the organization of genes and regulatory elements, leading to changes in gene expression patterns and genome structure.
3. **Genome Stability:** The activity of transposable elements can influence genome stability by creating insertions, deletions, and duplications. TEs can disrupt gene function, regulatory networks, and chromatin structure, leading to changes in genome stability and integrity.

Transposable Elements and Gene Regulation:

Transposable elements play a significant role in gene regulation by providing regulatory elements that influence gene expression and chromatin structure. Their ability to create new regulatory elements and disrupt existing networks contributes to the complexity of gene regulation.

1. **Regulatory Elements:** Transposable elements can create new regulatory elements, such as enhancers and silencers, that influence gene expression patterns. The presence of TEs can modulate the transcriptional activity of genes by providing binding sites for transcription factors and other regulatory proteins.
2. **Epigenetic Modifications:** Transposable elements can influence epigenetic modifications, such as DNA methylation and histone modifications, that alter chromatin structure and gene accessibility. TEs can create new epigenetic marks or disrupt existing modifications, leading to changes in gene expression and chromatin organization.
3. **Non-Coding RNAs:** Transposable elements can generate non-coding RNAs (ncRNAs) that play a role in gene regulation, RNA processing, and chromatin remodeling. TEs can create new ncRNA genes or influence the expression of existing ncRNAs, contributing to the complexity of regulatory networks.

Transposable Elements and Evolutionary Innovation:

Transposable elements contribute to evolutionary innovation by providing a reservoir of genetic variation that can drive adaptation and diversification. Their ability to create new genetic combinations and regulatory elements facilitates the exploration of novel traits and phenotypes.

1. **Genetic Variation:** Transposable elements contribute to genetic variation by creating insertions, deletions, and duplications that influence gene function and regulation. The presence of TEs can lead to the development of new genetic combinations and phenotypic diversity.
2. **Adaptive Evolution:** Transposable elements can drive adaptive evolution by creating novel regulatory elements and gene expression patterns that enhance fitness and adaptation. TEs can facilitate the evolution of complex traits and phenotypes, contributing to the diversification of species and the exploration of new ecological niches.
3. **Speciation:** Transposable elements can contribute to speciation by promoting reproductive isolation and genetic divergence between populations. The activity of TEs can create genetic barriers and incompatibilities that drive the evolution of new species and lineages.

The resolution of the C value paradox involves recognizing the functional significance of transposable elements in genome innovation and evolution. By exploring the roles of TEs in shaping genome architecture, regulation, and adaptation, researchers can gain insights into the dynamic nature of genomes and their role in shaping biodiversity.

Implications for Genomics and Biotechnology

The C value paradox has significant implications for genomics and biotechnology, influencing our understanding of genome evolution, regulation, and functionality. It challenges traditional notions of genome size and complexity, prompting researchers to explore the diverse elements within genomes that contribute to their size and organization. Understanding the C value paradox is essential for advancing genomics research and developing biotechnological applications that harness the potential of genetic material.

Genomics Research:

The C value paradox provides valuable insights into the complexity and dynamism of genomes, influencing the direction of genomics research and the development of new technologies.

1. **Genome Sequencing:** The C value paradox underscores the importance of sequencing and analyzing entire genomes to understand their complexity and diversity. Advances in genome sequencing technologies have enabled the comprehensive analysis of large and complex genomes, providing insights into the organization and function of genetic material.
2. **Functional Genomics:** The C value paradox highlights the need to explore the functional significance of non-coding DNA and repetitive sequences in shaping genome regulation and evolution. Functional genomics approaches, such as transcriptomics, proteomics, and epigenomics, provide insights into the roles of non-coding DNA in gene expression, regulation, and adaptation.
3. **Comparative Genomics:** The C value paradox emphasizes the importance of comparing genomes across different taxa to understand the evolutionary forces shaping genome

architecture and diversity. Comparative genomics approaches provide insights into the conservation and divergence of genetic elements, facilitating the identification of functional and regulatory elements within genomes.

Biotechnological Applications:

The C value paradox has significant implications for biotechnological applications, influencing the development of strategies to harness the potential of genetic material for various purposes.

1. **Crop Improvement:** The C value paradox provides insights into the genetic diversity and complexity of plant genomes, influencing the development of strategies for crop improvement and breeding. Understanding the roles of non-coding DNA, repetitive sequences, and transposable elements in shaping genome architecture and regulation can inform the development of crops with enhanced traits, such as yield, stress tolerance, and disease resistance.
2. **Gene Editing:** The C value paradox highlights the importance of non-coding DNA and regulatory elements in gene expression and regulation, influencing the development of gene editing technologies. Techniques such as CRISPR-Cas9 enable the precise modification of genetic elements within genomes, providing opportunities to manipulate gene expression, regulation, and adaptation for various purposes.
3. **Synthetic Biology:** The C value paradox provides insights into the complexity and diversity of genomes, influencing the development of synthetic biology approaches that harness the potential of genetic material for various applications. Synthetic biology approaches involve the design and construction of novel genetic circuits and pathways, enabling the creation of organisms with enhanced functions and capabilities.

Challenges and Opportunities:

The C value paradox presents challenges and opportunities for genomics and biotechnology, influencing the development of new technologies and approaches to explore the complexity and diversity of genomes.

1. **Data Analysis:** The complexity and size of genomes present challenges for data analysis and interpretation, requiring the development of new computational tools and algorithms to explore genome organization, regulation, and evolution.
2. **Functional Annotation:** The C value paradox highlights the need for comprehensive functional annotation of genomes to understand the roles of non-coding DNA and repetitive sequences in shaping genome function and evolution. Functional annotation approaches provide insights into the regulatory and evolutionary significance of genetic elements within genomes.
3. **Ethical Considerations:** The C value paradox raises ethical considerations related to the manipulation and modification of genetic material, influencing the development of guidelines and regulations for the responsible use of genomics and biotechnology. Ethical considerations include the potential impact of genetic modifications on biodiversity, ecosystems, and human health.

The resolution of the C value paradox involves recognizing the functional significance of non-coding DNA, repetitive sequences, and transposable elements in shaping genome evolution and complexity. By exploring the molecular and evolutionary processes underlying genome size variation, researchers can gain insights into the dynamic nature of genomes and their role in shaping biodiversity and advancing biotechnology.

E. Sequence complexity – Introns and Exons

Introns and exons are fundamental components of genes that contribute significantly to sequence complexity and gene regulation in eukaryotic organisms. These elements play essential roles in gene expression, mRNA processing, and the evolution of genetic diversity. The structure, function, and evolution of introns and exons is crucial for elucidating the mechanisms underlying genome complexity and the regulation of gene expression. Introns are non-coding sequences within a gene that are transcribed into RNA but are removed during the splicing process to produce mature mRNA. Exons are coding sequences that are transcribed and retained in the mature mRNA, ultimately translated into proteins. The interplay between introns and exons allows for the generation of diverse protein isoforms through alternative splicing, contributing to the functional diversity and adaptability of organisms.

Structure and Function of Introns and Exons

Introns and exons are key structural components of eukaryotic genes that play distinct roles in gene expression and regulation. Their unique characteristics and functions contribute to the complexity of the genome and the generation of diverse protein products.

Introns: Structure and Function

Introns are non-coding sequences found within genes that are transcribed into precursor mRNA (pre-mRNA) but are removed during RNA splicing to produce mature mRNA. Despite not encoding proteins, introns play crucial roles in regulating gene expression, alternative splicing, and genome evolution.

Structure of Introns:

Introns are interspersed between exons within a gene and can vary significantly in size and number across different organisms and genes. In humans, introns can range from a few hundred base pairs to over 100,000 base pairs in length, with some genes containing dozens of introns. The length and number of introns contribute to the overall complexity of gene structure and regulation. Introns typically contain specific sequence motifs that are recognized by the splicing machinery during RNA processing. These motifs include the 5' splice site (donor site), the 3' splice site (acceptor site), and the branch point sequence. The 5' splice site marks the beginning of the intron, while the 3' splice site marks the end. The branch point sequence, located upstream of the 3' splice site, plays a critical role in the splicing reaction.

Function of Introns:

Introns have several important functions that contribute to gene expression and regulation:

1. **Regulation of Gene Expression:** Introns can influence gene expression by affecting transcriptional efficiency, mRNA stability, and translation. The presence of introns can

enhance gene expression through a process known as intron-mediated enhancement (IME), where introns enhance the transcriptional activity of genes by facilitating the recruitment of transcriptional machinery and regulatory factors.

2. **Alternative Splicing:** One of the most significant roles of introns is their involvement in alternative splicing, a process that allows for the generation of multiple protein isoforms from a single gene. During alternative splicing, different combinations of exons and introns are selectively included or excluded from the final mRNA, resulting in diverse protein products with distinct functions and properties. Alternative splicing increases the functional diversity of proteins and allows organisms to adapt to changing environmental conditions and developmental cues.
3. **RNA Processing and Stability:** Introns play a role in RNA processing and stability by influencing mRNA splicing, export, and degradation. The presence of introns can affect the efficiency and accuracy of splicing, ensuring the proper removal of non-coding sequences and the generation of functional mRNA. Introns can also influence mRNA stability by regulating the binding of RNA-binding proteins and other factors that control mRNA degradation and turnover.
4. **Genomic Evolution and Innovation:** Introns contribute to genomic evolution and innovation by providing a substrate for genetic recombination, exon shuffling, and the generation of new genetic elements. The presence of introns allows for the rearrangement of exons, leading to the creation of novel gene structures and functions. Introns also serve as a reservoir of genetic variation that can drive evolutionary adaptation and diversification.

Biological Significance of Introns:

Introns are essential components of eukaryotic genomes that play critical roles in regulating gene expression and facilitating evolutionary innovation. Despite being non-coding sequences, introns contribute to the functional complexity and diversity of organisms by enabling alternative splicing, regulating gene expression, and promoting genomic evolution. The study of introns has revealed their importance in shaping genome architecture and function, challenging the traditional view of non-coding DNA as "**junk**" DNA. Understanding the roles of introns in gene regulation and evolution provides valuable insights into the mechanisms underlying genome complexity and the development of complex traits.

Exons: Structure and Function

Exons are coding sequences within genes that are transcribed into pre-mRNA and retained in the mature mRNA, ultimately translated into proteins. Exons are the building blocks of proteins, providing the genetic information required for the synthesis of functional polypeptides. Their structure and function contribute to the complexity of gene expression and the generation of diverse protein products.

Structure of Exons:

Exons are the coding regions of genes that are interspersed with non-coding introns. They can vary in size and number across different organisms and genes, contributing to the diversity of gene structures and functions. In humans, exons typically range from a few dozen to several hundred base pairs in length, with some genes containing dozens of exons. Exons contain specific sequence motifs that are recognized by the splicing machinery during RNA processing. These motifs include the 5' splice site (donor site) and the 3' splice site (acceptor site), which mark the boundaries between exons and introns. The recognition of these motifs ensures the accurate removal of introns and the precise joining of exons to generate functional mRNA.

The function of Exons:

Exons have several important functions that contribute to gene expression and protein synthesis:

1. **Protein Coding:** The primary function of exons is to provide the genetic information required for the synthesis of proteins. Exons contain codons, which are sequences of three nucleotides that specify the amino acids incorporated into a polypeptide chain. The sequence of exons determines the amino acid sequence of the protein, influencing its structure, function, and properties.
2. **Gene Expression Regulation:** Exons can influence gene expression by affecting transcriptional efficiency, mRNA stability, and translation. The presence of specific exonic sequences can regulate the binding of transcription factors, RNA-binding proteins, and other regulatory elements that control gene expression. Exons can also influence mRNA stability by affecting the binding of factors involved in mRNA degradation and turnover.
3. **Alternative Splicing:** Exons play a crucial role in alternative splicing, a process that allows for the generation of multiple protein isoforms from a single gene. During alternative splicing, different combinations of exons are selectively included or excluded from the final mRNA, resulting in diverse protein products with distinct functions and properties. Alternative splicing increases the functional diversity of proteins and allows organisms to adapt to changing environmental conditions and developmental cues.
4. **Exon Shuffling and Evolution:** Exons contribute to genomic evolution and innovation by providing a substrate for genetic recombination and exon shuffling. The presence of exons allows for the rearrangement of coding sequences, leading to the creation of novel gene structures and functions. Exon shuffling can result in the development of new protein domains and the evolution of complex traits.

Biological Significance of Exons:

Exons are essential components of eukaryotic genomes that play critical roles in encoding proteins and regulating gene expression. The structure and function of exons contribute to the functional complexity and diversity of organisms by enabling protein synthesis, alternative splicing, and genomic evolution. The study of exons has revealed their importance in shaping genome architecture and function, highlighting their role in gene regulation, protein diversity, and evolutionary innovation.

Evolution of Introns and Exons

The evolution of introns and exons is a complex process influenced by various factors, including genetic variation, selection, and recombination. The presence of introns and exons in eukaryotic genomes has significant implications for the evolution of gene structure, regulation, and function. Understanding the evolutionary dynamics of introns and exons provides insights into the mechanisms underlying genome complexity and diversity.

Origin and Evolution of Introns

The origin and evolution of introns have been subjects of extensive research and debate, with several theories proposed to explain their presence and significance in eukaryotic genomes. Introns are thought to have played a crucial role in the evolution of complex gene structures and regulatory networks, contributing to the diversity and adaptability of organisms.

Theories of Intron Origin:

Several theories have been proposed to explain the origin of introns in eukaryotic genomes, each with its implications for understanding genome evolution and complexity. The two main theories are the "introns-early" hypothesis and the "introns-late" hypothesis.

1. **Introns-Early Hypothesis:** The introns-early hypothesis suggests that introns were present in the ancestral genes of early life forms and played a role in the evolution of protein-coding sequences. According to this theory, introns facilitated the rearrangement and recombination of exonic sequences, allowing for the evolution of new protein domains and functions. The presence of introns in early ancestors may have contributed to the diversification of protein-coding genes and the development of complex traits.

2. **Introns-Late Hypothesis:** The introns-late hypothesis proposes that introns were introduced into eukaryotic genomes after the divergence of prokaryotes and eukaryotes. According to this theory, introns originated from the insertion of transposable elements or other non-coding sequences into genes. The introduction of introns may have facilitated the evolution of alternative splicing and complex regulatory networks, contributing to the adaptability and diversity of eukaryotic organisms.

Evolutionary Dynamics of Introns:

The evolution of introns is influenced by various factors, including mutation, selection, and recombination. Introns can undergo gain and loss events, leading to changes in gene structure and function. The evolutionary dynamics of introns are shaped by their functional roles in gene regulation and the selective pressures acting on them.

1. **Intron Gain:** Intron gain refers to the introduction of new introns into genes through processes such as transposable element insertion or the duplication of existing introns. Intron gain can lead to the development of new gene structures and the evolution of complex traits. The introduction of new introns can also facilitate alternative splicing and the generation of diverse protein isoforms.

2. **Intron Loss:** Intron loss refers to the removal of introns from genes through processes such as genomic deletion or exonization, where intronic sequences are incorporated into exons. Intron loss can lead to changes in gene structure and function, influencing gene expression and regulation. The loss of introns can result in the simplification of gene structures and the reduction of sequence complexity.

3. **Intron Conservation:** Introns can be conserved across different species, indicating their functional significance and evolutionary stability. Conserved introns often contain regulatory elements or sequence motifs that play essential roles in gene expression and regulation. The conservation of introns highlights their importance in maintaining gene function and genomic integrity.

Implications for Genome Evolution:

The evolution of introns has significant implications for understanding the complexity and diversity of eukaryotic genomes. Introns contribute to the evolution of gene structures and regulatory networks, facilitating the development of complex traits and phenotypes.

1. **Alternative Splicing:** Introns play a crucial role in alternative splicing, a process that allows for the generation of multiple protein isoforms from a single gene. The presence of introns enables the rearrangement and recombination of exonic sequences, leading to the diversification of protein functions and properties. Alternative splicing increases the functional diversity of proteins and allows organisms to adapt to changing environmental conditions and developmental cues.
2. **Genomic Innovation:** Introns contribute to genomic innovation by providing a substrate for genetic recombination, exon shuffling, and the generation of new genetic elements. The presence of introns allows for the rearrangement of exons, leading to the creation of novel gene structures and functions. Introns also serve as a reservoir of genetic variation that can drive evolutionary adaptation and diversification.
3. **Regulatory Complexity:** Introns play a role in regulating gene expression by influencing transcriptional efficiency, mRNA stability, and translation. The presence of introns can enhance gene expression through intron-mediated enhancement (IME), where introns facilitate the recruitment of transcriptional machinery and regulatory factors. Introns also contain regulatory elements that control the splicing and processing of mRNA, contributing to the complexity of gene regulation.

Origin and Evolution of Exons

The origin and evolution of exons are closely linked to the development of protein-coding genes and the diversification of eukaryotic genomes. Exons are thought to have played a crucial role in the evolution of complex gene structures and regulatory networks, contributing to the diversity and adaptability of organisms.

Theories of Exon Origin:

Several theories have been proposed to explain the origin of exons in eukaryotic genomes, each with its implications for understanding genome evolution and complexity. The main theories are the "exon theory of genes" and the "gene duplication and divergence" hypothesis.

1. **Exon Theory of Genes:** The exon theory of genes suggests that exons originated from the modular organization of ancestral genes, where distinct exonic sequences encoded specific protein domains. According to this theory, the modular organization of exons allowed for

the rearrangement and recombination of protein-coding sequences, facilitating the evolution of new protein functions and properties. The presence of exons in early ancestors may have contributed to the diversification of protein-coding genes and the development of complex traits.

2. **Gene Duplication and Divergence:** The gene duplication and divergence hypothesis proposes that exons originated from the duplication and divergence of existing genes. According to this theory, gene duplication events led to the creation of new copies of protein-coding genes, which subsequently underwent divergence to acquire new functions and properties. The duplication and divergence of exons contributed to the evolution of new protein domains and the diversification of eukaryotic genomes.

Evolutionary Dynamics of Exons:

The evolution of exons is influenced by various factors, including mutation, selection, and recombination. Exons can undergo duplication, loss, and shuffling events, leading to changes in gene structure and function. The evolutionary dynamics of exons are shaped by their functional roles in protein synthesis and the selective pressures acting on them.

1. **Exon Duplication:** Exon duplication refers to the creation of new copies of exons within a gene through processes such as tandem duplication or segmental duplication. Exon duplication can lead to the expansion of protein-coding genes and the evolution of new protein functions. The presence of duplicated exons can facilitate the development of novel protein domains and the diversification of protein properties.
2. **Exon Loss:** Exon loss refers to the removal of exons from genes through processes such as genomic deletion or alternative splicing, where exonic sequences are skipped or excluded from the final mRNA. Exon loss can lead to changes in gene structure and function, influencing protein synthesis and regulation. The loss of exons can result in the simplification of protein-coding genes and the reduction of sequence complexity.
3. **Exon Shuffling:** Exon shuffling refers to the rearrangement and recombination of exonic sequences within and between genes, leading to the creation of novel gene structures and functions. Exon shuffling can result in the development of new protein domains and the evolution of complex traits. The presence of introns facilitates exon shuffling by providing a substrate for genetic recombination and the rearrangement of exonic sequences.

Implications for Genome Evolution:

The evolution of exons has significant implications for understanding the complexity and diversity of eukaryotic genomes. Exons contribute to the evolution of protein-coding genes and regulatory networks, facilitating the development of complex traits and phenotypes.

1. **Protein Diversity:** Exons play a crucial role in the synthesis of proteins, providing the genetic information required for the production of functional polypeptides. The presence of diverse exonic sequences allows for the generation of multiple protein isoforms with distinct functions and properties. Exons contribute to the functional diversity of proteins and allow organisms to adapt to changing environmental conditions and developmental cues.

2. **Genomic Innovation:** Exons contribute to genomic innovation by providing a substrate for genetic recombination, exon shuffling, and the generation of new protein-coding sequences. The presence of exons allows for the rearrangement of coding sequences, leading to the creation of novel protein domains and functions. Exon shuffling can result in the development of new protein domains and the evolution of complex traits.
3. **Regulatory Complexity:** Exons play a role in regulating gene expression by influencing transcriptional efficiency, mRNA stability, and translation. The presence of specific exonic sequences can regulate the binding of transcription factors, RNA-binding proteins, and other regulatory elements that control gene expression. Exons can also influence mRNA stability by affecting the binding of factors involved in mRNA degradation and turnover.

The evolutionary dynamics of exons provide valuable insights into the mechanisms underlying genome complexity and diversity. By exploring the roles of exons in gene expression and evolution, researchers can gain insights into the dynamic nature of genomes and their role in shaping biodiversity.

Significance of Introns and Exons in Gene Regulation

Introns and exons play essential roles in gene regulation, contributing to the complexity and diversity of gene expression and protein synthesis. Their unique characteristics and functions enable the generation of diverse protein products and the regulation of gene expression in response to developmental cues and environmental signals.

Role of Introns in Gene Regulation

Introns play a crucial role in gene regulation by influencing transcriptional efficiency, mRNA processing, and alternative splicing. Despite being non-coding sequences, introns contribute to the complexity and diversity of gene expression, allowing organisms to adapt to changing conditions and developmental cues.

Transcriptional Regulation:

Introns can influence transcriptional regulation by affecting the efficiency and accuracy of gene transcription. The presence of introns can enhance gene expression through a process known as intron-mediated enhancement (IME), where introns facilitate the recruitment of transcriptional machinery and regulatory factors.

1. **Intron-Mediated Enhancement (IME):** Intron-mediated enhancement is a phenomenon where the presence of introns within a gene enhances its transcriptional activity. IME is thought to occur through the interaction of introns with the transcriptional machinery and regulatory elements, facilitating the efficient initiation and elongation of transcription. The presence of introns can also influence the chromatin structure and accessibility of the gene, contributing to its transcriptional regulation.
2. **Regulatory Elements:** Introns often contain regulatory elements, such as enhancers, silencers, and insulators, that modulate gene expression in response to developmental cues and environmental signals. These elements interact with transcription factors and other regulatory proteins to control the transcriptional activity of genes. The presence of regulatory

elements within introns allows for the fine-tuning of gene expression and the development of complex traits.

Alternative Splicing:

One of the most significant roles of introns is their involvement in alternative splicing, a process that allows for the generation of multiple protein isoforms from a single gene. Alternative splicing increases the functional diversity of proteins and allows organisms to adapt to changing conditions and developmental cues.

1. **Splice Site Selection:** Introns contain specific sequence motifs, such as the 5' splice site, 3' splice site, and branch point sequence, that are recognized by the splicing machinery during RNA processing. The selection of splice sites determines the inclusion or exclusion of exons and introns in the final mRNA, influencing the generation of protein isoforms. The presence of multiple splice sites within introns allows for the generation of diverse splicing patterns and protein products.
2. **Regulation of Alternative Splicing:** The regulation of alternative splicing involves the interaction of introns with splicing factors and regulatory elements that control the selection of splice sites. Splicing factors, such as serine/arginine-rich (SR) proteins and heterogeneous nuclear ribonucleoproteins (hnRNPs), bind to specific sequences within introns and exons, influencing splice site selection and alternative splicing outcomes. The regulation of alternative splicing allows for the generation of diverse protein isoforms and the adaptation to changing conditions and developmental cues.

RNA Processing and Stability:

Introns play a role in RNA processing and stability by influencing mRNA splicing, export, and degradation. The presence of introns can affect the efficiency and accuracy of splicing, ensuring the proper removal of non-coding sequences and the generation of functional mRNA.

1. **mRNA Splicing:** Introns are removed from pre-mRNA through a process known as splicing, where the splicing machinery recognizes specific splice sites and catalyzes the excision of introns. The presence of introns can influence the efficiency and accuracy of splicing, ensuring the generation of functional mRNA and protein products. The regulation of splicing is essential for maintaining the integrity of the transcriptome and proteome.
2. **mRNA Export and Stability:** Introns can influence the export and stability of mRNA by regulating the binding of RNA-binding proteins and other factors involved in mRNA processing. The presence of introns can enhance mRNA stability by influencing the binding of factors involved in mRNA degradation and turnover. Introns can also regulate mRNA export by influencing the interaction of mRNA with nuclear export machinery.

Biological Significance of Introns in Gene Regulation:

Introns are essential components of eukaryotic genomes that play critical roles in regulating gene expression and facilitating evolutionary innovation. Despite being non-coding sequences, introns contribute to the functional complexity and diversity of organisms by enabling alternative splicing, regulating gene expression, and promoting genomic evolution. The study of introns has revealed their importance in shaping genome architecture and function, challenging the traditional view of non-coding DNA as "junk" DNA.

Role of Exons in Gene Regulation

Exons play a crucial role in gene regulation by influencing transcriptional efficiency, mRNA stability, and protein synthesis. As coding sequences, exons provide the genetic information required for the synthesis of functional proteins, contributing to the complexity and diversity of gene expression and regulation.

Protein Coding:

The primary function of exons is to provide the genetic information required for the synthesis of proteins. Exons contain codons, which are sequences of three nucleotides that specify the amino acids incorporated into a polypeptide chain. The sequence of exons determines the amino acid sequence of the protein, influencing its structure, function, and properties.

1. **Amino Acid Sequence:** The sequence of exons determines the amino acid sequence of the protein, which in turn influences its structure and function. The arrangement of codons within exons specifies the order of amino acids in the polypeptide chain, contributing to the protein's three-dimensional structure and functional properties. The presence of specific exonic sequences can influence protein folding, stability, and activity.
2. **Protein Domains:** Exons often encode distinct protein domains, which are functional units within a protein that perform specific tasks or interact with other molecules. The modular organization of exons allows for the combination and rearrangement of protein domains, facilitating the evolution of new protein functions and properties. The presence of diverse exonic sequences contributes to the functional diversity of proteins and allows organisms to adapt to changing conditions and developmental cues.

Gene Expression Regulation:

Exons can influence gene expression by affecting transcriptional efficiency, mRNA stability, and translation. The presence of specific exonic sequences can regulate the binding of transcription factors, RNA-binding proteins, and other regulatory elements that control gene expression.

1. **Transcriptional Efficiency:** Exons can influence transcriptional efficiency by affecting the interaction of the gene with transcriptional machinery and regulatory elements. The presence of specific exonic sequences can enhance or repress transcriptional activity, influencing the level of gene expression. The regulation of transcriptional efficiency allows for the fine-tuning of gene expression in response to developmental cues and environmental signals.
2. **mRNA Stability:** Exons can influence mRNA stability by regulating the binding of factors involved in mRNA degradation and turnover. The presence of specific exonic sequences can enhance mRNA stability by protecting it from degradation or facilitating its interaction with stabilizing proteins. The regulation of mRNA stability ensures the proper translation of genetic information into functional proteins.

3. **Translation Regulation:** Exons can influence translation regulation by affecting the interaction of mRNA with ribosomes and translation factors. The presence of specific exonic sequences can modulate the efficiency and accuracy of translation, influencing protein synthesis and activity. The regulation of translation allows for the precise control of protein production in response to cellular and environmental conditions.

Alternative Splicing:

Exons play a crucial role in alternative splicing, a process that allows for the generation of multiple protein isoforms from a single gene. Alternative splicing increases the functional diversity of proteins and allows organisms to adapt to changing conditions and developmental cues.

1. **Exon Inclusion and Exclusion:** During alternative splicing, different combinations of exons are selectively included or excluded from the final mRNA, resulting in diverse protein products. The presence of multiple exons within a gene allows for the generation of diverse splicing patterns and protein isoforms. The regulation of exon inclusion and exclusion is essential for the generation of functional diversity and adaptation to changing conditions.
2. **Splicing Factors:** The regulation of alternative splicing involves the interaction of exons with splicing factors and regulatory elements that control the selection of splice sites. Splicing factors, such as serine/arginine-rich (SR) proteins and heterogeneous nuclear ribonucleoproteins (hnRNPs), bind to specific sequences within exons and introns, influencing splice site selection and alternative splicing outcomes. The regulation of alternative splicing allows for the generation of diverse protein isoforms and the adaptation to changing conditions and developmental cues.

Biological Significance of Exons in Gene Regulation:

Exons are essential components of eukaryotic genomes that play critical roles in encoding proteins and regulating gene expression. The structure and function of exons contribute to the functional complexity and diversity of organisms by enabling protein synthesis, alternative splicing, and genomic evolution. The study of exons has revealed their importance in shaping genome architecture and function, highlighting their role in gene regulation, protein diversity, and evolutionary innovation. The roles of exons in gene expression and evolution provide valuable insights into the mechanisms underlying genome complexity and the development of complex traits.

F. Repetitive sequences, Role of repetitive sequence

Repetitive sequences are a significant component of eukaryotic genomes, characterized by the presence of DNA sequences that are repeated multiple times within the genome. These sequences, which can vary in size and complexity, account for a substantial portion of the genetic material in many organisms. Despite their repetitive nature, these sequences play crucial roles in genome organization, stability, and evolution. Repetitive sequences are categorized into several types, including tandem repeats, interspersed repeats, and satellite DNA. Each type of repetitive sequence has unique structural and functional properties that contribute to the overall complexity and functionality of the genome. Understanding the nature and role of repetitive sequences provides valuable insights into the dynamic nature of genomes and their evolutionary history.

ypes of Repetitive Sequences

Repetitive sequences are classified into various types based on their structure, organization, and mode of repetition within the genome. These types include tandem repeats, interspersed repeats, and satellite DNA, each playing distinct roles in genome architecture and function.

Tandem Repeats

Tandem repeats are sequences of DNA in which identical or nearly identical sequences are arranged consecutively in a head-to-tail manner. They are commonly found in the genome and can vary in length and complexity. Tandem repeats are further divided into several categories based on the size of the repeated unit, including microsatellites, minisatellites, and macrosatellites.

Microsatellites:

Microsatellites, also known as short tandem repeats (STRs) or simple sequence repeats (SSRs), are tandem repeats consisting of short sequences of 1 to 6 base pairs. They are highly polymorphic and widely distributed throughout the genome, often found in non-coding regions, such as introns and intergenic spaces. The number of repeat units in microsatellites can vary between individuals, making them valuable markers for genetic studies.

1. **Structure and Distribution:** Microsatellites are composed of short repeating units, such as mono-, di-, tri-, or tetranucleotides. Common examples include (A)n, (CA)n, and (GATA) n repeats. These sequences are abundant in eukaryotic genomes, with humans having approximately 1 million microsatellite loci scattered throughout their DNA. Their distribution is often random, but certain genomic regions may be enriched with specific microsatellites.
2. **Function and Importance:** Microsatellites play a crucial role in genome evolution and diversity due to their high mutation rate, which is primarily driven by DNA replication slippage. This leads to variation in the number of repeat units among individuals, contributing to genetic diversity within populations. Microsatellites are widely used as molecular markers in genetic mapping, population genetics, and forensics due to their high polymorphism and ease of analysis.
3. **Applications:** The variability of microsatellites makes them useful tools in various fields, including forensic science, where they are employed for DNA fingerprinting and paternity testing. In conservation biology, microsatellites help assess genetic diversity and population structure of endangered species. Additionally, they are utilized in medical genetics to study linkage disequilibrium and identify disease-associated loci.

Minisatellites:

Minisatellites are tandem repeats consisting of longer sequences, typically 10 to 100 base pairs, repeated 10 to 100 times. They are found in the genome, often in telomeric regions and other specific loci. Minisatellites exhibit moderate polymorphism and play a role in genome stability and recombination.

1. **Structure and Distribution:** Minisatellites are composed of longer repeat units compared to microsatellites, and they often form clustered arrays in specific genomic regions. They are

commonly found in telomeres, where they contribute to chromosomal stability and integrity. The variable number of tandem repeats (VNTRs) observed in minisatellites makes them valuable genetic markers.

2. **Function and Importance:** Minisatellites are involved in various genomic functions, including the regulation of gene expression, chromatin structure, and telomere maintenance. They are also hotspots for homologous recombination, contributing to genetic diversity and evolution. The variation in minisatellite length among individuals can affect gene function and expression, potentially influencing phenotype and disease susceptibility.
3. **Applications:** Minisatellites have applications in DNA profiling, particularly in forensic investigations, where they are used to establish identity and familial relationships. They are also employed in evolutionary studies to assess genetic variation and divergence between species. In medical research, minisatellites serve as markers for studying genetic disorders and disease susceptibility.

Macrosatellites:

Macrosatellites are large tandem repeats, typically consisting of hundreds to thousands of base pairs repeated over extended regions. They are less common than microsatellites and minisatellites but play significant roles in chromosomal structure and function.

1. **Structure and Distribution:** Macrosatellites consist of long repeat units that can span several kilobases. They are often found in centromeric and pericentromeric regions of chromosomes, contributing to the organization and stability of these critical genomic areas. The large size and repetitive nature of macrosatellites make them challenging to study using traditional sequencing methods.
2. **Function and Importance:** Macrosatellites play crucial roles in maintaining chromosomal integrity and facilitating proper chromosome segregation during cell division. They contribute to the formation of heterochromatin, a tightly packed form of DNA that regulates gene expression and protects against genomic instability. The repetitive nature of macrosatellites also makes them targets for recombination events, influencing genome evolution and variation.
3. **Applications:** While macrosatellites are less commonly used as genetic markers due to their size and complexity, they have potential applications in studying chromosomal abnormalities and genetic diseases associated with centromeric and pericentromeric regions. Advances in sequencing technologies may enable more detailed analyses of macrosatellites and their role in genome dynamics.

Interspersed Repeats

Interspersed repeats are sequences that are dispersed throughout the genome rather than being arranged consecutively. They include transposable elements (TEs) and other repetitive sequences that can move within the genome, contributing to genome expansion, rearrangement, and evolution.

Transposable Elements (TEs):

Transposable elements, also known as "jumping genes," are DNA sequences that can change their position within the genome. They are classified into two main types: retrotransposons and DNA transposons, based on their mechanism of transposition. TEs are a major component of interspersed repeats and play significant roles in genome dynamics and evolution.

1. **Retrotransposons:** Retrotransposons move within the genome via an RNA intermediate. They are transcribed into RNA, which is then reverse-transcribed into DNA and inserted into a new genomic location. Retrotransposons are further divided into long terminal repeat (LTR) retrotransposons and non-LTR retrotransposons, such as long interspersed nuclear elements (LINEs) and short interspersed nuclear elements (SINEs).
   - **LTR Retrotransposons:** LTR retrotransposons are characterized by the presence of long terminal repeats at both ends of the element. They are similar to retroviruses in structure and replication and can contribute to genome expansion through replication and insertion events.
   - **Non-LTR Retrotransposons:** Non-LTR retrotransposons, including LINEs and SINEs, lack long terminal repeats and replicate via a mechanism known as target-primed reverse transcription. LINEs are autonomous elements that encode the proteins required for their replication, while SINEs are non-autonomous elements that rely on the replication machinery of LINEs.
2. **DNA Transposons:** DNA transposons move within the genome via a DNA intermediate. They are excised from their original location and inserted into a new site within the genome. DNA transposons encode a transposase enzyme that catalyzes their excision and insertion, allowing them to move within the genome.

Function and Importance of TEs:

Transposable elements play significant roles in genome evolution and dynamics by influencing genome structure, stability, and function. They can contribute to genome expansion through replication and insertion events, facilitating the rearrangement of genetic material and the generation of genetic diversity.

1. **Genome Expansion:** Transposable elements can increase genome size by creating new copies and inserting into new genomic locations. Their activity can lead to the creation of new genetic elements and the expansion of existing ones, contributing to the C value paradox and the variation in genome size among different organisms.
2. **Gene Regulation:** TEs can influence gene regulation by inserting into or near genes, regulatory elements, or other functional sequences. They can create new regulatory elements, such as enhancers and silencers, or disrupt existing regulatory networks, leading to changes in gene expression patterns.
3. **Genomic Instability:** The activity of transposable elements can influence genome stability by creating insertions, deletions, and duplications. TEs can disrupt gene function, regulatory networks, and chromatin structure, leading to changes in genome stability and integrity.

Applications of TEs:

Transposable elements have significant implications for various fields, including genomics, biotechnology, and evolutionary biology. Understanding their role in genome dynamics and evolution provides valuable insights into the mechanisms underlying genetic variation and adaptation.

1. **Genomic Studies:** TEs are used as markers in genomic studies to investigate genome structure, function, and evolution. Their presence and activity can provide insights into the dynamics of genome organization and the evolutionary history of species.
2. **Biotechnology:** TEs have potential applications in biotechnology, where they are used as tools for genetic manipulation and engineering. Their ability to move within the genome can be harnessed to introduce or remove genetic elements, facilitating the study of gene function and regulation.
3. **Evolutionary Biology:** TEs are important drivers of evolutionary change and adaptation, influencing genome structure and function. Their activity can lead to the development of new traits and phenotypes, contributing to the diversification of species and the exploration of new ecological niches.

Satellite DNA

Satellite DNA consists of highly repetitive sequences that form large arrays in specific genomic regions, such as centromeres and telomeres. These sequences are often organized into tandem repeats and play crucial roles in maintaining genome structure and stability.

Centromeric Satellite DNA:

Centromeric satellite DNA is a type of repetitive sequence found in the centromeric regions of chromosomes, where it plays a critical role in chromosome segregation and stability. These sequences are organized into large arrays of tandem repeats and contribute to the formation of heterochromatin, a tightly packed form of DNA that regulates gene expression and protects against genomic instability.

1. **Structure and Distribution:** Centromeric satellite DNA consists of long arrays of tandemly repeated sequences, often spanning millions of base pairs. The repeat units are typically short, ranging from a few to several hundred base pairs, and are highly conserved within a species. Centromeric satellite DNA is located at the primary constriction of chromosomes, where it forms the centromere, a specialized structure that ensures proper chromosome segregation during cell division.
2. **Function and Importance:** Centromeric satellite DNA plays a crucial role in maintaining chromosomal integrity and facilitating proper chromosome segregation during mitosis and meiosis. The repetitive nature of centromeric satellite DNA allows for the formation of kinetochore, a protein complex that attaches to spindle fibers and ensures accurate chromosome segregation. Centromeric satellite DNA also contributes to the formation of heterochromatin, a tightly packed form of DNA that regulates gene expression and protects against genomic instability.
3. **Applications:** Centromeric satellite DNA has potential applications in the study of

chromosomal abnormalities and genetic diseases associated with centromeric regions. Advances in sequencing technologies may enable more detailed analyses of centromeric satellite DNA and its role in genome dynamics and evolution.

Telomeric Satellite DNA:

Telomeric satellite DNA is a type of repetitive sequence found in the telomeric regions of chromosomes, where it plays a critical role in maintaining genome stability and protecting chromosome ends from degradation. These sequences are organized into tandem repeats and contribute to the formation of telomeres, specialized structures that regulate chromosome length and stability.

1. **Structure and Distribution:** Telomeric satellite DNA consists of long arrays of tandemly repeated sequences, often spanning several kilobases. The repeat units are typically short, with the most common sequence being the hexanucleotide repeat (TTAGGG)n in vertebrates. Telomeric satellite DNA is located at the ends of chromosomes, where it forms the telomeres, specialized structures that protect chromosome ends from degradation and recombination.
2. **Function and Importance:** Telomeric satellite DNA plays a crucial role in maintaining genome stability and protecting chromosome ends from degradation. The repetitive nature of telomeric satellite DNA allows for the formation of telomeres, specialized structures that regulate chromosome length and stability. Telomeres protect chromosome ends from degradation and recombination, ensuring proper chromosome function and integrity. Telomeric satellite DNA also plays a role in regulating cell division and aging, as the shortening of telomeres is associated with cellular senescence and aging.
3. **Applications:** Telomeric satellite DNA has potential applications in the study of aging and age-related diseases, as well as cancer and other genetic disorders associated with telomere dysfunction. Advances in sequencing technologies may enable more detailed analyses of telomeric satellite DNA and its role in genome dynamics and evolution.

Role of Repetitive Sequences

Repetitive sequences play crucial roles in genome organization, stability, and evolution, contributing to the complexity and functionality of the genome. Despite their repetitive nature, these sequences have diverse functions that influence gene regulation, chromatin structure, and genomic diversity.

Genome Organization and Stability

Repetitive sequences contribute significantly to genome organization and stability by influencing chromatin structure, chromosome dynamics, and genome integrity. Their presence and distribution play essential roles in maintaining the structural and functional organization of the genome.

Chromatin Structure:

Repetitive sequences contribute to the formation and maintenance of chromatin structure, influencing gene expression and genome stability. Chromatin is the complex of DNA and proteins that forms the structural framework of chromosomes, regulating the accessibility and activity of genetic material.

1. **Heterochromatin Formation:** Repetitive sequences, particularly satellite DNA, are often found in heterochromatic regions of the genome, where they contribute to the formation of tightly packed chromatin. Heterochromatin is characterized by its dense structure, which restricts access to the underlying DNA and silences gene expression. The presence of repetitive sequences in heterochromatin plays a critical role in regulating gene expression and maintaining genome stability.
2. **Epigenetic Regulation:** Repetitive sequences influence epigenetic modifications, such as DNA methylation and histone modifications, that alter chromatin structure and gene accessibility. These modifications are essential for regulating gene expression, maintaining cellular identity, and facilitating developmental processes. The interaction between repetitive sequences and epigenetic marks contributes to the dynamic nature of chromatin and its regulatory functions.
3. **Nuclear Organization:** Repetitive sequences play a role in organizing the genome within the nucleus, influencing the spatial arrangement of chromosomes and the formation of nuclear compartments. The presence of repetitive sequences can facilitate the clustering of specific genomic regions, contributing to the formation of nuclear domains and the regulation of gene expression.

Chromosome Dynamics:

Repetitive sequences are involved in various aspects of chromosome dynamics, influencing chromosome segregation, recombination, and repair processes. Their presence and activity play essential roles in maintaining chromosome stability and integrity.

1. **Centromere Function:** Repetitive sequences, particularly centromeric satellite DNA, are crucial for centromere function and chromosome segregation during cell division. The repetitive nature of centromeric sequences allows for the formation of the kinetochore, a protein complex that attaches to spindle fibers and ensures accurate chromosome segregation. The proper function of centromeres is essential for maintaining chromosome stability and preventing aneuploidy.
2. **Telomere Maintenance:** Repetitive sequences, particularly telomeric satellite DNA, are essential for telomere maintenance and chromosome stability. Telomeres protect chromosome ends from degradation and recombination, ensuring proper chromosome function and integrity. The repetitive nature of telomeric sequences allows for the regulation of telomere length and the prevention of genomic instability.
3. **Homologous Recombination:** Repetitive sequences play a role in homologous recombination, a process that facilitates the exchange of genetic material between homologous chromosomes. The presence of repetitive sequences can promote recombination events, contributing to genetic diversity and genome evolution. Homologous recombination is essential for DNA repair, genetic diversity, and adaptation.

Genome Integrity:

Repetitive sequences contribute to genome integrity by influencing DNA repair, replication, and stability. Their presence and activity play essential roles in maintaining the structural and functional integrity of the genome.

1. **DNA Repair:** Repetitive sequences play a role in DNA repair processes, such as homologous recombination and non-homologous end joining. The presence of repetitive sequences can facilitate the repair of DNA double-strand breaks, ensuring genome stability and integrity. DNA repair processes are essential for maintaining genome integrity and preventing mutations and genomic instability.
2. **DNA Replication:** Repetitive sequences influence DNA replication, affecting the efficiency and accuracy of genome duplication. The presence of repetitive sequences can pose challenges for DNA replication, leading to replication slippage and the generation of genetic variation. The regulation of DNA replication is essential for maintaining genome integrity and preventing genomic instability.
3. **Genome Stability:** Repetitive sequences play a role in maintaining genome stability by influencing chromatin structure, chromosome dynamics, and DNA repair processes. Their presence and activity contribute to the structural and functional integrity of the genome, ensuring proper gene expression, chromosome segregation, and DNA repair.

Gene Regulation

Repetitive sequences play significant roles in gene regulation by influencing transcriptional activity, mRNA processing, and chromatin structure. Their presence and activity contribute to the complexity and diversity of gene expression and regulation.

Transcriptional Regulation:

Repetitive sequences influence transcriptional regulation by affecting the efficiency and accuracy of gene transcription. Their presence and distribution can modulate the interaction of genes with transcriptional machinery and regulatory elements.

1. **Enhancers and Silencers:** Repetitive sequences can function as enhancers or silencers, regulatory elements that modulate gene expression in response to developmental cues and environmental signals. Enhancers increase transcriptional activity, while silencers repress it. The presence of repetitive sequences in enhancers and silencers can influence the transcriptional regulation of genes, contributing to the complexity of gene expression networks.
2. **Transcription Factor Binding Sites:** Repetitive sequences can serve as binding sites for transcription factors, proteins that regulate gene expression by interacting with specific DNA sequences. The presence of repetitive sequences in transcription factor binding sites can influence the recruitment of transcriptional machinery and the regulation of gene expression. The regulation of transcription factor binding is essential for controlling gene expression in response to cellular and environmental conditions.
3. **Promoter Regions:** Repetitive sequences can be located in promoter regions, the upstream sequences that regulate the initiation of gene transcription. The presence of repetitive sequences in promoters can influence the efficiency and accuracy of transcription initiation, affecting

the level of gene expression. The regulation of promoter activity is essential for controlling gene expression and ensuring proper cellular function.

mRNA Processing:

Repetitive sequences play a role in mRNA processing, influencing the splicing, export, and stability of mRNA molecules. Their presence and activity contribute to the regulation of gene expression and the generation of diverse protein products.

1. **Alternative Splicing:** Repetitive sequences can influence alternative splicing, a process that allows for the generation of multiple protein isoforms from a single gene. The presence of repetitive sequences in introns can affect the selection of splice sites, influencing the inclusion or exclusion of exons in the final mRNA. The regulation of alternative splicing is essential for generating diverse protein products and adapting to changing conditions.
2. **mRNA Stability:** Repetitive sequences can influence mRNA stability by regulating the binding of factors involved in mRNA degradation and turnover. The presence of repetitive sequences in mRNA can enhance or repress its stability, affecting the level of gene expression. The regulation of mRNA stability is essential for controlling protein production and ensuring proper cellular function.
3. **mRNA Export:** Repetitive sequences can influence mRNA export by regulating the interaction of mRNA with nuclear export machinery. The presence of repetitive sequences in mRNA can enhance or repress its export from the nucleus to the cytoplasm, affecting the level of gene expression. The regulation of mRNA export is essential for controlling protein production and ensuring proper cellular function.

Chromatin Structure:

Repetitive sequences influence chromatin structure, affecting the accessibility and activity of genetic material. Their presence and activity contribute to the regulation of gene expression and the maintenance of genome stability.

1. **Heterochromatin Formation:** Repetitive sequences, particularly satellite DNA, are often found in heterochromatic regions of the genome, where they contribute to the formation of tightly packed chromatin. Heterochromatin restricts access to the underlying DNA and silences gene expression, playing a critical role in regulating gene expression and maintaining genome stability.
2. **Epigenetic Modifications:** Repetitive sequences influence epigenetic modifications, such as DNA methylation and histone modifications, that alter chromatin structure and gene accessibility. These modifications are essential for regulating gene expression, maintaining cellular identity, and facilitating developmental processes. The interaction between repetitive sequences and epigenetic marks contributes to the dynamic nature of chromatin and its regulatory functions.
3. **Nuclear Organization:** Repetitive sequences play a role in organizing the genome within the nucleus, influencing the spatial arrangement of chromosomes and the formation of nuclear

compartments. The presence of repetitive sequences can facilitate the clustering of specific genomic regions, contributing to the formation of nuclear domains and the regulation of gene expression.

Evolutionary Significance

Repetitive sequences play significant roles in genome evolution and adaptation, contributing to the complexity and diversity of genetic material. Their presence and activity influence genome dynamics, genetic variation, and evolutionary innovation.

Genome Dynamics:

Repetitive sequences contribute to genome dynamics by influencing genome structure, stability, and function. Their presence and activity play essential roles in shaping the architecture and evolution of genomes.

1. **Genome Expansion:** Repetitive sequences, particularly transposable elements, can increase genome size by creating new copies and inserting into new genomic locations. Their activity can lead to the creation of new genetic elements and the expansion of existing ones, contributing to the C value paradox and the variation in genome size among different organisms.
2. **Genome Rearrangement:** Repetitive sequences facilitate genome rearrangements by promoting homologous recombination and genome shuffling. Their presence can influence the organization of genes and regulatory elements, leading to changes in gene expression patterns and genome structure. The regulation of genome rearrangement is essential for maintaining genome stability and facilitating evolutionary adaptation.
3. **Genomic Instability:** The activity of repetitive sequences can influence genome stability by creating insertions, deletions, and duplications. Their presence can disrupt gene function, regulatory networks, and chromatin structure, leading to changes in genome stability and integrity. The regulation of genomic instability is essential for maintaining genome integrity and preventing mutations and genomic disorders.

Genetic Variation:

Repetitive sequences contribute to genetic variation by influencing the generation and maintenance of genetic diversity within populations. Their presence and activity play essential roles in shaping the genetic landscape and evolutionary potential of organisms.

1. **Mutation and Polymorphism:** Repetitive sequences, particularly microsatellites, are prone to mutations, such as replication slippage, that lead to variations in the number of repeat units. These mutations contribute to genetic diversity and polymorphism within populations, providing a reservoir of genetic variation for evolutionary processes.
2. **Gene Duplication and Divergence:** Repetitive sequences play a role in gene duplication and divergence, processes that contribute to the expansion and diversification of gene families. The presence of repetitive sequences can facilitate the duplication of genetic elements, leading to the development of new gene functions and properties. The regulation of gene duplication and divergence is essential for evolutionary innovation and adaptation.

3. **Recombination Hotspots:** Repetitive sequences, particularly transposable elements, can serve as recombination hotspots, regions of the genome with increased rates of homologous recombination. Their presence can promote genetic recombination and the generation of novel genetic combinations, contributing to genetic diversity and evolutionary adaptation.

Evolutionary Innovation:

Repetitive sequences contribute to evolutionary innovation by providing a reservoir of genetic variation and facilitating the exploration of novel traits and phenotypes. Their presence and activity play essential roles in shaping the evolutionary trajectory of organisms.

1. **Adaptive Evolution:** Repetitive sequences play a role in adaptive evolution by influencing the generation and maintenance of genetic variation and adaptation. Their activity can lead to the development of new traits and phenotypes, enhancing fitness and adaptation to changing environmental conditions and ecological niches.
2. **Speciation:** Repetitive sequences contribute to speciation by promoting reproductive isolation and genetic divergence between populations. Their presence and activity can create genetic barriers and incompatibilities that drive the evolution of new species and lineages. The regulation of speciation is essential for maintaining biodiversity and facilitating evolutionary diversification.
3. **Functional Innovation:** Repetitive sequences play a role in functional innovation by facilitating the evolution of new gene functions and regulatory networks. Their presence and activity can influence gene expression, regulation, and adaptation, contributing to the development of complex traits and phenotypes. The regulation of functional innovation is essential for maintaining genome complexity and evolutionary potential.

Implications for Genomics and Biotechnology

Repetitive sequences have significant implications for genomics and biotechnology, influencing our understanding of genome evolution, regulation, and functionality. Their presence and activity play essential roles in shaping the genetic landscape and providing opportunities for technological advancements.

Genomics Research:

Repetitive sequences provide valuable insights into the complexity and dynamics of genomes, influencing the direction of genomics research and the development of new technologies.

1. **Genome Sequencing:** Repetitive sequences pose challenges for genome sequencing and assembly due to their repetitive nature and complex organization. Advances in sequencing technologies have enabled the comprehensive analysis of repetitive sequences, providing insights into their structure, function, and evolution.
2. **Functional Genomics:** Repetitive sequences influence functional genomics approaches, such as transcriptomics, proteomics, and epigenomics, that provide insights into gene expression, regulation, and adaptation. Understanding the role of repetitive sequences in functional

genomics is essential for elucidating the mechanisms underlying genome complexity and the development of complex traits.

3. **Comparative Genomics:** Repetitive sequences provide valuable insights into the conservation and divergence of genetic elements across different taxa, influencing the development of comparative genomics approaches. Comparative genomics provides insights into the evolutionary history and dynamics of repetitive sequences, facilitating the identification of functional and regulatory elements within genomes.

Biotechnological Applications:

Repetitive sequences have significant implications for biotechnological applications, influencing the development of strategies to harness the potential of genetic material for various purposes.

1. **Genetic Engineering:** Repetitive sequences provide opportunities for genetic engineering and manipulation, enabling the modification of genetic elements within genomes. Techniques such as CRISPR-Cas9 allow for the precise modification of repetitive sequences, facilitating the study of gene function and regulation.
2. **Synthetic Biology:** Repetitive sequences play a role in synthetic biology approaches that harness the potential of genetic material for various applications. Synthetic biology involves the design and construction of novel genetic circuits and pathways, enabling the creation of organisms with enhanced functions and capabilities.
3. **Genetic Markers:** Repetitive sequences serve as valuable genetic markers for studying genetic diversity, population structure, and evolutionary dynamics. Their variability and abundance make them useful tools for genetic mapping, population genetics, and forensic investigations.

Challenges and Opportunities:

Repetitive sequences present challenges and opportunities for genomics and biotechnology, influencing the development of new technologies and approaches to explore the complexity and diversity of genomes.

1. **Data Analysis:** The complexity and abundance of repetitive sequences present challenges for data analysis and interpretation, requiring the development of new computational tools and algorithms to explore genome organization, regulation, and evolution.
2. **Functional Annotation:** The presence and activity of repetitive sequences highlight the need for comprehensive functional annotation of genomes to understand their roles in shaping genome regulation and evolution. Functional annotation approaches provide insights into the regulatory and evolutionary significance of repetitive sequences within genomes.
3. **Ethical Considerations:** The manipulation and modification of repetitive sequences raise ethical considerations related to the responsible use of genomics and biotechnology. Ethical considerations include the potential impact of genetic modifications on biodiversity, ecosystems, and human health.

Chapter - 4

Karyotyping

A. Karyotyping – Chromosome Banding and Chromosome Painting

Karyotyping is a fundamental technique in cytogenetics that involves the examination of chromosomes within a cell to assess their number, size, shape, and structural integrity. This technique provides critical insights into the genetic makeup of an organism, allowing for the detection of chromosomal abnormalities and the study of genetic diseases. Karyotyping is essential for understanding genetic disorders, cancer genetics, chromosomal evolution, and various research applications.

Two advanced methodologies that have significantly enhanced the resolution and accuracy of karyotyping are **chromosome banding** and **chromosome painting**. These techniques allow for the precise identification of individual chromosomes and the detection of structural abnormalities, making them invaluable tools in genetic research and clinical diagnostics.

1. Chromosome Banding

Chromosome banding is a technique that produces characteristic patterns of light and dark bands on chromosomes, facilitating their identification and analysis. This method enables researchers to study the structural organization of chromosomes and identify abnormalities such as deletions, duplications, inversions, and translocations. Chromosome banding has revolutionized cytogenetics by providing detailed maps of chromosomal architecture and enhancing our understanding of genomic disorders.

Principles of Chromosome Banding

The principle behind chromosome banding involves the differential staining of chromosomes to produce a series of distinct bands along their length. Each band corresponds to a specific region of the chromosome, allowing for accurate identification and analysis. The banding pattern is consistent for each chromosome, enabling differentiation between homologous chromosomes and the identification of structural changes.

Types of Chromosome Banding:

Several chromosome banding techniques have been developed, each utilizing specific staining methods to produce distinct banding patterns. The most commonly used techniques include G-banding,

DOI: 10.1201/9781003717775-4

Q-banding, R-banding, and C-banding.

1. **G-Banding (Giemsa Banding):** G-banding is the most widely used chromosome banding technique and involves treating chromosomes with trypsin, an enzyme that partially digests proteins, followed by staining with Giemsa dye. This process results in a pattern of dark and light bands along the chromosome. Dark bands, known as G-bands, are rich in adenine-thymine (AT) base pairs and are composed of tightly packed chromatin, while light bands, known as R-bands, are rich in guanine-cytosine (GC) base pairs and contain less condensed chromatin. G-banding is instrumental in karyotyping and detecting structural abnormalities such as translocations, inversions, and deletions.
2. **Q-Banding (Quinacrine Banding):** Q-banding involves staining chromosomes with quinacrine, a fluorescent dye that binds to AT-rich regions. Under a fluorescence microscope, Q-banding produces bright bands corresponding to AT-rich regions and dark bands corresponding to GC-rich regions. This technique provides a similar banding pattern to G-banding but with enhanced fluorescence visualization. Q-banding is particularly useful for studying heteromorphic chromosomes and identifying specific chromosomal regions.
3. **R-Banding (Reverse Banding):** R-banding is the reverse of G-banding and produces an opposite banding pattern. The technique involves heating chromosomes before staining, which denatures AT-rich regions and enhances the staining of GC-rich regions. R-banding results in dark bands corresponding to GC-rich regions and light bands corresponding to AT-rich regions. This technique is particularly useful for studying telomeric regions and areas of chromosomes not well resolved by G-banding.
4. **C-Banding (Centromeric Banding):** C-banding specifically stains constitutive heterochromatin regions, primarily found at centromeres and other specific chromosomal regions. The technique involves treating chromosomes with an alkali solution followed by staining with Giemsa dye. C-banding produces dark bands at centromeric regions, highlighting the heterochromatin and facilitating the identification of centromeric abnormalities. This technique is essential for identifying chromosomal polymorphisms and studying centromeric heterochromatin.

Applications of Chromosome Banding:

Chromosome banding techniques have numerous applications in genetic research, clinical diagnostics, and evolutionary biology. Some of the key applications include:

1. **Clinical Diagnostics:** Chromosome banding is essential for diagnosing genetic disorders and chromosomal abnormalities such as Down syndrome, Turner syndrome, and Klinefelter syndrome. By analyzing banding patterns, clinicians can identify aneuploidies, translocations, deletions, and other structural abnormalities that contribute to genetic diseases. This information is critical for accurate diagnosis, prognosis, and genetic counseling.
2. **Cancer Research:** Chromosome banding is a valuable tool in cancer research, as many cancers are characterized by specific chromosomal aberrations, such as translocations, deletions, or amplifications. These chromosomal changes can be detected through banding techniques, providing insights into cancer genetics and contributing to the development of targeted therapies.

3. **Prenatal Screening:** Chromosome banding is used in prenatal screening to detect chromosomal abnormalities in developing fetuses. Techniques such as amniocentesis and chorionic villus sampling (CVS) provide fetal cells for karyotyping, allowing for the detection of genetic disorders and informing reproductive decision-making.

4. **Evolutionary Studies:** Chromosome banding techniques are used to compare karyotypes across different species, providing insights into chromosomal evolution and phylogenetic relationships. Analyzing banding patterns can reveal conserved and divergent chromosomal regions, contributing to our understanding of evolutionary processes and the history of chromosomal rearrangements.

5. **Genetic Research:** Chromosome banding is used in genetic research to study chromosomal structure and function, identify genetic markers, and investigate the mechanisms underlying genomic disorders. It is also employed in the mapping of disease-associated loci and the analysis of genomic architecture.

Advantages and Limitations of Chromosome Banding:

Chromosome banding offers several advantages, including the ability to identify specific chromosomes, detect structural abnormalities, and analyze chromosomal organization. However, it also has limitations, such as the inability to detect small genetic changes or sequence-level variations.

1. **Advantages:** Chromosome banding provides a reliable method for chromosome identification and analysis, allowing for the detection of structural abnormalities and aneuploidies. The distinct banding patterns facilitate the identification of homologous chromosomes and the detection of translocations, inversions, and other rearrangements. Banding techniques are relatively simple and cost-effective, making them accessible for various applications.

2. **Limitations:** Chromosome banding techniques have limitations in terms of resolution and sensitivity. They cannot detect small genetic changes, such as point mutations or small deletions, that occur at the sequence level. Additionally, banding techniques may not resolve complex chromosomal rearrangements or subtle structural variations, requiring complementary techniques for comprehensive analysis.

2. Chromosome Painting

Chromosome painting is a molecular cytogenetic technique that utilizes fluorescently labeled DNA probes to visualize specific chromosomes or chromosomal regions in different colors. This method provides a clear and detailed representation of chromosomes, allowing for precise identification and characterization. Chromosome painting enhances the resolution of karyotyping, facilitating the detection of structural abnormalities and chromosomal rearrangements with high accuracy.

Principles of Chromosome Painting

The principle of chromosome painting involves the hybridization of fluorescently labeled DNA probes to complementary sequences on target chromosomes. These probes are designed to be specific to particular chromosomes or chromosomal regions, enabling the visualization of distinct patterns of fluorescence. Chromosome painting allows for the differentiation of individual chromosomes and the detection of complex rearrangements.

Fluorescent In Situ Hybridization (FISH):

Chromosome painting is accomplished through a technique known as fluorescent in situ hybridization (FISH). FISH involves the hybridization of fluorescently labeled DNA probes to denatured chromosomal DNA on a microscope slide. The hybridized probes emit fluorescence when exposed to specific wavelengths of light, allowing for the visualization of chromosomes under a fluorescence microscope.

1. **DNA Probes:** DNA probes used in chromosome painting are derived from chromosome-specific libraries or synthesized to target specific chromosomal regions. These probes are labeled with fluorescent dyes, such as fluorescein, rhodamine, or cyanine, which emit distinct colors when excited by specific wavelengths of light. The choice of probe and fluorescent dye determines the specificity and resolution of the painting.
2. **Hybridization Process:** The hybridization process involves denaturing chromosomal DNA to single-stranded form and incubating it with labeled DNA probes. The probes hybridize to complementary sequences on the target chromosomes, forming stable DNA-DNA hybrids. After washing away unbound probes, the hybridized chromosomes are visualized under a fluorescence microscope.
3. **Visualization and Analysis:** The hybridized chromosomes are visualized using a fluorescence microscope equipped with filters that match the emission wavelengths of the fluorescent dyes. The distinct fluorescence patterns allow for the identification and characterization of individual chromosomes and chromosomal regions. Digital imaging systems are often used to capture and analyze fluorescence images, providing quantitative data on chromosomal organization and abnormalities.

Applications of Chromosome Painting:

Chromosome painting has a wide range of applications in genetic research, clinical diagnostics, and evolutionary studies. Some of the key applications include:

1. **Detection of Chromosomal Rearrangements:** Chromosome painting is used to detect chromosomal rearrangements, such as translocations, inversions, and insertions. By labeling specific chromosomes or regions, researchers can visualize and characterize complex rearrangements that may be associated with genetic disorders or cancers.
2. **Analysis of Chromosomal Aberrations:** Chromosome painting is employed in the analysis of chromosomal aberrations associated with genetic diseases and malignancies. It provides a high-resolution method for identifying specific chromosomal changes, facilitating the diagnosis and characterization of genetic conditions.
3. **Comparative Genomics:** Chromosome painting is used in comparative genomics to study chromosomal evolution and phylogenetic relationships between species. By comparing painted karyotypes across different taxa, researchers can identify conserved and divergent chromosomal regions, providing insights into evolutionary processes.
4. **Mapping of Genetic Loci:** Chromosome painting is used to map genetic loci and identify specific genes or genomic regions associated with particular traits or diseases. This information

is valuable for understanding the genetic basis of complex traits and identifying potential therapeutic targets.

5. **Research on Chromosomal Dynamics:** Chromosome painting is employed to study chromosomal dynamics, such as chromosome territory organization, replication timing, and nuclear architecture. This research provides insights into the spatial organization of the genome and its impact on gene regulation.

Advantages and Limitations of Chromosome Painting:

Chromosome painting offers several advantages, including high-resolution visualization, precise identification of chromosomal rearrangements, and versatility in various applications. However, it also has limitations, such as the requirement for specific probes and the inability to detect small genetic changes.

1. **Advantages:** Chromosome painting provides a high-resolution method for visualizing and analyzing chromosomes, allowing for the detection of complex rearrangements and structural abnormalities. The use of fluorescent probes enables the precise identification of individual chromosomes and chromosomal regions, facilitating the analysis of chromosomal organization and variation. Chromosome painting is versatile and can be applied to a wide range of research and clinical applications.
2. **Limitations:** Chromosome painting requires the use of specific DNA probes, which may not be available for all chromosomes or regions. The technique may not detect small genetic changes, such as point mutations or small deletions, that occur at the sequence level. Additionally, chromosome painting may not resolve subtle structural variations or complex rearrangements, requiring complementary techniques for comprehensive analysis.

Significance of Chromosome Banding and Painting

Chromosome banding and painting are powerful cytogenetic techniques that have revolutionized the study of chromosomes and genetic analysis. Their applications span clinical diagnostics, genetic research, evolutionary studies, and biotechnology, providing valuable insights into the structure, function, and evolution of genomes.

Clinical Diagnostics

Chromosome banding and painting play crucial roles in clinical diagnostics, enabling the detection and characterization of chromosomal abnormalities associated with genetic disorders and diseases. Their use in clinical settings provides valuable information for diagnosis, prognosis, and treatment planning.

Diagnosis of Genetic Disorders:

Chromosome banding and painting are used to diagnose a wide range of genetic disorders, including aneuploidies, structural abnormalities, and syndromic conditions. By analyzing the banding patterns and fluorescence signals, clinicians can identify specific chromosomal changes that contribute to genetic diseases.

1. **Aneuploidies:** Chromosome banding and painting are used to detect aneuploidies, conditions characterized by an abnormal number of chromosomes. Common examples include Down syndrome (trisomy 21), Turner syndrome (monosomy X), and Klinefelter syndrome (XXY). These techniques provide a reliable method for identifying aneuploidies and assessing their clinical significance.
2. **Structural Abnormalities:** Chromosome banding and painting are employed to detect structural abnormalities, such as translocations, deletions, duplications, and inversions. These abnormalities can disrupt gene function and regulation, leading to a variety of genetic disorders. By characterizing structural abnormalities, clinicians can provide accurate diagnoses and inform treatment decisions.
3. **Syndromic Conditions:** Chromosome banding and painting are used to diagnose syndromic conditions associated with specific chromosomal changes. These conditions may involve complex rearrangements or microdeletions that contribute to characteristic phenotypes. By analyzing chromosomal patterns, clinicians can identify the genetic basis of syndromic conditions and provide targeted care.

Cancer Diagnostics:

Chromosome banding and painting are valuable tools in cancer diagnostics, providing insights into the genetic changes associated with malignancies. Many cancers are characterized by specific chromosomal aberrations that can be detected through these techniques.

1. **Identification of Chromosomal Aberrations:** Chromosome banding and painting are used to identify chromosomal aberrations associated with cancer, such as translocations, amplifications, and deletions. These techniques provide a high-resolution method for characterizing chromosomal changes that contribute to cancer development and progression.
2. **Prognostic Assessment:** Chromosomal aberrations identified through banding and painting can provide valuable prognostic information, helping to assess the severity and progression of cancer. Specific chromosomal changes may be associated with particular cancer subtypes or stages, informing treatment decisions and outcomes.
3. **Monitoring of Treatment Response:** Chromosome banding and painting can be used to monitor the response to cancer treatment by assessing changes in chromosomal patterns. By analyzing chromosomal aberrations, clinicians can evaluate the effectiveness of therapy and make informed decisions about treatment strategies.

Reproductive Medicine:

Chromosome banding and painting have applications in reproductive medicine, where they are used to assess chromosomal integrity in gametes and embryos. These techniques provide valuable information for fertility assessments, assisted reproductive technologies, and prenatal diagnostics.

1. **Fertility Assessments:** Chromosome banding and painting are used to evaluate chromosomal integrity in sperm and oocytes, providing insights into fertility and reproductive health. Abnormalities in chromosomal structure or number can impact fertility and contribute to reproductive challenges.

2. **Assisted Reproductive Technologies:** Chromosome banding and painting are employed in assisted reproductive technologies, such as in vitro fertilization (IVF), to assess chromosomal integrity in embryos. Preimplantation genetic diagnosis (PGD) and screening (PGS) involve karyotyping techniques to identify chromosomal abnormalities in embryos before implantation.
3. **Prenatal Diagnostics:** Chromosome banding and painting are used in prenatal diagnostics to assess chromosomal integrity in fetuses. Techniques such as amniocentesis and chorionic villus sampling (CVS) provide fetal cells for karyotyping, allowing for the detection of genetic disorders and informing reproductive decision-making.

Genetic Research

Chromosome banding and painting are essential tools in genetic research, providing insights into genome organization, function, and evolution. Their applications in research settings contribute to our understanding of genetic diversity and the mechanisms underlying genetic processes.

Genome Organization and Function:

Chromosome banding and painting provide valuable insights into the organization and function of genomes, revealing the structural and functional relationships between chromosomes and genes.

1. **Chromosomal Mapping:** Chromosome banding and painting are used to map chromosomes and identify specific genomic regions associated with particular traits or functions. By analyzing chromosomal patterns, researchers can identify conserved and divergent regions, providing insights into the organization and function of genomes.
2. **Gene Regulation:** Chromosome banding and painting provide insights into the regulation of gene expression by revealing the spatial and functional relationships between genes and regulatory elements. By analyzing chromosomal patterns, researchers can identify regions of active and inactive chromatin, contributing to our understanding of gene regulation and expression.
3. **Functional Genomics:** Chromosome banding and painting contribute to functional genomics approaches, providing insights into the roles of specific genomic regions in shaping gene expression and phenotypes. By analyzing chromosomal patterns, researchers can identify functional elements within genomes and their contributions to complex traits and phenotypes.

Evolutionary Studies:

Chromosome banding and painting are used in evolutionary studies to compare karyotypes across different species, providing insights into chromosomal evolution and phylogenetic relationships.

1. **Comparative Genomics:** Chromosome banding and painting are employed in comparative genomics to study the conservation and divergence of chromosomal regions across different taxa. By comparing painted karyotypes, researchers can identify conserved and divergent regions, contributing to our understanding of chromosomal evolution and phylogenetic relationships.

2. **Chromosomal Evolution:** Chromosome banding and painting provide insights into the mechanisms underlying chromosomal evolution, such as rearrangements, duplications, and deletions. By analyzing chromosomal patterns, researchers can identify evolutionary changes that have contributed to the diversification of species and the development of complex traits.
3. **Phylogenetic Relationships:** Chromosome banding and painting are used to infer phylogenetic relationships between species by comparing chromosomal patterns and karyotypes. By analyzing chromosomal similarities and differences, researchers can construct phylogenetic trees and explore the evolutionary history of species.

Genetic Diversity and Adaptation:

Chromosome banding and painting are used to study genetic diversity and adaptation in populations, providing insights into the genetic basis of complex traits and evolutionary processes.

1. **Population Genetics:** Chromosome banding and painting are employed in population genetics to assess genetic diversity and structure within and between populations. By analyzing chromosomal patterns, researchers can identify genetic markers associated with specific traits or adaptations, contributing to our understanding of population dynamics and evolution.
2. **Adaptation and Speciation:** Chromosome banding and painting provide insights into the genetic basis of adaptation and speciation, revealing the chromosomal changes associated with evolutionary processes. By analyzing chromosomal patterns, researchers can identify regions of the genome that have undergone adaptive changes, contributing to the development of new species and lineages.
3. **Conservation Biology:** Chromosome banding and painting are used in conservation biology to assess genetic diversity and structure in endangered species. By analyzing chromosomal patterns, researchers can identify genetic markers associated with specific traits or adaptations, informing conservation efforts and management strategies.

Biotechnology

Chromosome banding and painting have significant implications for biotechnology, providing valuable tools for genetic engineering, synthetic biology, and other applications.

Genetic Engineering:

Chromosome banding and painting provide opportunities for genetic engineering and manipulation, enabling the modification of genetic elements within genomes.

1. **Gene Mapping and Targeting:** Chromosome banding and painting are used to map genes and identify specific genomic regions for genetic engineering. By analyzing chromosomal patterns, researchers can target specific genes or regions for modification, facilitating the study of gene function and regulation.
2. **Gene Editing:** Chromosome banding and painting contribute to gene editing approaches, such as CRISPR-Cas9, by providing insights into the organization and function of genomes. By analyzing chromosomal patterns, researchers can design targeted gene editing strategies to modify genetic elements within genomes.

3. **Bioproduction:** Chromosome banding and painting have potential applications in bioproduction, where genetic engineering is used to produce valuable compounds and materials. By analyzing chromosomal patterns, researchers can optimize the genetic pathways and regulatory networks involved in bioproduction, enhancing the efficiency and yield of desired products.

Synthetic Biology:

Chromosome banding and painting play a role in synthetic biology approaches that harness the potential of genetic material for various applications.

1. **Design and Construction of Genetic Circuits:** Chromosome banding and painting provide insights into the organization and function of genomes, facilitating the design and construction of novel genetic circuits and pathways. By analyzing chromosomal patterns, researchers can identify functional elements and regulatory networks for synthetic biology applications.
2. **Development of Engineered Organisms:** Chromosome banding and painting contribute to the development of engineered organisms with enhanced functions and capabilities. By analyzing chromosomal patterns, researchers can design and construct organisms with specific traits and phenotypes for various applications, such as bioremediation, agriculture, and medicine.
3. **Biological Computation:** Chromosome banding and painting have potential applications in biological computation, where genetic circuits and pathways are used to perform computational tasks. By analyzing chromosomal patterns, researchers can design and construct genetic systems for information processing and decision-making.

Pharmaceutical Applications:

Chromosome banding and painting have applications in pharmaceutical research and development, where they are used to study the genetic basis of diseases and identify potential therapeutic targets.

1. **Drug Discovery:** Chromosome banding and painting are used in drug discovery to identify genetic markers and pathways associated with specific diseases. By analyzing chromosomal patterns, researchers can identify potential therapeutic targets and develop targeted interventions for various conditions.
2. **Pharmacogenomics:** Chromosome banding and painting contribute to pharmacogenomics, the study of how genetic variation affects drug response. By analyzing chromosomal patterns, researchers can identify genetic markers associated with drug metabolism and efficacy, informing personalized medicine and treatment strategies.
3. **Biomarker Identification:** Chromosome banding and painting are used to identify biomarkers associated with specific diseases and conditions. By analyzing chromosomal patterns, researchers can identify genetic markers that can be used for diagnostic and prognostic purposes, enhancing the accuracy and effectiveness of clinical interventions.

Challenges and Opportunities:

Chromosome banding and painting present challenges and opportunities for genomics and biotechnology, influencing the development of new technologies and approaches to explore the complexity and diversity of genomes.

1. **Resolution and Sensitivity:** Chromosome banding and painting techniques have limitations in terms of resolution and sensitivity, requiring the development of new methods and technologies to enhance their capabilities. Advances in microscopy, imaging, and sequencing technologies may enable more detailed analyses of chromosomal patterns and abnormalities.
2. **Functional Annotation:** The analysis of chromosomal patterns highlights the need for comprehensive functional annotation of genomes to understand the roles of specific regions in shaping gene expression and regulation. Functional annotation approaches provide insights into the regulatory and evolutionary significance of chromosomal elements within genomes.
3. **Ethical Considerations:** The manipulation and modification of genetic material raise ethical considerations related to the responsible use of genomics and biotechnology. Ethical considerations include the potential impact of genetic modifications on biodiversity, ecosystems, and human health.

B. Tracking introgressions using FISH

Introgression is the incorporation of genetic material from one species into the genome of another through repeated backcrossing and hybridization. It plays a significant role in evolutionary biology, plant and animal breeding, and the study of genetic diversity. Tracking introgressions is crucial for elucidating the mechanisms of gene flow between species, the evolution of new traits, and the development of improved crop and livestock varieties.

Fluorescent *In Situ* Hybridization (FISH) is a powerful cytogenetic technique that can be used to track introgressions by visualizing specific DNA sequences within chromosomes. FISH employs fluorescently labeled DNA probes that bind to complementary sequences on chromosomes, allowing researchers to identify and characterize introgressed genetic material at a high resolution. This technique has become an indispensable tool in genetic research, breeding programs, and conservation biology.

Principles of FISH for Tracking Introgressions

The principle of FISH is based on the hybridization of fluorescently labeled DNA probes to specific target sequences on chromosomes. By designing probes specific to the introgressed genetic material, FISH enables the visualization and localization of these sequences within the recipient genome. This allows researchers to track the presence and distribution of introgressed genes and assess their impact on genome structure and function.

Components of FISH:

1. **DNA Probes:** DNA probes are synthesized to match specific target sequences of interest, such as introgressed genes or regions. These probes are labeled with fluorescent dyes that

emit distinct colors when excited by specific wavelengths of light. The choice of probe and dye determines the specificity and resolution of the FISH analysis.

2. **Chromosome Preparation:** Chromosome spreads are prepared from cells, typically metaphase cells, to allow for optimal visualization and analysis. The chromosomes are fixed onto microscope slides and denatured to expose the single-stranded DNA for probe hybridization.
3. **Hybridization and Detection:** The fluorescently labeled probes are applied to the denatured chromosomes, where they hybridize to complementary target sequences. After washing away unbound probes, the hybridized chromosomes are visualized under a fluorescence microscope. The fluorescent signals indicate the presence and location of the target sequences within the chromosomes.

Steps in FISH Procedure:

1. **Probe Design and Labeling:** Probes are designed to target specific introgressed sequences and are labeled with fluorescent dyes for detection. The selection of probes depends on the genomic region of interest and the desired resolution of the analysis.
2. **Chromosome Preparation:** Cells are harvested, and metaphase chromosome spreads are prepared on microscope slides. The chromosomes are treated with enzymes and fixatives to preserve their structure and expose the DNA for hybridization.
3. **Denaturation and Hybridization:** The chromosome spreads are denatured to separate the DNA strands, allowing the probes to access the target sequences. The labeled probes are then applied to the slides and allowed to hybridize to the complementary sequences.
4. **Washing and Detection:** Unbound probes are removed by washing the slides with buffer solutions. The hybridized chromosomes are then visualized under a fluorescence microscope, where the fluorescent signals indicate the presence of the target sequences.
5. **Analysis and Interpretation:** The fluorescent signals are analyzed to determine the presence, location, and distribution of the introgressed sequences within the recipient genome. Digital imaging systems can be used to capture and analyze the fluorescence images, providing quantitative data on the introgressions.

Advantages of FISH for Tracking Introgressions:

1. **High Resolution:** FISH provides high-resolution visualization of specific DNA sequences within chromosomes, allowing for precise identification and characterization of introgressed material.
2. **Specificity:** The use of fluorescently labeled probes allows for the specific detection of target sequences, enabling researchers to differentiate between introgressed and native genetic material.
3. **Versatility:** FISH can be applied to a wide range of organisms and cell types, making it a versatile tool for studying introgressions in various contexts.
4. **Real-Time Analysis:** FISH provides real-time visualization of introgressed sequences, allowing researchers to track their presence and distribution throughout the genome.

5. **Multiplexing Capability:** FISH can be used to simultaneously detect multiple target sequences by using probes labeled with different fluorescent dyes, providing a comprehensive analysis of introgressions.

Applications of FISH in Tracking Introgressions

FISH has a wide range of applications in tracking introgressions across different fields, including plant and animal breeding, evolutionary biology, and conservation genetics. By providing detailed insights into the presence and impact of introgressed genetic material, FISH contributes to our understanding of genetic diversity, adaptation, and evolution.

Plant Breeding and Genetics

In plant breeding and genetics, introgressions are commonly used to introduce desirable traits, such as disease resistance, drought tolerance, and improved yield, from wild relatives into cultivated varieties. FISH is a valuable tool for tracking these introgressions and assessing their effects on plant genomes and phenotypes.

Applications in Plant Breeding:

1. **Identification of Introgressed Regions:** FISH is used to identify and characterize introgressed regions in plant genomes, providing insights into the transfer and incorporation of beneficial traits. By using probes specific to the introgressed material, researchers can visualize the presence and distribution of these sequences within the recipient genome.
2. **Monitoring of Hybridization Events:** FISH is employed to monitor hybridization events and track the movement of introgressed genes across generations. This information is valuable for assessing the stability and inheritance of introgressed traits and for optimizing breeding strategies.
3. **Assessment of Genome Integrity:** FISH is used to assess the impact of introgressions on genome integrity and structure. By analyzing the chromosomal distribution of introgressed sequences, researchers can evaluate potential genomic disruptions and ensure the stability of breeding lines.
4. **Development of Marker-Assisted Selection:** FISH is used to develop marker-assisted selection strategies by identifying chromosomal markers associated with introgressed traits. These markers can be used to select individuals with desirable traits, enhancing the efficiency and precision of breeding programs.

Case Studies in Plant Breeding:

- **Wheat Breeding:** In wheat breeding, FISH has been used to track the introgression of disease-resistance genes from wild relatives into cultivated wheat varieties. By visualizing the chromosomal location of resistance genes, researchers can assess the success of introgression events and optimize breeding strategies.
- **Rice Improvement:** FISH has been employed to study introgressions in rice breeding programs, where wild rice species are used to introduce traits such as pest resistance and

stress tolerance into cultivated rice varieties. FISH provides insights into the genomic impact of these introgressions and aids in the development of improved rice lines.

Animal Breeding and Genetics

In animal breeding and genetics, introgressions are used to enhance desirable traits, such as milk production, disease resistance, and growth rate, in livestock populations. FISH is a valuable tool for tracking these introgressions and understanding their effects on animal genomes and phenotypes.

Applications in Animal Breeding:

1. **Detection of Introgressed Genes:** FISH is used to detect and characterize introgressed genes in animal genomes, providing insights into the transfer and incorporation of beneficial traits. By using probes specific to the introgressed material, researchers can visualize the presence and distribution of these sequences within the recipient genome.
2. **Evaluation of Hybridization Success:** FISH is employed to evaluate the success of hybridization events and track the movement of introgressed genes across generations. This information is valuable for assessing the stability and inheritance of introgressed traits and for optimizing breeding strategies.
3. **Analysis of Genomic Impact:** FISH is used to analyze the impact of introgressions on genome structure and integrity. By examining the chromosomal distribution of introgressed sequences, researchers can evaluate potential genomic disruptions and ensure the stability of breeding lines.
4. **Enhancement of Breeding Programs:** FISH is used to enhance breeding programs by identifying chromosomal markers associated with introgressed traits. These markers can be used to select individuals with desirable traits, improving the efficiency and precision of breeding efforts.

Case Studies in Animal Breeding:

- **Cattle Breeding:** In cattle breeding, FISH has been used to track the introgression of traits such as disease resistance and milk production from wild relatives or different breeds into commercial cattle populations. FISH provides insights into the genomic impact of these introgressions and aids in the development of improved cattle lines.
- **Pig Improvement:** FISH has been employed to study introgressions in pig breeding programs, where wild pig species or different breeds are used to introduce traits such as growth rate and feed efficiency into commercial pig populations. FISH helps researchers understand the genomic changes associated with these introgressions and optimize breeding strategies.

Evolutionary Biology

In evolutionary biology, introgressions provide valuable insights into the processes of gene flow, hybridization, and adaptation. FISH is a powerful tool for studying introgressions in natural populations and understanding their impact on evolutionary dynamics and genetic diversity.

Applications in Evolutionary Biology:

1. **Analysis of Hybrid Zones:** FISH is used to analyze hybrid zones, where individuals from different species or populations interbreed, resulting in introgressions. By visualizing the distribution of introgressed sequences, researchers can study the dynamics of gene flow and hybridization in natural populations.
2. **Investigation of Adaptive Introgressions:** FISH is employed to investigate adaptive introgressions, where introgressed genes confer advantageous traits that enhance fitness and adaptation. By tracking the presence and impact of adaptive introgressions, researchers can gain insights into the role of gene flow in shaping evolutionary processes.
3. **Study of Speciation Events:** FISH is used to study speciation events, where introgressions may contribute to the development of new species. By analyzing the genomic changes associated with introgressions, researchers can explore the genetic mechanisms underlying speciation and divergence.
4. **Examination of Genomic Stability:** FISH is used to examine the impact of introgressions on genomic stability and structure. By visualizing the chromosomal distribution of introgressed sequences, researchers can assess potential genomic disruptions and their implications for evolutionary processes.

Case Studies in Evolutionary Biology:

» **Hybridization in Fish:** FISH has been used to study introgressions in fish populations, where hybridization between different species or populations occurs. By visualizing the distribution of introgressed sequences, researchers can gain insights into the dynamics of gene flow and adaptation in aquatic environments.

» **Adaptive Introgressions in Plants:** FISH has been employed to investigate adaptive introgressions in plant species, where introgressed genes confer traits such as drought tolerance or disease resistance. FISH provides insights into the role of introgressions in shaping plant evolution and adaptation.

Conservation Genetics

In conservation genetics, introgressions are studied to assess their impact on genetic diversity and population viability. FISH is a valuable tool for tracking introgressions in endangered species and informing conservation efforts.

Applications in Conservation Genetics:

1. **Assessment of Genetic Diversity:** FISH is used to assess genetic diversity in endangered species, providing insights into the presence and impact of introgressions. By visualizing the distribution of introgressed sequences, researchers can evaluate genetic variation and its implications for population viability.
2. **Monitoring of Hybridization Events:** FISH is employed to monitor hybridization events between endangered species and their relatives, providing insights into the potential impact

of introgressions on genetic integrity. This information is valuable for informing conservation strategies and management decisions.

3. **Evaluation of Adaptive Introgressions:** FISH is used to evaluate the presence and impact of adaptive introgressions in endangered species, where introgressed genes may confer advantageous traits that enhance survival and adaptation. By tracking the presence and impact of adaptive introgressions, researchers can gain insights into the role of gene flow in shaping population dynamics.

4. **Development of Conservation Strategies:** FISH is used to develop conservation strategies by identifying genetic markers associated with introgressed traits. These markers can be used to guide breeding and management efforts, enhancing the genetic resilience and viability of endangered populations.

Case Studies in Conservation Genetics:

» **Hybridization in Birds:** FISH has been used to study introgressions in bird populations, where hybridization between endangered species and their relatives occurs. By visualizing the distribution of introgressed sequences, researchers can gain insights into the dynamics of gene flow and its implications for conservation.

» **Conservation of Rare Plant Species:** FISH has been employed to investigate introgressions in rare plant species, where hybridization with related species may impact genetic diversity and adaptation. FISH provides insights into the role of introgressions in shaping plant conservation and management strategies.

Significance of Tracking Introgressions Using FISH

Tracking introgressions using FISH is of great significance in various fields, as it provides valuable insights into genetic diversity, adaptation, and evolution. The ability to visualize and characterize introgressed sequences offers numerous benefits for genetic research, breeding programs, conservation efforts, and evolutionary studies.

Genetic Research and Breeding

Tracking introgressions using FISH is crucial for genetic research and breeding programs, where it facilitates the identification and incorporation of beneficial traits into populations. By providing detailed insights into the presence and impact of introgressions, FISH contributes to the development of improved crop and livestock varieties.

Benefits for Genetic Research:

1. **Enhanced Trait Improvement:** Tracking introgressions using FISH allows researchers to identify and incorporate beneficial traits into populations, enhancing trait improvement efforts. By visualizing the presence and distribution of introgressed sequences, researchers can assess the success of breeding programs and optimize strategies for trait enhancement.

2. **Increased Breeding Efficiency:** FISH enables researchers to track introgressions and assess

their impact on genomic structure and function, increasing breeding efficiency and precision. By identifying chromosomal markers associated with introgressed traits, researchers can select individuals with desirable characteristics, enhancing breeding efforts.

3. **Improved Understanding of Gene Flow:** Tracking introgressions using FISH provides insights into the dynamics of gene flow and hybridization, improving our understanding of genetic processes. By visualizing the presence and impact of introgressions, researchers can explore the mechanisms of gene flow and their implications for genetic diversity and adaptation.

Conservation and Evolution

Tracking introgressions using FISH is essential for conservation efforts and evolutionary studies, as it provides valuable insights into the processes of hybridization, adaptation, and speciation. By visualizing the presence and impact of introgressions, FISH contributes to our understanding of genetic diversity, adaptation, and evolution.

Benefits for Conservation and Evolution:

1. **Enhanced Conservation Strategies:** Tracking introgressions using FISH provides valuable insights into genetic diversity and hybridization in endangered species, enhancing conservation strategies. By visualizing the presence and impact of introgressions, researchers can develop targeted conservation efforts that promote genetic resilience and population viability.
2. **Increased Understanding of Adaptation:** Tracking introgressions using FISH improves our understanding of adaptation processes and the role of gene flow in shaping evolutionary dynamics. By visualizing the presence and impact of introgressions, researchers can explore the genetic mechanisms underlying adaptation and their implications for evolutionary change.
3. **Improved Knowledge of Speciation:** Tracking introgressions using FISH provides insights into the genetic mechanisms of speciation, improving our knowledge of evolutionary processes. By visualizing the presence and impact of introgressions, researchers can explore the role of gene flow in shaping speciation and divergence.

Technological Advancements

Tracking introgressions using FISH represents a significant advancement in cytogenetics and genetic research, as it provides high-resolution visualization and analysis of genetic material. The ability to track introgressions using FISH offers numerous benefits for technological advancements in genomics and biotechnology.

Benefits for Technological Advancements:

1. **Improved Genomic Analysis:** Tracking introgressions using FISH enhances genomic analysis by providing high-resolution visualization and characterization of genetic material. The ability to visualize introgressed sequences offers insights into the organization and function of genomes, facilitating the development of new technologies and approaches.

2. **Increased Precision in Genetic Engineering:** Tracking introgressions using FISH increases precision in genetic engineering efforts, enabling the targeted modification of genetic elements within genomes. By visualizing the presence and distribution of introgressed sequences, researchers can design and implement targeted gene editing strategies that improve genetic engineering outcomes.
3. **Enhanced Understanding of Genetic Diversity:** Tracking introgressions using FISH improves our understanding of genetic diversity and its implications for genetic processes. By visualizing the presence and impact of introgressions, researchers can explore the genetic basis of diversity and its role in shaping biological processes and phenotypes.

C. Tracking Introgressions Using GISH

Introgression, the integration of genetic material from one species into another through hybridization and backcrossing, is a key mechanism in evolutionary biology and breeding. It allows for the transfer of beneficial traits between species and can drive adaptation and speciation.

Genomic *In Situ* Hybridization (GISH) is a sophisticated cytogenetic technique that enables researchers to track introgressions by using whole genomic DNA from one species as a probe to hybridize with the chromosomes of another species. This method provides a high-resolution view of the genetic landscape, allowing for precise visualization and characterization of introgressed sequences. GISH has proven to be an invaluable tool in fields such as plant and animal breeding, evolutionary biology, and conservation genetics.

Principles of GISH

The core principle of GISH lies in its ability to distinguish between genomic DNA from different species through hybridization, leveraging fluorescent labeling to visualize these differences on chromosomes. This process involves using labeled genomic DNA as probes to detect and map specific regions of interest within the genome of a recipient species, enabling the tracking of introgressed genetic material.

Components of GISH:

1. **Genomic DNA Probes:** In GISH, genomic DNA from the donor species is extracted and labeled with fluorescent dyes to serve as probes. This labeled DNA hybridizes specifically to complementary sequences in the recipient species, allowing for the visualization of introgressed regions. The choice of donor species and labeling technique is crucial for ensuring specificity and resolution in detecting the targeted genetic material.
2. **Blocking DNA:** To enhance the specificity of the hybridization, unlabeled blocking DNA from the recipient species is used. This excess of blocking DNA binds to non-target sequences in the recipient genome, preventing non-specific binding of the labeled probe DNA and ensuring that the fluorescent signals observed are specific to the introgressed material.
3. **Chromosome Preparation:** High-quality metaphase chromosome spreads from the recipient species are prepared to provide a clear platform for hybridization. The chromosomes are treated to preserve their structural integrity and denatured to expose single-stranded DNA, allowing for effective probe binding.

4. **Hybridization and Detection:** The labeled genomic probes are applied to the denatured chromosomes, where they hybridize to their complementary sequences. After washing away unbound probes, the hybridized chromosomes are observed under a fluorescence microscope. The fluorescent signals indicate the presence, location, and extent of the introgressed sequences within the recipient genome.

Steps in GISH Procedure:

1. **Preparation of Probes and Blocking DNA:** Genomic DNA is extracted from the donor species and labeled with fluorescent dyes, such as fluorescein or rhodamine, which emit light at specific wavelengths. Simultaneously, blocking DNA is extracted from the recipient species to suppress background hybridization.
2. **Chromosome Preparation:** Metaphase chromosome spreads are prepared from the recipient species by arresting cells in metaphase and spreading them onto microscope slides. The chromosomes are fixed to preserve their structure and then denatured to facilitate probe hybridization.
3. **Hybridization Process:** The chromosome spreads are incubated with the labeled genomic probes and blocking DNA under specific conditions that promote hybridization. The temperature, time, and concentration of the components are optimized to achieve specific and strong binding of the probes to the target sequences.
4. **Post-Hybridization Washing:** After hybridization, slides are washed to remove excess and unbound probes, ensuring that only specific hybridization events remain. This washing step is critical for reducing background noise and enhancing the clarity of the fluorescent signals.
5. **Visualization and Analysis:** The hybridized chromosomes are viewed under a fluorescence microscope, where the labeled probes fluoresce at specific wavelengths, allowing for the identification and analysis of introgressed sequences. Digital imaging systems can be used to capture and analyze these signals, providing quantitative data on the extent and distribution of introgressions.

Advantages of GISH for Tracking Introgressions:

1. **High Specificity and Resolution:** GISH provides high specificity in detecting introgressed sequences by using whole genomic DNA as probes. This allows for the precise identification of alien genetic material within the recipient genome and provides detailed insights into the structure and organization of introgressions.
2. **Versatility Across Species:** GISH can be applied to a wide range of species and hybridization events, making it a versatile tool for studying genetic interactions across diverse taxa. It is particularly useful in species with complex genomes or where other techniques may fail to provide clear resolution.
3. **Detailed Visualization of Introgressions:** The fluorescent labeling in GISH enables detailed visualization of introgressed sequences, providing insights into their chromosomal location, distribution, and structural impact on the genome. This visualization is crucial for

understanding the genetic basis of introgressed traits and their effects on phenotype and adaptation.

4. **Non-Destructive Method:** GISH allows for non-destructive analysis of chromosomes, preserving the integrity of the genetic material and enabling further studies. This is particularly beneficial in studies requiring repeated observations or analyses of the same genetic material.

5. **Integration with Other Techniques:** GISH can be combined with other cytogenetic methods, such as FISH or chromosome banding, to provide comprehensive analyses of genomic interactions and rearrangements. This integration enhances the overall understanding of genetic complexity and evolutionary dynamics.

Applications of GISH in Tracking Introgressions

GISH has a wide range of applications in tracking introgressions, spanning fields such as plant and animal breeding, evolutionary biology, and conservation genetics. It provides detailed insights into the presence and impact of introgressed genetic material, contributing to our understanding of genetic diversity, adaptation, and evolutionary processes.

Plant Breeding and Genetics

In plant breeding and genetics, introgressions are commonly used to introduce desirable traits from wild relatives into cultivated varieties, such as disease resistance, stress tolerance, and improved yield. GISH is an essential tool for tracking these introgressions and assessing their effects on plant genomes and phenotypes.

Applications in Plant Breeding:

1. **Identification of Introgressed Regions:** GISH is employed to identify and characterize introgressed regions in plant genomes, providing insights into the transfer and incorporation of beneficial traits. By using genomic probes specific to the donor species, researchers can visualize the presence and distribution of introgressed sequences within the recipient genome, facilitating the precise mapping of introgressed genes and regions.

2. **Monitoring of Hybridization Events:** GISH is used to monitor hybridization events and track the movement of introgressed genes across generations. This information is invaluable for assessing the stability and inheritance of introgressed traits and for optimizing breeding strategies to ensure the successful incorporation of desirable traits into breeding lines.

3. **Assessment of Genome Integrity:** GISH is used to assess the impact of introgressions on genome integrity and structure. By analyzing the chromosomal distribution of introgressed sequences, researchers can evaluate potential genomic disruptions and ensure the stability of breeding lines. This assessment helps in maintaining the genetic balance and integrity of breeding populations.

4. **Development of Marker-Assisted Selection:** GISH is instrumental in developing marker-assisted selection strategies by identifying chromosomal markers associated with introgressed traits. These markers can be used to select individuals with desirable traits, enhancing the

efficiency and precision of breeding programs. Marker-assisted selection helps accelerate the development of improved plant varieties with enhanced traits.

Case Studies in Plant Breeding:

» **Wheat Breeding:** In wheat breeding, GISH has been used to track the introgression of disease resistance genes from wild relatives into cultivated wheat varieties. By visualizing the chromosomal location of resistance genes, researchers can assess the success of introgression events and optimize breeding strategies to enhance disease resistance in wheat lines.

» **Rice Improvement:** GISH has been employed to study introgressions in rice breeding programs, where wild rice species are used to introduce traits such as pest resistance and stress tolerance into cultivated rice varieties. GISH provides insights into the genomic impact of these introgressions, aiding in the development of improved rice lines with enhanced agronomic performance.

» **Tomato Cultivar Enhancement:** In tomato breeding, GISH has been used to identify and track the introgression of traits such as improved fruit quality and stress tolerance from wild tomato relatives. By visualizing the integration of these traits at the chromosomal level, researchers can develop tomato cultivars with enhanced characteristics tailored to specific growing conditions.

Animal Breeding and Genetics

In animal breeding and genetics, introgressions are used to enhance desirable traits such as disease resistance, productivity, and growth rate in livestock populations. GISH serves as a valuable tool for tracking these introgressions and understanding their effects on animal genomes and phenotypes.

Applications in Animal Breeding:

1. **Detection of Introgressed Genes:** GISH is utilized to detect and characterize introgressed genes in animal genomes, providing insights into the transfer and incorporation of beneficial traits. By using genomic probes specific to the donor species, researchers can visualize the presence and distribution of introgressed sequences within the recipient genome, facilitating the mapping of valuable genetic material.

2. **Evaluation of Hybridization Success:** GISH is employed to evaluate the success of hybridization events and track the movement of introgressed genes across generations. This information is crucial for assessing the stability and inheritance of introgressed traits, ensuring the desired genetic improvements are successfully integrated into breeding populations.

3. **Analysis of Genomic Impact:** GISH is used to analyze the impact of introgressions on genome structure and integrity. By examining the chromosomal distribution of introgressed sequences, researchers can evaluate potential genomic disruptions and ensure the stability of breeding lines. This analysis helps maintain genetic balance and mitigate any adverse effects of introgression on animal health and productivity.

4. **Enhancement of Breeding Programs:** GISH enhances breeding programs by identifying

chromosomal markers associated with introgressed traits. These markers can be used to select individuals with desirable traits, improving the efficiency and precision of breeding efforts. Marker-assisted selection allows breeders to target specific genetic improvements, accelerating the development of superior livestock breeds.

Case Studies in Animal Breeding:

» **Cattle Breeding:** In cattle breeding, GISH has been used to track the introgression of traits such as disease resistance and milk production from wild relatives or different breeds into commercial cattle populations. GISH provides insights into the genomic impact of these introgressions, aiding in the development of improved cattle lines with enhanced productivity and resilience.

» **Pig Improvement:** GISH has been employed to study introgressions in pig breeding programs, where wild pig species or different breeds are used to introduce traits such as growth rate and feed efficiency into commercial pig populations. GISH helps researchers understand the genomic changes associated with these introgressions, optimizing breeding strategies to develop pigs with improved growth performance and feed conversion.

» **Poultry Genetic Enhancement:** In poultry breeding, GISH has been used to track the introgression of traits such as disease resistance and egg production from different poultry breeds into commercial lines. By visualizing the integration of these traits, researchers can develop poultry breeds with enhanced disease resistance and improved production traits, catering to specific market demands.

Evolutionary Biology

In evolutionary biology, introgressions provide valuable insights into the processes of gene flow, hybridization, and adaptation. GISH is a powerful tool for studying introgressions in natural populations and understanding their impact on evolutionary dynamics and genetic diversity.

Applications in Evolutionary Biology:

1. **Analysis of Hybrid Zones:** GISH is used to analyze hybrid zones, where individuals from different species or populations interbreed, resulting in introgressions. By visualizing the distribution of introgressed sequences, researchers can study the dynamics of gene flow and hybridization in natural populations, providing insights into the genetic interactions and evolutionary outcomes of hybrid zones.
2. **Investigation of Adaptive Introgressions:** GISH is employed to investigate adaptive introgressions, where introgressed genes confer advantageous traits that enhance fitness and adaptation. By tracking the presence and impact of adaptive introgressions, researchers can gain insights into the role of gene flow in shaping evolutionary processes and the development of novel adaptations.
3. **Study of Speciation Events:** GISH is used to study speciation events, where introgressions may contribute to the development of new species. By analyzing the genomic changes associated with introgressions, researchers can explore the genetic mechanisms underlying

speciation and divergence, shedding light on the evolutionary pathways leading to the emergence of new species.

4. **Examination of Genomic Stability:** GISH is used to examine the impact of introgressions on genomic stability and structure. By visualizing the chromosomal distribution of introgressed sequences, researchers can assess potential genomic disruptions and their implications for evolutionary processes, providing insights into the maintenance of genetic integrity in the face of introgression.

Case Studies in Evolutionary Biology:

» **Hybridization in Fish:** GISH has been used to study introgressions in fish populations, where hybridization between different species or populations occurs. By visualizing the distribution of introgressed sequences, researchers can gain insights into the dynamics of gene flow and adaptation in aquatic environments, contributing to our understanding of fish evolution and biodiversity.

» **Adaptive Introgressions in Plants:** GISH has been employed to investigate adaptive introgressions in plant species, where introgressed genes confer traits such as drought tolerance or disease resistance. GISH provides insights into the role of introgressions in shaping plant evolution and adaptation, highlighting the genetic basis of adaptive traits and their evolutionary significance.

» **Introgression in Avian Species:** In avian species, GISH has been used to track introgressions resulting from hybridization events between different bird species. By visualizing the genomic integration of introgressed material, researchers can study the evolutionary implications of hybridization in birds, including the potential for adaptive introgressions and their impact on species divergence.

Conservation Genetics

In conservation genetics, introgressions are studied to assess their impact on genetic diversity and population viability. GISH is a valuable tool for tracking introgressions in endangered species and informing conservation efforts.

Applications in Conservation Genetics:

1. **Assessment of Genetic Diversity:** GISH is used to assess genetic diversity in endangered species, providing insights into the presence and impact of introgressions. By visualizing the distribution of introgressed sequences, researchers can evaluate genetic variation and its implications for population viability, informing conservation strategies aimed at preserving genetic diversity.
2. **Monitoring of Hybridization Events:** GISH is employed to monitor hybridization events between endangered species and their relatives, providing insights into the potential impact of introgressions on genetic integrity. This information is valuable for informing conservation strategies and management decisions, ensuring that hybridization does not compromise the genetic health of endangered populations.

3. **Evaluation of Adaptive Introgressions:** GISH is used to evaluate the presence and impact of adaptive introgressions in endangered species, where introgressed genes may confer advantageous traits that enhance survival and adaptation. By tracking the presence and impact of adaptive introgressions, researchers can gain insights into the role of gene flow in shaping population dynamics and enhancing the resilience of endangered species.
4. **Development of Conservation Strategies:** GISH is used to develop conservation strategies by identifying genetic markers associated with introgressed traits. These markers can be used to guide breeding and management efforts, enhancing the genetic resilience and viability of endangered populations and contributing to the development of effective conservation plans.

Case Studies in Conservation Genetics:

» **Hybridization in Birds:** GISH has been used to study introgressions in bird populations, where hybridization between endangered species and their relatives occurs. By visualizing the distribution of introgressed sequences, researchers can gain insights into the dynamics of gene flow and its implications for conservation, informing strategies to manage hybridization and preserve genetic integrity.

» **Conservation of Rare Plant Species:** GISH has been employed to investigate introgressions in rare plant species, where hybridization with related species may impact genetic diversity and adaptation. GISH provides insights into the role of introgressions in shaping plant conservation and management strategies, guiding efforts to preserve genetic diversity and resilience in rare plant populations.

» **Conservation of Mammalian Species:** In mammals, GISH has been used to track introgressions resulting from hybridization events between endangered species and their relatives. By visualizing the genomic integration of introgressed material, researchers can study the impact of hybridization on mammalian conservation, informing strategies to preserve genetic diversity and ensure population viability.

D. Localization and Mapping of Genes/Genomic Segments

Localization and mapping of genes and genomic segments are crucial processes in genetics that aim to determine the precise locations of genes or DNA sequences within the genome. These processes are fundamental for understanding genetic organization, function, and evolution and play a vital role in identifying genes associated with traits, diseases, and biological functions. Gene mapping is essential for advancing research in genetics, genomics, and biotechnology and has a wide range of applications in medicine, agriculture, and evolutionary biology.

Gene mapping involves various techniques to pinpoint the exact locations of genes or DNA segments within the genome. This information is critical for elucidating the genetic basis of traits and diseases, developing genetic markers, and facilitating the study of complex genetic interactions. The localization and mapping of genes are foundational to numerous applications, including medical genetics, plant and animal breeding, evolutionary studies, and personalized medicine.

1. Types of Gene Mapping

Gene mapping can be broadly classified into three main types: genetic mapping, physical mapping, and comparative mapping. Each type utilizes different principles and techniques to determine the positions of genes or DNA segments within the genome.

1. Genetic Mapping

Genetic mapping, also known as linkage mapping, involves determining the relative positions of genes based on their inheritance patterns. This approach uses genetic markers, such as microsatellites or single nucleotide polymorphisms (SNPs), to estimate the distances between genes based on recombination frequencies during meiosis. Genetic maps provide insights into the arrangement of genes on chromosomes and are instrumental in identifying loci associated with traits or diseases.

Principles of Genetic Mapping:

Genetic mapping is based on the principle of genetic linkage, which refers to the tendency of genes or genetic markers located close to each other on a chromosome to be inherited together during meiosis. By analyzing the patterns of inheritance in families or populations, researchers can estimate the genetic distances between markers and map the locations of genes or loci associated with traits.

Techniques in Genetic Mapping:

» **Linkage Analysis:** Linkage analysis involves studying the co-segregation of genetic markers with traits or diseases in families or populations. By analyzing recombination events and calculating recombination frequencies, researchers can construct linkage maps that show the relative positions of genes and markers on chromosomes.

» **Pedigree Analysis:** Pedigree analysis is used to trace the inheritance of genetic markers and traits through generations in families. By analyzing pedigrees, researchers can identify linkage relationships and map the locations of genes associated with hereditary conditions.

» **Genome-Wide Association Studies (GWAS):** GWAS involve scanning the entire genome for associations between genetic markers and traits in large populations. This approach identifies regions of the genome linked to complex traits and diseases, providing insights into their genetic architecture.

Applications of Genetic Mapping:

» **Identification of Disease Genes:** Genetic mapping is used to identify genes associated with inherited diseases and complex disorders. By analyzing linkage or association patterns, researchers can pinpoint the loci responsible for diseases such as cystic fibrosis, Huntington's disease, and diabetes.

» **Marker-Assisted Selection in Breeding:** Genetic mapping facilitates marker-assisted selection by identifying genetic markers linked to desirable traits. These markers are used to select individuals with favorable characteristics, enhancing the efficiency of breeding programs in plants and animals.

- **Understanding Genetic Architecture:** Genetic mapping provides insights into the genetic architecture of complex traits, revealing the contributions of multiple loci to phenotypic variation. This information is valuable for studying the genetic basis of traits and developing strategies for genetic improvement.

Case Studies in Genetic Mapping:

- **Mapping of BRCA1 and BRCA2:** The BRCA1 and BRCA2 genes, associated with hereditary breast and ovarian cancer, were mapped using linkage analysis and positional cloning. These discoveries have led to the development of genetic tests for cancer risk assessment and prevention.
- **QTL Mapping in Agriculture:** Quantitative trait loci (QTL) mapping has been used to identify genomic regions associated with traits such as yield, disease resistance, and drought tolerance in crops. This information is used to enhance crop improvement efforts and increase agricultural productivity.

2. Physical Mapping

Physical mapping involves determining the absolute positions of genes or DNA sequences on chromosomes by measuring physical distances, often in base pairs. Techniques such as restriction mapping, fluorescence in situ hybridization (FISH), and sequencing are used to construct physical maps that provide high-resolution views of genomic architecture. Physical maps are crucial for understanding genome structure and organization.

Principles of Physical Mapping:

Physical mapping is based on the direct measurement of DNA sequences and their arrangement on chromosomes. This approach provides a detailed view of the genome's structural organization, allowing researchers to precisely locate genes and genomic segments.

Techniques in Physical Mapping:

- **Fluorescence In Situ Hybridization (FISH):** FISH uses fluorescently labeled DNA probes to visualize specific genomic regions on chromosomes. By hybridizing probes to denatured chromosomal DNA, FISH allows researchers to localize genes or DNA segments within the genome, providing insights into their chromosomal positions and structural context.
- **Restriction Mapping:** Restriction mapping involves cutting DNA with specific restriction enzymes and analyzing the resulting fragments to determine the locations of restriction sites. This technique provides information about the physical distances between genetic elements.
- **Radiation Hybrid Mapping:** Radiation hybrid mapping uses irradiated cell lines to map the relative positions of genes or DNA segments. This technique leverages the retention of genetic markers in irradiated hybrids to estimate distances between loci and construct maps of genomic regions.
- **Sequencing-Based Approaches:** Sequencing technologies, such as whole-genome sequencing

(WGS) and targeted sequencing, provide high-resolution data for physical mapping. By sequencing entire genomes or specific regions, researchers can precisely determine the positions of genes or DNA segments, facilitating the construction of detailed genomic maps.

Applications of Physical Mapping:

- **Genome Assembly and Annotation:** Physical mapping is essential for genome assembly and annotation, providing the foundation for understanding genome structure and function. By determining the positions of genes and genomic segments, researchers can annotate genomes and identify functional elements.
- **Chromosomal Aberration Analysis:** Physical mapping is used to analyze chromosomal aberrations, such as deletions, duplications, and translocations, in genetic disorders and cancers. By localizing these aberrations, researchers can identify the underlying genetic causes and inform diagnostic and therapeutic strategies.
- **Comparative Genomics:** Physical mapping supports comparative genomics by providing a reference framework for comparing genomic organization across species. This information is valuable for studying evolutionary relationships and identifying conserved genetic elements.

Case Studies in Physical Mapping:

- **Human Genome Project:** The Human Genome Project involved the construction of a high-resolution physical map of the human genome, enabling the sequencing and annotation of the entire genome. This project has revolutionized our understanding of human genetics and disease.
- **Mapping of Disease Loci in Cancer:** Physical mapping techniques, such as FISH, have been used to identify chromosomal rearrangements and amplifications in cancer. This information has led to the discovery of oncogenes and tumor suppressor genes, advancing cancer research and treatment.

3. Comparative Mapping

Comparative mapping involves comparing the genomic organization of different species to identify conserved regions and evolutionary relationships. This approach leverages known genetic or physical maps from one species to infer the genomic organization in another species, facilitating the identification of homologous genes and evolutionary conserved sequences.

Principles of Comparative Mapping:

Comparative mapping is based on the concept of synteny, which refers to the conservation of gene order and arrangement between different species. By comparing genomic maps, researchers can identify conserved regions and gain insights into the evolutionary history and functional relationships of genes.

Techniques in Comparative Mapping:

- **Synteny Analysis:** Synteny analysis involves comparing the order and arrangement of genes between different species to identify conserved regions. This approach uses genomic data from multiple species to construct comparative maps and explore evolutionary relationships.

- **Cross-Species Mapping:** Cross-species mapping involves using genetic or physical maps from one species to guide the mapping efforts in another species. This approach leverages the conservation of genetic elements to identify homologous genes and genomic regions.
- **Comparative Genomic Hybridization (CGH):** CGH is a technique used to compare genomic content between different species or individuals. By hybridizing labeled DNA from two sources, researchers can identify genomic gains, losses, and rearrangements, providing insights into evolutionary changes.

Applications of Comparative Mapping:

- **Evolutionary Studies:** Comparative mapping provides valuable insights into evolutionary relationships and processes by identifying conserved genetic elements and syntenic regions. This information is used to study speciation, adaptation, and the evolution of complex traits.
- **Functional Genomics:** Comparative mapping supports functional genomics by identifying homologous genes and regulatory elements across species. This information is used to infer gene functions and explore the genetic basis of phenotypic diversity.
- **Conservation Biology:** Comparative mapping is used in conservation biology to assess genetic diversity and evolutionary relationships among species. By identifying conserved regions, researchers can develop strategies for preserving genetic diversity and managing endangered populations.

Case Studies in Comparative Mapping:

- **Comparative Mapping of Model Organisms:** Comparative mapping between model organisms, such as mice and humans, has provided insights into the conservation of genetic elements and pathways. This information is used to study human diseases and develop therapeutic interventions.
- **Evolutionary Conservation in Plants:** Comparative mapping in plants has identified conserved syntenic regions between different species, providing insights into the evolution of plant genomes and the genetic basis of traits such as stress tolerance and disease resistance.

2. Techniques for Gene Localization and Mapping

Various techniques are employed for the localization and mapping of genes and genomic segments, each offering unique advantages and applications. These techniques provide the tools necessary for constructing detailed genomic maps and elucidating the genetic basis of traits and diseases.

Linkage Analysis

Linkage analysis is a genetic mapping technique that involves studying the co-segregation of genetic markers with traits or diseases in families or populations. By analyzing recombination events and calculating recombination frequencies, researchers can construct linkage maps that show the relative positions of genes and markers on chromosomes.

Principles of Linkage Analysis:

Linkage analysis is based on the principle of genetic linkage, which refers to the tendency of genes or genetic markers located close to each other on a chromosome to be inherited together during meiosis. By analyzing the patterns of inheritance, researchers can estimate the genetic distances between markers and map the locations of genes or loci associated with traits.

Steps in Linkage Analysis:

1. **Selection of Genetic Markers:** Genetic markers, such as microsatellites or SNPs, are selected based on their polymorphism and informativeness. These markers serve as reference points for mapping the positions of genes or loci.
2. **Collection of Pedigree Data:** Pedigree data is collected from families or populations, including information on the inheritance of genetic markers and traits. This data is used to trace the co-segregation of markers and traits through generations.
3. **Calculation of Recombination Frequencies:** Recombination frequencies between genetic markers and traits are calculated based on the observed inheritance patterns. These frequencies provide estimates of the genetic distances between markers and loci.
4. **Construction of Linkage Maps:** Linkage maps are constructed by arranging genetic markers based on their recombination frequencies. These maps reveal the relative positions of genes and markers on chromosomes and provide insights into the genetic architecture of traits.

Applications of Linkage Analysis:

» **Identification of Disease Genes:** Linkage analysis is used to identify genes associated with inherited diseases and complex disorders. By analyzing linkage patterns, researchers can pinpoint the loci responsible for diseases such as cystic fibrosis, Huntington's disease, and Alzheimer's disease.

» **Mapping of Quantitative Trait Loci (QTL):** Linkage analysis is used to map QTLs, regions of the genome associated with complex traits, such as yield or disease resistance. This information is valuable for studying the genetic basis of traits and developing strategies for genetic improvement.

» **Genetic Counseling and Risk Assessment:** Linkage analysis provides valuable information for genetic counseling and risk assessment. By identifying the locations of disease-associated genes, clinicians can assess genetic risk, inform reproductive decision-making, and guide preventive measures.

Case Studies in Linkage Analysis:

» **Mapping of Cystic Fibrosis Gene:** Linkage analysis was used to identify and map the CFTR gene, responsible for cystic fibrosis. This discovery enabled the development of diagnostic tests and targeted therapies for cystic fibrosis, improving patient outcomes and quality of life.

» **Identification of Alzheimer's Disease Genes:** Linkage analysis has been used to identify genes associated with Alzheimer's disease, such as APP and PSEN1. These findings have contributed to understanding the genetic basis of Alzheimer's disease and informing therapeutic strategies.

Sequencing-Based Approaches

Sequencing technologies, such as whole-genome sequencing (WGS) and targeted sequencing, provide high-resolution data for physical mapping. By sequencing entire genomes or specific regions, researchers can precisely determine the positions of genes or DNA segments, facilitating the construction of detailed genomic maps.

Principles of Sequencing-Based Approaches:

Sequencing-based approaches involve determining the nucleotide sequences of DNA molecules to provide detailed information about the genome's structure and organization. These approaches enable researchers to precisely locate genes and genomic segments within the genome.

Techniques in Sequencing-Based Approaches:

» **Whole-Genome Sequencing (WGS):** WGS involves sequencing the entire genome of an organism, providing comprehensive data on the positions and arrangements of genes and DNA segments. This approach is used to construct high-resolution genomic maps and identify genetic variations.

» **Targeted Sequencing:** Targeted sequencing focuses on sequencing specific regions of the genome, such as exomes or candidate genes, to provide detailed information about specific genetic elements. This approach is used to identify and map genes associated with traits or diseases.

» **Exome Sequencing:** Exome sequencing involves sequencing the coding regions of the genome, providing insights into the genetic basis of traits and diseases. This approach is used to identify mutations and variations that affect gene function.

Applications of Sequencing-Based Approaches:

» **Genome Assembly and Annotation:** Sequencing-based approaches are essential for genome assembly and annotation, providing the foundation for understanding genome structure and function. By determining the positions of genes and genomic segments, researchers can annotate genomes and identify functional elements.

» **Identification of Genetic Variations:** Sequencing-based approaches are used to identify genetic variations, such as SNPs and indels, that contribute to phenotypic diversity and disease susceptibility. This information is valuable for studying the genetic basis of traits and developing personalized medicine strategies.

» **Comparative Genomics:** Sequencing-based approaches support comparative genomics by providing a reference framework for comparing genomic organization across species. This information is valuable for studying evolutionary relationships and identifying conserved genetic elements.

Case Studies in Sequencing-Based Approaches:

- **Human Genome Project:** The Human Genome Project involved sequencing the entire human genome, enabling the construction of a high-resolution genomic map and revolutionizing our understanding of human genetics and disease.
- **Cancer Genomics:** Sequencing-based approaches have been used to identify genetic mutations and variations in cancer, providing insights into the genetic basis of tumor development and informing the development of targeted therapies.

Radiation Hybrid Mapping

Radiation hybrid mapping is a technique used to map the relative positions of genes or DNA segments by using irradiated cell lines. This technique leverages the retention of genetic markers in irradiated hybrids to estimate distances between loci and construct maps of genomic regions.

Principles of Radiation Hybrid Mapping:

Radiation hybrid mapping involves using radiation to fragment the genome into smaller pieces, which are then incorporated into hybrid cell lines. By analyzing the retention of genetic markers in these hybrids, researchers can estimate the distances between loci and construct maps of genomic regions.

Steps in Radiation Hybrid Mapping:

1. **Generation of Radiation Hybrids:** Cells from a donor species are irradiated to create random genomic fragments. These fragments are fused with cells from a recipient species to create hybrid cell lines that contain fragments of the donor genome.
2. **Selection of Genetic Markers:** Genetic markers, such as microsatellites or SNPs, are selected based on their polymorphism and informativeness. These markers serve as reference points for mapping the positions of genes or loci.
3. **Analysis of Marker Retention:** The retention of genetic markers in radiation hybrids is analyzed to estimate the distances between loci. By calculating the frequency of marker retention, researchers can construct maps that reveal the relative positions of genes and markers.
4. **Construction of Radiation Hybrid Maps:** Radiation hybrid maps are constructed by arranging genetic markers based on their retention frequencies. These maps provide insights into the genetic architecture of traits and the structural organization of genomic regions.

Applications of Radiation Hybrid Mapping:

- **Genome Assembly and Annotation:** Radiation hybrid mapping is used to support genome assembly and annotation by providing high-resolution maps of genomic regions. This information is valuable for understanding genome structure and function.
- **Identification of Genetic Variations:** Radiation hybrid mapping is used to identify genetic variations, such as deletions and rearrangements, that contribute to phenotypic diversity and disease susceptibility. This information is valuable for studying the genetic basis of traits and developing personalized medicine strategies.

» **Comparative Genomics:** Radiation hybrid mapping supports comparative genomics by providing a reference framework for comparing genomic organization across species. This information is valuable for studying evolutionary relationships and identifying conserved genetic elements.

Case Studies in Radiation Hybrid Mapping:

» **Mapping of Disease Loci in Cancer:** Radiation hybrid mapping has been used to identify chromosomal rearrangements and amplifications in cancer. This information has led to the discovery of oncogenes and tumor suppressor genes, advancing cancer research and treatment.

» **Comparative Mapping in Model Organisms:** Radiation hybrid mapping has been used to compare the genomic organization of different species, providing insights into the conservation of genetic elements and pathways. This information is used to study human diseases and develop therapeutic interventions.

3. Chromosome Walking and Jumping

Chromosome walking and jumping are techniques used to clone and map genomic regions. Chromosome walking involves progressively isolating overlapping DNA fragments to traverse a genomic region, while chromosome jumping allows researchers to skip over large intervening sequences to map distant loci.

Principles of Chromosome Walking and Jumping:

Chromosome walking and jumping are based on the systematic isolation and analysis of DNA fragments to construct maps of genomic regions. These techniques provide detailed information about the arrangement and organization of genes and DNA segments within the genome.

Steps in Chromosome Walking and Jumping:

1. **Isolation of DNA Fragments:** DNA fragments are isolated from genomic libraries using specific probes or primers. These fragments serve as starting points for walking or jumping along the chromosome.
2. **Construction of Overlapping Clones:** Overlapping DNA clones are constructed by isolating and analyzing successive fragments that span the genomic region of interest. This process is repeated to traverse the region and construct a contiguous map.
3. **Analysis of Cloned Sequences:** The cloned sequences are analyzed to identify genes, markers, or loci of interest. This information is used to construct maps that reveal the organization and arrangement of genomic elements.
4. **Extension and Refinement of Maps:** The maps are extended and refined by isolating additional fragments and analyzing their sequences. This process provides detailed insights into the genomic architecture of the region and the relationships between genetic elements.

Applications of Chromosome Walking and Jumping:

» **Gene Cloning and Characterization:** Chromosome walking and jumping are used to clone and characterize genes of interest, providing insights into their structure, function, and regulation. This information is valuable for studying the genetic basis of traits and diseases.

» **Mapping of Disease Loci:** Chromosome walking and jumping are used to map disease loci and identify candidate genes associated with inherited disorders. By traversing genomic regions, researchers can pinpoint the locations of disease-associated genes and inform diagnostic and therapeutic strategies.

» **Functional Genomics:** Chromosome walking and jumping support functional genomics by providing insights into the organization and function of genomic regions. This information is used to explore the genetic basis of phenotypic diversity and identify regulatory elements.

Case Studies in Chromosome Walking and Jumping:

» **Cloning of Huntington's Disease Gene:** Chromosome walking was used to clone the HTT gene, responsible for Huntington's disease. This discovery has contributed to understanding the genetic basis of the disease and developing potential therapies.

» **Mapping of BRCA1 Gene:** Chromosome jumping was used to map the BRCA1 gene, associated with hereditary breast and ovarian cancer. This discovery has led to the development of genetic tests for cancer risk assessment and prevention.

4. Significance of Gene Localization and Mapping

The localization and mapping of genes and genomic segments are of great significance in various fields, providing valuable insights into genetic organization, function, and evolution. These processes are instrumental in advancing research and development in genetics, genomics, medicine, agriculture, and evolutionary biology.

Medical Genetics and Genomics

Gene localization and mapping are essential for identifying genes associated with diseases and traits. By pinpointing the locations of disease-associated genes, researchers can elucidate the genetic basis of conditions, develop diagnostic tools, and inform therapeutic strategies.

Benefits in Medical Genetics:

» **Identification of Disease Genes:** Gene mapping is used to identify genes associated with inherited diseases and complex disorders. By analyzing linkage or association patterns, researchers can pinpoint the loci responsible for diseases such as cystic fibrosis, Huntington's disease, and Alzheimer's disease.

» **Genetic Counseling and Risk Assessment:** Gene localization and mapping provide valuable information for genetic counseling and risk assessment. By identifying the locations of disease-associated genes, clinicians can assess genetic risk, inform reproductive decision-making, and guide preventive measures.

» **Development of Genetic Tests:** Mapping the locations of disease-associated genes facilitates the development of genetic tests for diagnostic and prognostic purposes. These tests enable the detection of genetic variants or mutations associated with diseases, aiding in early diagnosis and personalized treatment planning.

» **Pharmacogenomics and Personalized Medicine:** Gene mapping contributes to pharmacogenomics by identifying genetic variants that influence drug response and metabolism. By localizing genes involved in drug pathways, researchers can develop personalized medicine approaches that tailor treatments based on an individual's genetic profile, optimizing therapeutic outcomes.

Plant Breeding and Agricultural Biotechnology

Gene localization and mapping are used to identify and manipulate genes associated with desirable traits in plants, such as yield, disease resistance, and stress tolerance. These processes are essential for developing improved crop varieties and enhancing agricultural productivity.

Benefits in Plant Breeding:

» **Marker-Assisted Selection:** Gene mapping facilitates marker-assisted selection by identifying genetic markers linked to desirable traits. These markers are used to select plants with favorable characteristics, accelerating the breeding process and enhancing the efficiency of crop improvement programs.

» **Development of Transgenic Crops:** Localization and mapping of genes associated with beneficial traits enable the development of transgenic crops with enhanced properties. By identifying and introducing specific genes, researchers can create crops with improved resistance to pests, diseases, and environmental stresses, increasing agricultural productivity and sustainability.

» **Quantitative Trait Loci (QTL) Mapping:** Gene mapping is used to identify QTLs, regions of the genome associated with complex traits, such as yield or drought tolerance. By localizing these loci, researchers can gain insights into the genetic architecture of traits and inform breeding strategies for trait improvement.

» **Genomic Selection:** Gene mapping supports genomic selection, a breeding approach that uses genomic information to predict the genetic value of individuals. By mapping genes associated with traits, researchers can develop predictive models that inform selection decisions, enhancing the accuracy and efficiency of breeding programs.

Animal Breeding and Livestock Improvement

Gene localization and mapping are used to identify genes associated with desirable traits in livestock, such as milk production, growth rate, and disease resistance. These processes are essential for developing improved livestock breeds and enhancing animal productivity.

Benefits in Animal Breeding:

» **Identification of Trait-Associated Genes:** Gene mapping is used to identify genes associated with economically important traits in livestock, such as meat quality, reproductive performance, and disease resistance. By pinpointing the locations of these genes, researchers can enhance breeding programs and improve livestock production.

- **Genomic Selection and Marker-Assisted Breeding:** Gene mapping supports genomic selection and marker-assisted breeding by identifying genetic markers linked to desirable traits. These markers are used to select animals with favorable characteristics, accelerating genetic improvement and increasing breeding efficiency.
- **Development of Disease-Resistant Breeds:** Localization and mapping of genes associated with disease resistance enable the development of disease-resistant livestock breeds. By identifying and incorporating specific genes, researchers can enhance the resilience of animals to infectious diseases, reducing morbidity and mortality.
- **Enhancement of Reproductive Traits:** Gene mapping is used to identify genes associated with reproductive traits, such as fertility and litter size. By localizing these genes, researchers can develop strategies to improve reproductive performance and enhance the productivity of livestock populations.

Evolutionary Biology

Gene localization and mapping provide valuable insights into evolutionary relationships and processes by identifying conserved genetic elements and syntenic regions. This information is used to study speciation, adaptation, and the evolution of complex traits.

Benefits in Evolutionary Biology:

- **Evolutionary Studies:** Gene mapping provides valuable insights into evolutionary relationships and processes by identifying conserved genetic elements and syntenic regions. This information is used to study speciation, adaptation, and the evolution of complex traits.
- **Functional Genomics:** Gene mapping supports functional genomics by identifying homologous genes and regulatory elements across species. This information is used to infer gene functions and explore the genetic basis of phenotypic diversity.
- **Comparative Genomics:** Gene mapping supports comparative genomics by providing a reference framework for comparing genomic organization across species. This information is valuable for studying evolutionary relationships and identifying conserved genetic elements.

Chapter - 5

Pre-breeding and Applications of Cytogenetical Methods for Crop Improvement

Introduction to Pre-breeding in Crop Improvement

Pre-breeding, often referred to as "germplasm enhancement," is an essential precursor to traditional breeding activities that aim to enhance the genetic base of cultivated crops. It involves the identification and incorporation of desirable traits from wild relatives, landraces, or exotic germplasm into cultivated varieties, creating improved breeding lines for further development. Pre-breeding serves as a crucial foundation for the continuous improvement of crops, allowing breeders to respond to the evolving challenges of agriculture, such as pest resistance, climate change, and the need for increased food production.

The importance of pre-breeding has grown significantly in recent years, as agricultural scientists recognize the limitations of traditional breeding methods that rely on existing commercial varieties. These limitations include reduced genetic diversity, making crops more susceptible to diseases and environmental stresses. By tapping into the rich genetic pool of wild and underutilized species, pre-breeding provides the raw material necessary for breeding programs to develop resilient and high-yielding crop varieties.

The global challenge of food security, exacerbated by a rapidly growing population and changing climate, underscores the need for innovative approaches to crop improvement. Pre-breeding plays a pivotal role in addressing these challenges by expanding the genetic base of crops and allowing for the development of varieties that can thrive under diverse and changing conditions. As such, pre-breeding is a scientific endeavor and a strategic approach to ensuring food security and sustainable agriculture worldwide.

Definition and Importance

Pre-breeding is a strategic approach in plant breeding that involves the use of genetic resources from wild relatives, landraces, and other non-cultivated species to introduce desirable traits into cultivated crops. It serves as an intermediary step between the conservation of genetic resources and the development of commercial varieties, providing breeders with enhanced genetic material that can be further refined through traditional breeding techniques. The core objective of pre-breeding is to enrich

DOI: 10.1201/9781003717775-5

the genetic pool available for breeding, addressing the limitations of modern cultivars that often suffer from narrow genetic bases due to extensive selection and breeding for specific traits. The process of pre-breeding begins with the identification of valuable traits in non-cultivated germplasm, such as disease resistance, drought tolerance, or nutritional quality. This is followed by the incorporation of these traits into a compatible genetic background through methods such as hybridization, backcrossing, and selection. The resulting pre-bred lines serve as a bridge, transferring genetic diversity from wild or exotic sources into the genetic makeup of cultivated crops. This process often involves complex genetic analyses, phenotypic evaluations, and rigorous selection procedures to ensure that the desired traits are effectively integrated without compromising the agronomic performance of the resulting lines.

The importance of pre-breeding lies in its ability to overcome the genetic bottlenecks present in modern agriculture. Over the past few decades, the focus on developing high-yielding varieties has led to a reduction in genetic diversity, making crops more vulnerable to pests, diseases, and environmental changes. Pre-breeding offers a solution by reintroducing genetic variability into breeding programs, thereby enhancing the adaptability and resilience of crops. This increased genetic diversity is crucial for developing varieties that can withstand the challenges posed by climate change, such as extreme weather conditions, water scarcity, and shifting pest and disease pressures.

Moreover, pre-breeding is essential for tapping into the vast genetic potential of wild relatives and landraces, which often possess traits that have been lost in modern cultivars due to selective breeding. For instance, wild relatives of wheat, rice, and maize have shown resistance to diseases and pests that are prevalent in cultivated varieties. By incorporating these traits into breeding lines, pre-breeding enables the development of crops that are not only high-yielding but also resilient to biotic and abiotic stresses. Pre-breeding is also crucial for addressing the nutritional needs of a growing global population. By introducing traits related to nutritional quality, such as increased protein content, enhanced micronutrient levels, or improved digestibility, pre-breeding contributes to the development of crops that can provide more balanced and nutritious diets. This aspect of pre-breeding aligns with the goals of improving food security and nutrition, particularly in regions where malnutrition and food scarcity are pressing concerns.

Role in enhancing genetic diversity and addressing global food security challenges.

The role of pre-breeding in enhancing genetic diversity and addressing global food security challenges is multifaceted, encompassing various aspects of crop improvement that contribute to agricultural sustainability and resilience. One of the primary benefits of pre-breeding is its ability to broaden the genetic base of crops, thereby enhancing their adaptability to changing environmental conditions and resistance to biotic stresses.

Enhancing Genetic Diversity

Genetic diversity is a critical factor in the success of breeding programs, as it provides the raw material necessary for developing improved crop varieties. However, modern agricultural practices have often led to a narrowing of the genetic base of many crops, resulting in increased vulnerability to pests, diseases, and environmental stresses. Pre-breeding addresses this issue by introducing novel genetic variation from wild relatives and underutilized germplasm into breeding lines. This process involves the careful selection and incorporation of desirable traits, such as disease resistance, drought tolerance, and nutrient efficiency, from genetically diverse sources. For example, wild relatives of staple crops like wheat, rice, and maize possess a wealth of genetic diversity that has been lost in cultivated varieties due to intensive breeding for specific traits. Pre-breeding enables the transfer of these valuable traits into modern breeding lines, thereby enhancing the genetic diversity of cultivated crops. This increased diversity not only improves the adaptability of crops to diverse environments but also reduces their susceptibility to emerging pests and diseases. Moreover, pre-breeding facilitates the exploration of new genetic combinations that may not be present in existing breeding populations. By introducing novel alleles and recombining them with desirable traits, pre-breeding expands the genetic potential of crops, allowing breeders to develop varieties with improved performance and resilience. This increased genetic diversity is essential for meeting the demands of a rapidly changing agricultural landscape, where crops must adapt to fluctuating temperatures, unpredictable rainfall patterns, and evolving pest and disease pressures.

Addressing Global Food Security Challenges

Pre-breeding plays a pivotal role in addressing the global food security challenges posed by a growing population, climate change, and limited natural resources. As the world population continues to rise, the demand for food is expected to increase significantly, necessitating the development of crop varieties that can produce higher yields with fewer inputs. Pre-breeding contributes to this goal by providing breeders with the genetic tools needed to enhance crop productivity and efficiency. One of the key ways pre-breeding addresses food security is by developing crop varieties that are resilient to climate change. As global temperatures rise and weather patterns become more erratic, crops face increased stress from heat, drought, and flooding. Pre-breeding enables the identification and incorporation of traits that enhance a crop's tolerance to these abiotic stresses, ensuring stable yields under adverse conditions. For instance, pre-breeding efforts have led to the development of drought-tolerant maize varieties that can thrive in water-scarce environments, thereby supporting food production in regions vulnerable to climate change. Additionally, pre-breeding addresses food security by enhancing the nutritional quality of crops. By incorporating traits related to nutrient content, pre-breeding contributes to the development of biofortified varieties that provide essential vitamins and minerals to populations suffering from malnutrition. For example, pre-breeding efforts have resulted in rice varieties with higher iron and zinc content, addressing micronutrient deficiencies in regions where rice is a staple food. These biofortified crops play a crucial role in improving public health and reducing the burden of malnutrition-related diseases. Furthermore, pre-breeding supports sustainable agriculture by developing crop varieties that require fewer chemical inputs, such as pesticides and fertilizers. By incorporating traits for pest resistance and nutrient efficiency, pre-breeding reduces the reliance on chemical interventions, minimizing environmental impact and promoting ecological balance. This aligns with the principles of sustainable agriculture, which aim to optimize resource use while preserving biodiversity and ecosystem services.

Historical Background

The historical development of pre-breeding techniques in agriculture is a testament to the continuous quest for innovation in crop improvement. As the demands on agriculture have grown, pre-breeding has evolved as a critical scientific endeavor aimed at enhancing genetic diversity and addressing the challenges posed by biotic and abiotic stresses. This section delves into the historical context of pre-breeding, exploring its origins, key milestones, and how it has evolved to become a cornerstone of modern plant breeding.

Origins and Early Recognition of Genetic Diversity

The roots of pre-breeding can be traced back to the early 20th century when the limitations of traditional breeding approaches began to emerge. At that time, plant breeders primarily relied on selecting superior individuals from cultivated varieties, often overlooking the vast genetic potential present in wild relatives and landraces. This approach led to a narrowing of the genetic base, resulting in increased vulnerability to pests, diseases, and environmental changes. One of the pioneering figures in recognizing the importance of genetic diversity was Nikolai Vavilov, a Russian botanist and geneticist. In the 1920s and 1930s, Vavilov embarked on extensive expeditions to collect plant genetic resources from around the world, identifying centers of origin for major crops. His work laid the foundation for the development of genebanks, which became repositories for preserving the genetic diversity of crops and their wild relatives. Vavilov's vision emphasized the need to conserve and utilize genetic diversity as a means to improve crop resilience and productivity.

The Emergence of Pre-breeding Concepts

The concept of pre-breeding began to take shape as plant scientists recognized the limitations of existing breeding practices and the potential benefits of introducing novel genetic material into cultivated varieties. The early 20th century saw the establishment of agricultural research institutions and botanical gardens that focused on collecting and conserving plant genetic resources. These efforts were driven by the understanding that wild relatives and landraces possessed traits that could enhance the resilience and performance of crops. During this period, the concept of introgression, which involves the transfer of genetic material from wild relatives to cultivated varieties, gained prominence. Introgression offered a way to incorporate beneficial traits such as disease resistance, drought tolerance, and improved nutritional quality into breeding lines. However, the process was often hindered by the genetic incompatibility between distantly related species and the complexity of hybridization.

Advancements in Genetics and Cytogenetics

The mid-20th century marked a turning point in pre-breeding with the advent of modern genetics and cytogenetics. The discovery of the structure of DNA by James Watson and Francis Crick in 1953 provided a molecular basis for understanding genetic inheritance and variation. This breakthrough paved the way for the development of genetic tools and techniques that facilitated pre-breeding efforts.

Chromosome manipulation techniques, such as polyploidy induction and aneuploidy, became instrumental in pre-breeding programs. Induced polyploidy, achieved through the use of colchicine, allowed breeders to create polyploid plants with increased genetic variability and improved agronomic traits. Aneuploidy, the presence of extra or missing chromosomes, provided insights into the genetic control of specific traits and facilitated the transfer of desired genes from wild relatives. The field of cytogenetics also advanced significantly during this time, enabling the visualization and analysis of chromosomes. Techniques such as chromosome banding and karyotyping allowed breeders to identify chromosomal rearrangements and monitor the introgression of genetic material. These cytogenetic tools provided valuable information on the genetic architecture of crops and their wild relatives, guiding pre-breeding efforts.

The Green Revolution and Hybrid Varieties

The Green Revolution of the 1960s and 1970s brought about a paradigm shift in agriculture, driven by the development of high-yielding hybrid varieties and the widespread adoption of modern agricultural practices. Pre-breeding played a pivotal role in this transformation by providing the genetic resources needed to develop hybrid varieties with enhanced performance. Norman Borlaug, often referred to as the father of the Green Revolution, demonstrated the power of pre-breeding in his work on wheat improvement. By introducing dwarfing genes and disease-resistance traits into wheat, Borlaug developed semi-dwarf varieties that significantly increased wheat yields. These varieties, coupled with improved agronomic practices, contributed to food security in many developing countries and showcased the potential of pre-breeding to address global agricultural challenges.

Integration of Molecular Genetics and Biotechnology

The latter half of the 20th century witnessed significant advancements in molecular genetics and biotechnology, which revolutionized pre-breeding by providing precise tools for genetic analysis and manipulation. The development of molecular markers, such as RFLPs (restriction fragment length polymorphisms), SSRs (simple sequence repeats), and SNPs (single nucleotide polymorphisms), enabled breeders to identify and track specific genes associated with desirable traits.

Marker-assisted selection (MAS) became a powerful tool in pre-breeding, allowing breeders to select individuals with favorable alleles based on their genetic profiles. This approach accelerated the introgression of traits from wild relatives and reduced the time required to develop improved breeding lines. The integration of molecular markers with traditional breeding methods enhanced the efficiency and precision of pre-breeding programs. The advent of genetic engineering and biotechnology further expanded the scope of pre-breeding by enabling the direct transfer of genes across species barriers. Transgenic approaches allowed breeders to introduce genes conferring resistance to pests, diseases, and herbicides into crops, enhancing their resilience and productivity. While transgenic methods have faced regulatory and public acceptance challenges, they have demonstrated the potential to address complex agricultural issues.

Emergence of Genomic Technologies and Precision Breeding

The 21st century has witnessed a rapid evolution in pre-breeding techniques driven by the integration of genomic technologies and precision breeding approaches. Next-generation sequencing (NGS) has revolutionized the way breeders access and analyze genetic information, providing insights into the entire genome of crops and their wild relatives. Genome-wide association studies (GWAS) have facilitated the identification of genomic regions associated with complex traits, enabling breeders to target specific loci for improvement.

Genomic selection (GS) has emerged as a powerful tool in pre-breeding, allowing breeders to predict the performance of breeding lines based on their genomic information. GS leverages large-scale genomic data to identify individuals with favorable genetic profiles, enabling more accurate and efficient selection. This approach is particularly valuable for traits influenced by multiple genes and environmental interactions, where traditional selection methods may be less effective. The development of gene-editing technologies, such as CRISPR-Cas9, has further expanded the possibilities for pre-breeding by enabling precise modification of specific genes. Gene editing allows breeders to introduce or enhance desirable traits with unprecedented accuracy, overcoming the limitations of conventional breeding methods. These technologies have the potential to accelerate the development of crop varieties that are resilient to climate change, have improved nutritional quality, and require fewer inputs.

Integration of Traditional Knowledge and Participatory Approaches

Throughout its history, pre-breeding has benefited from the integration of traditional knowledge and participatory approaches. Indigenous communities and local farmers have long practiced informal pre-breeding by selecting and cultivating landraces with traits suited to their specific environments. Recognizing the value of this knowledge, modern pre-breeding programs have increasingly involved farmers and local communities in the selection and evaluation of breeding lines. Participatory plant breeding (PPB) has emerged as a collaborative approach that combines scientific expertise with farmers' insights, leading to the development of varieties that meet the needs and preferences of local communities. PPB emphasizes the importance of involving stakeholders in the breeding process, ensuring that the resulting varieties are well-adapted to local conditions and socio-economic contexts. This approach has been particularly successful in developing countries, where smallholder farmers play a crucial role in agricultural production.

Evolution Over Time

The evolution of pre-breeding techniques over time reflects the dynamic nature of agriculture and the need for continuous innovation. As global challenges such as climate change, population growth, and resource scarcity intensify, pre-breeding remains a critical component of sustainable crop improvement. The historical development of pre-breeding has paved the way for a more resilient and adaptable agricultural system, capable of meeting the demands of the 21st century and beyond.

Early Challenges and Innovations

In the early stages of pre-breeding, the primary challenges faced by breeders were the complexity of genetic crosses and the limited understanding of genetic inheritance. The introduction of chromosome manipulation techniques, such as colchicine-induced polyploidy, allowed breeders to overcome some of these challenges by creating new genetic combinations. The use of cytogenetic techniques to visualize chromosomal changes further advanced pre-breeding efforts, enabling the identification of chromosomal rearrangements associated with desired traits.

Collaboration and International Efforts

The recognition of the global nature of agricultural challenges led to increased collaboration and international efforts in pre-breeding. Organizations such as the Consultative Group on International Agricultural Research (CGIAR) and its network of research centers have played a pivotal role in coordinating pre-breeding programs and facilitating the exchange of genetic resources. These collaborative efforts have resulted in the development of improved crop varieties that address regional and global challenges, such as drought-tolerant maize for Sub-Saharan Africa and disease-resistant rice for Southeast Asia.

Enhancing Genetic Diversity

Enhancing genetic diversity is a fundamental goal of pre-breeding, aiming to address the genetic limitations found in many modern crop varieties. The narrowing of the genetic base in cultivated crops, due to extensive selection for specific traits, has resulted in increased vulnerability to pests, diseases, and environmental changes. Pre-breeding seeks to counteract this by introducing new alleles and genetic combinations from wild relatives, landraces, and exotic germplasm, thereby enriching the genetic pool available for breeding programs.

Importance of Broadening the Genetic Base of Crops

1. Adaptability to Changing Environments

The broadening of the genetic base is essential for improving the adaptability of crops to changing environmental conditions. As climate change continues to impact global agriculture, crops must be equipped to endure extreme weather events such as droughts, floods, and heat waves. Pre-breeding plays a vital role in this by introducing traits from wild relatives and landraces that enhance a crop's tolerance to these abiotic stresses. For example, wild relatives of rice possess genes that confer drought tolerance, which can be introgressed into cultivated varieties to improve their resilience in water-scarce environments. Wild relatives often thrive in diverse and challenging environments, making them valuable sources of genes for abiotic stress tolerance. By tapping into this genetic diversity, pre-breeding helps create crop varieties that can maintain productivity and yield stability in the face of adverse conditions. This adaptability is crucial for ensuring food security in regions vulnerable to climate change, where traditional crops may struggle to survive.

2. Increased Resilience to Pests and Diseases

A narrow genetic base can make crops more susceptible to pests and diseases, as uniform populations are more easily targeted by pathogens. Pre-breeding addresses this vulnerability by introducing genetic variability that enhances disease resistance. For instance, the wild relatives of wheat have been shown to possess resistance to rust diseases, which can be transferred to commercial varieties through pre-breeding efforts. This genetic diversity acts as a buffer against the emergence of new or virulent strains of pests and diseases. The introduction of resistance genes from wild relatives not only improves crop resilience but also reduces the reliance on chemical control measures such as pesticides and fungicides. This shift towards more sustainable agricultural practices benefits the environment and reduces the risk of developing pesticide-resistant pest populations. Additionally, diverse genetic backgrounds provide a broader range of defense mechanisms, enabling crops to resist multiple pests and diseases simultaneously.

3. Potential for Yield Improvement

Genetic diversity is essential for yield improvement, as it provides the raw material for selecting and developing high-yielding varieties. By broadening the genetic base, pre-breeding introduces novel alleles that can contribute to traits such as increased biomass, improved photosynthetic efficiency, and enhanced nutrient use efficiency. These traits collectively contribute to higher yields and better resource utilization, supporting the goal of increasing agricultural productivity. Moreover, diverse genetic backgrounds enable breeders to explore new combinations of alleles that may result in synergistic effects on yield and performance. This exploration can lead to the discovery of unique traits and gene interactions that were previously unavailable in cultivated varieties. The resulting high-yielding varieties are better equipped to meet the growing demand for food, particularly in regions with limited arable land.

4. Preservation of Genetic Resources

Pre-breeding plays a vital role in the preservation and utilization of genetic resources. Many wild relatives and landraces are at risk of extinction due to habitat loss, changes in land use, and shifts in agricultural practices. By incorporating these genetic resources into breeding programs, pre-breeding ensures their conservation and sustainable use. This preservation of genetic diversity is critical for maintaining the long-term viability of crop improvement efforts and for safeguarding valuable genetic traits for future generations. The conservation of genetic resources also provides a reservoir of genes that can be accessed in response to emerging challenges, such as new pests, diseases, or climatic conditions. By maintaining a diverse genetic base, breeders can adapt quickly to unforeseen threats and opportunities, ensuring the continued success and resilience of agricultural systems.

5. Facilitating Breeding for Complex Traits

Complex traits, such as yield, drought tolerance, and disease resistance, are often controlled by multiple genes and influenced by environmental interactions. Pre-breeding facilitates the breeding of complex traits by introducing a wider range of genetic variation, allowing breeders to explore new genetic combinations and interactions. This expanded genetic base provides more opportunities for selecting individuals with superior performance and adaptability. Breeding for complex traits often requires a holistic approach that considers multiple genes and environmental factors. Pre-breeding helps lay the groundwork for such an approach by enriching the genetic diversity available for selection. This diversity enables breeders to develop varieties that exhibit stable performance across diverse environments and management practices, increasing their utility and effectiveness in real-world agricultural settings.

6. Supporting Sustainable Agriculture

Broadening the genetic base of crops through pre-breeding contributes to sustainable agriculture by enhancing the resilience and stability of cropping systems. Genetically diverse populations are better equipped to withstand environmental fluctuations and biological threats, reducing the need for external inputs such as pesticides and fertilizers. This aligns with the principles of sustainable agriculture, which emphasize the importance of resource conservation, ecological balance, and the reduction of negative environmental impacts. By fostering genetic diversity, pre-breeding supports the development of crop varieties that are resilient, high-yielding, and capable of thriving under sustainable management practices. These varieties contribute to the long-term sustainability of agricultural systems, promoting food security and environmental stewardship in the face of growing challenges.

Incorporating Resistance Traits

Incorporating resistance traits is another key objective of pre-breeding, aimed at developing crop varieties that can withstand biotic stresses such as pests and diseases. As agriculture intensifies and ecosystems change, crops face increasing pressure from pathogens and pests, necessitating the development of resistant varieties. Pre-breeding plays a crucial role in identifying and transferring resistance traits from wild relatives and exotic germplasm into cultivated crops, enhancing their resilience and reducing reliance on chemical inputs.

1. Biotic Stress Resistance

Biotic stresses, including diseases caused by bacteria, fungi, viruses, and nematodes, as well as insect and pest infestations, pose significant threats to crop productivity. Traditional breeding methods have often focused on selecting resistant individuals from existing populations, but this approach can be limited by the narrow genetic base of cultivated varieties. Pre-breeding addresses this limitation by exploring the genetic diversity found in wild relatives, which often possess unique resistance genes that have been lost in modern crops.

Disease Resistance

Pre-breeding efforts have been instrumental in introducing disease resistance traits into crops. For example, the wild relatives of potatoes have been used to introduce resistance to late blight, a devastating disease caused by the oomycete *Phytophthora infestans*. Similarly, pre-breeding has led to the development of rice varieties with resistance to blast disease, utilizing resistance genes from wild rice species. These resistant varieties reduce the need for fungicides and contribute to more sustainable crop production systems. The introduction of disease resistance traits also enhances the durability of resistance, as diverse genetic backgrounds provide a broader range of defense mechanisms. This diversity reduces the likelihood of resistance breakdown due to the evolution of new pathogen strains, ensuring long-term protection against diseases.

Insect and Pest Resistance

Insect pests are a major challenge in agriculture, causing significant yield losses and requiring extensive use of chemical insecticides. Pre-breeding provides an opportunity to incorporate insect resistance traits from wild relatives, reducing the reliance on chemical control measures. For instance, the wild relatives of maize have been used to introduce resistance to the corn rootworm, a major pest in maize production. This genetic resistance provides an environmentally friendly alternative to chemical control and contributes to integrated pest management strategies. The introduction of pest resistance traits also reduces the environmental impact of agriculture, as it minimizes the use of chemical insecticides that can harm non-target organisms and disrupt ecosystems. By promoting biological control and natural pest resistance, pre-breeding supports the development of sustainable and ecologically sound cropping systems.

2. Abiotic Stress Resistance

Abiotic stresses, such as drought, salinity, and extreme temperatures, are becoming increasingly prevalent due to climate change. Pre-breeding plays a vital role in incorporating traits that enhance a crop's ability to withstand these stresses, ensuring stable yields under adverse conditions.

Drought Tolerance

Drought is one of the most significant abiotic stresses affecting agriculture, particularly in arid and semi-arid regions. Pre-breeding efforts have focused on identifying drought tolerance traits in wild relatives and landraces, allowing for the development of varieties that can maintain productivity under water-limited conditions. For example, pre-breeding has contributed to the development of drought-tolerant wheat and maize varieties, which are essential for food security in regions prone to water scarcity. The introduction of drought tolerance traits often involves the transfer of genes that enhance water use efficiency, root architecture, and stomatal regulation. These traits collectively improve a crop's ability to survive and produce under drought conditions, reducing the risk of yield loss and crop failure.

Salinity Tolerance

Soil salinity is a major constraint to agricultural productivity, particularly in irrigated areas. Pre-breeding efforts have identified salinity tolerance traits in wild relatives of crops such as rice and barley, enabling the development of varieties that can thrive in saline environments. These salt-tolerant varieties are crucial for maintaining productivity in areas affected by soil salinization, which is expected to worsen with climate change and unsustainable irrigation practices. The incorporation of salinity tolerance traits often involves the transfer of genes that regulate ion transport, osmotic adjustment, and salt exclusion. These mechanisms enable crops to maintain cellular homeostasis and metabolic function under saline conditions, supporting growth and yield in challenging environments.

3. Nutrient Use Efficiency

Pre-breeding also focuses on incorporating traits that enhance nutrient use efficiency, reducing the need for chemical fertilizers and improving environmental sustainability. Wild relatives often possess traits that enable more efficient uptake and utilization of nutrients, which can be introgressed into cultivated varieties through pre-breeding.

Nitrogen Use Efficiency

Nitrogen is a critical nutrient for plant growth, but its excessive use can lead to environmental pollution and contribute to greenhouse gas emissions. Pre-breeding efforts have identified traits related to nitrogen use efficiency in wild relatives of crops such as wheat and rice. By incorporating these traits, breeders can develop varieties that require less nitrogen fertilizer, reducing the environmental impact of agriculture and promoting sustainable nutrient management. The introduction of nitrogen use efficiency traits often involves the transfer of genes that enhance nitrogen uptake, assimilation, and remobilization. These traits improve a crop's ability to utilize nitrogen efficiently, reducing fertilizer inputs and minimizing environmental pollution.

Phosphorus Use Efficiency

Phosphorus is another essential nutrient that is often limited in soils, leading to the need for phosphorus fertilizers. Pre-breeding has focused on identifying traits that enhance phosphorus uptake and utilization in crops, reducing the reliance on phosphorus fertilizers. For example, wild relatives of soybean have been used to introduce traits related to phosphorus use efficiency, contributing to more sustainable crop production systems. The incorporation of phosphorus use efficiency traits often involves the transfer of genes that regulate root architecture, phosphate transport, and organic acid secretion. These mechanisms enable crops to access and utilize phosphorus more efficiently, reducing fertilizer inputs and improving soil health.

4. Nutritional Quality Improvement

In addition to resistance traits, pre-breeding also aims to enhance the nutritional quality of crops, addressing malnutrition and improving public health. By incorporating traits related to nutritional content, pre-breeding contributes to the development of biofortified varieties that provide essential vitamins and minerals.

Micronutrient Enhancement

Micronutrient deficiencies, such as iron, zinc, and vitamin A deficiencies, are prevalent in many developing countries, leading to health issues such as anemia and impaired immune function. Pre-breeding has played a crucial role in developing biofortified varieties with increased micronutrient content. For example, pre-breeding efforts have led to the development of iron- and zinc-rich rice and wheat varieties, addressing micronutrient deficiencies, and improving dietary quality. The introduction of micronutrient enhancement traits often involves the transfer of genes that regulate micronutrient uptake, transport, and accumulation. These traits increase the nutritional content of crops, contributing to improved health and well-being for consumers.

Protein Quality Improvement

Protein quality is an important aspect of nutritional quality, particularly in regions where plant-based diets are the primary source of protein. Pre-breeding has focused on improving the amino acid composition and protein content of crops such as legumes and cereals. By introducing traits related to protein quality from wild relatives, pre-breeding contributes to the development of varieties that provide balanced and nutritious diets.

The incorporation of protein quality improvement traits often involves the transfer of genes that regulate protein synthesis, amino acid composition, and digestibility. These traits enhance the nutritional value of crops, supporting dietary diversity and health.

Improving Yield and Quality

Improving yield and quality in crops is one of the most critical goals of pre-breeding programs, as it directly impacts food security, economic viability, and consumer satisfaction. Yield improvement focuses on maximizing the productivity of crops, while quality enhancement aims to improve the nutritional value and market appeal of agricultural products. Through pre-breeding, which involves the strategic use of genetic resources from wild relatives, landraces, and exotic germplasm, breeders can introduce new traits and alleles that boost both yield and quality, thereby addressing the diverse challenges faced by modern agriculture.

Enhancing Yield Potential

Enhancing yield potential involves increasing the amount of produce that can be obtained from a crop under optimal conditions. This goal is essential for meeting the growing global demand for food, especially in regions with limited arable land. Yield potential is influenced by a range of genetic and environmental factors, and pre-breeding provides the tools needed to exploit genetic diversity to achieve higher productivity.

1. Genetic Basis of Yield Potential

The genetic basis of yield potential is complex, involving multiple genes and their interactions with environmental factors. Pre-breeding programs aim to identify and incorporate beneficial alleles from wild relatives and landraces that can enhance yield. These efforts focus on key traits such as photosynthetic efficiency, resource allocation, and plant architecture, which collectively determine the productivity of a crop.

» **Photosynthetic Efficiency**

Photosynthesis is the process by which plants convert light energy into chemical energy, driving growth and biomass accumulation. Improving photosynthetic efficiency is a crucial strategy for enhancing yield potential, as it allows plants to convert sunlight into biomass more effectively. Wild relatives and landraces often possess genetic variations that can improve photosynthesis, such as enhanced chlorophyll content, optimized leaf structure, and efficient carbon fixation mechanisms. Pre-breeding efforts focus on identifying and introgressing these beneficial traits into cultivated varieties. For example, researchers have explored the potential of transferring C4 photosynthesis traits from wild relatives to C3 crops like rice, which could significantly increase photosynthetic efficiency and yield. Additionally, genetic variations that improve the efficiency of photosystem II, a key component of the photosynthetic machinery, have been targeted for introgression.

» **Resource Allocation and Plant Architecture**

Resource allocation refers to the distribution of resources such as nutrients, water, and carbohydrates within a plant. Optimizing resource allocation can lead to improved growth and yield by ensuring that resources are directed towards organs that contribute most to productivity. Plant architecture, including traits such as leaf area, root system development, and plant height, plays a critical role in determining resource allocation. Pre-breeding efforts focus on identifying and incorporating traits that optimize plant architecture and resource allocation. For example, traits such as increased leaf area can enhance light capture and photosynthesis, while robust root systems improve water and nutrient uptake. Optimized plant height can reduce lodging risk, improving yield stability.

2. Overcoming Yield Plateaus

Many modern crop varieties have reached yield plateaus, where further improvements are challenging to achieve using conventional breeding methods. This plateau is often due to the limited genetic diversity available within cultivated germplasm. Pre-breeding offers a way to break these plateaus by introducing novel genetic variations that can unlock new yield potential.

» **Heterosis (Hybrid Vigor)**

Heterosis, or hybrid vigor, refers to the phenomenon where hybrid offspring exhibit superior performance compared to their parents. This effect is often observed in hybrid varieties, which can show increased yield, vigor, and resilience. Pre-breeding plays a crucial role in creating hybrid varieties that capitalize on heterosis by introducing diverse genetic backgrounds. By incorporating genetic diversity from wild relatives and landraces, pre-breeding enhances the potential for heterosis in hybrid breeding programs. These hybrids often exhibit traits such as increased biomass, improved stress tolerance, and enhanced nutrient use efficiency, contributing to higher yields. For example, hybrid rice varieties developed through pre-breeding have shown significant yield advantages over conventional varieties.

» **Exploring Novel Genetic Combinations**

Pre-breeding enables breeders to explore new genetic combinations that may not be present in existing breeding populations. By introducing alleles from wild relatives, breeders can create novel gene interactions that result in improved yield traits. These new combinations can lead to synergistic effects, where the combined action of multiple alleles results in greater yield potential.

For instance, pre-breeding efforts in wheat have focused on introducing novel alleles for grain size and number, leading to increased yield potential. Similarly, genetic variations that enhance plant growth and development under stress conditions have been targeted for introgression into maize and soybean.

3. Enhancing Stress Tolerance for Yield Stability

Yield potential is often compromised by biotic and abiotic stresses, which can significantly reduce productivity. Pre-breeding contributes to yield stability by incorporating stress tolerance traits that enable crops to maintain high yields under challenging conditions.

» **Abiotic Stress Tolerance**

Abiotic stresses, such as drought, salinity, and extreme temperatures, are becoming increasingly prevalent due to climate change. These stresses can severely impact yield potential, necessitating the development of crop varieties that can withstand adverse conditions. Pre-breeding plays a vital role in incorporating traits that enhance abiotic stress tolerance. Drought tolerance, for example, is a critical trait for maintaining yield stability in water-limited environments. Pre-breeding efforts have focused on identifying and introgressing genes that improve drought tolerance from wild relatives. Traits such as deeper root systems, improved water use efficiency, and osmotic adjustment have been targeted for introgression into cultivated varieties. Similarly, salinity tolerance is essential for crops grown in saline soils or irrigated with saline water. Pre-breeding efforts have identified salinity tolerance traits in wild relatives of crops such as rice and barley, enabling the development of varieties that can thrive in saline environments. These salt-tolerant varieties are crucial for maintaining productivity in areas affected by soil salinization.

» **Biotic Stress Resistance**

Pests and diseases pose significant threats to yield potential, leading to yield losses and increased reliance on chemical control measures. Pre-breeding helps develop varieties that can withstand these biotic challenges by incorporating resistance traits from wild relatives.

Disease resistance, for example, is a key focus of pre-breeding efforts. Wild relatives often possess resistance genes that have been lost in cultivated varieties due to selective breeding. By introgressing these genes, pre-breeding enables the development of varieties that are resistant to diseases such as rust, blight, and mildew. These resistant varieties not only maintain high yields but also reduce the need for fungicides, promoting sustainable agricultural practices. Similarly, insect resistance is a critical trait for reducing yield losses caused by pests. Pre-breeding efforts have focused on introducing genes that confer resistance to major pests, such as the corn rootworm in maize and the bollworm in cotton. These resistant varieties reduce the reliance on chemical insecticides and support integrated pest management strategies.

Improving Crop Quality Traits

In addition to yield, crop quality is a critical factor that determines the market value and nutritional content of agricultural produce. Quality traits include nutritional composition, taste, texture, and appearance, all of which are important for consumer acceptance and marketability.

1. Nutritional Quality Enhancement

Improving the nutritional quality of crops is essential for addressing malnutrition and ensuring food security. Pre-breeding efforts focus on enhancing the content of essential nutrients such as proteins, vitamins, and minerals, thereby contributing to the development of biofortified varieties that provide enhanced nutritional benefits.

» **Biofortification**

Biofortification is the process of increasing the nutritional value of crops through breeding. It involves the incorporation of genes that enhance the content of micronutrients like iron, zinc, and vitamin A, which are essential for human health. Pre-breeding plays a pivotal role in biofortification by identifying and introgressing genes from wild relatives and landraces that contribute to improved nutritional content. For example, biofortified rice varieties have been developed with higher iron and zinc levels, addressing micronutrient deficiencies in regions where rice is a staple food. Similarly, pre-breeding efforts in maize have focused on increasing the content of provitamin A carotenoids, resulting in varieties that can help combat vitamin A deficiency. The success of biofortification efforts hinges on the genetic diversity available for breeding. Wild relatives and landraces often possess genetic variations that enhance the uptake, transport, and accumulation of micronutrients in plants. By tapping into this diversity, pre-breeding provides the genetic resources needed to develop nutritionally enriched crop varieties.

» **Protein Quality and Content**

Improving protein quality and content is particularly important for crops that serve as major protein sources, such as legumes and cereals. Protein quality is determined by the amino acid composition of the crop, with a focus on essential amino acids that are crucial for human health. Pre-breeding efforts aim to enhance protein quality and content by leveraging the genetic diversity found in wild relatives. For example, wild relatives of maize have been used to introduce high-lysine genes, resulting in varieties with improved protein quality. Similarly, pre-breeding efforts in legumes such as chickpeas and lentils have focused on increasing protein content and enhancing the amino acid profile. By improving protein quality and content, pre-breeding contributes to the development of crop varieties that can provide balanced and nutritious diets. These varieties are particularly valuable in regions where plant-based diets are the primary source of protein, addressing protein malnutrition and supporting dietary diversity.

2. Organoleptic Traits

Organoleptic traits, which include taste, aroma, and texture, significantly influence consumer preferences and marketability. Pre-breeding contributes to improving these traits by introducing genetic diversity that enhances sensory qualities, making crops more appealing to consumers and increasing their market potential.

» **Flavor and Aroma**

Flavor and aroma are important determinants of consumer acceptance. They are influenced by a variety of compounds, including sugars, acids, and volatile compounds, which contribute to the overall sensory experience of a crop. Pre-breeding efforts focus on identifying genes that influence flavor profiles and incorporating them into cultivated varieties. For example, wild relatives of tomatoes have been used to introduce genes that enhance flavor and aroma, resulting in varieties that are more appealing to consumers. These efforts have focused on increasing the content of flavor-enhancing compounds such as sugars and organic acids, as well as optimizing the balance of flavor compounds to create a more desirable taste profile. Flavor and aroma are also important for crops such as fruits, herbs, and spices, where sensory qualities play a crucial role in determining market value. Pre-breeding efforts in these crops have focused on enhancing flavor profiles by introducing genetic variations that improve taste and aroma, resulting in varieties with superior sensory qualities.

» **Texture and Appearance**

Texture and appearance are also critical factors in determining the quality of crops. Traits such as fruit firmness, color, and size can impact consumer preferences and market value. Pre-breeding helps improve these traits by introducing genetic variations that enhance texture and appearance, making crops more attractive to consumers. For example, pre-breeding efforts in strawberries have focused on improving fruit firmness, which is important for shelf life and transportability. By introducing genes that enhance cell wall structure and composition, breeders have developed varieties with improved texture and extended shelf life. Similarly, pre-breeding efforts in carrots have focused on improving color and appearance by introducing genes that enhance carotenoid content. These efforts have resulted in varieties with vibrant colors that are more appealing to consumers and have higher market value.

3. Post-harvest Quality and Shelf Life

Improving post-harvest quality and shelf life is essential for reducing food waste and extending the availability of fresh produce. Pre-breeding efforts focus on incorporating traits that enhance storage properties and reduce post-harvest losses, ensuring that crops remain safe and marketable during storage and transport.

» **Shelf Life Extension**

Extending the shelf life of crops can significantly reduce post-harvest losses and improve food availability. Pre-breeding contributes to this goal by identifying genes that delay ripening and senescence, leading to varieties with longer shelf lives. For example, the introduction of delayed-ripening genes from wild relatives of fruits like bananas and tomatoes has resulted in varieties with extended freshness. The extension of shelf life is particularly important for perishable crops such as fruits and vegetables, which are prone to spoilage during storage and transport. By incorporating genetic variations that enhance shelf life, pre-breeding helps reduce food waste and improve the efficiency of supply chains.

» **Disease and Pest Resistance During Storage**

Post-harvest losses are often caused by diseases and pests that attack stored produce. Pre-breeding efforts focus on developing varieties with resistance to storage pests and diseases, ensuring that crops remain safe and marketable during storage and transport.

For example, pre-breeding efforts in cereals have focused on introducing genes that confer resistance to storage pests such as weevils and moths. By incorporating these resistance traits, breeders have developed varieties that are less susceptible to pest damage during storage. Similarly, pre-breeding efforts in fruits have focused on introducing genes that confer resistance to post-harvest diseases such as fungal infections. These resistant varieties reduce the need for chemical treatments during storage, promoting sustainable and safe storage practices.

3. Cytogenetical Methods in Pre-breeding

Cytogenetical methods in pre-breeding are pivotal in advancing crop improvement strategies. By focusing on the manipulation and understanding of chromosomes, these methods enable breeders to introduce desirable traits, enhance genetic diversity, and improve crop resilience and productivity. Cytogenetics combines classical genetics with modern molecular biology techniques, providing breeders with powerful tools to manipulate plant genomes at the chromosomal level.

Chromosome Manipulation Techniques

Chromosome manipulation techniques are essential tools in pre-breeding, allowing breeders to modify the structure and number of chromosomes in plants. These techniques enable the transfer of beneficial traits from wild relatives and other exotic germplasm into cultivated varieties, facilitating the development of improved crop lines. The primary methods of chromosome manipulation include chromosome engineering, polyploidy induction, and aneuploidy, each offering unique advantages and applications in crop improvement.

. Overview of Chromosome Engineering and Its Applications in Crop Improvement

Chromosome engineering is a sophisticated technique that involves the intentional modification of chromosome structures to introduce desirable traits into plant genomes. This method enables breeders to manipulate chromosomes to incorporate specific genes or entire chromosomal segments from one species into another, providing a means to overcome the limitations of traditional breeding methods. Chromosome engineering has become a cornerstone in pre-breeding efforts, offering precise control over genetic material and facilitating the development of crop varieties with enhanced traits.

Structural Chromosomal Changes

Chromosome engineering involves various structural changes, including translocations, inversions, and deletions, each playing a critical role in crop improvement. These structural modifications allow breeders to introduce new genetic material into a plant's genome, thereby facilitating the expression of beneficial traits and increasing genetic diversity.

» **Translocations**

Translocations are structural chromosomal changes involving the exchange of segments between non-homologous chromosomes. This process enables breeders to transfer specific genes associated with desirable traits, such as disease resistance or abiotic stress tolerance, from one chromosome to another. Translocations can occur naturally but are often induced through techniques such as irradiation or chemical treatments.

In crop improvement, translocations have been successfully used to introduce resistance genes from wild relatives into cultivated varieties. For instance, translocations involving wheat and its wild relative *Aegilops tauschii* have facilitated the transfer of genes conferring resistance to wheat rust diseases, resulting in wheat varieties with enhanced disease resistance. This approach has significantly reduced the reliance on chemical fungicides and contributed to more sustainable agricultural practices.

» **Inversions**

Inversions are structural changes that involve reversing the orientation of a chromosome segment. This modification can affect gene expression by altering the arrangement of genes along the chromosome. Inversions can also suppress recombination within the inverted region, preserving desirable gene combinations during meiosis. In crop improvement, inversions have been utilized to stabilize beneficial traits by preventing recombination between homologous chromosomes. For example, inversions in maize have been exploited to maintain advantageous trait combinations related to yield and stress tolerance. By preserving these trait combinations, breeders can ensure that the desired characteristics are consistently expressed in subsequent generations, contributing to increased productivity and crop resilience.

» **Deletions and Duplications**

Deletions involve the removal of a chromosome segment, while duplications involve the repetition of a segment. Both structural changes can impact gene dosage and expression, influencing traits such as growth, development, and stress response. In crop improvement, deletions have been strategically used to eliminate undesirable traits or create new gene combinations. For example, deletions in tomato have been employed to remove genes associated with susceptibility to specific diseases, resulting in disease-resistant varieties. Similarly, duplications have been utilized to increase the expression of beneficial genes, such as those involved in nutrient uptake or photosynthesis, thereby enhancing crop performance.

Alien Chromosome Introgression

Alien chromosome introgression is a technique that involves the transfer of entire chromosomes or chromosome segments from wild relatives into cultivated crops. This approach is used to introduce beneficial traits that are absent in the gene pool of cultivated varieties, thereby expanding genetic diversity and improving crop performance.

» **Chromosome Substitution Lines**

Chromosome substitution lines are developed by replacing one or more chromosomes of a cultivated variety with chromosomes from a wild relative. This substitution allows breeders to evaluate and select desirable traits associated with the alien chromosomes, facilitating the introgression of beneficial genes. In wheat breeding, chromosome substitution lines have been instrumental in introducing traits such as disease resistance, abiotic stress tolerance, and improved grain quality. For example, substitution lines involving chromosomes from wild wheat species have been used to enhance resistance to wheat stem rust, a major disease affecting wheat production globally. This approach has contributed to the development of wheat varieties with improved resilience and productivity, supporting food security and sustainable agriculture.

» **Chromosome Addition Lines**

Chromosome addition lines involve the introduction of additional chromosomes from a wild relative into a cultivated variety. This approach allows for the expression of specific traits controlled by genes located on the added chromosome, thereby enhancing crop performance. In barley breeding, chromosome addition lines have been used to introduce resistance to barley yellow dwarf virus, a significant pathogen affecting barley crops. By incorporating chromosomes from resistant wild relatives, breeders have developed varieties with improved resistance, reducing yield losses caused by the virus. This method has proven effective in enhancing crop resilience and ensuring stable yields, even under challenging conditions.

Polyploidy Induction

Polyploidy induction is a chromosome manipulation technique that involves increasing the number of chromosome sets in a plant. Polyploidy can result in increased genetic diversity and improved traits, making it a valuable tool in crop improvement.

» **Autopolyploidy**

Autopolyploidy occurs when a plant has multiple sets of chromosomes derived from a single species. This increase in chromosome number can enhance traits such as vigor, size, and stress tolerance. Autopolyploids often exhibit increased biomass and improved adaptability to environmental stresses, making them valuable in crop improvement efforts. In crops like alfalfa, potato, and strawberry, autopolyploidy has been used to enhance traits such as yield, fruit size, and stress tolerance. For example, autopolyploid potato varieties often exhibit larger tubers and increased resistance to environmental stresses, contributing to improved agricultural productivity. Similarly, autopolyploid strawberries demonstrate enhanced fruit size and quality, making them desirable for commercial production.

» **Allopolyploidy**

Allopolyploidy occurs when a plant has multiple sets of chromosomes derived from different species. This type of polyploidy combines the genetic traits of both parent species, resulting in new and improved varieties. Allopolyploids often exhibit increased genetic diversity and resilience to biotic and abiotic stresses, making them important for crop improvement. Allopolyploidy has been successfully utilized in the development of crops such as wheat, cotton, and canola. Allopolyploid wheat varieties, for example, combine the traits of several wheat species, resulting in increased yield, disease resistance, and adaptability to diverse environments. This genetic diversity is crucial for developing crop varieties that can thrive under various conditions and meet the demands of modern agriculture.

Aneuploidy and Its Applications

Aneuploidy involves changes in the number of individual chromosomes, resulting in plants with extra or missing chromosomes. While aneuploidy can cause developmental issues, it can also be exploited in pre-breeding to study gene dosage effects and identify beneficial traits.

» **Gene Dosage Effects**

Aneuploidy can alter the expression of genes by changing their dosage. By studying aneuploids, researchers can identify genes that have significant effects on traits such as growth, development, and stress response. This knowledge can inform breeding strategies and guide the selection of desirable traits. In tomato breeding, aneuploidy has been used to study the effects of gene dosage on fruit size and shape. By analyzing aneuploid lines with different chromosome complements, researchers have identified genes that control these traits, enabling the development of varieties with improved fruit characteristics. This understanding of gene dosage effects has been instrumental in optimizing crop performance and enhancing yield and quality.

» **Mapping and Identifying Traits**

Aneuploidy can also be used to map and identify genes associated with specific traits. By comparing the phenotypes of aneuploid lines with different chromosome complements, researchers can pinpoint the chromosomal location of genes controlling important traits. This mapping process facilitates the identification of candidate genes and supports the development of molecular markers for trait selection. In maize breeding, aneuploidy has been used to map genes associated with yield and stress tolerance. By analyzing aneuploid lines with different chromosome complements, breeders have identified chromosomal regions that contribute to these traits, facilitating the development of improved varieties. This approach has enhanced the efficiency of breeding programs and accelerated the development of high-yielding, resilient crops.

Applications in Crop Improvement

Chromosome manipulation techniques have numerous applications in crop improvement, enabling the development of varieties with enhanced traits and increased genetic diversity. By harnessing the power of chromosome engineering, breeders can overcome the limitations of traditional breeding methods and create crops that are more resilient, productive, and adaptable to changing environments.

» **Disease Resistance**

Chromosome manipulation techniques are used to introduce resistance genes from wild relatives into cultivated varieties, enhancing their ability to withstand diseases. For example, translocations and chromosome addition lines have been used to transfer resistance genes to crops such as wheat, rice, and barley, resulting in varieties that are more resilient to diseases. This genetic resistance reduces the reliance on chemical control measures and supports sustainable agricultural practices.

» **Stress Tolerance**

Chromosome manipulation techniques are also used to introduce traits that enhance abiotic stress tolerance, such as drought and salinity resistance. By incorporating genes from wild relatives, breeders can develop varieties that can thrive in challenging environments, ensuring stable yields under adverse conditions. This resilience is crucial for maintaining food security in the face of climate change and other environmental challenges.

» **Nutritional Quality**

Chromosome manipulation techniques can be used to improve the nutritional quality of crops by introducing genes that enhance the content of essential nutrients. For example, polyploidy induction has been used to increase the nutritional value of crops such as wheat and potatoes, resulting in varieties with higher protein and micronutrient content. This improvement in nutritional quality supports efforts to combat malnutrition and promote health and well-being.

» **Yield Enhancement**

Chromosome manipulation techniques contribute to yield enhancement by introducing traits that improve productivity. For example, allopolyploidy has been used to create new wheat varieties with increased yield potential, combining the genetic traits of several wheat species. These high-yielding varieties are essential for meeting the growing demand for food and ensuring the economic viability of agricultural production.

b. Polyploidy in Crop Improvement

Polyploidy, the condition of having more than two complete sets of chromosomes, plays a significant role in plant evolution and crop improvement. It is a powerful tool in pre-breeding and breeding programs, providing genetic diversity and enhancing crop traits. Polyploidy can occur naturally or be induced artificially, offering breeders the opportunity to create new varieties with superior characteristics. This section explores the role of polyploidy in enhancing crop traits and the mechanisms through which it contributes to the development of improved varieties.

The Role of Polyploidy in Enhancing Crop Traits and Creating New Varieties

Polyploidy contributes to crop improvement by providing several advantages, including increased genetic variation, enhanced stress tolerance, and improved yield and quality traits. These advantages arise from the duplication of genetic material, which allows for the expression of beneficial alleles, increased heterozygosity, and the potential for novel gene interactions.

1. Genetic Diversity and Novelty

Polyploidy leads to an increase in genetic diversity, which is crucial for crop improvement. The duplication of entire genomes results in multiple copies of each gene, providing a larger pool of genetic material for selection. This increased genetic diversity can lead to the emergence of new traits and the enhancement of existing ones, offering breeders more options for improving crop performance.

Polyploid plants often exhibit novel traits that are not present in their diploid counterparts. These traits can arise from gene dosage effects, where the expression of certain genes is amplified due to the presence of additional gene copies. For example, polyploidy can lead to increased size and vigor, as observed in crops like wheat and strawberries. The larger cell size and increased biomass associated with polyploidy can result in higher yields and improved stress tolerance.

The genetic novelty introduced by polyploidy can also lead to the development of new crop varieties with unique characteristics. For instance, polyploidy has been instrumental in the domestication and improvement of crops such as cotton, where polyploid varieties have contributed to enhanced fiber quality and disease resistance. By leveraging the genetic diversity and novelty provided by polyploidy, breeders can create crop varieties that meet specific needs and preferences, thereby expanding the genetic base and adaptability of agricultural systems.

2. Stress Tolerance and Adaptability

Polyploidy enhances stress tolerance in crops, allowing them to better withstand environmental challenges such as drought, salinity, and temperature extremes. The presence of additional gene copies in polyploids can provide a buffer against adverse conditions, enabling the expression of stress-responsive genes and pathways.

Polyploid plants often exhibit increased tolerance to abiotic stresses due to their enhanced ability to regulate gene expression and maintain cellular homeostasis. For example, polyploidy in wheat has been associated with improved drought tolerance, enabling the development of varieties that can maintain productivity under water-limited conditions. Similarly, polyploidy in barley has contributed to salinity tolerance, allowing the cultivation of barley in saline environments where diploid varieties struggle to thrive.

In addition to abiotic stress tolerance, polyploidy can also enhance resistance to biotic stresses, such as pests and diseases. The increased genetic diversity in polyploids can lead to the expression of multiple resistance genes, providing a broader range of defense mechanisms. This diversity reduces the likelihood of resistance breakdown and enhances the durability of resistance, contributing to more sustainable pest and disease management.

The adaptability of polyploid crops to diverse environments is a key advantage in crop improvement. Polyploidy allows plants to exploit different ecological niches and expand their geographic range, increasing their resilience and adaptability to changing climates. This adaptability is particularly important in the context of climate change, where crops must be able to cope with unpredictable environmental conditions and maintain stable yields.

3. Yield Improvement and Agronomic Traits

Polyploidy is associated with yield improvement and the enhancement of agronomic traits, making it a valuable tool in crop improvement programs. Polyploid plants often exhibit increased biomass, larger organ size, and improved resource use efficiency, contributing to higher yields and better performance.

The increase in cell size and biomass observed in polyploids can lead to higher productivity in crops such as cereals, fruits, and vegetables. For example, polyploid wheat varieties have been developed with increased grain size and yield potential, contributing to improved food security and agricultural productivity. Similarly, polyploid strawberries and grapes are known for their larger fruit size and enhanced flavor, making them more appealing to consumers and increasing their market value.

Polyploidy can also improve agronomic traits such as flowering time, plant architecture, and nutrient use efficiency. These traits are important for optimizing crop performance and ensuring efficient resource use. For instance, polyploidy in rice has been associated with improved nutrient uptake and utilization, leading to higher yields and reduced input requirements. This improvement in resource use efficiency supports sustainable agricultural practices and reduces the environmental impact of crop production. In addition to yield improvement, polyploidy can enhance the quality of agricultural products, including nutritional content, flavor, and post-harvest characteristics. Polyploidy in crops such as tomatoes and potatoes has been associated with increased nutrient content, improved flavor profiles, and extended shelf life. These quality improvements contribute to better consumer acceptance and increased marketability, supporting the economic viability of agricultural production.

4. Creating New Varieties and Crop Evolution

Polyploidy has played a significant role in the evolution and domestication of crops, leading to the creation of new varieties with desirable traits. Polyploidy-induced changes in genetic and phenotypic diversity have been instrumental in the development of many modern crops.

The allopolyploidy of wheat, for example, involved the hybridization of different species, resulting in new genetic combinations that contributed to the evolution of modern bread wheat (*Triticum aestivum*). This polyploidization event led to the development of a species with increased yield, adaptability, and resistance to various stresses. The success of wheat as a global staple crop is largely attributed to its polyploid nature, which has allowed it to thrive in diverse environments and meet the demands of a growing population. Polyploidy has also been crucial in the development of new fruit and vegetable varieties. The induction of polyploidy in crops such as strawberries, bananas, and watermelons has resulted in the creation of seedless or improved-quality varieties, enhancing consumer appeal and marketability. These new varieties often exhibit increased size, flavor, and nutritional content, making them desirable for commercial production and consumption. In addition to its role in crop evolution, polyploidy continues to be a valuable tool in modern plant breeding. The deliberate induction of polyploidy in crops allows breeders to exploit the benefits of polyploidy and create new varieties with enhanced traits. Techniques such as colchicine treatment and tissue culture are commonly used to induce polyploidy, enabling the development of improved crop varieties that can meet specific agricultural and market needs.

c. Gene Mapping and Marker-Assisted Selection

Gene mapping and marker-assisted selection (MAS) are essential components of modern plant breeding, providing powerful tools for identifying and selecting desirable traits in crops. These techniques enable breeders to accelerate the development of improved varieties by using genetic markers to track the inheritance of important traits.

Application of Gene Mapping Techniques in Identifying Desirable Traits

Gene mapping is a critical process in plant breeding that involves locating genes or genetic markers associated with specific traits on a chromosome. This technique allows breeders to understand the genetic basis of traits and identify the chromosomal regions that control them. By mapping genes associated with desirable traits, breeders can facilitate the selection and introgression of these traits into cultivated varieties.

1. Understanding the Genetic Architecture of Traits

Gene mapping provides insights into the genetic architecture of traits, helping breeders understand the complex interactions between genes and their influence on phenotype. This understanding is crucial for identifying the genetic basis of traits and developing strategies for their improvement.

Quantitative trait loci (QTL) mapping is a common approach used to identify regions of the genome associated with complex traits, such as yield, disease resistance, and stress tolerance. QTL mapping involves the analysis of genetic linkage between markers and phenotypic traits in a segregating population. By identifying QTL associated with desirable traits, breeders can pinpoint the chromosomal locations of genes that contribute to these traits. In addition to QTL mapping, genome-wide association studies (GWAS) are used to identify genetic variants associated with traits in diverse populations. GWAS involves scanning the entire genome for associations between genetic markers and phenotypic traits, allowing for the identification of candidate genes and loci. This approach is particularly useful for studying traits controlled by multiple genes and environmental interactions. The genetic architecture of traits through gene mapping provides breeders with valuable information for developing targeted breeding strategies. By identifying the genes and loci associated with desirable traits, breeders can focus their efforts on selecting and introgressing these traits into breeding lines, ultimately accelerating the development of improved crop varieties.

2. Identifying Genes for Complex Traits

Complex traits, such as yield, stress tolerance, and nutritional quality, are often controlled by multiple genes and influenced by environmental factors. Gene mapping is essential for identifying the genetic basis of these complex traits and understanding the underlying mechanisms that control them.

In rice breeding, for example, gene mapping has been used to identify QTL associated with drought tolerance, a critical trait for maintaining productivity in water-limited environments. By mapping the genetic regions associated with drought tolerance, researchers have identified candidate genes involved in water uptake, osmotic adjustment, and root development. This knowledge has facilitated the development of drought-tolerant rice varieties through marker-assisted selection. Similarly, gene mapping has been used to identify genes associated with disease resistance in crops such as wheat and maize. By mapping the loci associated with resistance to pathogens like rust, blight, and mildew, breeders have been able to identify resistance genes and incorporate them into breeding programs. This approach has led to the development of disease-resistant varieties that reduce the reliance on chemical control measures and support sustainable agriculture. The identification of genes for complex traits through gene mapping provides breeders with the tools needed to develop crop varieties with enhanced performance and resilience. By targeting specific genes and loci associated with desirable traits, breeders can efficiently select and introgress these traits into breeding lines, ultimately improving crop productivity and quality.

3. Marker Development and Validation

The development and validation of genetic markers are essential components of gene mapping and marker-assisted selection. Genetic markers are DNA sequences associated with specific traits, and they serve as tools for tracking the inheritance of these traits in breeding populations. Marker development involves identifying DNA sequences that are tightly linked to desirable traits, allowing breeders to use these markers as proxies for the traits themselves. The most common types of markers used in plant breeding include simple sequence repeats (SSRs), single nucleotide polymorphisms (SNPs), and amplified fragment length polymorphisms (AFLPs). Each marker type has its advantages and limitations, and the choice of marker depends on factors such as the trait of interest, the breeding population, and the available resources. Validation of genetic markers is crucial to ensure their reliability and accuracy in predicting trait inheritance. Marker validation involves testing the association between markers and traits in diverse populations and environments, confirming that the markers accurately reflect the presence or absence of the traits. This validation process is essential for establishing the utility of markers in breeding programs and ensuring their effectiveness in selecting desirable traits. The development and validation of genetic markers provide breeders with powerful tools for tracking and selecting desirable traits in breeding populations. By using markers to guide selection, breeders can accelerate the development of improved crop varieties and enhance the efficiency of breeding programs.

4. Mapping and Utilizing Novel Traits

Gene mapping also plays a crucial role in identifying and utilizing novel traits from wild relatives and exotic germplasm. These traits often represent untapped genetic resources that can contribute to crop improvement by introducing new variations and enhancing existing traits. Wild relatives of cultivated crops often possess traits that have been lost in modern breeding, such as resistance to pests and diseases, tolerance to abiotic stresses, and improved nutritional quality. Gene mapping allows breeders to identify the genetic basis of these novel traits and facilitate their introgression into breeding lines.

In tomato breeding, for example, gene mapping has been used to identify loci associated with resistance to late blight, a devastating disease caused by the oomycete *Phytophthora infestans*. By mapping the genetic regions associated with resistance in wild tomato relatives, breeders have identified resistance genes and introgressed them into cultivated varieties. This approach has led to the development of late blight-resistant tomato varieties, reducing yield losses and supporting sustainable production. Similarly, gene mapping has been used to identify novel traits associated with nutritional quality in crops such as maize and rice. By mapping the loci associated with enhanced nutrient content, researchers have identified candidate genes involved in nutrient uptake, transport, and accumulation. This knowledge has facilitated the development of biofortified varieties with improved nutritional profiles, contributing to efforts to combat malnutrition and promote health. Mapping and utilizing novel traits through gene mapping provide breeders with valuable genetic resources for crop improvement. By identifying and introgressing these traits into breeding lines, breeders can develop crop varieties that are more resilient, productive, and adaptable to changing environments.

d. Importance of Marker-Assisted Selection in Accelerating Pre-breeding Efforts

Marker-assisted selection (MAS) is a powerful tool in modern plant breeding, enabling the efficient selection and introgression of desirable traits into breeding lines. MAS utilizes genetic markers to track the inheritance of specific traits, allowing breeders to accelerate the development of improved crop varieties. This technique offers several advantages over traditional breeding methods, including increased selection accuracy, reduced breeding cycles, and enhanced efficiency in managing complex traits.

1. Increasing Selection Accuracy

Marker-assisted selection increases selection accuracy by providing breeders with precise information about the genetic makeup of breeding lines. By using genetic markers to track the inheritance of desirable traits, breeders can accurately identify individuals that carry the target traits and select them for further breeding. In traditional breeding, selection is often based on phenotypic traits, which can be influenced by environmental factors and may not accurately reflect the genetic potential of a plant. MAS overcomes this limitation by directly targeting the genetic markers associated with the traits of interest, ensuring that the selected individuals possess the desired genetic makeup. The increased selection accuracy provided by MAS is particularly valuable for traits that are difficult to assess phenotypically, such as disease resistance, stress tolerance, and quality traits. By using markers to guide selection, breeders can efficiently identify individuals with the desired traits and accelerate the development of improved varieties.

2. Reducing Breeding Cycles

Marker-assisted selection reduces the time required for breeding by allowing for the early identification of individuals with desirable traits. By using markers to select for target traits in the early generations of a breeding program, breeders can reduce the number of generations needed to achieve the desired outcomes.In traditional breeding, the selection process often involves multiple generations of testing and evaluation to identify individuals with the desired traits. This process can be time-consuming and resource-intensive, particularly for traits that require extensive field testing or multiple years of evaluation. MAS streamlines the breeding process by providing breeders with tools to select for target traits in the early stages of a breeding program. This reduction in breeding cycles not only accelerates the development of improved varieties but also reduces the costs associated with breeding, allowing for more efficient use of resources.

3. Managing Complex Traits

Marker-assisted selection is particularly effective in managing complex traits that are controlled by multiple genes and influenced by environmental factors. These traits, such as yield, stress tolerance, and quality, often involve complex genetic interactions that are challenging to address through traditional breeding methods.

MAS provides breeders with the tools needed to dissect the genetic architecture of complex traits and identify the genetic components that contribute to them. By using markers to track the inheritance of multiple genes and loci associated with a trait, breeders can efficiently select individuals with the desired genetic makeup and manage the complexity of these traits. The ability to manage complex traits through MAS is particularly valuable in pre-breeding efforts, where the goal is to introduce new genetic variations and enhance the performance of breeding lines. By using markers to guide selection, breeders can efficiently integrate complex traits into breeding programs and develop improved crop varieties that exhibit superior performance and resilience.

4. Enhancing Breeding Efficiency

Marker-assisted selection enhances the efficiency of breeding programs by providing breeders with tools to make informed decisions and optimize resource allocation. By using markers to guide selection, breeders can focus their efforts on individuals with the highest genetic potential and reduce the need for extensive phenotypic testing.

The efficiency gains provided by MAS are particularly valuable in pre-breeding efforts, where the goal is to evaluate and select a diverse array of genetic resources. By using markers to track the inheritance of target traits, breeders can efficiently assess the genetic potential of breeding lines and prioritize individuals for further development. MAS also enhances breeding efficiency by enabling the simultaneous selection of multiple traits, allowing breeders to address multiple breeding objectives within a single program. This capability is particularly important in the context of modern agriculture, where crops must meet a range of requirements related to yield, quality, stress tolerance, and sustainability.

4. Integration of Wild Relatives and Exotic Germplasm

The integration of wild relatives and exotic germplasm into breeding programs is a fundamental strategy for enhancing crop diversity, resilience, and productivity. Wild relatives of domesticated crops harbor a wealth of genetic diversity that has been shaped by natural selection over millennia. This genetic reservoir provides invaluable resources for breeding programs aimed at improving traits such as stress tolerance, disease resistance, and yield potential. In recent years, the importance of utilizing wild relatives and exotic germplasm has gained prominence as breeders seek to overcome the limitations of modern cultivars and address global challenges such as climate change, food security, and biodiversity loss.

Importance of Wild Relatives

Wild relatives of crops are species that are genetically related to domesticated crops but have not undergone the same level of artificial selection. They possess a rich array of genetic diversity that can be harnessed to improve cultivated varieties. The importance of wild relatives in crop improvement cannot be overstated, as they offer a source of novel traits that can enhance genetic diversity, improve resilience, and increase productivity.

a. How Wild Relatives Contribute to Genetic Diversity and Trait Improvement

Wild relatives contribute to crop improvement by offering a broad spectrum of genetic diversity and a repository of traits that can enhance the adaptability and performance of cultivated crops. Their integration into breeding programs allows for the introduction of new genetic material, the enhancement of existing traits, and the development of novel characteristics that can address the challenges faced by modern agriculture.

1. Genetic Diversity and Novel Alleles

Wild relatives of crops are reservoirs of genetic diversity, harboring a wide range of alleles that have been shaped by natural selection in diverse environments. This genetic variation is critical for breeding programs seeking to expand the genetic base of cultivated crops and enhance their adaptability to changing conditions.

» **Introduction of Novel Alleles**

The genetic diversity found in wild relatives includes novel alleles that are not present in cultivated varieties. These alleles can introduce new traits or enhance existing ones, providing breeders with valuable genetic resources for crop improvement. By incorporating novel alleles from wild relatives, breeders can increase the genetic variability within breeding populations, creating opportunities for the selection of superior traits.

For example, wild relatives of wheat, such as *Aegilops* species, have been used to introduce novel alleles for disease resistance, drought tolerance, and nutrient use efficiency. These alleles have expanded the genetic base of wheat breeding programs, enabling the development of varieties with enhanced performance and resilience. The introduction of novel alleles from wild relatives also provides a means to overcome genetic bottlenecks in cultivated crops. Many modern varieties have experienced a reduction in genetic diversity due to intensive selection for specific traits, leading to increased vulnerability to pests, diseases, and environmental stresses. By incorporating genetic material from wild relatives, breeders can break these bottlenecks and increase the genetic diversity available for selection, enhancing the adaptability and resilience of crops.

» **Enhancement of Genetic Diversity**

Wild relatives contribute to the enhancement of genetic diversity in breeding programs by providing a source of alleles that can be introgressed into cultivated varieties. This increased genetic diversity is crucial for developing crop varieties that can adapt to changing environments and meet the demands of modern agriculture.

The enhancement of genetic diversity through the integration of wild relatives allows breeders to explore new combinations of alleles and create novel gene interactions that can result in improved traits. For instance, the introgression of genetic material from wild relatives into tomato breeding programs has increased the genetic diversity of the crop, leading to the development of varieties with improved yield, quality, and stress tolerance. The enhanced genetic diversity provided by wild relatives also supports the development of varieties that are more resilient to environmental changes and biotic stresses. By increasing the genetic variability within breeding populations, breeders can select individuals with superior performance under diverse conditions, ensuring the stability and productivity of crops in the face of climate change and other challenges.

2. Trait Improvement and Adaptation

Wild relatives possess a wide array of traits that have been shaped by natural selection, enabling them to survive and thrive in diverse environments. These traits can be harnessed to improve the performance and adaptability of cultivated crops, addressing the challenges posed by biotic and abiotic stresses.

» **Disease Resistance**

One of the most significant contributions of wild relatives to crop improvement is the introduction of disease-resistance traits. Wild relatives often harbor resistance genes that have been lost in cultivated varieties due to selective breeding. These genes can provide crops with resistance to a wide range of pathogens, reducing the reliance on chemical control measures and promoting sustainable agricultural practices. For example, wild relatives of rice, such as *Oryza nivara* and *Oryza rufipogon*, have been used to introduce resistance genes against rice blast, bacterial blight, and other diseases. The introgression of these genes into cultivated rice varieties has led to the development of disease-resistant cultivars that can maintain high yields and productivity in the presence of pathogens. Similarly, wild relatives of potatoes, such as *Solanum bulbocastanum*, have been utilized to transfer resistance genes against late blight, a devastating disease caused by the oomycete *Phytophthora infestans*. The incorporation of these resistance genes into cultivated potato varieties has resulted in the development of blight-resistant cultivars, reducing yield losses, and supporting sustainable production.

» **Abiotic Stress Tolerance**

Wild relatives also contribute to the improvement of abiotic stress tolerance in crops, enhancing their ability to withstand adverse environmental conditions such as drought, salinity, and temperature extremes. The genetic diversity present in wild relatives provides a source of alleles that can confer tolerance to these stresses, enabling the development of resilient crop varieties. For instance, wild relatives of wheat, such as *Triticum urartu* and *Triticum dicoccoides*, have been used to introduce genes associated with drought tolerance. These genes enhance the ability of wheat varieties to maintain productivity under water-limited conditions, supporting food security in regions prone to drought. Similarly, wild relatives of barley, such as *Hordeum spontaneum*, have been utilized to transfer genes associated with salinity tolerance. The incorporation of these genes into cultivated barley varieties has led to the development of salt-tolerant cultivars that can thrive in saline environments, expanding the potential growing areas for barley production.

The improvement of abiotic stress tolerance through the integration of wild relatives is crucial for ensuring crop resilience and productivity in the face of climate change and environmental challenges. By incorporating stress tolerance traits from wild relatives, breeders can develop varieties that can maintain stable yields under diverse conditions, contributing to sustainable agriculture and food security.

3. Quality Traits and Nutritional Improvement

Wild relatives also play a vital role in the improvement of quality traits and nutritional content in crops. The genetic diversity present in wild relatives can be harnessed to enhance the nutritional profile, flavor, and other quality attributes of cultivated varieties.

» Nutritional Quality Enhancement

The improvement of nutritional quality is a critical objective in crop breeding, as it addresses malnutrition and promotes human health. Wild relatives offer a source of genetic variation that can be used to enhance the content of essential nutrients in crops, contributing to the development of biofortified varieties. For example, wild relatives of maize, such as *Zea mexicana*, have been used to increase the content of essential amino acids, vitamins, and minerals in cultivated maize varieties. The introgression of these traits has led to the development of nutritionally enhanced maize varieties that provide a more balanced and nutritious diet.

Similarly, wild relatives of wheat, such as *Aegilops tauschii*, have been utilized to improve the protein content and nutritional quality of wheat. The incorporation of these traits into cultivated wheat varieties has resulted in the development of biofortified wheat cultivars that address protein malnutrition and support dietary diversity. The enhancement of nutritional quality through the integration of wild relatives is essential for combating micronutrient deficiencies and promoting health and well-being. By leveraging the genetic diversity present in wild relatives, breeders can develop crop varieties that provide enhanced nutritional benefits, contributing to efforts to improve food security and nutrition.

» Flavor, Aroma, and Other Quality Traits

Wild relatives also contribute to the improvement of flavor, aroma, and other quality traits in crops. These traits are important for consumer acceptance and marketability, and the genetic diversity present in wild relatives can be used to enhance the sensory attributes of cultivated varieties. For instance, wild relatives of tomatoes, such as *Solanum pimpinellifolium*, have been used to improve the flavor and aroma of cultivated tomato varieties. The introgression of genes associated with flavor-enhancing compounds has led to the development of tomato varieties with superior sensory qualities, increasing their appeal to consumers. Similarly, wild relatives of grapes, such as *Vitis vinifera sylvestris*, have been utilized to enhance the flavor and aroma of cultivated grape varieties. The incorporation of these traits has resulted in the development of grape cultivars with improved taste and quality, supporting the production of high-quality wines and other grape products. The improvement of flavor, aroma, and other quality traits through the integration of wild relatives is crucial for meeting consumer preferences and enhancing the marketability of agricultural products. By harnessing the genetic diversity present in wild relatives, breeders can develop crop varieties that offer superior quality and appeal, supporting the economic viability of agricultural production.

4. Breeding Strategies and Challenges

The integration of wild relatives into breeding programs involves various strategies and challenges that must be addressed to achieve successful outcomes. These strategies include the identification and characterization of wild relatives, the development of pre-breeding lines, and the use of advanced breeding techniques to introgress desirable traits.

» Identification and Characterization of Wild Relatives

The identification and characterization of wild relatives are essential steps in the integration process, as they provide information on the genetic diversity and traits present in these species. This information is crucial for selecting wild relatives that possess desirable traits and can contribute to crop improvement.

Genetic characterization involves the use of molecular markers and genomic tools to assess the genetic diversity and relationships between wild relatives and cultivated crops. This characterization helps breeders identify the genetic resources that offer the greatest potential for improving specific traits, guiding the selection of wild relatives for breeding programs. The identification and characterization of wild relatives also involve the evaluation of their phenotypic traits, such as disease resistance, stress tolerance, and quality attributes. This evaluation provides insights into the traits that can be introgressed into cultivated varieties, supporting the development of pre-breeding lines with enhanced performance and adaptability.

» Development of Pre-breeding Lines

The development of pre-breeding lines is a critical step in the integration of wild relatives, as it involves the introgression of desirable traits into cultivated varieties. Pre-breeding lines serve as intermediaries that facilitate the transfer of genetic material from wild relatives to commercial crops. The development of pre-breeding lines involves the use of hybridization and backcrossing techniques to introduce genetic material from wild relatives into cultivated varieties. This process often requires multiple generations of crossing and selection to achieve the desired level of introgression and stabilize the expression of target traits. The development of pre-breeding lines also involves the use of advanced breeding techniques, such as marker-assisted selection and genomic selection, to accelerate the introgression process and enhance the efficiency of breeding programs. These techniques enable breeders to track the inheritance of target traits and select individuals with the desired genetic makeup, ensuring the successful integration of wild relatives into cultivated varieties.

» Challenges and Considerations

The integration of wild relatives into breeding programs presents several challenges and considerations that must be addressed to achieve successful outcomes. These challenges include the potential for genetic incompatibility, the presence of undesirable traits, and the need for extensive evaluation and selection. Genetic incompatibility between wild relatives and cultivated crops can pose a significant barrier to the introgression of desirable traits. This incompatibility may result in reduced fertility, hybrid breakdown, or other reproductive barriers that hinder the successful transfer of genetic material. To overcome these challenges, breeders may employ techniques such as embryo rescue, chromosome engineering, or bridge crosses to facilitate the introgression process. The presence of undesirable traits in wild relatives, such as low yield or poor quality, may also pose challenges for breeding programs. The introgression of desirable traits from wild relatives may inadvertently introduce undesirable traits that require additional selection and evaluation to eliminate. Breeders must carefully manage the introgression process to ensure that the benefits of integrating wild relatives outweigh the potential drawbacks. The need for extensive evaluation and selection is another challenge associated with the integration of wild relatives. Breeders must thoroughly evaluate pre-breeding lines for the expression of target traits and assess their performance under diverse conditions. This evaluation requires significant time, resources, and expertise, but it is essential for ensuring the success of breeding programs and the development of improved crop varieties.

b. Challenges and Strategies for Germplasm Integration

The integration of wild relatives and exotic germplasm into crop breeding programs is essential for enhancing genetic diversity and improving crop resilience. However, this integration process is fraught with various challenges that must be addressed to realize the potential benefits. These challenges include genetic incompatibility, undesirable traits, and the complexities of introgressing beneficial alleles into cultivated varieties. Effective strategies are needed to overcome these obstacles and successfully harness the genetic potential of wild relatives for crop improvement.

Challenges Faced in Using Wild Relatives

The utilization of wild relatives in breeding programs presents several challenges that can hinder the successful integration of genetic material into cultivated varieties. These challenges arise from the inherent differences between wild relatives and domesticated crops, as well as the complexities of breeding and selection processes.

1. Genetic Incompatibility

Genetic incompatibility between wild relatives and cultivated crops is one of the primary challenges faced in germplasm integration. Wild relatives often belong to different species or subspecies, leading to reproductive barriers that complicate the crossing and hybridization processes.

» **Reproductive Barriers**

Reproductive barriers can prevent successful hybridization between wild relatives and cultivated crops. These barriers may be pre-zygotic, such as differences in flowering time or pollen-pistil incompatibility, or post-zygotic, such as hybrid inviability or sterility. Pre-zygotic barriers hinder the formation of hybrid embryos, while post-zygotic barriers result in hybrids that are unable to survive or reproduce.

For example, crossing cultivated wheat (*Triticum aestivum*) with its wild relatives, such as *Aegilops tauschii*, often faces challenges due to differences in ploidy levels and genomic composition. These differences can lead to hybrid sterility, where the resulting hybrids are unable to produce viable seeds, limiting the transfer of desirable traits. Reproductive barriers pose significant challenges for breeders attempting to introgress beneficial alleles from wild relatives into cultivated varieties. Overcoming these barriers requires innovative breeding techniques and a thorough understanding of the genetic and reproductive biology of both wild and cultivated species.

» **Genomic Incompatibility**

Genomic incompatibility can also arise from differences in chromosome structure and number between wild relatives and cultivated crops. These differences can lead to issues such as chromosome misalignment during meiosis, resulting in reduced fertility and hybrid breakdown.

For instance, the integration of wild relatives into tomato breeding programs can be complicated by genomic differences between species, such as variations in chromosome number and structure. These differences can lead to challenges in maintaining stable and fertile hybrids, limiting the successful transfer of genetic material. Genomic incompatibility presents a barrier to the successful introgression of traits from wild relatives, requiring breeders to develop strategies to overcome these challenges and facilitate the exchange of genetic material.

2. Undesirable Traits

Wild relatives often possess undesirable traits that can be introgressed into cultivated varieties along with beneficial alleles. These traits may include low yield, poor quality, or susceptibility to pests and diseases, which can negatively impact the performance and acceptability of cultivated crops.

» **Low Yield and Poor Quality**

Many wild relatives have not undergone the same level of artificial selection as cultivated crops, resulting in traits that are less favorable for agricultural production. For example, wild relatives may exhibit low yield potential, poor fruit or grain quality, or undesirable growth habits.

In the case of rice breeding, wild relatives such as *Oryza rufipogon* may possess traits such as shattering, low seed weight, and poor grain quality. While these wild relatives offer valuable genetic resources for traits such as disease resistance and stress tolerance, the presence of undesirable agronomic traits poses challenges for their integration into cultivated varieties. The introgression of undesirable traits from wild relatives can compromise the performance and marketability of cultivated crops, necessitating careful selection and breeding strategies to retain beneficial traits while minimizing negative impacts.

» **Susceptibility to Pests and Diseases**

Wild relatives may also possess susceptibility to pests and diseases that can be transferred to cultivated crops during the introgression process. This susceptibility can increase the vulnerability of crops to biotic stresses, reducing yield and productivity. For example, while wild relatives of potatoes such as *Solanum demissum* offer valuable resistance genes to late blight, they may also introduce susceptibility to other pests and diseases. The presence of these undesirable traits can complicate the breeding process, requiring additional selection and evaluation to ensure that the introgressed lines exhibit enhanced resistance without increased susceptibility. Managing the introgression of undesirable traits is a critical challenge in the integration of wild relatives, requiring breeders to carefully evaluate and select for beneficial traits while minimizing the negative impacts of undesirable traits.

3. Complexity of Trait Introgression

The introgression of beneficial traits from wild relatives into cultivated crops is a complex process that requires careful management of genetic resources and breeding strategies. This complexity arises from the need to identify, transfer, and stabilize desirable traits while maintaining the overall performance and adaptability of cultivated varieties.

» **Identifying Desirable Traits**

Identifying desirable traits in wild relatives is a crucial step in the introgression process. This identification involves the evaluation of wild relatives for traits such as disease resistance, stress tolerance, and quality attributes, as well as the characterization of the genetic basis of these traits. For example, identifying disease resistance traits in wild relatives of wheat involves screening for resistance to pathogens such as rusts, smuts, and blights. This screening process requires the use of molecular markers and genomic tools to map the loci associated with resistance and identify candidate genes for introgression. The identification of desirable traits is often complicated by the complex genetic architecture of these traits, which may be controlled by multiple genes and influenced by environmental factors. This complexity requires breeders to employ advanced breeding techniques and genomic tools to accurately identify and select for target traits.

» **Transferring Beneficial Traits**

Transferring beneficial traits from wild relatives into cultivated crops involves the use of hybridization and backcrossing techniques to introduce the desired genetic material. This transfer process requires multiple generations of crossing and selection to achieve the desired level of introgression and stabilize the expression of target traits. The transfer of beneficial traits is often complicated by genetic linkage, where desirable traits are linked to undesirable traits on the same chromosome. This linkage can hinder the separation of beneficial and undesirable traits during the breeding process, requiring breeders to use techniques such as marker-assisted selection and genomic selection to break the linkage and select for the desired traits. The successful transfer of beneficial traits requires careful management of genetic resources and breeding strategies to ensure that the introgressed lines exhibit enhanced performance and adaptability while maintaining the overall integrity of the cultivated variety.

» **Stabilizing Trait Expression**

Stabilizing the expression of introgressed traits is a critical step in the integration process, as it ensures that the desired traits are consistently expressed in subsequent generations. This stabilization involves the selection and evaluation of introgressed lines to confirm the stable inheritance and expression of target traits.

The stabilization of trait expression can be complicated by genetic and environmental factors that influence the expression of introgressed traits. For example, the expression of stress tolerance traits may be influenced by environmental conditions such as temperature and moisture levels, requiring breeders to evaluate the performance of introgressed lines under diverse conditions. The stabilization of trait expression also involves the assessment of potential interactions between introgressed traits and the existing genetic background of the cultivated variety. These interactions can affect the overall performance and adaptability of the introgressed lines, requiring breeders to carefully evaluate and select for stable and consistent trait expression.

Strategies to Overcome Challenges

To overcome the challenges associated with the integration of wild relatives and exotic germplasm, breeders employ a range of strategies that leverage advanced breeding techniques, genomic tools, and innovative approaches to facilitate the successful introgression of beneficial traits.

1. Advanced Breeding Techniques

Advanced breeding techniques are essential for overcoming the challenges of genetic incompatibility and trait introgression, providing breeders with tools to facilitate the exchange of genetic material and enhance the efficiency of breeding programs.

» Hybridization and Backcrossing

Hybridization and backcrossing are fundamental techniques used to introduce genetic material from wild relatives into cultivated crops. These techniques involve the crossing of wild and cultivated species to produce hybrids, followed by repeated backcrossing to the cultivated parent to stabilize the expression of target traits.

The use of hybridization and backcrossing requires careful management of genetic resources and breeding strategies to ensure the successful transfer of beneficial traits. Breeders employ techniques such as controlled pollination, embryo rescue, and bridge crosses to facilitate hybridization and overcome reproductive barriers, ensuring the successful production of hybrid progeny.

Backcrossing is a critical step in the introgression process, allowing breeders to recover the genetic background of the cultivated variety while retaining the desired traits from the wild relative. This process involves multiple generations of crossing and selection to achieve the desired level of introgression and stabilize the expression of target traits.

» Embryo Rescue and Tissue Culture

Embryo rescue and tissue culture are advanced breeding techniques used to overcome reproductive barriers and facilitate the production of hybrid progeny. These techniques involve the culture of embryos or tissues in vitro to promote growth and development, allowing for the successful production of hybrids that may not survive under natural conditions.

Embryo rescue is particularly useful in overcoming pre-zygotic and post-zygotic barriers, such as hybrid inviability and sterility. By culturing embryos in vitro, breeders can bypass these barriers and produce viable hybrids, enabling the successful transfer of genetic material from wild relatives to cultivated crops.

Tissue culture techniques, such as somatic hybridization and protoplast fusion, provide additional tools for overcoming genetic incompatibility and facilitating the exchange of genetic material. These techniques involve the fusion of somatic cells or protoplasts to create hybrids with combined genetic material, allowing for the introgression of desirable traits from wild relatives.

The use of advanced breeding techniques such as hybridization, backcrossing, embryo rescue, and tissue culture provides breeders with the tools needed to overcome genetic incompatibility and facilitate the successful integration of wild relatives into breeding programs. These techniques enable the efficient transfer of beneficial traits and support the development of improved crop varieties.

2. Genomic Tools and Molecular Markers

Genomic tools and molecular markers play a crucial role in overcoming the challenges associated with trait introgression and genetic diversity, providing breeders with the tools needed to identify, track, and select for desirable traits.

» **Marker-Assisted Selection**

Marker-assisted selection (MAS) is a powerful tool used to facilitate the introgression of target traits by providing breeders with precise information about the genetic makeup of breeding lines. MAS utilizes molecular markers to track the inheritance of desirable traits, allowing breeders to accurately identify individuals that carry the target traits and select them for further breeding. The use of MAS enhances the efficiency of breeding programs by reducing the need for extensive phenotypic testing and allowing for the early selection of individuals with the desired genetic makeup. This approach is particularly valuable for traits that are difficult to assess phenotypically, such as disease resistance and stress tolerance. By using markers to guide selection, breeders can efficiently introgress desirable traits from wild relatives into breeding lines, ensuring the successful integration of genetic material and the development of improved crop varieties.

» **Genomic Selection**

Genomic selection (GS) is an advanced breeding technique that uses genomic information to predict the breeding value of individuals and guide selection decisions. GS involves the analysis of genome-wide markers to estimate the genetic potential of breeding lines, allowing breeders to select individuals with the highest predicted performance.

The use of GS provides breeders with a comprehensive understanding of the genetic architecture of traits and allows for the simultaneous selection of multiple traits. This capability is particularly important in pre-breeding efforts, where the goal is to introduce new genetic variations and enhance the performance of breeding lines. By leveraging genomic information, GS enables breeders to overcome the challenges associated with complex traits and accelerate the development of improved crop varieties. This approach supports the efficient integration of wild relatives and exotic germplasm, contributing to the enhancement of genetic diversity and trait improvement. The use of genomic tools and molecular markers provides breeders with the tools needed to overcome the challenges associated with trait introgression and genetic diversity. By leveraging these tools, breeders can efficiently identify, track, and select for desirable traits, ensuring the successful integration of wild relatives and exotic germplasm into breeding programs.

3. Pre-breeding and Evaluation

Pre-breeding and evaluation are essential components of germplasm integration, providing a framework for the identification, assessment, and selection of desirable traits from wild relatives and exotic germplasm.

» **Pre-breeding Programs**

Pre-breeding programs serve as a bridge between the exploration of genetic resources and the development of commercial crop varieties, facilitating the transfer of genetic material from wild relatives to breeding lines. These programs involve the identification and characterization of wild relatives, the development of pre-breeding lines, and the evaluation of introgressed traits.

The goal of pre-breeding programs is to identify and introgress beneficial traits from wild relatives into cultivated varieties, creating breeding lines with enhanced performance and adaptability. This process involves the use of advanced breeding techniques and genomic tools to facilitate the introgression of target traits and ensure their stable expression in breeding lines. Pre-breeding programs also involve the evaluation of pre-breeding lines for the expression of introgressed traits and their performance under diverse conditions. This evaluation provides insights into the potential of pre-breeding lines for further development and selection, supporting the integration of wild relatives and exotic germplasm into breeding programs.

» **Evaluation and Selection**

Evaluation and selection are critical steps in the integration process, providing breeders with the information needed to assess the performance and adaptability of introgressed lines. This evaluation involves the assessment of introgressed lines for traits such as yield, quality, stress tolerance, and disease resistance, as well as their performance under diverse environmental conditions. The evaluation process requires the use of phenotypic and genotypic data to assess the expression of introgressed traits and their interactions with the existing genetic background of the cultivated variety. This assessment provides insights into the potential of introgressed lines for further development and selection, guiding the breeding process and supporting the successful integration of wild relatives. The evaluation and selection of introgressed lines are crucial for ensuring the successful integration of wild relatives and exotic germplasm into breeding programs. By carefully assessing the performance and adaptability of introgressed lines, breeders can identify individuals with the greatest potential for crop improvement and develop improved crop varieties that meet the demands of modern agriculture.

Chapter - 6

Location and Mapping of Genes on Chromosomes

A. Location and Mapping of Genes on Chromosomes: Deficiency Method

Mapping genes on chromosomes is a fundamental component of molecular cytogenetics, essential for understanding genetic inheritance and facilitating crop improvement. The deficiency method, a classical approach in genetics, involves using chromosomal deletions to determine the locations of genes. This method has been instrumental in advancing our understanding of gene function and has significant implications for agricultural genetics.

Introduction to Gene Mapping

Gene mapping is the process of identifying the specific locations of genes on chromosomes. This technique is crucial for linking genetic markers with phenotypic traits, allowing researchers to understand the genetic basis of various characteristics. Gene mapping provides insights into gene function, genetic variation, and evolutionary relationships, enabling targeted breeding and biotechnological interventions to improve crops.

Historical Background of Gene Mapping

The history of gene mapping dates back to the early 20th century when geneticists began exploring the relationship between chromosomal structure and gene function. The development of gene mapping techniques was propelled by the pioneering work of researchers like Thomas Hunt Morgan and Calvin Bridges. Their studies on *Drosophila melanogaster* (fruit flies) revealed that genes are arranged linearly on chromosomes and that specific genetic traits could be linked to particular chromosomal locations.

The concept of gene mapping evolved from classical genetics to incorporate molecular techniques, leading to more precise and efficient mapping methods. The deficiency method emerged as a powerful tool in this context, enabling the identification of gene locations based on the phenotypic effects of chromosomal deletions. This method was initially applied to model organisms such as *Drosophila* but was later adapted for use in plants and other organisms. In plants, gene mapping became increasingly important for understanding the genetic basis of complex traits such as yield, disease resistance, and stress tolerance. The deficiency method played a crucial role in these efforts, allowing researchers to pinpoint the chromosomal regions associated with important agricultural traits. As technology

DOI: 10.1201/9781003717775-6

advanced, gene mapping techniques evolved to incorporate molecular markers, genomic data, and bioinformatics tools, enhancing the accuracy and efficiency of gene location studies.

Evolution of the Deficiency Method

The deficiency method has evolved significantly since its inception, with several key developments contributing to its advancement. Initially, this method relied on the natural occurrence of chromosomal deletions or deficiencies, which were identified through visible phenotypic changes in organisms. These natural deficiencies provided a starting point for understanding the relationship between chromosomal structure and gene function. With the advent of mutagenesis techniques, researchers gained the ability to induce chromosomal deletions artificially. Chemical mutagens, radiation, and genetic engineering tools such as CRISPR-Cas9 allowed for the targeted creation of chromosomal deficiencies, expanding the scope of the deficiency method. This enabled researchers to study the effects of specific deletions on phenotypes and map genes more precisely. The integration of molecular markers and genomic tools further enhanced the deficiency method. Molecular markers, such as restriction fragment length polymorphisms (RFLPs), simple sequence repeats (SSRs), and single nucleotide polymorphisms (SNPs), provided additional information for mapping genes in relation to chromosomal deficiencies. These markers allowed for the construction of high-resolution genetic maps and facilitated the identification of candidate genes within deleted regions. Modern advancements in bioinformatics and genomic data analysis have revolutionized the deficiency method. Researchers can now combine deletion mapping with whole-genome sequencing and transcriptomics to refine gene locations and identify functional elements within chromosomal regions. This integration of data from multiple sources enhances the accuracy and efficiency of gene mapping, making it a powerful tool for studying complex traits in crops.

Fundamental Principles of the Deficiency Method

The deficiency method is based on the principle that the absence of certain chromosomal segments leads to the loss of specific genes, resulting in observable phenotypic changes. By analyzing these changes, researchers can determine which chromosomal regions contain the genes responsible for particular traits. The deficiency method is particularly useful for identifying recessive genes and studying gene interactions, as the loss of dominant alleles can reveal the effects of hidden genetic factors. One of the core principles of the deficiency method is the concept of complementation. Complementation occurs when two individuals with different deficiencies are crossed, and the resulting offspring exhibit a phenotype not present in either parent. This indicates that the missing genes in one parent are present in the other, allowing researchers to infer the locations of genes within the deleted regions. Another key principle is the use of overlapping deficiencies to refine gene locations. By creating a series of overlapping chromosomal deletions, researchers can pinpoint the exact position of a gene within a chromosomal region. This approach involves comparing the phenotypes of individuals with different deficiencies and identifying the smallest region that must be present for a specific trait to be expressed. The deficiency method also relies on the analysis of quantitative traits, which are controlled by multiple genes and influenced by environmental factors. By examining the effects of chromosomal deletions on quantitative traits, researchers can identify chromosomal regions that contribute to complex traits such as yield, stress tolerance, and disease resistance. This information is invaluable for breeding programs aimed at improving crop performance.

Mechanisms and Procedures of the Deficiency Method

The deficiency method involves several key steps, including the creation of chromosomal deficiencies, the analysis of phenotypic effects, and the inference of gene locations through deletion mapping. Each step is critical for accurately mapping genes and understanding their functions.

Creating Chromosomal Deficiencies

Creating chromosomal deficiencies is the first step in the deficiency method. This process involves inducing deletions in specific chromosomal regions to study their effects on phenotypes. Several techniques are employed to create chromosomal deficiencies, each with its advantages and limitations.

1. **Radiation-Induced Mutagenesis**: Exposure to ionizing radiation, such as X-rays or gamma rays, can cause chromosomal breakage, leading to deletions. This method has been extensively used in model organisms like *Drosophila* and plants such as maize. The extent of deletions can vary, providing a range of deficiencies for mapping studies.
2. **Chemical Mutagenesis**: Chemical agents like ethyl methanesulfonate (EMS) are used to induce mutations and deletions in chromosomes. This approach is particularly useful in plants and allows for targeted deletions in specific chromosomal regions. Chemical mutagenesis offers the advantage of creating precise deletions for gene mapping.
3. **Genetic Engineering**: Advances in genetic engineering, particularly CRISPR-Cas9 technology, have revolutionized the creation of chromosomal deficiencies. Researchers can now design specific deletions with precision, facilitating the study of gene function in crops. This approach is highly efficient and enables the creation of targeted deficiencies for mapping purposes.

The creation of chromosomal deficiencies is followed by the identification and characterization of the deletions. Researchers use cytogenetic techniques, such as fluorescence in situ hybridization (FISH), to visualize chromosomal deficiencies and confirm their locations. FISH involves using fluorescent probes that bind to specific DNA sequences, allowing researchers to detect and map deletions on chromosomes. The use of molecular markers is also essential for characterizing chromosomal deficiencies. Markers that are deleted in individuals with specific deficiencies can be used to identify the boundaries of the deletions. This information is critical for determining the extent of the deleted regions and for correlating deletions with phenotypic effects.

Analyzing Phenotypic Effects

Once chromosomal deficiencies are created, the next step is to analyze the phenotypic effects of these deletions. By observing changes in phenotypes, researchers can identify the chromosomal regions associated with specific traits.

1. **Quantitative and Qualitative Trait Analysis**: Researchers assess the impact of chromosomal deficiencies on both quantitative traits (e.g., yield, height, stress tolerance) and qualitative traits (e.g., flower color, leaf shape). By comparing the phenotypes of individuals with different deficiencies, scientists can identify the chromosomal regions influencing these traits.
2. **Complementation Tests**: Complementation tests are conducted to confirm the association

between a chromosomal deficiency and a specific trait. By crossing individuals with different deficiencies, researchers can determine whether the presence of a specific chromosomal region restores the trait. Complementation tests provide valuable information about the genetic basis of traits and help refine gene mapping.

3. **Phenotypic Assays**: Phenotypic assays involve measuring the effects of chromosomal deficiencies on various traits. This may include assessing plant growth, development, and reproductive success, as well as evaluating resistance to diseases and environmental stresses. Phenotypic assays provide a comprehensive understanding of the impact of chromosomal deletions on organismal fitness and trait expression.

The analysis of phenotypic effects is complemented by statistical methods that quantify the relationship between chromosomal deficiencies and trait variation. Quantitative trait locus (QTL) mapping is a powerful technique used to identify chromosomal regions associated with quantitative traits. QTL mapping involves statistical analysis of the phenotypic data from individuals with different deficiencies to pinpoint the chromosomal regions contributing to trait variation.

Inferring Gene Locations

The final step in the deficiency method is inferring the locations of genes based on deletion mapping. By analyzing the phenotypic data from individuals with various deficiencies, researchers can create a map of gene locations on chromosomes.

1. **Deletion Mapping**: Deletion mapping involves comparing the phenotypes of individuals with overlapping deficiencies. By identifying the minimal chromosomal region required for a specific trait, researchers can infer the location of the gene responsible for that trait. This process is repeated for multiple traits, resulting in a comprehensive map of gene locations on the chromosome.

2. **Gene Clustering**: Gene clustering is a technique used to identify groups of genes located within the same chromosomal region. By analyzing the phenotypic effects of overlapping deficiencies, researchers can identify clusters of genes that contribute to similar traits. Gene clustering provides insights into the organization of genes on chromosomes and aids in the identification of candidate genes for further study.

3. **Fine Mapping**: Fine mapping is the process of narrowing down the location of a gene to a specific chromosomal region. This involves using additional genetic markers and molecular techniques to refine the gene location and identify candidate genes. Fine mapping is critical for understanding the genetic basis of traits and for developing targeted breeding strategies.

The integration of molecular data, such as whole-genome sequencing and transcriptomics, enhances the accuracy and efficiency of gene mapping. By combining deletion mapping with genomic data, researchers can refine gene locations and identify functional elements within chromosomal regions. This integration of data from multiple sources provides a comprehensive view of gene organization and function, making it a powerful tool for studying complex traits in crops.

Applications of the Deficiency Method in Crop Improvement

The deficiency method has numerous applications in crop improvement, enabling researchers to map genes associated with important agronomic traits and develop improved crop varieties.

Mapping Agronomic Traits

Mapping agronomic traits is a key application of the deficiency method. By identifying the chromosomal regions associated with traits like yield, disease resistance, and stress tolerance, researchers can develop crop varieties with enhanced performance.

1. **Yield Improvement**: Yield is a complex trait influenced by multiple genes and environmental factors. The deficiency method allows researchers to identify chromosomal regions associated with yield and develop varieties with improved productivity. By selecting for specific deficiencies, breeders can enhance yield potential and develop high-yielding crop varieties.
2. **Disease Resistance**: Disease resistance is a crucial trait for crop improvement, as it reduces the impact of pathogens on crop yield and quality. The deficiency method enables researchers to map genes associated with disease resistance and develop resistant varieties. By identifying chromosomal regions linked to resistance, breeders can incorporate these regions into new varieties through marker-assisted selection.
3. **Stress Tolerance**: Environmental stresses, such as drought and salinity, pose significant challenges to crop production. The deficiency method helps researchers map genes associated with stress tolerance and develop resilient crop varieties. By identifying chromosomal regions linked to stress tolerance, breeders can select for these regions and enhance the ability of crops to withstand adverse conditions.

The deficiency method also aids in understanding the genetic basis of other important traits, such as nutrient use efficiency, quality attributes, and adaptability to changing climates. By mapping genes associated with these traits, researchers can develop crop varieties that are better suited to meet the demands of modern agriculture.

Identifying Candidate Genes

The deficiency method is also valuable for identifying candidate genes responsible for specific traits. By narrowing down the chromosomal regions associated with a trait, researchers can identify potential candidate genes for further study.

1. **Gene Annotation**: Once candidate genes are identified, researchers can annotate these genes to determine their function and role in the trait of interest. Gene annotation involves analyzing the gene sequence, structure, and expression patterns to gain insights into its biological function. This information is essential for understanding the genetic basis of traits and developing targeted breeding strategies.
2. **Functional Validation**: Functional validation is a critical step in confirming the role of candidate genes in trait expression. This involves using techniques like gene knockouts, overexpression, and RNA interference to assess the impact of candidate genes on the trait. Functional validation provides evidence for the involvement of specific genes in trait expression and supports their use in crop improvement.

3. **Comparative Genomics**: Comparative genomics involves comparing the genomes of different species to identify conserved genes and regions associated with specific traits. By integrating data from the deficiency method with comparative genomics, researchers can identify candidate genes that are conserved across species, providing valuable insights into the genetic basis of traits and their evolution.

The identification of candidate genes is facilitated by advances in bioinformatics and genomic data analysis. Researchers can use computational tools to predict gene function, identify regulatory elements, and assess the impact of genetic variation on gene expression. This integration of genomic data with the deficiency method enhances the efficiency of gene identification and characterization, accelerating the development of improved crop varieties.

Enhancing Genetic Diversity

The deficiency method contributes to enhancing genetic diversity in crops by facilitating the introduction of new genetic material and traits.

1. **Introgression of Novel Traits**: The deficiency method allows researchers to introduce novel traits from wild relatives and exotic germplasm into cultivated varieties. By identifying chromosomal regions associated with desirable traits, breeders can introgress these regions into new varieties through cross-breeding and selection. This process enhances genetic diversity and expands the genetic base of cultivated crops.
2. **Genetic Recombination**: Genetic recombination is a powerful tool for creating new combinations of genes and traits. The deficiency method facilitates genetic recombination by identifying chromosomal regions that can be recombined to produce novel trait combinations. This process enhances the adaptability and resilience of crops to changing environmental conditions.
3. **Conservation of Genetic Resources**: The deficiency method can also be used to conserve and utilize genetic resources in crop breeding. By identifying and mapping genes associated with important traits, researchers can develop strategies for preserving genetic diversity and incorporating valuable genetic material into breeding programs. This approach ensures the sustainable use of genetic resources and supports the development of resilient crop varieties.

Challenges and Limitations

While the deficiency method has numerous applications, it also presents challenges and limitations that researchers must address.

Difficulty in Creating Specific Deficiencies

Creating specific chromosomal deficiencies can be challenging due to the random nature of mutagenesis and the complexity of chromosome structure.

1. **Randomness of Mutagenesis**: Radiation and chemical mutagenesis can result in random chromosomal deletions, making it difficult to create specific deficiencies. Researchers must carefully select and screen for desired deletions, which can be time-consuming and labor-intensive.

2. **Complex Chromosome Structure**: Chromosomes have complex structures, with regions that are difficult to delete without affecting adjacent genes. This complexity can complicate the creation of specific deficiencies and limit the precision of gene mapping.
3. **Technical Limitations**: The creation of chromosomal deficiencies may also be limited by technical constraints, such as the availability of suitable mutagenesis techniques and the capacity to analyze and interpret phenotypic data. These limitations can impact the efficiency and accuracy of the deficiency method and require the development of improved techniques and methodologies.

Researchers are continually working to overcome these challenges by developing new mutagenesis techniques, refining deletion mapping methods, and integrating data from multiple sources to enhance the precision and efficiency of gene mapping. Advances in technology and bioinformatics are expected to address many of the limitations associated with the deficiency method, enabling more accurate and efficient mapping of genes in crops.

Potential for Lethal Effects

Chromosomal deficiencies can have lethal effects, particularly when essential genes are deleted. This presents challenges in studying certain traits and requires careful consideration of experimental design.

1. **Essential Genes**: Deleting essential genes can result in lethality or severe phenotypic effects, making it difficult to study specific traits. Researchers must identify non-lethal deficiencies and use complementary techniques to study essential genes.
2. **Compensatory Mechanisms**: Organisms may have compensatory mechanisms that mask the effects of certain deficiencies. These mechanisms can complicate the interpretation of phenotypic data and require additional experiments to confirm gene function.
3. **Phenotypic Variability**: The phenotypic effects of chromosomal deficiencies may vary depending on the genetic background and environmental conditions. This variability can complicate the analysis and interpretation of phenotypic data and requires careful consideration of experimental design and data analysis.

To address these challenges, researchers are developing strategies to identify and study essential genes, such as using conditional knockouts and inducible expression systems. These approaches allow for the controlled study of essential genes and their functions, minimizing the risk of lethality and enabling a more comprehensive understanding of gene function and regulation.

Limited Resolution

The deficiency method has limited resolution compared to other gene mapping techniques, making it difficult to pinpoint the exact location of genes within a chromosomal region.

1. **Resolution Constraints**: The resolution of the deficiency method is limited by the size and number of available chromosomal deletions. Researchers may need to create and analyze a large number of overlapping deficiencies to achieve high-resolution mapping, which can be time-consuming and labor-intensive.
2. **Integration with Other Techniques**: The limited resolution of the deficiency method can

be addressed by integrating it with other gene mapping techniques, such as fine mapping, association mapping, and genome-wide association studies (GWAS). These techniques provide complementary information that can enhance the resolution and accuracy of gene mapping.

3. **Advancements in Technology**: Advances in genomic technologies, such as whole-genome sequencing and high-throughput genotyping, are improving the resolution and efficiency of gene mapping. These technologies enable researchers to identify and characterize smaller deletions and integrate data from multiple sources, enhancing the precision and accuracy of gene mapping.

Despite its limitations, the deficiency method remains a valuable tool for gene mapping and studying gene function. Researchers continue to refine and enhance this method, integrating it with other techniques and technologies to overcome its limitations and expand its applications in crop improvement.

B. Interchange: Genetic Consequences

Interchanges, also known as reciprocal translocations, are chromosomal rearrangements involving the exchange of segments between non-homologous chromosomes. These genetic alterations can have profound consequences, affecting gene expression, chromosomal stability, and the overall genetic diversity of plant populations. In the context of crop improvement, understanding the genetic consequences of interchanges is crucial for utilizing them in developing superior crop varieties with enhanced traits. This section delves into the mechanisms of interchange, their genetic consequences, and their implications for crop improvement.

Mechanism of Interchange

Interchanges involve the breakage and rejoining of chromosome segments between non-homologous chromosomes. This process can occur naturally or be induced through external factors such as radiation or chemicals. The mechanism of interchange is intricate, involving several key steps:

Chromosome Breakage and Rejoining

The interchange process begins with the breakage of chromosomes at specific sites. These breaks can occur spontaneously due to errors in DNA replication or repair mechanisms or be induced by mutagenic agents. Once the breaks occur, the broken ends of the chromosomes undergo rejoining, leading to the exchange of segments between non-homologous chromosomes. This rejoining process is mediated by cellular repair machinery, which can sometimes lead to incorrect rejoining, resulting in an interchange.

Multivalent Formation During Meiosis

Interchanges lead to the formation of complex multivalents during meiosis, the process of gamete formation. Normally, homologous chromosomes pair and segregate during meiosis, but interchanges result in the pairing of non-homologous chromosomes. This altered pairing leads to the formation of multivalents, which are complex structures involving multiple chromosomes. The presence of multivalents can disrupt the normal segregation of chromosomes, leading to genetic imbalances in the resulting gametes.

Segregation Patterns and Aneuploidy

The presence of interchanges affects the segregation patterns of chromosomes during meiosis. As the multivalents attempt to segregate, they can do so in several ways, leading to different outcomes. These segregation patterns can result in balanced or unbalanced gametes. Balanced gametes have the correct chromosomal complement, while unbalanced gametes have duplications or deletions of genetic material. The formation of unbalanced gametes can lead to aneuploidy, a condition where the zygote has an abnormal number of chromosomes, potentially resulting in reduced fertility or viability.

Genetic Consequences of Interchange

The genetic consequences of interchange are multifaceted, impacting various aspects of an organism's genetic makeup. These consequences can have both beneficial and detrimental effects, influencing plant breeding and crop improvement strategies.

Gene Disruption and Expression

Interchanges can disrupt genes at the breakpoints, leading to loss of function or altered gene expression. This disruption can result in phenotypic changes, affecting traits such as growth, development, and stress response. For example, if an interchange disrupts a gene responsible for disease resistance, the plant may become more susceptible to pathogens. Conversely, if an interchange activates a previously silent beneficial gene, it may enhance the plant's performance or adaptability.

Generation of Novel Gene Combinations

Interchanges contribute to genetic diversity by creating novel gene combinations. The exchange of chromosomal segments can bring together genes from different regions, leading to new combinations that may confer adaptive advantages. This increased genetic variability provides raw material for natural selection and evolution, enhancing the resilience of plant populations to environmental changes.

Impacts on Chromosomal Stability and Fertility

While interchanges can increase genetic diversity, they can also compromise chromosomal stability and fertility. The presence of multivalents during meiosis can lead to the formation of unbalanced gametes, resulting in reduced fertility. Plants carrying interchanges may exhibit decreased seed set or produce non-viable offspring. This can be a significant limitation in breeding programs, as reduced fertility may hinder the propagation of desirable traits.

Applications in Crop Improvement

Interchanges have significant implications for crop improvement, offering both opportunities and challenges for plant breeders. By understanding and manipulating interchanges, breeders can develop crop varieties with enhanced traits and improved adaptability.

Enhancing Disease Resistance

One of the primary goals of crop improvement is enhancing disease resistance. Interchanges can play a crucial role in this endeavor by introducing novel resistance genes or modifying existing ones. For instance, if an interchange disrupts a susceptibility gene, it may confer resistance to a particular pathogen. Alternatively, an interchange may activate a latent resistance gene, providing the plant with an additional defense mechanism. By selecting plants with interchange-induced resistance, breeders can develop crop varieties that are more resilient to diseases, reducing the need for chemical pesticides and increasing agricultural sustainability.

Improving Yield and Quality

Interchanges can also improve crop yield and quality by reshuffling genes involved in plant growth, development, and metabolism. By introducing new gene combinations, interchanges can enhance traits such as plant height, seed size, and fruit quality. For example, an interchange may lead to increased expression of genes involved in nutrient uptake, resulting in improved plant vigor and higher yields. Additionally, interchanges can modify genes related to flavor, texture, or nutritional content, enhancing the quality of agricultural products. These improvements can increase crop value and meet consumer demands for high-quality produce.

Challenges and Considerations

While interchanges offer significant opportunities for crop improvement, they also pose challenges and require careful management. One of the primary challenges is the potential for reduced fertility in plants carrying interchanges. This can limit the propagation of desirable traits and hinder the development of new crop varieties. Breeders must carefully assess the fertility and viability of plants with interchanges to ensure successful breeding outcomes. Additionally, the effects of interchanges can be unpredictable, leading to unintended phenotypic changes or compromised plant performance. Therefore, thorough evaluation and selection are essential to harness the benefits of interchanges while minimizing potential drawbacks.

Case Studies in Crop Improvement

Several case studies highlight the successful utilization of interchanges in crop improvement programs. These examples demonstrate the potential of interchanges to enhance crop performance, adaptability, and resilience.

Wheat

In wheat (*Triticum aestivum*), interchanges have been used to introduce disease resistance genes from wild relatives. By crossing wheat with wild species and selecting plants with favorable interchanges, breeders have developed wheat varieties with enhanced resistance to rust diseases. These interchanges have enabled the transfer of resistance genes that were not present in cultivated wheat, providing a valuable source of genetic diversity for breeding programs.

Maize

In maize (*Zea mays*), interchanges have been utilized to improve yield and stress tolerance. By selecting plants with beneficial interchange-induced traits, breeders have developed maize varieties with increased drought tolerance and improved nutrient use efficiency. These interchanges have reshuffled genes involved in stress response and metabolic pathways, enhancing the plant's ability to thrive in challenging environments.

Rice

In rice (*Oryza sativa*), interchanges have been employed to enhance grain quality and yield. By introducing interchanges between chromosomes carrying yield-related genes, breeders have developed rice varieties with increased grain size and improved milling quality. These interchanges have modified the expression of genes involved in grain development, resulting in superior rice cultivars that meet consumer preferences and market demands.

Techniques for Identifying Interchanges

Identifying and characterizing interchanges is crucial for understanding their genetic consequences and harnessing their potential in crop improvement. Several techniques are employed to detect and analyze interchanges at the chromosomal and molecular levels.

Cytogenetic Techniques

Cytogenetic techniques, such as chromosome banding and fluorescence in situ hybridization (FISH), are used to visualize and identify interchanges. Chromosome banding techniques, such as G-banding and C-banding, allow researchers to distinguish individual chromosomes and identify structural abnormalities. FISH involves labeling specific DNA sequences with fluorescent probes and hybridizing them to metaphase chromosomes. This technique can detect interchanges by visualizing the altered positions of chromosome segments.

Molecular Techniques

Molecular techniques, such as polymerase chain reaction (PCR) and next-generation sequencing (NGS), are used to analyze the genetic consequences of interchanges at the DNA level. PCR can amplify specific regions of DNA to detect the presence of interchanges and analyze breakpoint sequences. NGS allows for high-throughput sequencing of entire genomes, providing comprehensive information on the genetic changes associated with interchanges. These molecular techniques complement cytogenetic methods and provide detailed insights into the genetic consequences of interchanges.

Applications in Biotechnology

The understanding of interchange genetic consequences has significant applications in biotechnology and genetic engineering. By leveraging interchanges, scientists can manipulate plant genomes to introduce desirable traits and enhance crop performance.

Genetic Engineering and Transformation

Interchanges can be used in genetic engineering to introduce foreign genes into plant genomes. By creating interchanges between chromosomes carrying target genes and vectors carrying desired traits, scientists can integrate new genes into the plant's genetic material. This approach allows for precise gene insertion and expression, enabling the development of genetically modified crops with improved traits, such as herbicide resistance or enhanced nutritional content.

Gene Editing Technologies

Gene editing technologies, such as CRISPR-Cas9, can be used to induce targeted interchanges and study their genetic consequences. By creating specific interchanges at desired genomic locations, researchers can investigate the effects of structural rearrangements on gene function and expression. This approach provides valuable insights into the roles of specific genes and their interactions, facilitating the development of crops with optimized traits.

Future Prospects and Challenges

While interchanges offer significant potential for crop improvement and biotechnology, several challenges and future prospects must be considered.

Ethical and Regulatory Considerations

The use of interchanges in crop improvement and genetic engineering raises ethical and regulatory considerations. The development and release of genetically modified crops require rigorous testing and evaluation to ensure safety and environmental sustainability. Regulatory frameworks must be established to oversee the use of interchanges and other genetic engineering techniques, balancing innovation with public safety and environmental protection.

Advancements in Technology

Advancements in technology, such as high-throughput sequencing and gene editing, will continue to enhance our understanding of interchange genetic consequences. These technologies will enable more precise manipulation of plant genomes, allowing for the development of crops with tailored traits. Continued research and innovation in this field will drive progress in crop improvement and biotechnology.

Integration with Traditional Breeding

Interchanges should be integrated with traditional breeding methods to maximize their potential in crop improvement. Combining interchange-based approaches with conventional breeding techniques can enhance genetic diversity and improve breeding outcomes. Collaboration between researchers, breeders, and farmers will be essential to harness the full potential of interchanges and address the challenges of modern agriculture.

C. Identification of chromosomes involved and gene location

Chromosome Identification in Interchanges

Chromosome identification is a critical step in studying interchanges, as it allows researchers to determine which chromosomes are involved in the rearrangement. Various techniques are employed to achieve accurate identification and characterization of chromosomes.

Cytogenetic Techniques

Cytogenetic techniques are fundamental tools for chromosome identification in interchanges. These methods provide a visual representation of chromosomes, allowing researchers to detect structural abnormalities and identify specific chromosomes involved in interchanges.

» **Chromosome Banding:** Chromosome banding techniques, such as G-banding, C-banding, and R-banding, are widely used to distinguish individual chromosomes and identify structural rearrangements. These techniques involve staining chromosomes with specific dyes that produce distinct banding patterns, allowing for the identification of chromosomes and the detection of interchanges.

» **Fluorescence In Situ Hybridization (FISH):** FISH is a powerful technique used to identify specific chromosomes and visualize interchanges at a high resolution. This method involves labeling DNA probes with fluorescent dyes and hybridizing them to metaphase chromosomes. The fluorescent signals allow researchers to identify the location of specific DNA sequences and detect interchanges involving those sequences. FISH is particularly useful for identifying chromosomes in complex interchanges and mapping gene locations.

Molecular Techniques

Molecular techniques complement cytogenetic methods by providing detailed information about the genetic sequences involved in interchanges. These techniques are essential for identifying chromosomes and locating genes at the DNA level.

- **Polymerase Chain Reaction (PCR):** PCR is a widely used molecular technique for amplifying specific DNA regions. This method can be employed to detect the presence of interchanges by amplifying breakpoint regions and analyzing the sequences involved. PCR is useful for identifying chromosomes involved in interchanges and locating genes near the breakpoints.
- **Next-Generation Sequencing (NGS):** NGS is a high-throughput sequencing technology that allows for the comprehensive analysis of entire genomes. This technique can be used to identify chromosomes involved in interchanges by sequencing the genomic regions surrounding the breakpoints. NGS provides detailed information about the genetic changes associated with interchanges, facilitating the identification of chromosomes and gene locations.

Gene Location and Mapping

Determining the location of genes involved in interchanges is crucial for understanding the genetic consequences of these rearrangements and harnessing their potential in crop improvement. Several techniques are employed to map gene locations and analyze their interactions with interchanges.

Genetic Mapping

Genetic mapping involves the identification of specific genes and their locations on chromosomes. This process is essential for understanding how interchanges affect gene expression and contribute to phenotypic changes.

- **Linkage Mapping:** Linkage mapping is a traditional method used to map genes based on their inheritance patterns. This technique involves analyzing the segregation of genetic markers and phenotypic traits in a population to determine the relative positions of genes on chromosomes. Linkage mapping is useful for identifying genes involved in interchanges and understanding their genetic interactions.
- **QTL Mapping:** Quantitative trait locus (QTL) mapping is a method used to identify the chromosomal regions associated with quantitative traits. This technique involves analyzing the association between genetic markers and phenotypic variation to locate QTLs on chromosomes. QTL mapping is particularly useful for identifying genes involved in complex traits influenced by interchanges.

Physical Mapping

Physical mapping provides a more precise locatiòn of genes by analyzing the physical distance between genetic markers on chromosomes. This approach is essential for understanding the structural organization of chromosomes and the effects of interchanges on gene location.

- **Radiation Hybrid Mapping:** Radiation hybrid mapping is a technique used to determine the physical distance between genes on chromosomes. This method involves creating hybrid cell lines containing fragments of donor DNA and analyzing the segregation of genetic markers. Radiation hybrid mapping provides detailed information about the physical location of genes and can be used to map genes involved in interchanges.

» **Comparative Genomic Hybridization (CGH):** CGH is a technique used to detect genomic imbalances and map gene locations on chromosomes. This method involves comparing the DNA of different individuals or species to identify variations in copy number and chromosomal structure. CGH is useful for identifying genes involved in interchanges and analyzing their effects on genome organization.

Applications in Crop Improvement

The identification of chromosomes involved in interchanges and the precise location of genes have significant implications for crop improvement. By understanding the genetic consequences of interchanges, researchers and breeders can develop crop varieties with enhanced traits and improved adaptability.

Marker-Assisted Selection

Marker-assisted selection (MAS) is a breeding technique that uses genetic markers to select plants with desirable traits. By identifying chromosomes involved in interchanges and mapping genes, MAS can be used to select plants with beneficial interchange-induced traits. This approach allows breeders to efficiently develop crop varieties with improved yield, quality, and disease resistance.

Gene Editing and Transformation

The identification of chromosomes and gene locations involved in interchanges facilitates gene editing and transformation efforts. By understanding the genetic changes associated with interchanges, researchers can use gene editing technologies, such as CRISPR-Cas9, to induce targeted interchanges and introduce desirable traits into plant genomes. This approach allows for precise manipulation of plant genomes, enabling the development of genetically modified crops with enhanced traits.

Conservation of Genetic Diversity

Understanding the genetic consequences of interchanges is essential for conserving genetic diversity in crop populations. By identifying chromosomes and genes involved in interchanges, researchers can assess the impact of these rearrangements on genetic diversity and develop strategies to preserve valuable genetic resources. This knowledge is crucial for maintaining the resilience of crop populations in the face of environmental challenges and changing agricultural demands.

Challenges and Future Prospects

While significant progress has been made in identifying chromosomes and gene locations involved in interchanges, several challenges and future prospects must be considered.

Complexity of Interchanges

Interchanges can be complex, involving multiple chromosomes and intricate rearrangements. The complexity of these interchanges can make it challenging to accurately identify chromosomes and map gene locations. Continued research and technological advancements are needed to improve the resolution and accuracy of chromosome identification and gene mapping techniques.

Integration with Genomic Data

The integration of chromosome identification and gene mapping data with genomic information is essential for understanding the broader genetic context of interchanges. Advances in bioinformatics and computational biology will facilitate the integration of genomic data with cytogenetic and molecular techniques, providing a comprehensive understanding of interchanges and their genetic consequences.

Ethical and Environmental Considerations

The use of interchanges in crop improvement and genetic engineering raises ethical and environmental considerations. The development and release of genetically modified crops require careful assessment of their potential impact on ecosystems and biodiversity. Regulatory frameworks and ethical guidelines must be established to ensure the responsible use of interchanges in agriculture and biotechnology.

D. Balanced lethal systems, their maintenance, and utility

Balanced lethal systems are intriguing genetic mechanisms where lethality is balanced within a population, ensuring the maintenance of specific genetic configurations over generations. These systems play a crucial role in maintaining genetic stability and diversity, especially in controlled breeding programs and cytogenetic research. The mechanics, maintenance, and utility of balanced lethal systems provide valuable insights into genetic inheritance and offer practical applications in crop improvement and breeding strategies.

Balanced Lethal Systems

Balanced lethal systems occur when two or more alleles at different loci are mutually lethal when homozygous but survive in heterozygous combinations. In these systems, the lethal alleles are balanced, such that the survival of the organism relies on the presence of both alleles in a heterozygous state. This ensures that only heterozygous individuals are viable, while homozygous individuals do not survive. Such systems are often observed in diploid organisms and can be artificially maintained in laboratory or breeding settings.

Mechanism of Balanced Lethal Systems

Balanced lethal systems typically involve two or more alleles at different loci that are lethal when present in a homozygous state. These alleles can arise due to mutations, chromosomal rearrangements, or other genetic alterations. The system is maintained through heterozygous individuals that carry one copy of each lethal allele, ensuring the continued presence of the alleles within the population. During reproduction, only the heterozygous combinations are viable, while homozygous combinations result in lethality.

- **Allelic Lethality:** The alleles involved in balanced lethal systems are often recessive lethal alleles. This means that the lethal effect is masked in heterozygotes but becomes evident in homozygotes. For example, in a simple two-locus system with alleles A1 and A2 at one locus and B1 and B2 at another, only the A1B2 and A2B1 heterozygotes survive, while A1B1 and A2B2 homozygotes are lethal.

» **Chromosomal Rearrangements:** Balanced lethal systems can also result from chromosomal rearrangements, such as translocations or inversions, that create lethal alleles or disrupt essential genes. These rearrangements can lead to the formation of lethal combinations in homozygotes, maintaining the system's balance in heterozygotes.

Maintenance of Balanced Lethal Systems

Maintaining balanced lethal systems requires careful management to ensure the continued presence of heterozygous individuals. In nature, these systems are relatively rare, but they can be artificially maintained in laboratory or breeding settings to achieve specific genetic outcomes.

Natural Maintenance

In natural populations, balanced lethal systems can be maintained through natural selection and genetic drift. The presence of lethal alleles can be beneficial if they confer an advantage in a heterozygous state. For example, a lethal allele may provide resistance to a particular pathogen or environmental stress, ensuring its persistence in the population.

» **Heterozygote Advantage:** The key to maintaining balanced lethal systems in nature is the heterozygote advantage, where heterozygous individuals have a selective advantage over homozygous individuals. This advantage can result from increased fitness, resistance to diseases, or adaptation to specific environmental conditions.

» **Genetic Drift:** In small populations, genetic drift can play a significant role in maintaining balanced lethal systems. Random fluctuations in allele frequencies can lead to the persistence of lethal alleles, even if they do not confer a significant advantage. This can occur in isolated populations or in species with restricted gene flow.

Artificial Maintenance

In controlled settings, such as laboratories or breeding programs, balanced lethal systems can be artificially maintained to achieve specific genetic objectives. This involves carefully managing breeding strategies and selecting individuals to ensure the continued presence of heterozygotes.

» **Breeding Strategies:** Breeders can use specific breeding strategies to maintain balanced lethal systems, such as controlled crosses between heterozygous individuals. By selecting and breeding only heterozygotes, breeders can ensure the persistence of the system and prevent the emergence of lethal homozygotes.

» **Genetic Markers:** The use of genetic markers can aid in maintaining balanced lethal systems by allowing breeders to identify and select heterozygous individuals. Molecular markers, such as single nucleotide polymorphisms (SNPs) or microsatellites, can be used to track the presence of lethal alleles and guide breeding decisions.

Utility of Balanced Lethal Systems

Balanced lethal systems have several applications in genetics and plant breeding, offering unique opportunities for research and crop improvement.

Research Applications

Balanced lethal systems provide valuable tools for genetic research, allowing scientists to study gene interactions, chromosomal rearrangements, and evolutionary processes.

- **Gene Mapping:** Balanced lethal systems can be used for gene mapping by analyzing the segregation patterns of lethal alleles. By studying the inheritance of these alleles, researchers can identify linked genes and map their locations on chromosomes. This information is crucial for understanding genetic architecture and identifying genes associated with specific traits.
- **Evolutionary Studies:** Balanced lethal systems offer insights into evolutionary processes by revealing how lethal alleles are maintained in populations. By studying these systems, researchers can gain a better understanding of genetic drift, selection pressures, and the role of lethal alleles in adaptation and speciation.

Plant Breeding Applications

In plant breeding, balanced lethal systems can be harnessed to develop crop varieties with desirable traits and improved performance.

- **Hybrid Vigor:** Balanced lethal systems can contribute to hybrid vigor, or heterosis, by promoting the maintenance of heterozygous individuals with enhanced traits. By selecting and breeding heterozygotes, breeders can develop hybrids with improved yield, disease resistance, and stress tolerance.
- **Trait Selection:** Balanced lethal systems can be used for trait selection by maintaining specific genetic configurations that confer desirable traits. For example, breeders can use balanced lethal systems to select for resistance to diseases, pests, or environmental stresses, improving crop performance and sustainability.
- **Conservation of Genetic Diversity:** Balanced lethal systems can help conserve genetic diversity by maintaining rare alleles that might otherwise be lost in a population. This is particularly important in crop improvement, where maintaining genetic diversity is crucial for resilience to changing environmental conditions and emerging threats.

Challenges and Considerations

While balanced lethal systems offer several advantages, they also present challenges and require careful management to avoid potential drawbacks.

Risk of Inbreeding

One of the primary challenges of maintaining balanced lethal systems is the risk of inbreeding. By selecting and breeding only heterozygous individuals, there is a risk of reducing genetic diversity and increasing the likelihood of inbreeding depression. This can lead to a decline in fitness and performance, reducing the effectiveness of the system.

- **Managing Inbreeding:** To mitigate the risk of inbreeding, breeders must carefully manage the genetic diversity within a population. This can be achieved by introducing new genetic material, maintaining a diverse gene pool, and using genetic markers to monitor genetic variation.

Lethal Allele Management

The presence of lethal alleles in a population can pose challenges for breeding programs, as they can lead to the production of non-viable offspring. Managing these alleles requires careful breeding strategies and genetic monitoring.

» **Genetic Monitoring:** The use of genetic markers and molecular techniques can aid in managing lethal alleles by allowing breeders to identify and track their presence. This information can guide breeding decisions and help prevent the emergence of lethal homozygotes.

» **Breeding Strategies:** Breeders must develop specific breeding strategies to manage lethal alleles, such as controlled crosses and selection of heterozygous individuals. By carefully managing breeding practices, breeders can ensure the continued maintenance of balanced lethal systems and avoid the production of non-viable offspring.

Ethical Considerations

The use of balanced lethal systems in breeding and research raises ethical considerations, particularly regarding the maintenance of lethal alleles and the potential impact on animal welfare and biodiversity.

» **Ethical Guidelines:** Ethical guidelines and regulatory frameworks must be established to ensure the responsible use of balanced lethal systems in breeding and research. These guidelines should consider the welfare of organisms, the conservation of biodiversity, and the potential impact on ecosystems.

» **Public Perception:** The use of balanced lethal systems may also face public scrutiny and concerns about the manipulation of genetic material. Researchers and breeders must engage with stakeholders and the public to communicate the benefits and risks of balanced lethal systems and address any ethical concerns.

Future Prospects

Balanced lethal systems hold promise for advancing genetic research and crop improvement, with several future prospects for their application.

Advances in Genomic Technologies

Advances in genomic technologies, such as high-throughput sequencing and gene editing, will enhance our understanding and utilization of balanced lethal systems. These technologies will enable more precise identification and manipulation of lethal alleles, facilitating the development of improved breeding strategies and crop varieties.

» **Gene Editing:** Gene editing technologies, such as CRISPR-Cas9, offer the potential to modify lethal alleles and create new balanced lethal systems. By targeting specific genes and creating novel genetic configurations, researchers can develop crops with enhanced traits and improved performance.

» **Genomic Selection:** Genomic selection, which uses genome-wide markers to predict breeding values, can be integrated with balanced lethal systems to enhance breeding efficiency. This approach allows breeders to select individuals with the desired genetic configurations and accelerate the development of improved crop varieties.

Integration with Breeding Programs

The integration of balanced lethal systems with traditional breeding programs offers new opportunities for crop improvement. By combining the benefits of balanced lethal systems with conventional breeding techniques, breeders can develop crops with enhanced traits and improved adaptability.

» **Hybrid Breeding:** Balanced lethal systems can be used in hybrid breeding programs to promote heterosis and improve hybrid performance. By selecting and breeding heterozygous individuals, breeders can develop hybrids with enhanced yield, quality, and stress tolerance.

» **Crossbreeding:** Balanced lethal systems can also be integrated into crossbreeding programs to introduce novel traits and maintain genetic diversity. By incorporating lethal alleles into crossbreeding efforts, breeders can develop crops with new genetic combinations and improved resilience.

E. Multiple interchanges-use in producing inbreds

Multiple interchanges refer to complex chromosomal rearrangements involving the exchange of segments between several non-homologous chromosomes. These interchanges can significantly impact genetic architecture and are utilized strategically in plant breeding, particularly in producing inbred lines. The role of multiple interchanges in generating inbred lines can provide valuable insights into genetic inheritance and breeding strategies.

Multiple Interchanges

Multiple interchanges occur when segments from two or more non-homologous chromosomes are reciprocally exchanged, leading to complex genetic configurations. These rearrangements can arise naturally or be induced through mutagenic agents, and they result in structural chromosomal changes that have profound implications for genetic diversity and trait expression.

Mechanism of Multiple Interchanges

Multiple interchanges involve intricate chromosomal breakage and rejoining events, resulting in a reshuffling of genetic material across several chromosomes. This process can be described as follows:

» **Chromosomal Breakage:** The interchange process begins with chromosomal breakage at specific sites. These breaks can be spontaneous, due to errors in DNA replication or repair mechanisms, or induced by external agents like radiation or chemicals.

» **Rejoining and Exchange:** Following breakage, the broken chromosome segments undergo rejoining, leading to the exchange of genetic material between non-homologous chromosomes. This rejoining is mediated by cellular repair mechanisms, which can sometimes misalign chromosome segments, leading to interchanges.

» **Multivalent Formation:** During meiosis, the presence of multiple interchanges can result in the formation of multivalents, where non-homologous chromosomes pair with each other. This altered pairing disrupts normal segregation, potentially leading to genetic imbalances in the resulting gametes.

Genetic Consequences of Multiple Interchanges

Multiple interchanges can have several genetic consequences, affecting gene expression, chromosomal stability, and genetic diversity. These consequences play a critical role in plant breeding and the development of inbred lines.

Gene Disruption and Novel Gene Combinations

The rearrangement of chromosome segments in multiple interchanges can lead to gene disruption at the breakpoints. This disruption may result in the loss of gene function or altered gene expression, affecting phenotypic traits.

» **Novel Gene Combinations:** Multiple interchanges create novel gene combinations by bringing together genetic material from different chromosomal regions. These new combinations can lead to the expression of unique traits, contributing to genetic diversity within a population. This increased variability provides raw material for selection and breeding.

Chromosomal Stability and Segregation Patterns

The presence of multiple interchanges affects chromosomal stability and segregation patterns during meiosis. Multivalents formed by interchanges can lead to irregular segregation, resulting in aneuploid gametes with missing or extra chromosomes.

» **Balanced vs. Unbalanced Gametes:** The segregation of multivalents can produce both balanced and unbalanced gametes. Balanced gametes maintain the correct chromosomal complement, while unbalanced gametes exhibit duplications or deletions of genetic material, potentially leading to reduced fertility or viability.

Utilization in Producing Inbreds

Multiple interchanges are strategically utilized in plant breeding to produce inbred lines, which are homozygous for specific traits. Inbred lines are essential in hybrid breeding, providing a stable genetic foundation for developing superior crop varieties.

Advantages of Inbred Lines

Inbred lines offer several advantages in plant breeding, contributing to the development of uniform and stable crop varieties.

» **Homozygosity:** Inbred lines are homozygous for specific traits, ensuring uniformity and stability in the resulting progeny. This homozygosity allows breeders to fix desirable traits and eliminate unwanted variations, leading to consistent crop performance.

» **Genetic Stability:** The genetic stability of inbred lines makes them valuable in breeding programs, as they provide a reliable genetic base for developing hybrids and new cultivars. This stability reduces the risk of genetic drift and ensures the reproducibility of desired traits.

» **Trait Fixation:** Inbred lines facilitate the fixation of desirable traits, such as disease resistance, yield, and quality. By selecting and breeding individuals with advantageous traits, breeders can develop inbred lines that exhibit superior performance and adaptability.

Role of Multiple Interchanges in Inbred Production

Multiple interchanges play a crucial role in the development of inbred lines by enhancing genetic diversity and facilitating the fixation of desirable traits.

- **Genetic Variation:** Multiple interchanges introduce genetic variation by creating novel gene combinations, providing breeders with a diverse pool of genetic material to select from. This variation is essential for identifying and fixing beneficial traits in inbred lines.
- **Selection and Fixation:** The genetic reshuffling caused by multiple interchanges allows breeders to select and fix desirable traits more effectively. By identifying individuals with advantageous gene combinations, breeders can develop inbred lines with improved traits and performance.
- **Resistance to Diseases and Pests:** Multiple interchanges can introduce new resistance genes or modify existing ones, enhancing the resistance of inbred lines to diseases and pests. This increased resistance contributes to the development of robust and resilient crop varieties.

Techniques for Inducing and Identifying Multiple Interchanges

To utilize multiple interchanges in producing inbred lines, it is essential to induce and identify these rearrangements accurately. Several techniques are employed to achieve this:

Inducing Multiple Interchanges

- Multiple interchanges can be induced using physical and chemical mutagens, which create chromosomal breakages and facilitate the exchange of genetic material.
- **Radiation:** Radiation, such as X-rays or gamma rays, is commonly used to induce chromosomal interchanges. The high-energy radiation causes DNA breaks, promoting the formation of multiple interchanges and increasing genetic variability.
- **Chemical Mutagens:** Chemical mutagens, such as ethyl methanesulfonate (EMS), can induce chromosomal rearrangements by altering DNA structure and causing breaks. These chemicals create opportunities for genetic recombination and interchange formation.

Identifying Multiple Interchanges

Accurate identification of multiple interchanges is crucial for understanding their genetic consequences and utilizing them in breeding programs.

- **Cytogenetic Techniques:** Cytogenetic techniques, such as chromosome banding and fluorescence in situ hybridization (FISH), are used to visualize and identify interchanges. These methods provide a visual representation of chromosomes, allowing researchers to detect structural abnormalities and characterize interchanges.
- **Molecular Techniques:** Molecular techniques, such as polymerase chain reaction (PCR) and next-generation sequencing (NGS), provide detailed information about the genetic sequences involved in interchanges. These methods allow for the precise identification of breakpoints and the characterization of genetic changes associated with interchanges.

Challenges and Considerations in Using Multiple Interchanges

While multiple interchanges offer significant potential for producing inbred lines, they also present challenges that must be carefully managed.

Genetic Complexity and Stability

The genetic complexity introduced by multiple interchanges can complicate breeding efforts and affect chromosomal stability.

- **Managing Complexity:** Breeders must carefully manage the complexity of genetic material resulting from multiple interchanges. This involves selecting individuals with desirable traits while ensuring genetic stability and avoiding potential negative effects of genetic imbalances.
- **Ensuring Stability:** Maintaining chromosomal stability is crucial for the success of inbred lines. Breeders must monitor and manage the segregation patterns of multivalents to prevent the emergence of unbalanced gametes and ensure the viability of progeny.

Ethical and Environmental Considerations

The use of multiple interchanges in breeding raises ethical and environmental considerations that must be addressed.

- **Regulatory Frameworks:** Regulatory frameworks and ethical guidelines are necessary to ensure the responsible use of multiple interchanges in plant breeding. These frameworks should consider the potential impact on ecosystems and biodiversity and address public concerns about genetic manipulation.
- **Conservation of Biodiversity:** The introduction of multiple interchanges must be balanced with the conservation of genetic diversity. Breeders should ensure that the use of interchanges does not lead to the loss of valuable genetic resources or negatively impact biodiversity.

Future Prospects of Multiple Interchanges in Breeding

The future of plant breeding and inbred production will be shaped by advances in genomic technologies and the strategic use of multiple interchanges.

Integration with Genomic Tools

Advances in genomic tools, such as gene editing and genomic selection, will enhance the utilization of multiple interchanges in breeding programs.

- **Gene Editing:** Gene editing technologies, such as CRISPR-Cas9, offer the potential to induce targeted interchanges and modify specific genes. This precision allows breeders to create novel genetic configurations and develop crops with enhanced traits.
- **Genomic Selection:** Genomic selection, which uses genome-wide markers to predict breeding values, can be integrated with multiple interchanges to enhance breeding efficiency. This approach allows breeders to select individuals with desired genetic configurations and accelerate the development of improved inbred lines.

Breeding for Climate Resilience

Multiple interchanges will play a crucial role in breeding for climate resilience, providing genetic variability and adaptability to changing environmental conditions.

» **Developing Resilient Varieties:** By utilizing multiple interchanges, breeders can develop crop varieties with enhanced tolerance to abiotic stresses, such as drought, heat, and salinity. This resilience will be essential for ensuring food security in the face of climate change.

» **Enhancing Genetic Diversity:** Multiple interchanges will continue to enhance genetic diversity, providing a diverse pool of genetic material for breeding resilient crops. This diversity will enable breeders to respond to emerging challenges and develop crops that thrive in diverse environments.

F. Transfer of Genes: Linked Marker Methods

Introduction

The transfer of genes using linked marker methods is a transformative approach in plant breeding and genetic research, offering precision and efficiency in developing crop varieties with desirable traits. Linked marker methods utilize genetic markers that are closely associated with genes of interest, allowing researchers and breeders to track and transfer these genes with greater accuracy.

The Concept of Linkage

The concept of linkage forms the foundation of linked marker methods. Linkage refers to the tendency of genes or genetic markers located near each other on the same chromosome to be inherited together during meiosis. This occurs because the physical proximity of linked genes reduces the likelihood of recombination events separating them.

Linkage and Recombination

During meiosis, homologous chromosomes pair and exchange genetic material through a process called recombination or crossing over. Recombination occurs at specific points along the chromosome, known as chiasmata, where segments of genetic material are exchanged between homologous chromatids. The probability of recombination separating two genes is inversely proportional to the distance between them on the chromosome. Genes that are close together are less likely to be separated by recombination, leading to linkage.

» **Genetic Distance:** Genetic distance is a measure of the likelihood of recombination occurring between two genes. It is often expressed in centimorgans (cM), where one centimorgan represents a 1% chance of recombination occurring between two loci during meiosis. Closely linked genes have a small genetic distance, indicating a low probability of recombination separating them.

» **Linkage Disequilibrium (LD):** Linkage disequilibrium refers to the non-random association of alleles at different loci within a population. LD is influenced by factors such as recombination rate, mutation rate, genetic drift, and selection. High LD between markers and target genes indicates strong linkage, making them useful for gene transfer.

Genetic Markers and Their Types

Genetic markers are essential tools in linked marker methods, serving as proxies for tracking and selecting genes of interest. These markers are DNA sequences with known locations on chromosomes and can be categorized into several types based on their nature and properties.

Morphological Markers

Morphological markers are phenotypic traits that can be easily observed and scored. These markers include characteristics such as flower color, seed shape, and plant height. While morphological markers are simple to identify, they are limited by their environmental dependence and the possibility of pleiotropic effects, where a single gene influences multiple traits.

Biochemical Markers

Biochemical markers involve proteins or enzymes that can be detected using biochemical assays. Isozymes, which are different forms of an enzyme that catalyze the same reaction, are a common type of biochemical marker. While biochemical markers provide more stability than morphological markers, they offer limited coverage of the genome and lower polymorphism.

Molecular Markers

Molecular markers are DNA sequences that exhibit variation between individuals and are not influenced by environmental factors. They offer several advantages over morphological and biochemical markers, including high polymorphism, genome-wide coverage, and the ability to differentiate between homozygous and heterozygous individuals. Key types of molecular markers include:

» **Simple Sequence Repeats (SSRs):** Also known as microsatellites, SSRs are short tandem repeats of DNA sequences, typically 1-6 base pairs in length. SSRs are highly polymorphic and distributed throughout the genome, making them valuable for genetic mapping and marker-assisted selection. They are co-dominant markers, allowing differentiation between homozygous and heterozygous individuals.

» **Single Nucleotide Polymorphisms (SNPs):** SNPs are single base pair variations in the DNA sequence and are the most abundant type of genetic variation in genomes. SNPs are highly stable and can be used to develop high-density genetic maps. They are particularly useful in high-throughput genotyping platforms, enabling the rapid screening of large populations.

» **Random Amplified Polymorphic DNA (RAPD):** RAPD markers are generated by amplifying random DNA segments using short, arbitrary primers. They are relatively simple and cost-effective to produce but are dominant markers, limiting their ability to distinguish between homozygous and heterozygous individuals.

» **Amplified Fragment Length Polymorphisms (AFLPs):** AFLPs involve the selective amplification of restriction fragments from a total digest of genomic DNA. They offer high reproducibility and the ability to detect multiple polymorphic loci simultaneously, making them useful for genetic mapping and diversity studies.

» **Restriction Fragment Length Polymorphisms (RFLPs):** RFLPs are based on variations in the length of DNA fragments produced by restriction enzyme digestion. They were among the first molecular markers developed and have been used extensively in genetic mapping and linkage analysis.

Techniques for Linked Marker Methods

Several techniques are employed to identify, develop, and utilize linked markers for gene transfer, each offering unique advantages and applications in genetic research and breeding.

Marker Discovery and Development

The discovery and development of genetic markers involve several steps, from identifying polymorphic regions in the genome to validating marker-trait associations.

- **Identification of Polymorphic Regions:** The first step in marker development is identifying polymorphic regions in the genome that exhibit variation between individuals. This can be achieved through whole-genome sequencing, transcriptome analysis, or targeted resequencing of candidate regions.
- **Marker Validation:** Once potential markers are identified, they must be validated for their association with target genes. This involves testing the markers in diverse populations and assessing their correlation with phenotypic traits. Statistical methods, such as quantitative trait locus (QTL) mapping and genome-wide association studies (GWAS), are used to validate marker-trait associations.
- **Development of Marker Assays:** Marker assays are developed to enable the efficient detection and genotyping of markers in breeding populations. These assays can be based on various techniques, such as PCR, multiplex PCR, or SNP genotyping platforms, depending on the type and number of markers being analyzed.

Genetic Mapping

Genetic mapping is a crucial step in identifying linked markers and understanding their relationship with target genes. This process involves analyzing the inheritance patterns of markers and traits in a population to determine their genetic distance.

- **Linkage Mapping:** Linkage mapping involves creating a genetic map that shows the relative positions of markers and genes on chromosomes. This map is constructed by analyzing the co-segregation of markers and traits in a population, allowing the identification of linked markers for gene transfer. Linkage maps provide valuable information about the genetic architecture of traits and facilitate the selection of markers for breeding.
- **QTL Mapping:** Quantitative trait locus (QTL) mapping is a method used to identify chromosomal regions associated with quantitative traits, such as yield, disease resistance, and stress tolerance. By analyzing the association between markers and phenotypic variation, researchers can locate QTLs and identify linked markers for complex traits. QTL mapping involves several steps, including the development of a mapping population, phenotyping for target traits, and statistical analysis of marker-trait associations.

- **Genome-Wide Association Studies (GWAS):** GWAS involve scanning the entire genome to identify markers associated with specific traits. This approach is particularly useful for studying complex traits influenced by multiple genes. GWAS utilizes high-density SNP arrays to analyze marker-trait associations in diverse populations, providing insights into the genetic basis of traits and identifying linked markers for gene transfer.

Marker-Assisted Selection (MAS)

Marker-assisted selection (MAS) is a powerful tool for transferring genes and improving crop traits by selecting individuals with desired genetic configurations.

- **Selection of Linked Markers:** In MAS, linked markers are used to select individuals carrying the desired alleles for specific traits. This selection process is based on the presence of markers that are closely linked to target genes, ensuring the transfer of beneficial traits. MAS enhances breeding efficiency by reducing the time and resources required to achieve desired genetic improvements.
- **Backcross Breeding:** Backcross breeding is a technique used to transfer genes from donor to recipient plants while retaining the recipient's genetic background. MAS facilitates this process by tracking linked markers, enabling the efficient transfer of target genes and reducing the number of backcross generations required. This approach is particularly useful for introgressing specific traits, such as disease resistance or stress tolerance, into elite cultivars.
- **Pyramiding of Traits:** MAS allows for the pyramiding of multiple genes or QTLs that confer desirable traits, such as disease resistance, yield, and quality. By selecting individuals with markers linked to multiple target genes, breeders can develop crop varieties with enhanced performance and adaptability.

Applications of Linked Marker Methods

Linked marker methods have transformed plant breeding and genetic research, offering numerous applications in crop improvement and biotechnology.

Development of Improved Crop Varieties

Linked marker methods enable the development of crop varieties with enhanced traits, such as yield, quality, and resistance to biotic and abiotic stresses.

- **Disease Resistance:** Linked markers are used to transfer and select genes for disease resistance, enabling the development of crop varieties that withstand pathogen attacks. By identifying markers linked to resistance genes, breeders can efficiently incorporate resistance traits into breeding programs, reducing the reliance on chemical pesticides and enhancing agricultural sustainability. For example, markers linked to rust resistance genes in wheat have been used to develop resistant cultivars, reducing yield losses and improving food security.
- **Abiotic Stress Tolerance:** Linked markers facilitate the transfer of genes conferring tolerance to abiotic stresses, such as drought, salinity, and temperature extremes. By selecting markers associated with stress tolerance traits, breeders can develop crops that thrive in challenging

environments, contributing to food security and climate resilience. For instance, markers linked to drought tolerance QTLs in maize have been used to develop varieties that perform well under water-limited conditions.

» **Nutritional Enhancement:** Linked markers are used to transfer genes involved in nutritional traits, such as vitamin content, mineral uptake, and protein quality. By incorporating markers linked to nutritional traits, breeders can develop biofortified crops that address nutritional deficiencies and improve human health. Examples include the use of markers linked to beta-carotene content in rice to develop Golden Rice, a biofortified variety that addresses vitamin A deficiency.

Genetic Engineering and Biotechnology

Linked marker methods play a vital role in genetic engineering and biotechnology, enabling the precise manipulation of plant genomes.

» **Transgenic Crop Development:** Linked markers are used in the development of transgenic crops, where genes of interest are introduced into plant genomes. By identifying markers linked to transgenes, researchers can track their integration and expression, ensuring the successful development of genetically modified crops with enhanced traits. For example, markers linked to herbicide resistance genes have been used to monitor the integration and expression of transgenes in genetically modified soybean.

» **Gene Stacking:** Gene stacking involves the introduction of multiple genes into a single plant genome to confer multiple traits. Linked markers facilitate gene stacking by tracking the presence of stacked genes, enabling the development of crops with combined resistance to pests, diseases, and environmental stresses. This approach has been used to develop maize varieties with stacked traits for insect resistance and herbicide tolerance.

» **Genome Editing:** Genome editing technologies, such as CRISPR-Cas9, can be combined with linked marker methods to achieve precise gene editing. Linked markers guide the identification and modification of target genes, allowing for the development of crops with tailored traits and improved performance. For instance, linked markers have been used to identify target sites for CRISPR-Cas9 editing in wheat, enabling the precise modification of genes associated with disease resistance.

Challenges and Considerations in Linked Marker Methods

While linked marker methods offer significant advantages, they also present challenges that must be addressed to ensure their successful application.

Linkage Disequilibrium

Linkage disequilibrium (LD) refers to the non-random association of alleles at different loci, which can complicate the identification of linked markers.

» **Managing LD:** Breeders must carefully manage LD to accurately identify linked markers and ensure the transfer of target genes. This involves selecting appropriate populations for mapping and using statistical methods to account for LD in marker-trait associations. Understanding the genetic architecture of target traits and considering population structure are essential for effective LD management.

Cost and Resource Requirements

The development and application of linked marker methods can be resource-intensive, requiring significant investments in technology and expertise.

» **Cost-Effective Approaches:** Breeders must develop cost-effective approaches for marker development and genotyping, such as using high-throughput platforms and optimizing marker panels. Collaboration and resource-sharing among research institutions can also help reduce costs and enhance efficiency. Public-private partnerships and international collaborations can facilitate access to advanced technologies and resources.

Ethical and Regulatory Considerations

The use of linked marker methods in genetic engineering and biotechnology raises ethical and regulatory considerations.

» **Regulatory Frameworks:** Regulatory frameworks must be established to ensure the safe and responsible use of linked marker methods in breeding and biotechnology. These frameworks should address potential environmental and health risks associated with genetically modified crops and ensure compliance with international standards. Transparent and evidence-based regulatory processes are essential for building public trust and facilitating the adoption of new technologies.

» **Public Engagement:** Engaging with stakeholders and the public is essential to address ethical concerns and build trust in the use of linked marker methods. Researchers and breeders should communicate the benefits and risks of these methods and involve the public in decision-making processes. Public engagement can foster informed discussions and promote the acceptance of innovative breeding technologies.

Future Prospects of Linked Marker Methods

The future of plant breeding and genetic research will be shaped by advances in genomic technologies and the strategic use of linked marker methods.

Integration with Genomic Tools

The integration of linked marker methods with genomic tools, such as high-throughput sequencing and bioinformatics, will enhance their application in breeding and research.

» **High-Throughput Genotyping:** Advances in high-throughput genotyping technologies will enable the rapid and cost-effective screening of large populations for linked markers. This will facilitate the identification and transfer of target genes, accelerating breeding programs and enhancing genetic gain. Platforms such as genotyping-by-sequencing (GBS) and SNP arrays offer opportunities for large-scale marker analysis and selection.

» **Bioinformatics and Data Analysis:** The use of bioinformatics and data analysis tools will enhance the interpretation of marker-trait associations and improve the accuracy of linked marker methods. This will enable breeders to make informed decisions and optimize breeding strategies. Machine learning and artificial intelligence algorithms can be leveraged to analyze complex datasets and identify marker-trait correlations.

Breeding for Climate Resilience

Linked marker methods will play a crucial role in breeding for climate resilience, providing tools to develop crops that withstand changing environmental conditions.

» **Developing Resilient Varieties:** By using linked markers to transfer genes for stress tolerance and adaptability, breeders can develop crop varieties that thrive in diverse environments. This resilience will be essential for ensuring food security in the face of climate change and environmental challenges. Breeding programs can focus on traits such as heat tolerance, water-use efficiency, and resistance to emerging pests and diseases.

» **Enhancing Genetic Diversity:** Linked marker methods will continue to enhance genetic diversity, providing a diverse pool of genetic material for breeding resilient crops. This diversity will enable breeders to respond to emerging threats and develop crops with improved performance and adaptability. Conservation of wild relatives and landraces can provide valuable genetic resources for enhancing crop diversity.

G. Duplication - production and use; Inversions and location of genes; B/ A chromosome translocations and gene location

Duplication: Production and Use

Introduction to Duplication

Duplication refers to a chromosomal segment's repeated occurrence within the genome. This duplication can involve a single gene, a large chromosomal region, or even entire genomes. The occurrence of duplications is a significant driver of genetic diversity and evolutionary change. Duplication events can lead to the development of new gene functions, influence gene expression, and contribute to genomic plasticity.

Production of Duplications

Duplications can arise through several mechanisms, both naturally and artificially. Understanding these mechanisms is crucial for harnessing the potential of duplications in crop improvement and genetic research.

Natural Mechanisms

1. **Unequal Crossing Over:** This is a primary natural mechanism through which duplications occur. During meiosis, homologous chromosomes pair and exchange genetic material. If misalignment occurs during crossing over, one chromosome may gain extra genetic material while the other loses some, leading to duplication on one of the chromatids.
2. **Replication Slippage:** This occurs during DNA replication when the DNA polymerase

slips and realigns at a different repeat unit on the template strand, causing a duplication of a segment. This mechanism is more common in regions with repetitive sequences.

3. **Retrotransposition:** Duplications can also occur through retrotransposition, where RNA is reverse-transcribed and integrated back into the genome at a new location. This can result in the duplication of genes or other genetic elements.
4. **Segmental Duplications:** Large-scale duplications involve segments of chromosomes, often leading to the duplication of multiple genes. These can occur due to errors in DNA replication or repair processes.

Artificial Induction

1. **Mutagenic Agents:** Physical or chemical mutagens can induce duplications by causing DNA breaks and misrepair during the cell cycle. Radiation and chemical agents like ethyl methanesulfonate (EMS) are commonly used to induce duplications in plant breeding.
2. **Genetic Engineering:** Advanced techniques, such as CRISPR-Cas9, can be employed to create targeted duplications. By designing specific guide RNAs, researchers can induce double-strand breaks at desired locations, facilitating the duplication of specific genomic regions.

Uses of Duplications

Duplications have significant implications for genetics and breeding, offering opportunities for developing improved crop varieties and advancing genetic research.

Gene Dosage Effects

1. **Increased Expression:** Duplications can lead to increased expression of duplicated genes due to higher gene dosage. This can enhance traits such as growth rate, yield, and resistance to biotic and abiotic stresses.
2. **Metabolic Pathways:** In crops, duplications of genes involved in metabolic pathways can increase the production of valuable metabolites, such as vitamins, antioxidants, and secondary metabolites with medicinal properties.
3. **Trait Enhancement:** By increasing the copy number of beneficial genes, duplications can enhance traits like drought tolerance, pest resistance, and nutritional quality. For example, duplication of genes involved in the biosynthesis of drought-responsive proteins can improve drought tolerance in crops.

Evolutionary Innovation

1. **Neofunctionalization:** Duplicated genes can acquire new functions, contributing to evolutionary innovation. Over time, mutations in one copy of the duplicated gene may lead to new functions, allowing organisms to adapt to changing environments.
2. **Subfunctionalization:** In some cases, duplicated genes may divide their original function between them, leading to subfunctionalization. This can help maintain essential functions while allowing for specialization.

3. **Genomic Plasticity:** Duplications contribute to genomic plasticity by increasing genetic diversity and providing raw material for evolution. This plasticity enables species to adapt to environmental changes and survive in diverse habitats.

Breeding Applications

1. **Crop Improvement:** Duplications can be harnessed in breeding programs to improve crop traits. By selecting plants with beneficial duplications, breeders can develop varieties with enhanced yield, quality, and stress resistance.
2. **Trait Mapping:** Duplications can serve as markers for trait mapping, helping researchers identify genomic regions associated with specific traits. This information can guide breeding efforts and facilitate the development of superior crop varieties.
3. **Genetic Buffering:** Duplications can buffer against deleterious mutations by providing redundant copies of essential genes. This redundancy increases the robustness of organisms and contributes to genetic stability.

Inversions and Location of Genes

Introduction to Inversions

Inversions are chromosomal rearrangements in which a segment of a chromosome is reversed end-to-end. This reversal changes the linear order of genes but does not affect the overall genetic content. Inversions can have significant effects on gene expression, recombination rates, and the evolution of species.

Types of Inversions

Inversions can be classified based on their size and the position of the centromere relative to the inverted segment:

1. **Paracentric Inversions:** These inversions occur when the inverted segment does not include the centromere. Paracentric inversions can affect recombination rates and lead to the formation of acentric and dicentric chromatids during meiosis.
2. **Pericentric Inversions:** These inversions involve the centromere, resulting in a change in the relative positions of the centromere and the inverted segment. Pericentric inversions can alter chromosome morphology and affect the distribution of genetic material.

Effects of Inversions on Gene Location

Inversions can influence gene location and expression, impacting genetic diversity and evolution.

Suppression of Recombination

1. **Reduced Recombination Rates:** Inversions suppress recombination within the inverted region, as misaligned pairing during meiosis leads to the formation of inversion loops. This suppression can maintain advantageous gene combinations and contribute to the maintenance of genetic diversity.

2. **Protection of Adapted Gene Complexes:** By reducing recombination, inversions can protect adapted gene complexes from being broken up, preserving beneficial combinations of alleles. This can be advantageous in environments where specific gene combinations confer a selective advantage.

Position Effects

1. **Position Effect Variegation:** Inversions can cause position effect variegation, where the expression of genes near the inversion breakpoint is influenced by their new chromosomal environment. This can lead to changes in gene expression and phenotypic variation.
2. **Altered Gene Regulation:** The relocation of genes to different chromosomal regions can alter their regulation due to changes in chromatin structure or the proximity to regulatory elements. This can impact developmental processes and trait expression.

Uses of Inversions

Inversions have practical applications in genetics and breeding, offering tools for studying gene function and developing improved crop varieties.

Genetic Mapping and Research

1. **Mapping Recombination Rates:** Inversions can be used to study recombination rates and genetic linkage. By analyzing the suppression of recombination within inverted regions, researchers can gain insights into the mechanisms of genetic recombination and the distribution of crossover events.
2. **Studying Gene Function:** Inversions provide opportunities to study gene function by analyzing the effects of altered gene order and regulation. Researchers can investigate the consequences of position effects and identify regulatory elements influencing gene expression.

Breeding Applications

1. **Development of Inversion Lines:** Inversion lines, where specific chromosomal segments are inverted, can be developed to study the effects of inversions on trait expression and genetic diversity. These lines serve as valuable resources for breeding and genetic research.
2. **Enhanced Trait Stability:** Inversions can enhance trait stability by maintaining favorable gene combinations and reducing the effects of deleterious recombination events. This stability is particularly important in breeding programs aiming to develop varieties with consistent performance across environments.

B/A Chromosome Translocations and Gene Location

Introduction to B/A Chromosome Translocations

B/A chromosome translocations involve the exchange of genetic material between B chromosomes (extra, non-essential chromosomes) and A chromosomes (normal autosomes). These translocations can impact gene location, expression, and the overall genomic architecture. Understanding B/A chromosome translocations is essential for studying chromosomal evolution and their potential applications in plant breeding.

Mechanism of B/A Chromosome Translocations

B/A chromosome translocations can occur through several mechanisms, resulting in the integration of B chromosome segments into A chromosomes or vice versa.

1. **Non-Homologous End Joining:** Double-strand breaks in DNA can lead to the exchange of segments between B and A chromosomes through non-homologous end joining. This process involves the repair of broken DNA ends without the need for homologous sequences, facilitating translocations.
2. **Illegitimate Recombination:** B/A translocations can result from illegitimate recombination events, where non-homologous chromosomes undergo recombination due to sequence similarities or chromatin interactions. This mechanism can lead to the incorporation of B chromosome segments into A chromosomes.
3. **Chromosome Fusions:** Chromosome fusions, where entire chromosomes or large segments fuse, can result in B/A translocations. This mechanism can lead to the integration of B chromosome material into A chromosomes, affecting gene location and expression.

Effects of B/A Translocations on Gene Location

B/A chromosome translocations can influence gene location and expression, impacting genetic diversity and the evolution of species.

Alteration of Gene Expression

1. **Position Effects:** B/A translocations can cause position effects, where the expression of genes near the translocation breakpoint is influenced by their new chromosomal environment. This can lead to changes in gene expression and phenotypic variation.
2. **Gene Silencing:** Translocations can lead to gene silencing due to changes in chromatin structure or the relocation of genes to heterochromatic regions. This can impact the regulation of genes involved in essential biological processes.

Impact on Genetic Diversity

1. **Introduction of Novel Variants:** B/A translocations can introduce novel genetic variants by incorporating B chromosome sequences into A chromosomes. This can increase genetic diversity and provide raw material for evolution.
2. **Evolutionary Significance:** The integration of B chromosome segments into A chromosomes can have evolutionary significance, contributing to the adaptation of species to changing environments. This process can lead to the development of new gene functions and the diversification of species.

Uses of B/A Chromosome Translocations

B/A chromosome translocations have practical applications in genetics and breeding, offering tools for studying gene function and developing improved crop varieties.

Genetic Research and Mapping

1. **Studying Chromosomal Evolution:** B/A translocations provide opportunities to study chromosomal evolution and the mechanisms of genomic rearrangements. Researchers can investigate the impact of translocations on gene function and the role of B chromosomes in genomic evolution.
2. **Mapping Genetic Traits:** B/A translocations can be used to map genetic traits by analyzing the effects of translocations on trait expression. This information can guide breeding efforts and facilitate the development of superior crop varieties.

Breeding Applications

1. **Development of Translocation Lines:** Translocation lines, where specific chromosomal segments are exchanged between B and A chromosomes, can be developed to study the effects of translocations on trait expression and genetic diversity. These lines serve as valuable resources for breeding and genetic research.
2. **Enhanced Genetic Diversity:** B/A translocations can enhance genetic diversity by introducing novel variants and facilitating the exchange of genetic material between chromosomes. This diversity is particularly important in breeding programs aiming to develop crops with improved performance and adaptability.

Chapter - 7

Trisomics

A. Trisomics: An Overview

Trisomics are individuals or organisms with an extra chromosome in addition to the normal diploid chromosome set, resulting in a 2n+1 chromosomal arrangement. This chromosomal condition leads to a variety of genetic and phenotypic consequences, crucial for understanding plant genetics, breeding, and crop improvement. Trisomics can arise naturally or be induced through mutagenic treatments, providing valuable tools for genetic studies and plant breeding.

1. Significance in Plant Genetics

The study of trisomics is significant in plant genetics because it allows researchers to identify the functions and effects of specific chromosomes and their associated genes. By analyzing the phenotypic and genotypic variations caused by the presence of an extra chromosome, scientists can gain insights into the role of individual chromosomes in growth, development, and trait expression. Trisomics also helps in understanding the mechanisms of chromosomal behavior during cell division, particularly meiosis, which is essential for breeding and genetic manipulation.

Mapping and Identifying Genes: Trisomics facilitates the mapping and identification of genes associated with specific traits. By comparing the phenotypes of trisomic individuals with those of normal diploids, researchers can determine which genes are present on the extra chromosome. This helps in associating specific traits with particular chromosomes, enabling the development of molecular markers for breeding programs.

Understanding Dosage Effects: Trisomics are also valuable for studying gene dosage effects, which occur when the number of copies of a gene affects the organism's phenotype. In trisomics, the extra chromosome leads to an increased dosage of genes, providing insights into how changes in gene dosage influence traits such as growth, development, and stress responses.

Applications in Breeding Programs: In applied plant breeding programs, trisomics are used to identify and transfer desirable traits. By creating and studying trisomic plants, breeders can identify chromosomes that carry genes for traits like disease resistance, stress tolerance, and improved yield.

DOI: 10.1201/9781003717775-7

This knowledge is used to develop new crop varieties through targeted breeding strategies.

Genetic Studies: Trisomics is essential in genetic studies for understanding chromosomal behavior during meiosis. By analyzing the behavior of the extra chromosome during cell division, researchers can gain insights into mechanisms of chromosomal segregation, recombination, and inheritance patterns.

2. Types of Trisomics

Trisomics can be classified based on the specific chromosome that is present in triplicate. Each type of trisomy has distinct characteristics and implications for plant genetics and breeding.

Primary Trisomics

Primary trisomics involve the addition of a single extra chromosome to the normal diploid complement. This is the most basic and frequently studied type of trisomy. Primary trisomics are valuable in genetic mapping and functional genomics because they help in associating phenotypic traits with specific chromosomes.

Trisomy 1

Genetic Implications: Trisomy 1 involves an extra copy of the first chromosome in the chromosome set. This addition leads to significant changes in gene expression and phenotypic traits due to the imbalance of genetic material. The genes located on chromosome 1 are present in three copies instead of the usual two, resulting in altered dosage effects and potential gene interactions. The presence of an extra chromosome can disrupt normal cellular processes by altering the dosage of genes involved in critical pathways. For instance, genes that regulate cell cycle progression, hormone biosynthesis, or metabolic pathways may be overexpressed, leading to developmental abnormalities. The extra chromosome can also affect gene silencing mechanisms, such as RNA interference (RNAi) and epigenetic modifications, further influencing gene expression patterns.

Phenotypic Effects: The phenotypic effects of trisomy 1 can vary widely depending on the species and the genes located on the first chromosome. Common effects include changes in plant height, leaf morphology, and flowering time. For example, in some crops, trisomy 1 might result in taller plants with larger leaves, while in others, it might cause delayed flowering or altered flower structures. These phenotypic variations provide insights into the functions of genes on chromosome 1. In addition to these visible traits, trisomy 1 can also impact physiological and biochemical processes. Plants with trisomy 1 may exhibit altered photosynthetic efficiency, nutrient uptake, and stress responses. For instance, the extra chromosome may contain genes that enhance drought tolerance or improve nutrient use efficiency, providing a selective advantage under specific environmental conditions. These physiological changes can be measured using techniques such as chlorophyll fluorescence, gas exchange analysis, and metabolomics.

Applications in Crop Improvement: Understanding trisomy 1 can help breeders identify genes responsible for desirable traits. By studying the phenotypic effects of an extra chromosome 1, researchers can pinpoint critical genes for crop improvement. For instance, if trisomy 1 is associated with increased resistance to a particular disease, breeders can focus on the genes located on chromosome 1 that contribute to this resistance. This information can be used to develop new crop varieties with enhanced traits through traditional breeding or genetic engineering. In addition to disease resistance, trisomy

1 can be leveraged to improve other agronomic traits such as yield, quality, and stress tolerance. For example, if trisomy 1 enhances grain size or fruit flavor, breeders can incorporate this knowledge into their breeding programs to develop superior cultivars. Furthermore, the identification of molecular markers linked to beneficial genes on chromosome 1 can facilitate marker-assisted selection (MAS), speeding up the breeding process and increasing its precision.

Secondary Trisomics

Secondary trisomics involve an extra chromosome that is an isochromosome or contains duplicated genetic material. Isochromosomes are chromosomes with identical arms, resulting in a duplication of certain genetic regions. Secondary trisomics provide unique insights into the effects of specific chromosomal regions and gene dosage.

Trisomy 2

Genetic Implications: Trisomy 2 features an additional copy of the second chromosome. This can lead to overexpression of genes on chromosome 2, affecting the plant's overall genetic balance. The presence of an extra chromosome 2 may disrupt normal regulatory mechanisms, leading to changes in metabolic pathways and developmental processes. The genetic implications of trisomy 2 extend beyond simple gene dosage effects. The additional chromosome can also lead to structural changes in the genome, such as chromosomal rearrangements and translocations. These changes can result in the creation of novel gene fusions or the disruption of existing genes, further complicating the genetic landscape. Additionally, the presence of an extra chromosome can affect the stability of the genome, leading to increased rates of mutation and recombination.

Phenotypic Effects: Plants with trisomy 2 may exhibit changes in root structure, stem robustness, and resistance to certain diseases. The specific phenotypic outcomes depend on the genes located on the second chromosome. For example, in some species, trisomy 2 might enhance root development and improve water and nutrient uptake, while in others, it might increase susceptibility to specific pathogens due to gene interactions. The phenotypic effects of trisomy 2 are not limited to physical traits. Plants with this chromosomal configuration may also show changes in physiological processes such as photosynthesis, respiration, and secondary metabolite production. For instance, trisomy 2 may lead to increased production of specific secondary metabolites that confer resistance to herbivores or pathogens. These changes can be measured using techniques such as metabolite profiling, enzyme activity assays, and bioassays for resistance traits.

Applications in Crop Improvement: Trisomy 2 studies can reveal genes involved in disease resistance and stress tolerance. These findings are valuable for breeding programs aiming to enhance crop resilience. By identifying and characterizing the genes on chromosome 2 that contribute to desirable traits, breeders can develop new varieties with improved performance under biotic and abiotic stresses. Additionally, understanding the genetic basis of these traits can facilitate marker-assisted selection and genomic selection in breeding programs. In addition to disease resistance and stress tolerance, trisomy 2 can be used to improve other important traits such as yield, quality, and adaptability. For example, if trisomy 2 enhances grain filling or fruit setting, breeders can incorporate this knowledge into their breeding strategies to develop high-yielding cultivars. Moreover, the identification of quantitative trait loci (QTLs) associated with beneficial traits on chromosome 2 can provide valuable markers for selection in breeding programs.

Tertiary Trisomics

Tertiary trisomics involves an extra chromosome that consists of segments from two different chromosomes. This type of trisomy results in a complex genetic configuration, with implications for gene expression and chromosomal behavior. Tertiary trisomics are particularly useful for studying the interactions between genes located on different chromosomes.

Trisomy 3

Genetic Implications: In trisomy 3, an extra chromosome combines segments from chromosome 3 and another chromosome. This results in a unique genetic makeup with implications for gene expression and chromosomal behavior. The presence of genetic material from two different chromosomes on the extra chromosome can lead to novel phenotypic effects and interactions. The genetic implications of trisomy 3 extend to the regulatory networks that control gene expression. The extra chromosome may contain regulatory elements, such as enhancers and silencers, that interact with genes on both chromosomes, leading to altered expression patterns. Additionally, the presence of an extra chromosome can affect chromatin structure and nuclear organization, influencing the accessibility of genes to transcriptional machinery.

Phenotypic Effects: The phenotypic effects of trisomy 3 can include alterations in flower color, seed size, and fruit development. These changes are due to the interaction between genes on the extra chromosome and the plant's existing genetic material. For example, trisomy 3 might result in flowers with new color patterns or fruits with modified shapes and sizes. These phenotypic variations provide insights into the functions and interactions of genes from different chromosomes. In addition to visible traits, trisomy 3 can impact physiological processes such as hormone biosynthesis, nutrient transport, and stress responses. For instance, the extra chromosome may contain genes that enhance hormone production or improve nutrient use efficiency, providing a selective advantage under specific environmental conditions. These physiological changes can be measured using techniques such as hormone assays, nutrient analysis, and stress tolerance tests. Plants with trisomy 3 might also exhibit alterations in their growth patterns and developmental timing. For example, the interaction of genes from the extra chromosome with the plant's endogenous genetic material might result in altered flowering times, changes in seed dormancy, or variations in vegetative growth. Such changes can significantly impact the plant's adaptation to its environment and its overall fitness.

Applications in Crop Improvement: Trisomy 3 can help breeders manipulate flower and fruit traits for commercial purposes. This knowledge is essential for developing new varieties with enhanced aesthetic and nutritional qualities. For instance, if trisomy 3 is associated with improved fruit flavor or increased nutritional content, breeders can focus on the genes involved in these traits to develop superior crop varieties. Additionally, the study of trisomy 3 can reveal genetic mechanisms underlying complex traits and epistatic interactions, which are important for breeding strategies. In addition to flower and fruit traits, trisomy 3 can be used to improve other important traits such as disease resistance, stress tolerance, and yield. For example, if trisomy 3 enhances resistance to a specific pathogen or improves tolerance to drought, breeders can incorporate this knowledge into their breeding programs to develop resilient cultivars. Furthermore, the identification of molecular markers linked to beneficial genes on chromosome 3 can facilitate marker-assisted selection (MAS) and genomic selection (GS) in breeding programs. The comprehensive understanding of trisomy 3 also aids in the development of crops that are better suited to diverse environmental conditions. By identifying genes that confer resilience to abiotic stresses such as salinity, drought, and extreme temperatures, breeders can develop crop varieties that maintain high yields and quality under adverse conditions. This is particularly important in the context of climate change and the need for sustainable agricultural practices.

Double Trisomics

Double trisomics involves the addition of two extra chromosomes to the normal diploid complement. This type of trisomy results in a 2n+2 chromosomal arrangement and provides insights into the combined effects of two extra chromosomes on gene expression and phenotypic traits.

Trisomy 4

Genetic Implications: Double trisomy involving chromosome 4 and another chromosome leads to complex genetic interactions and dosage effects. The presence of two extra chromosomes can disrupt normal regulatory mechanisms and result in significant changes in gene expression and phenotypic traits. The specific genetic implications depend on the chromosomes involved and the genes located on them. The genetic implications of double trisomy extend to the interactions between genes on different chromosomes. The presence of two extra chromosomes can lead to epistatic interactions, where the expression of one gene affects the expression of another gene on a different chromosome. Additionally, double trisomy can result in changes in chromatin structure and nuclear organization, influencing the accessibility of genes to transcriptional machinery. Double trisomy can also result in genomic instability, leading to increased rates of mutation and chromosomal rearrangements. This instability can have both positive and negative effects on the plant's fitness and adaptability. On one hand, it can generate genetic diversity that may be beneficial under certain environmental conditions. On the other hand, it can lead to deleterious mutations that reduce the plant's viability and reproductive success.

Phenotypic Effects: The phenotypic effects of double trisomy can include changes in plant morphology, growth rate, and reproductive success. For example, double trisomy might result in plants with abnormal leaf shapes, increased growth vigor, or reduced fertility. These phenotypic variations provide insights into the combined effects of multiple chromosomes on plant development and trait expression. Plants with double trisomy may also exhibit alterations in their physiological processes, such as photosynthesis, respiration, and secondary metabolite production. For instance, double trisomy may lead to increased production of specific secondary metabolites that confer resistance to herbivores or pathogens. These changes can be measured using techniques such as metabolite profiling, enzyme activity assays, and bioassays for resistance traits. In addition to visible traits, double trisomy can impact the plant's ability to cope with environmental stresses. For example, the interaction of genes from the extra chromosomes may enhance the plant's tolerance to abiotic stresses such as drought, salinity, and extreme temperatures. This can be particularly beneficial in developing crop varieties that can thrive under challenging environmental conditions.

Applications in Crop Improvement: Double trisomics are valuable for studying the combined effects of multiple chromosomes on complex traits. This knowledge is crucial for breeding programs aiming to develop new crop varieties with enhanced performance. By understanding the genetic basis of the combined effects of two extra chromosomes, breeders can identify key genes and regulatory pathways involved in desirable traits. This information can be used to develop superior crop varieties through traditional breeding or genetic engineering.

In addition to studying the combined effects of multiple chromosomes, double trisomics can be used to improve important traits such as yield, quality, and adaptability. For example, if double trisomy enhances grain filling or fruit setting, breeders can incorporate this knowledge into their breeding strategies to develop high-yielding cultivars. Moreover, the identification of quantitative trait loci (QTLs) associated with beneficial traits on the extra chromosomes can provide valuable markers for selection in breeding programs. Double trisomics can also be used to develop crop varieties with improved resistance to pests and diseases. By identifying genes that confer resistance to specific pathogens or pests, breeders can develop cultivars that are less reliant on chemical pesticides, thereby promoting sustainable agricultural practices. Additionally, understanding the genetic interactions involved in double trisomy can help breeders design strategies to combine multiple resistance genes in a single cultivar, providing broad-spectrum protection against a range of biotic stresses.

Segmental Trisomics

Segmental trisomics involves an extra chromosome segment rather than an entire chromosome. This type of trisomy results in a partial 2n+1 chromosomal arrangement and provides insights into the effects of specific chromosomal regions on gene expression and phenotypic traits.

Trisomy 5

Genetic Implications: Segmental trisomy involving chromosome 5 results in the duplication of specific genetic regions. This leads to dosage effects and potential gene interactions that can significantly impact gene expression and phenotypic traits. The specific genetic implications depend on the duplicated regions and the genes located within them. The genetic implications of segmental trisomy extend to the regulatory networks that control gene expression. The extra chromosome segment may contain regulatory elements, such as enhancers and silencers, that interact with genes on the duplicated regions, leading to altered expression patterns. Additionally, the presence of an extra chromosome segment can affect chromatin structure and nuclear organization, influencing the accessibility of genes to transcriptional machinery. Segmental trisomy can also result in changes in genomic stability, leading to increased rates of mutation and recombination. These changes can generate genetic diversity that may be beneficial for adaptation to changing environmental conditions. However, they can also result in deleterious mutations that reduce the plant's fitness and reproductive success.

Phenotypic Effects: The phenotypic effects of segmental trisomy can include changes in plant height, leaf morphology, and reproductive success. For example, segmental trisomy might result in taller plants with altered leaf shapes or reduced fertility due to disruptions in reproductive processes. These phenotypic variations provide insights into the functions and interactions of genes located within the duplicated regions. In addition to visible traits, segmental trisomy can impact physiological processes such as hormone biosynthesis, nutrient transport, and stress responses. For instance, the extra chromosome segment may contain genes that enhance hormone production or improve nutrient use efficiency, providing a selective advantage under specific environmental conditions. These physiological changes can be measured using techniques such as hormone assays, nutrient analysis, and stress tolerance tests. Segmental trisomy can also affect the plant's ability to respond to biotic and abiotic stresses. For example, the interaction of genes from the extra chromosome segment with the plant's existing genetic material might enhance resistance to pathogens or improve tolerance to environmental stresses such as drought and salinity. These traits can be evaluated using various assays and field trials to determine their potential for crop improvement.

Applications in Crop Improvement: Segmental trisomics are valuable for studying the effects of specific chromosomal regions on gene expression and phenotypic traits. This knowledge is crucial for breeding programs aiming to develop new crop varieties with enhanced performance. By understanding the genetic basis of the effects of specific chromosome segments, breeders can identify key genes and regulatory pathways involved in desirable traits. This information can be used to develop superior crop varieties through traditional breeding or genetic engineering. In addition to studying the effects of specific chromosome segments, segmental trisomics can be used to improve important traits such as yield, quality, and adaptability. For example, if segmental trisomy enhances grain filling or fruit setting, breeders can incorporate this knowledge into their breeding strategies to develop high-yielding cultivars. Moreover, the identification of quantitative trait loci (QTLs) associated with beneficial traits on the extra chromosome segment can provide valuable markers for selection in breeding programs. Segmental trisomics can also be used to develop crop varieties with improved resistance to pests and diseases. By identifying genes that confer resistance to specific pathogens or pests, breeders can develop cultivars that are less reliant on chemical pesticides, thereby promoting sustainable agricultural practices. Additionally, understanding the genetic interactions involved in segmental trisomy can help breeders design strategies to combine multiple resistance genes in a single cultivar, providing broad-spectrum protection against a range of biotic stresses.

B. Production, breeding behavior, and location of genes

Production of Trisomics

The production of trisomics is a fundamental aspect of cytogenetic research and plant breeding, providing valuable insights into chromosomal behavior, gene function, and the development of new crop varieties with desirable traits.

Methods of Producing Trisomics

Spontaneous Occurrence: Trisomics can occur spontaneously in nature due to errors in meiotic or mitotic cell division. These errors result in the nondisjunction of chromosomes, leading to the formation of gametes with an extra chromosome. When these gametes fuse with normal gametes during fertilization, the resulting zygote is trisomic. Spontaneous trisomics are rare and typically identified through phenotypic screening and cytogenetic analysis. In natural populations, spontaneous trisomics can arise as a result of environmental stresses, such as temperature fluctuations, chemical exposure, or radiation. These stresses can disrupt normal cell division, increasing the likelihood of nondisjunction events. Additionally, genetic factors such as mutations in genes involved in chromosome segregation can predispose certain individuals to produce trisomic offspring. The natural occurrence of trisomics helps researchers study the evolutionary significance of chromosomal abnormalities and their impact on genetic diversity.

Induced Mutagenesis: Trisomics can also be produced artificially through mutagenesis. Chemical mutagens such as colchicine, a known spindle fiber inhibitor, are commonly used to induce nondisjunction during cell division. Colchicine disrupts the normal separation of chromosomes, leading to the formation of cells with an abnormal number of chromosomes. These cells can then be propagated to produce trisomic plants. Other mutagens, such as ethyl methanesulfonate (EMS) and gamma radiation, can also induce chromosomal aberrations, including trisomy. The use of mutagenesis for producing trisomics involves careful optimization of treatment conditions to achieve the desired chromosomal abnormalities without causing excessive damage to the plant. Researchers must determine the appropriate concentration and duration of mutagen exposure, as well as the developmental stage at which treatment is most effective. Following mutagenesis, plants are screened for the presence of trisomic individuals using cytogenetic techniques such as chromosome counting and karyotyping. The identification and selection of trisomics are critical steps in establishing stable trisomic lines for further study.

Hybridization Techniques: Hybridization techniques involve crossing plants with known chromosomal abnormalities or using wide crosses between different species or genera. These crosses can produce hybrid offspring with trisomic conditions. Wide crosses often result in hybrid sterility or reduced fertility, but they can be valuable for generating unique chromosomal combinations, including trisomics. Hybridization followed by backcrossing to the parent species can help stabilize the trisomic condition. In addition to wide crosses, researchers may use interspecific or intergeneric hybridization to produce trisomics. This approach can introduce novel genetic variation and create trisomic individuals with unique chromosomal configurations. The success of hybridization techniques depends on the compatibility of the parent species, the ability to overcome reproductive barriers, and the fertility of the resulting hybrids. By carefully selecting and propagating trisomic hybrids, researchers can develop valuable genetic resources for breeding and cytogenetic studies.

Cytogenetic Techniques: Cytogenetic techniques, such as the use of aneuploid stocks, can facilitate the production of trisomics. Aneuploid stocks are plant lines with known chromosomal abnormalities, including monosomics, trisomics, and tetrasomics. By crossing these aneuploid stocks with normal diploid plants, researchers can generate progeny with specific chromosomal configurations, including trisomics. Cytogenetic analysis using chromosome staining and microscopy helps identify and select the desired trisomic individuals. The use of cytogenetic techniques requires a detailed understanding of chromosomal behavior and the ability to perform precise genetic crosses. Researchers must maintain and propagate aneuploid stocks, perform controlled crosses, and analyze the resulting progeny for the presence of trisomic individuals. Advanced cytogenetic methods, such as fluorescence in situ hybridization (FISH) and genomic in situ hybridization (GISH), can enhance the accuracy and efficiency of trisomic identification. By integrating cytogenetic techniques with traditional breeding methods, researchers can produce and study trisomics more effectively.

Challenges in Producing Trisomics

Low Fertility and Viability: One of the major challenges in producing trisomics is their typically low fertility and viability. The presence of an extra chromosome often disrupts normal meiosis, leading to reduced gamete formation and lower seed set. Trisomics may also exhibit reduced vigor and growth abnormalities, making it difficult to propagate and maintain stable trisomic lines.

The reduced fertility and viability of trisomics can result from various factors, including chromosomal imbalances, disrupted gene expression, and developmental abnormalities. Trisomics often exhibit abnormal meiosis, resulting in the production of aneuploid gametes with missing or extra chromosomes. These aneuploid gametes are less likely to participate in successful fertilization, leading to reduced seed set and lower fertility. Additionally, trisomics may experience reduced growth and vigor due to the presence of an extra chromosome, which can disrupt normal metabolic and developmental processes.

Identification and Confirmation: Accurate identification and confirmation of trisomics require detailed cytogenetic analysis. This involves chromosome counting and karyotyping using techniques such as Giemsa staining, fluorescence in situ hybridization (FISH), and genomic in situ hybridization (GISH). These techniques help visualize the extra chromosome and confirm the trisomic condition. However, cytogenetic analysis is labor-intensive and requires specialized equipment and expertise. In addition to traditional cytogenetic techniques, molecular approaches such as quantitative PCR (qPCR) and next-generation sequencing (NGS) can be used to identify and confirm trisomics. These methods allow researchers to detect specific chromosomal regions and quantify the number of copies present in the genome. By combining cytogenetic and molecular techniques, researchers can achieve a higher level of accuracy and reliability in identifying trisomics. The development of automated imaging and analysis systems also holds promise for improving the efficiency and throughput of trisomic identification.

Genetic and Phenotypic Variability: Trisomics often exhibit significant genetic and phenotypic variability due to the presence of an extra chromosome. This variability can complicate the selection and stabilization of trisomic lines. Researchers must carefully select and propagate individuals with consistent phenotypic traits and chromosomal configurations to develop stable trisomic stocks. The genetic and phenotypic variability observed in trisomics can result from various factors, including gene dosage effects, epigenetic modifications, and genetic interactions. The presence of an extra chromosome can lead to changes in gene expression, resulting in a wide range of phenotypic traits. Researchers must carefully evaluate and select trisomic individuals with stable and desirable traits to develop reliable genetic resources for breeding and research. Advanced phenotyping techniques, such as high-throughput imaging and machine learning algorithms, can assist in the accurate characterization and selection of trisomics.

Ethical and Regulatory Considerations: The production and use of trisomics in breeding programs must adhere to ethical and regulatory guidelines. This includes considerations related to genetic diversity, environmental impact, and the potential release of genetically modified organisms (GMOs). Researchers and breeders must ensure that their work complies with national and international regulations governing the use of genetic resources and biotechnology. Ethical considerations in the production and use of trisomics involve the responsible management of genetic resources and the potential impacts on biodiversity and ecosystem health. Researchers must carefully evaluate the potential risks and benefits of using trisomics in breeding programs, considering factors such as genetic diversity, environmental sustainability, and food safety. Regulatory frameworks, such as those established by the International Union for the Protection of New Varieties of Plants (UPOV) and the Cartagena Protocol on Biosafety, provide guidelines for the responsible use of genetic resources and the safe handling of GMOs. Compliance with these regulations ensures that the production and use of trisomics are conducted safely and ethically.

Breeding Behavior of Trisomics

The breeding behavior of trisomics is crucial for their effective use in genetic studies and crop improvement programs. The presence of an extra chromosome affects various aspects of plant breeding, including meiosis, inheritance patterns, and the expression of genetic traits.

Meiosis in Trisomics

Chromosomal Pairing: During meiosis in trisomics, the extra chromosome complicates the normal pairing of homologous chromosomes. Instead of forming the usual bivalent structures, trisomic cells may form trivalents, where three homologous chromosomes are associated. This can result in irregular segregation patterns and the formation of aneuploid gametes with abnormal chromosome numbers. The formation of trivalents during meiosis in trisomics can lead to various chromosomal configurations, including univalents, bivalents, and trivalents. These configurations result from the interactions between the extra chromosome and the homologous chromosomes, which can affect the stability and behavior of the chromosomes during cell division. The irregular segregation of trivalents can produce a wide range of aneuploid gametes with different chromosomal compositions, contributing to the genetic variability observed in trisomics.

Nondisjunction: Nondisjunction is a common occurrence during meiosis in trisomics, where the extra chromosome fails to segregate properly. This leads to the production of gametes with additional or missing chromosomes. Nondisjunction can occur during the first or second meiotic division, resulting in various chromosomal configurations in the resulting gametes. Nondisjunction can result in the formation of gametes with different chromosomal compositions, including trisomic, monosomic, and nullisomic gametes. The frequency and patterns of nondisjunction can vary depending on the genetic background of the trisomic individual, the specific chromosome involved, and the environmental conditions. Researchers must carefully analyze the chromosomal composition of gametes produced by trisomics to understand the mechanisms of nondisjunction and its impact on genetic variability.

Chromosomal Bridges and Lagging Chromosomes: Trisomics often exhibit chromosomal bridges and lagging chromosomes during cell division. Chromosomal bridges occur when the extra chromosome forms connections between homologous chromosomes, preventing proper segregation. Lagging chromosomes are chromosomes that fail to move to the poles during anaphase, leading to aneuploidy in the daughter cells. These anomalies contribute to the high frequency of aneuploid gametes in trisomic plants. The presence of chromosomal bridges and lagging chromosomes can result in various chromosomal abnormalities, including deletions, duplications, and translocations. These abnormalities can have significant effects on gene expression and phenotypic traits, contributing to the genetic variability observed in trisomics. Researchers must carefully analyze the chromosomal behavior of trisomics during meiosis to understand the mechanisms underlying these abnormalities and their impact on genetic variability.

Inheritance Patterns in Trisomics

Mendelian Inheritance: The presence of an extra chromosome in trisomics affects the inheritance patterns of genetic traits. Trisomics typically exhibit non-Mendelian inheritance, where the segregation ratios deviate from the expected Mendelian ratios. For example, a trisomic individual with an extra chromosome carrying a dominant allele may produce a higher proportion of offspring expressing the dominant trait due to the increased dosage of the dominant allele. The non-Mendelian inheritance patterns observed in trisomics can result from various factors, including gene dosage effects, epistatic interactions, and chromosomal rearrangements. Researchers must carefully analyze the inheritance patterns of traits in trisomics to understand the underlying genetic mechanisms and their impact on trait expression. Advanced genetic mapping techniques, such as quantitative trait locus (QTL) analysis and genome-wide association studies (GWAS), can help identify the genetic factors contributing to the non-Mendelian inheritance patterns observed in trisomics.

Dosage Compensation: Dosage compensation mechanisms play a crucial role in trisomics, where the extra chromosome's genes may be partially or fully silenced to maintain normal gene expression levels. Dosage compensation can occur through various mechanisms, including gene silencing, RNA interference, and epigenetic modifications. Understanding dosage compensation is essential for interpreting the phenotypic effects of trisomy and predicting the inheritance of specific traits. The mechanisms of dosage compensation in trisomics can vary depending on the species, the specific chromosome involved, and the genetic background of the individual. Researchers must carefully analyze the expression patterns of genes on the extra chromosome to understand the mechanisms of dosage compensation and their impact on phenotypic traits. Advanced genomic techniques, such as RNA sequencing (RNA-seq) and chromatin immunoprecipitation sequencing (ChIP-seq), can provide valuable insights into the mechanisms of dosage compensation in trisomics.

Epistatic Interactions: Epistatic interactions, where the expression of one gene affects the expression of another gene, are common in trisomics. The extra chromosome can introduce novel genetic interactions that influence the expression of multiple traits. These interactions can complicate the interpretation of inheritance patterns and require detailed genetic analysis to identify the underlying genetic mechanisms. The study of epistatic interactions in trisomics can provide valuable insights into the genetic architecture of complex traits and the mechanisms of gene regulation. Researchers must carefully analyze the genetic interactions between the extra chromosome and the endogenous genome to understand the impact of epistasis on phenotypic traits. Advanced genetic mapping techniques, such as epistasis mapping and gene interaction networks, can help identify the genetic factors contributing to the epistatic interactions observed in trisomics.

Breeding Strategies for Trisomics

Backcrossing: Backcrossing is a common breeding strategy used to stabilize trisomic traits and incorporate them into breeding programs. By repeatedly crossing trisomic individuals with a recurrent parent, breeders can select progeny that exhibit the desired traits while maintaining the trisomic condition. This approach helps stabilize the genetic background and reduce genetic variability. Backcrossing involves selecting and crossing individuals with the desired trisomic traits to the recurrent parent over multiple generations. Researchers must carefully monitor the chromosomal composition and phenotypic traits of the progeny to ensure the stable transmission of the trisomic condition. Advanced breeding techniques, such as marker-assisted selection (MAS) and genomic selection (GS), can enhance the efficiency and accuracy of backcrossing programs by identifying individuals with the desired chromosomal configurations and traits.

Marker-Assisted Selection: Marker-assisted selection (MAS) is a powerful tool for breeding trisomics. Molecular markers linked to the extra chromosome or specific traits can be used to identify and select trisomic individuals. MAS enables breeders to efficiently screen large populations and select individuals with the desired chromosomal configurations and traits, accelerating the breeding process. MAS involves the use of molecular markers, such as single nucleotide polymorphisms

(SNPs), simple sequence repeats (SSRs), and restriction fragment length polymorphisms (RFLPs), to identify individuals with the desired trisomic traits. Researchers must carefully develop and validate the molecular markers to ensure their accuracy and reliability in identifying trisomic individuals. The integration of MAS with traditional breeding methods can enhance the efficiency and precision of breeding programs, enabling the rapid development of new crop varieties with desirable traits.

Genomic Selection: Genomic selection (GS) involves using genome-wide markers to predict the breeding value of individuals. GS can be applied to trisomics to identify individuals with the highest potential for trait expression and genetic stability. This approach leverages high-throughput genotyping and advanced statistical models to enhance the efficiency and accuracy of breeding programs. GS involves the development of genomic prediction models that use genome-wide markers to estimate the breeding value of individuals. Researchers must carefully develop and validate the genomic prediction models to ensure their accuracy and reliability in predicting the breeding value of trisomic individuals. The integration of GS with traditional breeding methods can enhance the efficiency and precision of breeding programs, enabling the rapid development of new crop varieties with desirable traits.

Introgression of Desirable Traits: Introgression involves transferring desirable traits from one species or variety to another. Trisomics can be used as a bridge for introgressing traits from wild relatives or exotic germplasm into cultivated varieties. By crossing trisomics with the desired traits to cultivated varieties, breeders can introduce new genetic variation and enhance crop performance. Introgression involves the use of trisomics to transfer specific chromosomal regions or genes from wild relatives or exotic germplasm into cultivated varieties. Researchers must carefully monitor the chromosomal composition and phenotypic traits of the progeny to ensure the stable transmission of the desired traits. Advanced breeding techniques, such as marker-assisted selection (MAS) and genomic selection (GS), can enhance the efficiency and precision of introgression programs by identifying individuals with the desired chromosomal configurations and traits.

Location of Genes

The precise location of genes on chromosomes is critical for understanding their function and their role in trait expression. Various molecular cytogenetic techniques are used to map genes to specific chromosomal regions, facilitating the identification and manipulation of genes for crop improvement.

Techniques for Gene Mapping

Fluorescence In Situ Hybridization (FISH): FISH is a widely used technique for locating genes on chromosomes. It involves labeling specific DNA sequences with fluorescent probes and hybridizing them to chromosome spreads. The probes bind to their complementary sequences on the chromosomes, allowing visualization of the gene's location under a fluorescence microscope. FISH is highly precise and can be used to map single genes or entire chromosomal regions. FISH provides a high-resolution method for visualizing the location of specific genes or chromosomal regions within the genome. Researchers can use FISH to map the positions of multiple genes simultaneously, enabling the construction of detailed chromosomal maps. FISH can also be used to study the spatial organization of the genome within the nucleus, providing insights into the relationships between gene location and gene expression. The integration of FISH with other cytogenetic techniques, such as comparative genomic hybridization (CGH) and genomic in situ hybridization (GISH), can enhance the accuracy and resolution of gene mapping studies.

Comparative Genomic Hybridization (CGH): CGH is a technique used to detect and map chromosomal aberrations, including duplications and deletions. It involves hybridizing labeled DNA from a test sample and a reference sample to a microarray or chromosome spread. Differences in hybridization patterns indicate the presence and location of chromosomal abnormalities. CGH is useful for identifying gene duplications and deletions that may affect trait expression. CGH provides a comprehensive method for detecting and mapping chromosomal aberrations across the entire genome. Researchers can use CGH to identify regions of the genome that are duplicated or deleted, providing insights into the genetic basis of trait expression and disease susceptibility. The integration of CGH with other genomic techniques, such as next-generation sequencing (NGS) and quantitative PCR (qPCR), can enhance the accuracy and resolution of gene mapping studies.

Linkage Mapping: Linkage mapping involves analyzing the co-segregation of genetic markers and traits in a population of offspring. By determining the frequency of recombination between markers and traits, researchers can construct linkage maps that indicate the relative positions of genes on chromosomes. Linkage mapping is essential for identifying quantitative trait loci (QTLs) associated with complex traits. Linkage mapping provides a method for identifying the relative positions of genes on chromosomes based on their genetic linkage. Researchers can use linkage mapping to identify QTLs associated with complex traits, providing insights into the genetic architecture of trait expression. The integration of linkage mapping with other genetic techniques, such as genome-wide association studies (GWAS) and marker-assisted selection (MAS), can enhance the accuracy and resolution of gene mapping studies.

Genetic and Physical Mapping: Genetic mapping involves determining the relative positions of genes based on recombination frequencies, while physical mapping involves measuring the actual physical distances between genes on chromosomes. Techniques such as restriction fragment length polymorphism (RFLP), simple sequence repeats (SSRs), and single nucleotide polymorphisms (SNPs) are used for genetic mapping. Physical mapping techniques include chromosome walking, contig mapping, and optical mapping. Genetic and physical mapping provide complementary methods for determining the locations of genes within the genome. Researchers can use genetic mapping to identify regions of the genome that are associated with specific traits, while physical mapping provides a high-resolution method for determining the actual physical positions of genes. The integration of genetic and physical mapping techniques can enhance the accuracy and resolution of gene mapping studies.

Next-Generation Sequencing (NGS): NGS technologies, such as whole-genome sequencing (WGS) and RNA sequencing (RNA-seq), provide high-resolution data for gene mapping. NGS can identify the precise locations of genes, regulatory elements, and structural variations across the genome. This information is crucial for understanding the genetic architecture of traits and for identifying candidate genes for crop improvement. NGS provides a high-throughput method for sequencing and analyzing entire genomes, enabling researchers to identify the precise locations of genes and regulatory elements. Researchers can use NGS to study the genetic basis of trait expression, identify candidate genes for crop improvement, and investigate the relationships between gene location and gene expression. The integration of NGS with other genomic techniques, such as chromatin immunoprecipitation sequencing (ChIP-seq) and RNA sequencing (RNA-seq), can enhance the accuracy and resolution of gene mapping studies.

Applications of Gene Mapping in Crop Improvement

Marker-Assisted Breeding: Gene mapping facilitates the development of molecular markers linked to desirable traits. These markers can be used in marker-assisted breeding programs to select individuals with the desired traits, accelerating the breeding process and increasing precision. Marker-assisted breeding is particularly useful for traits that are difficult to measure or are controlled by multiple genes. Marker-assisted breeding involves the use of molecular markers to identify and select individuals with the desired traits. Researchers can use gene mapping to identify markers that are linked to specific traits, enabling the efficient and accurate selection of individuals with the desired chromosomal configurations and traits. The integration of marker-assisted breeding with traditional breeding methods can enhance the efficiency and precision of breeding programs, enabling the rapid development of new crop varieties with desirable traits.

Identification of Candidate Genes: Gene mapping helps identify candidate genes responsible for important agronomic traits. By mapping QTLs and fine-mapping candidate regions, researchers can pinpoint specific genes that contribute to traits such as yield, disease resistance, and stress tolerance. Identifying candidate genes enables targeted breeding and genetic engineering efforts to enhance crop performance. The identification of candidate genes provides valuable insights into the genetic basis of trait expression and the mechanisms of gene regulation. Researchers can use gene mapping to identify genes that are associated with specific traits, enabling the targeted breeding and genetic engineering of crop varieties with desirable traits. The integration of gene mapping with other genomic techniques, such as genome-wide association studies (GWAS) and functional genomics, can enhance the accuracy and resolution of candidate gene identification.

Genetic Architecture: Gene mapping provides insights into the genetic architecture of complex traits. By identifying the locations and effects of multiple genes and their interactions, researchers can understand how genetic variation influences trait expression. This knowledge is essential for designing effective breeding strategies and for predicting the outcomes of genetic crosses. The study of genetic architecture provides valuable insights into the genetic basis of trait expression and the mechanisms of gene regulation. Researchers can use gene mapping to identify the genetic factors that contribute to complex traits, enabling the development of breeding strategies that incorporate multiple genes and their interactions. The integration of gene mapping with other genomic techniques, such as quantitative trait locus (QTL) analysis and genome-wide association studies (GWAS), can enhance the accuracy and resolution of genetic architecture studies.

Functional Genomics: Mapping genes to specific chromosomal regions facilitates functional genomics studies. By associating genes with phenotypic traits, researchers can investigate the functions of individual genes and their roles in biological processes. Functional genomics approaches, such as gene knockouts, overexpression, and RNA interference (RNAi), can be used to validate the functions of mapped genes. Functional genomics provides valuable insights into the functions of individual genes and their roles in biological processes. Researchers can use gene mapping to identify genes that are associated with specific traits, enabling the investigation of their functions using various functional genomics approaches. The integration of functional genomics with other genomic techniques, such as RNA sequencing (RNA-seq) and chromatin immunoprecipitation sequencing (ChIP-seq), can enhance the accuracy and resolution of functional genomics studies.

Gene Editing: The precise location of genes is essential for gene editing technologies such as CRISPR/Cas9. By mapping genes to specific chromosomal regions, researchers can design targeted gene edits to introduce or remove specific genetic variants. Gene editing can be used to improve traits such as disease resistance, stress tolerance, and nutritional content, providing a powerful tool for crop improvement. Gene editing provides a powerful tool for the precise manipulation of genes within the genome. Researchers can use gene mapping to identify the precise locations of genes and regulatory elements, enabling the targeted editing of specific genetic variants. The integration of gene editing with other genomic techniques, such as next-generation sequencing (NGS) and functional genomics, can enhance the accuracy and efficiency of gene editing studies.

Conservation of Genetic Resources: Gene mapping helps conserve genetic resources by identifying and characterizing valuable genetic variation. By mapping genes and QTLs associated with desirable traits, researchers can identify and preserve germplasm with unique genetic profiles. This is crucial for maintaining genetic diversity and for developing resilient crop varieties that can adapt to changing environmental conditions. The conservation of genetic resources provides valuable insights into the genetic basis of trait expression and the mechanisms of gene regulation. Researchers can use gene mapping to identify and characterize valuable genetic variation, enabling the development of breeding strategies that incorporate multiple genes and their interactions. The integration of gene mapping with other genomic techniques, such as genome-wide association studies (GWAS) and quantitative trait locus (QTL) analysis, can enhance the accuracy and resolution of genetic resource conservation studies.

C. Use of balanced tertiary trisomics in hybrid seed production

Balanced tertiary trisomics, where an extra chromosome is formed by combining segments from two different chromosomes, offers unique advantages for hybrid seed production. These trisomics play a crucial role in genetic studies and breeding programs by facilitating the transfer and stabilization of desirable traits.

Methods for Producing Balanced Tertiary Trisomics

Hybridization and Cytogenetic Techniques: Balanced tertiary trisomics are often produced through hybridization techniques that involve crossing plants with known chromosomal rearrangements. This process can result in the formation of hybrid offspring with balanced tertiary trisomic conditions. Hybridization followed by cytogenetic analysis helps identify and confirm the presence of balanced tertiary trisomics. The production of balanced tertiary trisomics through hybridization requires careful selection of parent plants with specific chromosomal rearrangements. Researchers must ensure that the parental chromosomes are compatible and capable of forming balanced tertiary trisomics. Cytogenetic techniques such as fluorescence in situ hybridization (FISH) and genomic in situ hybridization (GISH) are used to visualize and confirm the presence of the extra chromosome segments in the progeny.

Induced Mutagenesis: Induced mutagenesis using chemical mutagens, such as colchicine, or physical mutagens, such as radiation, can also produce balanced tertiary trisomics. These mutagens induce chromosomal breaks and rearrangements, leading to the formation of tertiary trisomic configurations. The resulting plants are then screened for the presence of balanced tertiary trisomics using cytogenetic and molecular techniques. The use of induced mutagenesis for producing balanced tertiary trisomics involves optimizing treatment conditions to achieve the desired chromosomal rearrangements. Researchers must determine the appropriate concentration and duration of mutagen exposure, as well as the developmental stage at which treatment is most effective. Following mutagenesis, plants are screened for chromosomal rearrangements using techniques such as chromosome counting, karyotyping, and molecular marker analysis.

Utilizing Aneuploid Stocks: Aneuploid stocks, which are plant lines with known chromosomal abnormalities, can be crossed with normal diploid plants to produce balanced tertiary trisomics. By selecting and crossing specific aneuploid stocks, researchers can generate progeny with balanced tertiary trisomic configurations. This approach is facilitated by detailed cytogenetic analysis to confirm the presence of the desired chromosomal segments. The use of aneuploid stocks involves maintaining and propagating plant lines with specific chromosomal rearrangements. Researchers perform controlled crosses between aneuploid stocks and normal diploid plants, followed by cytogenetic analysis to identify balanced tertiary trisomics. Advanced cytogenetic techniques, such as FISH and GISH, enhance the accuracy and efficiency of identifying balanced tertiary trisomics.

Benefits of Using Balanced Tertiary Trisomics in Hybrid Seed Production

Enhanced Genetic Diversity: Balanced tertiary trisomics introduce novel genetic combinations by incorporating segments from different chromosomes. This enhances genetic diversity in the breeding population, which is crucial for developing new hybrid varieties with improved traits. The increased genetic diversity provides a broader genetic base for selecting desirable traits. The introduction of novel genetic combinations through balanced tertiary trisomics can lead to the expression of new phenotypic traits that are not present in the parent lines. This genetic diversity is beneficial for breeding programs as it increases the likelihood of identifying unique and valuable traits. The enhanced genetic diversity also provides a buffer against environmental fluctuations and biotic stresses, improving the overall resilience of the hybrid varieties.

Stabilization of Desirable Traits: Balanced tertiary trisomics help stabilize desirable traits by maintaining specific chromosomal segments in the progeny. This stabilization is particularly useful for traits that are controlled by genes located on the tertiary chromosome segments. By ensuring the consistent presence of these segments, breeders can develop stable hybrid varieties with reliable trait expression. The stabilization of desirable traits through balanced tertiary trisomics is achieved by selecting and propagating individuals with the desired chromosomal configurations. Researchers must carefully monitor the inheritance of the tertiary chromosome segments to ensure their stable transmission across generations. The use of molecular markers linked to the tertiary segments can facilitate the identification and selection of individuals with stable trait expression.

Facilitation of Gene Introgression: Balanced tertiary trisomics facilitate the introgression of genes from wild relatives or exotic germplasm into cultivated varieties. By incorporating specific chromosomal segments from donor plants, breeders can introduce valuable traits such as disease resistance, stress tolerance, and improved yield. This introgression process is essential for broadening the genetic base of hybrid varieties. The use of balanced tertiary trisomics for gene introgression involves crossing trisomic individuals with donor plants carrying the desired traits. Researchers must ensure the stable transmission of the tertiary chromosome segments that contain the target genes. Advanced breeding techniques, such as marker-assisted selection (MAS) and genomic selection (GS), can enhance the efficiency and precision of the introgression process.

Challenges in Using Balanced Tertiary Trisomics

Low Fertility and Viability: Similar to other trisomics, balanced tertiary trisomics often exhibit low fertility and viability due to chromosomal imbalances. The presence of extra chromosome segments can disrupt normal meiosis, leading to reduced gamete formation and lower seed set. This challenge can complicate the propagation and maintenance of balanced tertiary trisomic lines. The reduced fertility and viability of balanced tertiary trisomics can result from various factors, including abnormal chromosomal pairing, nondisjunction, and chromosomal rearrangements. Researchers must carefully monitor the reproductive performance of trisomic individuals and select those with higher fertility and viability. The use of advanced reproductive techniques, such as embryo rescue and tissue culture, can also assist in propagating and maintaining balanced tertiary trisomic lines.

Identification and Confirmation: Accurate identification and confirmation of balanced tertiary trisomics require detailed cytogenetic and molecular analysis. This involves chromosome counting, karyotyping, and the use of molecular markers to confirm the presence of the tertiary chromosome segments. The identification process is labor-intensive and requires specialized equipment and expertise. In addition to traditional cytogenetic techniques, molecular approaches such as quantitative PCR (qPCR) and next-generation sequencing (NGS) can be used to identify and confirm balanced tertiary trisomics. These methods provide high-resolution data on the chromosomal composition and help confirm the presence of specific tertiary segments. Automated imaging and analysis systems can also enhance the efficiency and accuracy of identifying balanced tertiary trisomics.

Genetic and Phenotypic Variability: Balanced tertiary trisomics often exhibit significant genetic and phenotypic variability due to the presence of extra chromosome segments. This variability can complicate the selection and stabilization of trisomic lines. Researchers must carefully select and propagate individuals with consistent phenotypic traits and chromosomal configurations to develop stable trisomic stocks. The genetic and phenotypic variability observed in balanced tertiary trisomics can result from gene dosage effects, epigenetic modifications, and genetic interactions. The presence of extra chromosome segments can lead to changes in gene expression, resulting in a wide range of phenotypic traits. Researchers must carefully evaluate and select trisomic individuals with stable and desirable traits to develop reliable genetic resources for breeding and research. Advanced phenotyping techniques, such as high-throughput imaging and machine learning algorithms, can assist in the accurate characterization and selection of balanced tertiary trisomics.

Applications of Balanced Tertiary Trisomics in Hybrid Seed Production

Development of Hybrid Varieties: Balanced tertiary trisomics are used to develop hybrid varieties with enhanced traits such as disease resistance, stress tolerance, and improved yield. By incorporating specific chromosomal segments from donor plants, breeders can introduce valuable traits into hybrid varieties. This approach is particularly useful for traits that are difficult to transfer through traditional breeding methods. The development of hybrid varieties using balanced tertiary trisomics involves crossing trisomic individuals with donor plants carrying the desired traits. Researchers must ensure the stable transmission of the tertiary chromosome segments that contain the target genes. The use of molecular markers linked to the tertiary segments can facilitate the identification and selection of individuals with stable trait expression. The resulting hybrid varieties can exhibit improved performance and adaptability, making them valuable for commercial cultivation.

Mapping and Cloning of Genes: Balanced tertiary trisomics are valuable tools for mapping and cloning genes associated with important agronomic traits. By analyzing the genetic and phenotypic effects of the tertiary chromosome segments, researchers can identify and isolate specific genes responsible for desirable traits. This information is crucial for targeted breeding and genetic engineering efforts. The use of balanced tertiary trisomics for gene mapping and cloning involves detailed genetic and cytogenetic analysis to identify the chromosomal regions associated with specific traits. Researchers can use techniques such as quantitative trait locus (QTL) mapping, genome-wide association studies (GWAS), and fine mapping to pinpoint the location of target genes. The integration of molecular markers and advanced genomic techniques can enhance the accuracy and resolution of gene mapping and cloning studies.

Functional Genomics Studies: Balanced tertiary trisomics provide a unique platform for functional genomics studies. By associating specific chromosomal segments with phenotypic traits, researchers can investigate the functions of individual genes and their roles in biological processes. Functional genomics approaches, such as gene knockouts, overexpression, and RNA interference (RNAi), can be used to validate the functions of mapped genes. Functional genomics studies using balanced tertiary trisomics involve associating specific chromosomal segments with phenotypic traits and investigating the functions of individual genes. Researchers can use techniques such as RNA sequencing (RNA-seq), chromatin immunoprecipitation sequencing (ChIP-seq), and gene editing to study the functions of target genes. The integration of functional genomics with other genomic techniques can enhance the accuracy and resolution of gene function studies.

Introgression of Multiple Traits: Balanced tertiary trisomics facilitate the introgression of multiple traits simultaneously by incorporating different chromosomal segments into the breeding population. This approach allows breeders to combine desirable traits such as disease resistance, stress tolerance, and improved yield into a single hybrid variety. The introgression of multiple traits is essential for developing resilient and high-performing hybrid varieties. The use of balanced tertiary trisomics for introgressing multiple traits involves crossing trisomic individuals with donor plants carrying the desired traits. Researchers must ensure the stable transmission of the tertiary chromosome segments that contain the target genes. Advanced breeding techniques, such as marker-assisted selection (MAS) and genomic selection (GS), can enhance the efficiency and precision of the introgression process. The resulting hybrid varieties can exhibit improved performance and adaptability, making them valuable for commercial cultivation.

D. Monosomics methods of production

Monosomics are individuals with one chromosome missing from an otherwise diploid chromosome set, resulting in a 2n-1 chromosomal arrangement. They play a significant role in genetic studies and breeding programs by helping to identify gene functions, map chromosomes, and develop new crop varieties. This section explores the methods for producing monosomics, along with their applications, challenges, and benefits.

Natural Occurrence

Spontaneous Chromosomal Nondisjunction: Monosomics can occur spontaneously in nature due to errors in meiotic or mitotic cell division. These errors result in the nondisjunction of chromosomes, leading to the formation of gametes with a missing chromosome. When these gametes fuse with normal gametes during fertilization, the resulting zygote is monosomic. Spontaneous monosomics are rare and typically identified through phenotypic screening and cytogenetic analysis. In natural populations, environmental stresses such as temperature fluctuations, chemical exposure, or radiation can increase the frequency of chromosomal nondisjunction events. These stresses disrupt normal cell division processes, leading to the formation of gametes with missing chromosomes. Genetic factors, such as mutations in genes involved in chromosome segregation, can also predispose certain individuals to produce monosomic offspring. Understanding the natural occurrence of monosomics helps researchers study the evolutionary significance of chromosomal abnormalities and their impact on genetic diversity.

Induced Mutagenesis

Chemical Mutagens: Chemical mutagens like colchicine, which is a known spindle fiber inhibitor, can induce chromosomal nondisjunction and lead to the production of monosomics. Colchicine disrupts the normal separation of chromosomes during cell division, causing some cells to lose a chromosome. These cells can then be propagated to produce monosomic plants. Other chemical mutagens, such as ethyl methanesulfonate (EMS) and nitrosoguanidine, can also induce chromosomal abnormalities, including monosomy. The use of chemical mutagens involves optimizing treatment conditions to achieve the desired chromosomal abnormalities without causing excessive damage to the plant. Researchers must determine the appropriate concentration and duration of mutagen exposure and the developmental stage at which treatment is most effective. Following mutagenesis, plants are screened for chromosomal abnormalities using techniques such as chromosome counting, karyotyping, and molecular marker analysis.

Physical Mutagens: Physical mutagens such as gamma radiation and X-rays can also induce monosomy by causing chromosomal breaks and subsequent loss of chromosomes during cell division. Radiation treatments are applied to seeds, seedlings, or plant tissues, and the resulting plants are screened for the presence of monosomics using cytogenetic and molecular techniques. The use of physical mutagens for producing monosomics involves careful control of radiation dose and exposure time to induce chromosomal breaks without causing excessive damage to the plant. Researchers must optimize the treatment conditions to achieve the desired chromosomal abnormalities while minimizing adverse effects on plant growth and development. Following radiation treatment, plants are screened for chromosomal abnormalities using cytogenetic techniques such as chromosome counting, karyotyping, and molecular marker analysis.

Hybridization Techniques

Crosses with Aneuploid Stocks: Monosomics can be produced by crossing diploid plants with known aneuploid stocks, such as nullisomics or other monosomics. These crosses result in the formation of progeny with specific chromosomal deficiencies, including monosomics. Detailed cytogenetic analysis is used to identify and confirm the presence of monosomics in the progeny. The use of aneuploid stocks involves maintaining and propagating plant lines with specific chromosomal abnormalities. Researchers perform controlled crosses between diploid plants and aneuploid stocks, followed by cytogenetic analysis to identify monosomic individuals. Advanced cytogenetic techniques, such as fluorescence in situ hybridization (FISH) and genomic in situ hybridization (GISH), enhance the accuracy and efficiency of identifying monosomics.

Interspecific and Intergeneric Hybridization: Wide crosses between different species or genera can also produce monosomics. These hybridizations often result in hybrids with chromosomal imbalances, including monosomics. The hybrid offspring are screened for monosomy using cytogenetic and molecular techniques. The use of interspecific and intergeneric hybridization involves crossing plants from different species or genera to produce hybrids with unique chromosomal configurations. Researchers must carefully select parent plants with compatible chromosomes and the potential to form monosomic hybrids. Following hybridization, the progeny are screened for chromosomal abnormalities using techniques such as chromosome counting, karyotyping, and molecular marker analysis.

Tissue Culture Techniques

Anther and Ovule Culture: Monosomics can be produced through tissue culture techniques such as anther and ovule culture. These methods involve culturing anthers or ovules in vitro to develop haploid plants. Haploids are then treated with colchicine to double their chromosome number, resulting in diploid plants with a missing chromosome, or monosomics. Anther and ovule culture techniques involve isolating and culturing anthers or ovules under sterile conditions to induce the formation of haploid embryos. These haploid embryos are then treated with colchicine to double their chromosome number, resulting in diploid plants with a missing chromosome. Researchers must optimize the culture conditions, including the nutrient medium, temperature, and light, to achieve successful haploid and monosomic production.

Somatic Cell Hybridization: Somatic cell hybridization, or protoplast fusion, can also be used to produce monosomics. This technique involves fusing protoplasts (plant cells without cell walls) from different plant species or genotypes. The resulting hybrid cells can exhibit chromosomal imbalances, including monosomy. These hybrid cells are cultured to regenerate whole plants, which are then screened for monosomics. The use of somatic cell hybridization involves isolating and fusing protoplasts from different plant species or genotypes under controlled conditions. Researchers must optimize the fusion conditions, including the use of chemical or electrical treatments, to achieve successful protoplast fusion. Following fusion, the hybrid cells are cultured to regenerate whole plants, which are then screened for chromosomal abnormalities using cytogenetic and molecular techniques.

Applications of Monosomics in Crop Improvement

Gene Mapping and Chromosomal Assignment: Monosomics are valuable tools for gene mapping and chromosomal assignment. By comparing the phenotypes of monosomic plants with those of normal diploid plants, researchers can determine which traits are associated with specific chromosomes. This information is crucial for constructing genetic linkage maps and identifying the chromosomal locations of important genes. The use of monosomics for gene mapping involves detailed genetic and phenotypic analysis of monosomic plants. Researchers must carefully compare the phenotypic traits of monosomics with those of normal diploids to identify the chromosomal locations of target genes. Advanced genetic mapping techniques, such as quantitative trait locus (QTL) mapping and genome-wide association studies (GWAS), can enhance the accuracy and resolution of gene mapping studies using monosomics.

Functional Genomics Studies: Monosomics provides a unique platform for functional genomics studies. By associating specific chromosomal deficiencies with phenotypic traits, researchers can investigate the functions of individual genes and their roles in biological processes. Functional genomics approaches, such as gene knockouts, overexpression, and RNA interference (RNAi), can be used to validate the functions of mapped genes. The use of monosomics for functional genomics studies involves associating specific chromosomal deficiencies with phenotypic traits and investigating the functions of individual genes. Researchers can use techniques such as RNA sequencing (RNA-seq), chromatin immunoprecipitation sequencing (ChIP-seq), and gene editing to study the functions of target genes. The integration of functional genomics with other genomic techniques can enhance the accuracy and resolution of gene function studies.

Breeding and Crop Improvement: Monosomics are used in breeding programs to transfer desirable traits from one species or variety to another. By incorporating specific chromosomal deficiencies, breeders can introduce valuable traits such as disease resistance, stress tolerance, and improved yield. Monosomics facilitate the introgression of genes from wild relatives or exotic germplasm into cultivated varieties, broadening the genetic base of crop populations. The use of monosomics in breeding programs involves crossing monosomic individuals with donor plants carrying the desired traits. Researchers must ensure the stable transmission of the chromosomal deficiencies that contain the target genes. Advanced breeding techniques, such as marker-assisted selection (MAS) and genomic selection (GS), can enhance the efficiency and precision of the breeding process. The resulting crop varieties can exhibit improved performance and adaptability, making them valuable for commercial cultivation.

Cytogenetic and Evolutionary Studies: Monosomics are valuable tools for cytogenetic and evolutionary studies. By analyzing the chromosomal behavior and genetic interactions of monosomic plants, researchers can gain insights into the mechanisms of chromosome segregation, recombination, and gene regulation. Monosomics also provides a platform for studying the evolutionary significance of chromosomal abnormalities and their impact on genetic diversity. The use of monosomics for cytogenetic and evolutionary studies involves a detailed analysis of chromosomal behavior and genetic interactions. Researchers must carefully monitor the chromosomal segregation and recombination patterns of monosomic plants to understand the mechanisms underlying these processes. Advanced cytogenetic techniques, such as FISH and GISH, can enhance the accuracy and resolution of cytogenetic and evolutionary studies using monosomics.

E. Breeding behavior, and location of genes

Breeding Behavior of Monosomics

Monosomics, individuals with one missing chromosome (2n-1), are valuable for genetic studies and breeding programs. Understanding their breeding behavior is crucial for utilizing them effectively in genetic research and crop improvement.

Meiosis in Monosomics

Chromosomal Pairing: During meiosis in monosomics, the missing chromosome disrupts normal pairing. Instead of forming typical bivalents, monosomic cells may form univalents, where one chromosome lacks a homologous partner. This can result in irregular segregation patterns and the formation of aneuploid gametes with abnormal chromosome numbers. In monosomics, the lack of a homologous chromosome leads to univalents that often fail to align properly on the metaphase plate. This misalignment can cause nondisjunction, where the univalent chromosome does not segregate correctly into daughter cells, leading to gametes with an abnormal chromosome number. Researchers must analyze the behavior of univalents during meiosis to understand the mechanisms of chromosomal segregation in monosomics.

Nondisjunction: Nondisjunction is common in monosomics during meiosis, where the unpaired chromosome fails to segregate properly. This leads to the production of gametes with additional or missing chromosomes. Nondisjunction can occur during the first or second meiotic division, resulting in various chromosomal configurations in the resulting gametes. Nondisjunction in monosomics can result in gametes with different chromosomal compositions, including nullisomic (lacking both copies of a chromosome) or disomic (having two copies of a chromosome) gametes. These gametes can fuse with normal gametes to produce zygotes with various chromosomal abnormalities. Researchers must carefully analyze the chromosomal composition of gametes produced by monosomics to understand the impact of nondisjunction on genetic variability.

Chromosomal Bridges and Lagging Chromosomes: Monosomics often exhibit chromosomal bridges and lagging chromosomes during cell division. Chromosomal bridges occur when the univalent chromosome forms connections between homologous chromosomes, preventing proper segregation. Lagging chromosomes are chromosomes that fail to move to the poles during anaphase, leading to aneuploidy in the daughter cells. These anomalies contribute to the high frequency of aneuploid gametes in monosomic plants. The presence of chromosomal bridges and lagging chromosomes can lead to various chromosomal abnormalities, including deletions, duplications, and translocations. These abnormalities can have significant effects on gene expression and phenotypic traits, contributing to the genetic variability observed in monosomics. Researchers must carefully analyze the chromosomal behavior of monosomics during meiosis to understand the mechanisms underlying these abnormalities and their impact on genetic variability.

Inheritance Patterns in Monosomics

Mendelian Inheritance: The absence of one chromosome in monosomics affects the inheritance patterns of genetic traits. Monosomics typically exhibit non-Mendelian inheritance, where the segregation ratios deviate from the expected Mendelian ratios. For example, a monosomic individual with a missing chromosome carrying a recessive allele may produce a higher proportion of offspring expressing the recessive trait due to the lack of the dominant allele on the missing chromosome. The non-Mendelian inheritance patterns observed in monosomics can result from various factors, including gene dosage effects, epistatic interactions, and chromosomal rearrangements. Researchers must carefully analyze the inheritance patterns of traits in monosomics to understand the underlying genetic mechanisms and their impact on trait expression. Advanced genetic mapping techniques, such as quantitative trait locus (QTL) analysis and genome-wide association studies (GWAS), can help identify the genetic factors contributing to the non-Mendelian inheritance patterns observed in monosomics.

Dosage Compensation: Dosage compensation mechanisms play a crucial role in monosomics, where the genes on the missing chromosome may be partially or fully silenced to maintain normal gene expression levels. Dosage compensation can occur through various mechanisms, including gene silencing, RNA interference, and epigenetic modifications. Understanding dosage compensation is essential for interpreting the phenotypic effects of monosomy and predicting the inheritance of specific traits. The mechanisms of dosage compensation in monosomics can vary depending on the species, the specific chromosome involved, and the genetic background of the individual. Researchers must carefully analyze the expression patterns of genes on the missing chromosome to understand the mechanisms of dosage compensation and their impact on phenotypic traits. Advanced genomic techniques, such as RNA sequencing (RNA-seq) and chromatin immunoprecipitation sequencing (ChIP-seq), can provide valuable insights into the mechanisms of dosage compensation in monosomics.

Epistatic Interactions: Epistatic interactions, where the expression of one gene affects the expression of another gene, are common in monosomics. The absence of one chromosome can introduce novel genetic interactions that influence the expression of multiple traits. These interactions can complicate the interpretation of inheritance patterns and require detailed genetic analysis to identify the underlying genetic mechanisms. The study of epistatic interactions in monosomics can provide valuable insights into the genetic architecture of complex traits and the mechanisms of gene regulation. Researchers must carefully analyze the genetic interactions between the missing chromosome and the endogenous genome to understand the impact of epistasis on phenotypic traits. Advanced genetic mapping techniques, such as epistasis mapping and gene interaction networks, can help identify the genetic factors contributing to the epistatic interactions observed in monosomics.

Breeding Strategies for Monosomics

Backcrossing: Backcrossing is a common breeding strategy used to stabilize monosomic traits and incorporate them into breeding programs. By repeatedly crossing monosomic individuals with a recurrent parent, breeders can select progeny that exhibit the desired traits while maintaining the monosomic condition. This approach helps stabilize the genetic background and reduce genetic variability. Backcrossing involves selecting and crossing individuals with the desired monosomic traits to the recurrent parent over multiple generations. Researchers must carefully monitor the chromosomal composition and phenotypic traits of the progeny to ensure the stable transmission of the monosomic condition. Advanced breeding techniques, such as marker-assisted selection (MAS) and genomic selection (GS), can enhance the efficiency and accuracy of backcrossing programs by identifying individuals with the desired chromosomal configurations and traits.

Marker-Assisted Selection: Marker-assisted selection (MAS) is a powerful tool for breeding monosomics. Molecular markers linked to the missing chromosome or specific traits can be used to identify and select monosomic individuals. MAS enables breeders to efficiently screen large populations and select individuals with the desired chromosomal configurations and traits, accelerating the breeding process. MAS involves the use of molecular markers, such as single nucleotide polymorphisms (SNPs), simple sequence repeats (SSRs), and restriction fragment length polymorphisms (RFLPs), to identify individuals with the desired monosomic traits. Researchers must carefully develop and validate the molecular markers to ensure their accuracy and reliability in identifying monosomic individuals. The integration of MAS with traditional breeding methods can enhance the efficiency and precision of breeding programs, enabling the rapid development of new crop varieties with desirable traits.

Genomic Selection: Genomic selection (GS) involves using genome-wide markers to predict the breeding value of individuals. GS can be applied to monosomics to identify individuals with the highest potential for trait expression and genetic stability. This approach leverages high-throughput genotyping and advanced statistical models to enhance the efficiency and accuracy of breeding programs. GS involves the development of genomic prediction models that use genome-wide markers to estimate the breeding value of individuals. Researchers must carefully develop and validate the genomic prediction models to ensure their accuracy and reliability in predicting the breeding value of monosomic individuals. The integration of GS with traditional breeding methods can enhance the efficiency and precision of breeding programs, enabling the rapid development of new crop varieties with desirable traits.

Introgression of Desirable Traits: Introgression involves transferring desirable traits from one species or variety to another. Monosomics can be used as a bridge for introgressing traits from wild relatives or exotic germplasm into cultivated varieties. By crossing monosomics with the desired traits to cultivated varieties, breeders can introduce new genetic variation and enhance crop performance. Introgression involves the use of monosomics to transfer specific chromosomal regions or genes from wild relatives or exotic germplasm into cultivated varieties. Researchers must carefully monitor the chromosomal composition and phenotypic traits of the progeny to ensure the stable transmission of the desired traits. Advanced breeding techniques, such as marker-assisted selection (MAS) and genomic selection (GS), can enhance the efficiency and precision of introgression programs by identifying individuals with the desired chromosomal configurations and traits.

Location of Genes

The precise location of genes on chromosomes is critical for understanding their function and their role in trait expression. Various molecular cytogenetic techniques are used to map genes to specific chromosomal regions, facilitating the identification and manipulation of genes for crop improvement.

Techniques for Gene Mapping

Fluorescence In Situ Hybridization (FISH): FISH is a widely used technique for locating genes on chromosomes. It involves labeling specific DNA sequences with fluorescent probes and hybridizing them to chromosome spreads. The probes bind to their complementary sequences on the chromosomes, allowing visualization of the gene's location under a fluorescence microscope. FISH is highly precise and can be used to map single genes or entire chromosomal regions. FISH provides a high-resolution method for visualizing the location of specific genes or chromosomal regions within the genome. Researchers can use FISH to map the positions of multiple genes simultaneously, enabling the construction of detailed chromosomal maps. FISH can also be used to study the spatial organization of the genome within the nucleus, providing insights into the relationships between gene location and gene expression. The integration of FISH with other cytogenetic techniques, such as comparative genomic hybridization (CGH) and genomic in situ hybridization (GISH), can enhance the accuracy and resolution of gene mapping studies.

Comparative Genomic Hybridization (CGH): CGH is a technique used to detect and map chromosomal aberrations, including duplications and deletions. It involves hybridizing labeled DNA from a test sample and a reference sample to a microarray or chromosome spread. Differences in hybridization patterns indicate the presence and location of chromosomal abnormalities. CGH is useful for identifying gene duplications and deletions that may affect trait expression. CGH provides a comprehensive method for detecting and mapping chromosomal aberrations across the entire genome. Researchers can use CGH to identify regions of the genome that are duplicated or deleted, providing insights into the genetic basis of trait expression and disease susceptibility. The integration of CGH with other genomic techniques, such as next-generation sequencing (NGS) and quantitative PCR (qPCR), can enhance the accuracy and resolution of gene mapping studies.

Linkage Mapping: Linkage mapping involves analyzing the co-segregation of genetic markers and traits in a population of offspring. By determining the frequency of recombination between markers and traits, researchers can construct linkage maps that indicate the relative positions of genes on chromosomes. Linkage mapping is essential for identifying quantitative trait loci (QTLs) associated with complex traits. Linkage mapping provides a method for identifying the relative positions of genes on chromosomes based on their genetic linkage. Researchers can use linkage mapping to identify QTLs associated with complex traits, providing insights into the genetic architecture of trait expression. The integration of linkage mapping with other genetic techniques, such as genome-wide association studies (GWAS) and marker-assisted selection (MAS), can enhance the accuracy and resolution of gene mapping studies.

Genetic and Physical Mapping: Genetic mapping involves determining the relative positions of genes based on recombination frequencies, while physical mapping involves measuring the actual physical distances between genes on chromosomes. Techniques such as restriction fragment length polymorphism (RFLP), simple sequence repeats (SSRs), and single nucleotide polymorphisms (SNPs) are used for genetic mapping. Physical mapping techniques include chromosome walking, contig mapping, and optical mapping. Genetic and physical mapping provide complementary methods for determining the locations of genes within the genome. Researchers can use genetic mapping to identify regions of the genome that are associated with specific traits, while physical mapping provides a high-resolution method for determining the actual physical positions of genes. The integration of genetic and physical mapping techniques can enhance the accuracy and resolution of gene mapping studies.

Next-Generation Sequencing (NGS): NGS technologies, such as whole-genome sequencing (WGS) and RNA sequencing (RNA-seq), provide high-resolution data for gene mapping. NGS can identify the precise locations of genes, regulatory elements, and structural variations across the genome. This information is crucial for understanding the genetic architecture of traits and for identifying candidate genes for crop improvement. NGS provides a high-throughput method for sequencing and analyzing entire genomes, enabling researchers to identify the precise locations of genes and regulatory elements. Researchers can use NGS to study the genetic basis of trait expression, identify candidate genes for crop improvement, and investigate the relationships between gene location and gene expression. The integration of NGS with other genomic techniques, such as chromatin immunoprecipitation sequencing (ChIP-seq) and RNA sequencing (RNA-seq), can enhance the accuracy and resolution of gene mapping studies.

Applications of Gene Mapping in Crop Improvement

Marker-Assisted Breeding: Gene mapping facilitates the development of molecular markers linked to desirable traits. These markers can be used in marker-assisted breeding programs to select individuals with the desired traits, accelerating the breeding process and increasing precision. Marker-assisted breeding is particularly useful for traits that are difficult to measure or are controlled by multiple genes. Marker-assisted breeding involves the use of molecular markers to identify and select individuals with the desired traits. Researchers can use gene mapping to identify markers that are linked to specific traits, enabling the efficient and accurate selection of individuals with the desired chromosomal configurations and traits. The integration of marker-assisted breeding with traditional breeding methods can enhance the efficiency and precision of breeding programs, enabling the rapid development of new crop varieties with desirable traits.

Identification of Candidate Genes: Gene mapping helps identify candidate genes responsible for important agronomic traits. By mapping QTLs and fine-mapping candidate regions, researchers can pinpoint specific genes that contribute to traits such as yield, disease resistance, and stress tolerance. Identifying candidate genes enables targeted breeding and genetic engineering efforts to enhance crop performance. The identification of candidate genes provides valuable insights into the genetic basis of trait expression and the mechanisms of gene regulation. Researchers can use gene mapping to identify genes that are associated with specific traits, enabling the targeted breeding and genetic engineering of crop varieties with desirable traits. The integration of gene mapping with other genomic techniques, such as genome-wide association studies (GWAS) and functional genomics, can enhance the accuracy and resolution of candidate gene identification.

Gene mapping provides insights into the genetic architecture of complex traits. By identifying the locations and effects of multiple genes and their interactions, researchers can understand how genetic variation influences trait expression. This knowledge is essential for designing effective breeding strategies and for predicting the outcomes of genetic crosses. The study of genetic architecture provides valuable insights into the genetic basis of trait expression and the mechanisms of gene regulation. Researchers can use gene mapping to identify the genetic factors that contribute to complex traits, enabling the development of breeding strategies that incorporate multiple genes and their interactions. The integration of gene mapping with other genomic techniques, such as quantitative trait locus (QTL) analysis and genome-wide association studies (GWAS), can enhance the accuracy and resolution of genetic architecture studies.

Functional Genomics: Mapping genes to specific chromosomal regions facilitates functional genomics studies. By associating genes with phenotypic traits, researchers can investigate the functions of individual genes and their roles in biological processes. Functional genomics approaches, such as gene knockouts, overexpression, and RNA interference (RNAi), can be used to validate the functions of mapped genes. Functional genomics provides valuable insights into the functions of individual genes and their roles in biological processes. Researchers can use gene mapping to identify genes that are associated with specific traits, enabling the investigation of their functions using various functional genomics approaches. The integration of functional genomics with other genomic techniques, such as RNA sequencing (RNA-seq) and chromatin immunoprecipitation sequencing (ChIP-seq), can enhance the accuracy and resolution of functional genomics studies.

Gene Editing: The precise location of genes is essential for gene editing technologies such as CRISPR/Cas9. By mapping genes to specific chromosomal regions, researchers can design targeted gene edits to introduce or remove specific genetic variants. Gene editing can be used to improve traits such as disease resistance, stress tolerance, and nutritional content, providing a powerful tool for crop improvement. Gene editing provides a powerful tool for the precise manipulation of genes within the genome. Researchers can use gene mapping to identify the precise locations of genes and regulatory elements, enabling the targeted editing of specific genetic variants. The integration of gene editing with other genomic techniques, such as next-generation sequencing (NGS) and functional genomics, can enhance the accuracy and efficiency of gene editing studies.

Conservation of Genetic Resources: Gene mapping helps conserve genetic resources by identifying and characterizing valuable genetic variation. By mapping genes and QTLs associated with desirable traits, researchers can identify and preserve germplasm with unique genetic profiles. This is crucial for maintaining genetic diversity and for developing resilient crop varieties that can adapt to changing environmental conditions. The conservation of genetic resources provides valuable insights into the genetic basis of trait expression and the mechanisms of gene regulation. Researchers can use gene mapping to identify and characterize valuable genetic variation, enabling the development of breeding strategies that incorporate multiple genes and their interactions. The integration of gene mapping with other genomic techniques, such as genome-wide association studies (GWAS) and quantitative trait locus (QTL) analysis, can enhance the accuracy and resolution of genetic resource conservation studies.

F. Intervarietal substitutions-allelic and nonallelic interactions

Intervarietal substitutions involve replacing a specific chromosome or chromosome segment from one variety (donor) with the corresponding chromosome or segment from another variety (recipient). This process is instrumental in studying genetic interactions, mapping genes, and improving crop varieties. Intervarietal substitutions can reveal the effects of allelic and nonallelic interactions on trait expression, providing valuable insights into genetic mechanisms.

Allelic Interactions

Allelic interactions, also known as intralocus interactions, occur between different alleles of the same gene. These interactions can affect trait expression in various ways, such as dominance, codominance, and incomplete dominance.

Dominance and Recessiveness: In allelic interactions, dominance and recessiveness describe the relationship between two alleles at a single locus. A dominant allele masks the expression of a recessive allele in heterozygous individuals. Intervarietal substitutions can help identify dominant and recessive alleles by comparing the phenotypes of homozygous and heterozygous individuals. For example, in maize (Zea mays), intervarietal substitutions of chromosomes carrying alleles for kernel color can reveal dominance relationships. If the donor variety has a dominant allele for red kernel color (R) and the recipient variety has a recessive allele for white kernel color (r), substituting the donor chromosome into the recipient variety will result in red kernels, demonstrating the dominance of the R allele over the r allele. Dominance and recessiveness play a critical role in breeding programs as they determine the expression of essential traits. By understanding the dominance relationships, breeders can predict the outcomes of crosses and select parents to achieve desired phenotypes in the progeny.

Codominance and Incomplete Dominance: Codominance occurs when both alleles at a locus contribute equally to the phenotype, resulting in a distinct expression of both alleles in heterozygous individuals. Incomplete dominance, or partial dominance, occurs when the phenotype of heterozygous individuals is intermediate between the phenotypes of homozygous individuals. Intervarietal substitutions can elucidate codominance and incomplete dominance by analyzing the phenotypes of hybrid individuals. For instance, in snapdragons (Antirrhinum majus), substituting chromosomes with alleles for flower color can reveal incomplete dominance. If the donor variety has an allele for red flowers (R) and the recipient variety has an allele for white flowers (r), substituting the donor chromosome may result in pink flowers in heterozygous individuals, indicating incomplete dominance. Understanding codominance and incomplete dominance is crucial for accurately predicting phenotypic outcomes in hybrid breeding programs and for harnessing these genetic interactions to create new varieties with unique and desirable traits.

Epistasis: Epistasis is a form of allelic interaction where one gene's allele masks or modifies the expression of alleles at another gene. Intervarietal substitutions can help identify epistatic interactions by introducing chromosomes with specific alleles into a different genetic background and observing changes in phenotype. For example, in rice (Oryza sativa), epistatic interactions between genes controlling plant height can be studied through intervarietal substitutions. If the donor variety has a dominant allele for tall stature (T) and the recipient variety has a recessive allele for dwarf stature (t), substituting the donor chromosome may result in tall plants. However, if another gene in the recipient background modifies the expression of T, resulting in intermediate height, it indicates an epistatic interaction. Epistasis can significantly impact breeding strategies as it reveals complex genetic relationships that need to be considered when selecting parent plants and predicting the inheritance of multiple traits.

Nonallelic Interactions

Nonallelic interactions, also known as interlocus interactions, occur between alleles of different genes. These interactions can influence trait expression and complicate genetic analysis. Intervarietal substitutions are valuable tools for studying nonallelic interactions by introducing specific chromosome segments from one variety into another and analyzing the resulting phenotypes.

Additive Gene Action: Additive gene action refers to the cumulative effect of alleles at different loci on a trait. Each allele contributes a specific amount to the overall phenotype. Intervarietal substitutions can help quantify additive gene effects by comparing the phenotypes of individuals with different combinations of substituted chromosomes. For instance, in wheat (Triticum aestivum), intervarietal substitutions can be used to study additive effects on grain yield. If the donor variety has alleles that increase yield on multiple chromosomes, substituting these chromosomes into the recipient variety can result in a cumulative increase in yield. The additive effects of these alleles can be quantified by analyzing the yield of hybrids with different combinations of substituted chromosomes. Understanding additive gene action is essential for breeding programs aiming to combine multiple beneficial alleles to achieve superior performance in complex traits like yield, quality, and stress tolerance.

Complementary Gene Action: Complementary gene action occurs when alleles at different loci interact to produce a specific phenotype that neither allele can produce alone. This type of interaction is crucial for understanding the genetic basis of certain traits and can be studied effectively through intervarietal substitutions. For example, in peas (*Pisum sativum*), complementary gene action for flower color can be studied through intervarietal substitutions. Suppose the donor variety has an allele for pigment production (A) and the recipient variety has an allele for pigment deposition (B). Substituting the donor chromosome into the recipient may result in colored flowers only if both alleles (A and B) are present, indicating complementary gene action. This interaction highlights the need for both genetic factors to be present to achieve the desired trait. Understanding complementary gene action allows breeders to strategically combine alleles from different varieties to create new plant varieties with enhanced traits that cannot be achieved by individual alleles alone.

Epistasis and Gene Suppression: Epistasis involves interactions where one gene's allele masks or modifies the expression of alleles at another gene. Gene suppression is a specific form of epistasis where one gene suppresses the expression of another gene. Intervarietal substitutions can help identify epistatic and suppressive interactions by introducing chromosomes with specific alleles into a different genetic background and observing changes in phenotype. For instance, in maize (*Zea mays*), intervarietal substitutions can be used to study epistasis and gene suppression in anthocyanin biosynthesis. If the donor variety has an allele for anthocyanin production (A) and the recipient variety has a suppressor allele (S), substituting the donor chromosome may result in a lack of anthocyanin production, indicating suppression. Conversely, if the recipient variety lacks the suppressor allele, substituting the donor chromosome may result in anthocyanin production, revealing epistatic interactions. Identifying these interactions is crucial for understanding the complex genetic networks that control important traits and for designing effective breeding strategies.

Gene Interaction Networks: Gene interaction networks involve complex interactions among multiple genes that contribute to a trait. Intervarietal substitutions can help unravel these networks by introducing specific chromosome segments and analyzing the resulting phenotypes to identify interacting genes. In *Arabidopsis thaliana*, for example, intervarietal substitutions can be used to study gene interaction networks involved in stress responses. If the donor variety has alleles for drought tolerance on multiple chromosomes, substituting these chromosomes into the recipient variety can help identify the interactions between these genes and their cumulative effect on drought tolerance. Researchers can use advanced genetic mapping techniques, such as quantitative trait locus (QTL) mapping and genome-wide association studies (GWAS), to identify the specific genes involved in these networks. Gene interaction networks enable breeders to manipulate multiple genetic pathways simultaneously, leading to the development of crop varieties with enhanced resilience and performance under various environmental conditions.

G. Telocentric method of mapping

The telocentric method of mapping is a cytogenetic technique used to determine the physical location of genes on chromosomes. Telocentric chromosomes are chromosomes where the centromere is located very close to one end, resulting in a chromosome with a single arm. This unique structure makes telocentric chromosomes valuable tools for gene mapping, as they allow for precise localization of genes along the chromosomal arm. This method involves the use of these telocentric chromosomes to map genes and genetic markers, providing insights into the genetic architecture of organisms.

Production of Telocentric Chromosomes

Spontaneous Formation: Telocentric chromosomes can form spontaneously due to structural rearrangements during cell division. These rearrangements can result in the loss of chromosomal segments, leading to the formation of a chromosome with the centromere near one end. In nature, this spontaneous formation can be influenced by environmental factors such as radiation, chemical exposure, or stress conditions that disrupt normal chromosomal segregation. Researchers studying organisms that naturally exhibit high rates of chromosomal rearrangements can utilize spontaneous telocentric chromosomes to identify naturally occurring genetic variations. This natural variation provides a rich source of genetic diversity, which is crucial for evolutionary studies and for understanding the mechanisms of chromosomal stability and instability. By analyzing these spontaneous telocentric chromosomes, scientists can gain insights into the processes that drive chromosomal changes and their impacts on genetic traits.

Induced by Radiation: Radiation, such as X-rays or gamma rays, can induce chromosomal breaks and rearrangements, resulting in the formation of telocentric chromosomes. This method involves exposing cells to radiation and then screening for the presence of telocentric chromosomes using cytogenetic techniques. Radiation-induced telocentric chromosomes are particularly useful for creating specific chromosomal alterations in a controlled manner. The process of inducing telocentric chromosomes with radiation requires careful calibration of exposure levels to balance the induction of chromosomal breaks with the survival of the cells. Researchers must optimize the radiation dose to create a sufficient number of telocentric chromosomes without causing excessive damage to the cell's overall genome. Following radiation exposure, cells undergo rigorous screening using techniques such as fluorescence in situ hybridization (FISH) to identify and confirm the presence of telocentric chromosomes. This method provides a powerful tool for generating specific chromosomal changes for detailed genetic studies.

Chromosome Manipulation: Telocentric chromosomes can also be produced through chromosome manipulation techniques, such as the fusion of centric fragments from different chromosomes or the selective elimination of chromosomal arms. These techniques require precise control and expertise in cytogenetics. Chromosome manipulation allows researchers to design specific chromosomal configurations to study the effects of genetic changes. Chromosome manipulation involves advanced techniques such as microdissection and microinjection to physically alter chromosomes. Researchers can create telocentric chromosomes by selectively removing or fusing chromosomal segments, resulting in chromosomes with the centromere positioned near one end. These manipulated chromosomes are then propagated in cell cultures or whole organisms for further analysis. The ability to manipulate chromosomes with such precision allows scientists to investigate the functional consequences of specific chromosomal rearrangements and to identify genes located near the centromere.

Application of Telocentric Chromosomes in Mapping

Gene Localization: The primary application of telocentric chromosomes in gene mapping is the precise localization of genes. By comparing the phenotypes of individuals with different telocentric chromosomes, researchers can determine the position of genes along the chromosomal arm. This process involves crossing individuals with known genetic markers to those with telocentric chromosomes and analyzing the segregation patterns in the progeny. Gene localization using telocentric chromosomes provides high-resolution mapping, enabling researchers to pinpoint the exact position of genes relative to the centromere. This precision is crucial for identifying genes involved in specific traits and for understanding the genetic basis of complex phenotypes. By utilizing telocentric chromosomes, scientists can dissect the genetic architecture of traits and identify candidate genes for further functional studies.

Mapping Genetic Markers: Telocentric chromosomes are also used to map genetic markers. By identifying the position of markers relative to the centromere, researchers can construct detailed genetic maps. This information is crucial for understanding the genetic architecture of traits and for marker-assisted breeding programs. Genetic markers located on telocentric chromosomes provide reference points for linking phenotypic traits to specific chromosomal regions. Mapping genetic markers with telocentric chromosomes enhances the resolution of genetic maps, facilitating the identification of quantitative trait loci (QTLs) and other genetic elements associated with important traits. These high-resolution maps are invaluable for breeding programs, as they enable the precise selection of individuals with desirable genetic profiles. By integrating marker data with phenotypic analysis, breeders can develop new crop varieties with enhanced traits, improving agricultural productivity and sustainability.

Studying Chromosomal Behavior: Telocentric chromosomes provide insights into chromosomal behavior during cell division. By observing the segregation patterns of telocentric chromosomes, researchers can study the mechanisms of chromosome movement and alignment during meiosis and mitosis. Understanding these mechanisms is fundamental for unraveling the processes that ensure accurate chromosome segregation and genetic stability. The study of chromosomal behavior using telocentric chromosomes reveals important information about the dynamics of centromere function and kinetochore attachment. Researchers can investigate how telocentric chromosomes interact with the spindle apparatus during cell division and identify factors that influence chromosomal stability. This knowledge is essential for understanding the causes of chromosomal abnormalities, such as nondisjunction and aneuploidy, which are associated with various genetic disorders and diseases.

Telocentric Mapping in Practice

Creating Telocentric Lines: The first step in telocentric mapping is to create lines or stocks with telocentric chromosomes. This involves inducing or selecting for telocentric chromosomes and then propagating these lines for further study. Cytogenetic techniques, such as fluorescence in situ hybridization (FISH) and chromosome painting, are used to confirm the presence and structure of telocentric chromosomes. Creating telocentric lines requires careful screening and selection to ensure the stability and integrity of the chromosomes. Researchers propagate these lines through multiple generations to establish stable stocks for genetic analysis. Maintaining telocentric lines involves rigorous monitoring to prevent chromosomal rearrangements and ensure the consistency of the genetic material. These stable lines serve as valuable resources for mapping genes and studying chromosomal behavior.

Crossing and Phenotypic Analysis: Telocentric lines are crossed with lines carrying known genetic markers or traits. The progeny are then analyzed for the presence of these markers or traits, and their segregation patterns are compared to the structure of the telocentric chromosomes. This analysis helps determine the relative position of genes and markers along the chromosomal arm. The crossing and phenotypic analysis process involves detailed genetic and cytogenetic examination of the progeny. Researchers analyze the inheritance patterns of the genetic markers and traits about the telocentric chromosomes. By correlating the phenotypic data with the chromosomal structure, scientists can construct high-resolution genetic maps that provide precise information about gene location. This process requires meticulous record-keeping and statistical analysis to ensure the accuracy and reliability of the mapping results.

Constructing Genetic Maps: The data obtained from the phenotypic analysis of the progeny are used to construct genetic maps. These maps indicate the position of genes and markers relative to the centromere and each other. Genetic mapping software and statistical tools are often used to analyze the data and generate accurate maps. Constructing genetic maps with telocentric chromosomes involves integrating phenotypic data with chromosomal information to create detailed representations of the genetic landscape. Researchers use advanced computational tools to analyze the segregation patterns and calculate the genetic distances between markers. These maps provide a visual representation of the chromosomal regions associated with specific traits, facilitating the identification of candidate genes and genetic elements. High-resolution genetic maps are essential for guiding breeding programs and for advancing our understanding of genetic architecture.

Challenges and Considerations

Chromosomal Stability: One of the challenges in using telocentric chromosomes is their stability. Telocentric chromosomes can be prone to breakage and rearrangement, which can complicate mapping efforts. Researchers must carefully monitor the stability of telocentric lines and take measures to minimize chromosomal instability. Maintaining chromosomal stability in telocentric lines requires rigorous monitoring and management practices. Researchers use cytogenetic techniques to regularly assess the integrity of the chromosomes and identify any structural changes. Strategies such as minimizing environmental stress, optimizing growth conditions, and implementing robust propagation methods are employed to enhance chromosomal stability. Ensuring the stability of telocentric chromosomes is crucial for obtaining reliable and reproducible mapping results.

Resolution and Accuracy: The resolution and accuracy of telocentric mapping depend on the number and distribution of genetic markers along the chromosomal arm. Dense marker coverage is necessary for high-resolution maps, but achieving this can be challenging. Researchers must balance the need for resolution with the practical limitations of marker availability and phenotypic analysis. Achieving high-resolution mapping requires the identification and utilization of numerous genetic markers distributed along the telocentric chromosome. Researchers employ techniques such as whole-genome sequencing, marker discovery, and genotyping to enhance marker density. Additionally, phenotypic analysis must be conducted with precision to accurately correlate genetic markers with specific traits. By optimizing marker coverage and phenotypic analysis, researchers can achieve high-resolution and accurate genetic maps that provide valuable insights into gene location and function.

Cytogenetic Expertise: Telocentric mapping requires expertise in cytogenetics and chromosome manipulation. Researchers must be proficient in techniques such as FISH, chromosome painting, and cytogenetic analysis. Collaboration with cytogeneticists and access to specialized equipment are often necessary for successful telocentric mapping projects. Conducting telocentric mapping studies involves collaboration with cytogenetic experts who possess specialized knowledge and skills. Researchers must be trained in advanced cytogenetic techniques to accurately identify and manipulate telocentric chromosomes. Access to specialized equipment, such as fluorescence microscopes and chromosome analysis software, is essential for conducting detailed cytogenetic studies. Collaboration with cytogeneticists ensures the accuracy and reliability of the mapping results and enhances the overall success of telocentric mapping projects.

Advantages of Telocentric Mapping

Precision and Specificity: Telocentric mapping offers high precision and specificity in gene localization. By focusing on a single chromosomal arm, researchers can achieve more accurate mapping compared to techniques that involve entire chromosomes. The unique structure of telocentric chromosomes allows for detailed analysis of specific chromosomal regions, providing precise information about gene location and function. The precision and specificity of telocentric mapping are particularly valuable for identifying genes associated with complex traits. By analyzing specific chromosomal regions, researchers can dissect the genetic architecture of traits and identify candidate genes for further study. This level of precision enhances our understanding of gene function and regulation, facilitating the development of targeted breeding strategies and genetic interventions.

Utility in Diverse Organisms: Telocentric mapping can be applied to a wide range of organisms, including plants, animals, and fungi. This versatility makes it a valuable tool for genetic research across different species. The ability to produce and utilize telocentric chromosomes in various organisms expands the scope of genetic studies and enables cross-species comparisons. The utility of telocentric mapping in diverse organisms allows researchers to explore genetic mechanisms that are conserved across species. This cross-species approach provides insights into fundamental biological processes and enhances our understanding of evolutionary relationships. By applying telocentric mapping to different organisms, scientists can uncover universal principles of gene function and regulation, advancing the field of genetics.

Complementary to Other Techniques: Telocentric mapping can be used in conjunction with other genetic mapping techniques, such as linkage mapping and physical mapping. By integrating data from multiple methods, researchers can obtain a more comprehensive understanding of the genetic architecture of traits. This complementary approach enhances the accuracy and resolution of genetic maps, providing valuable insights into gene location and function. Integrating telocentric mapping with other genetic techniques allows researchers to leverage the strengths of each method. For example, linkage mapping provides information about the relative positions of genes based on genetic recombination, while telocentric mapping offers precise localization of genes along the chromosomal arm. Combining these approaches enhances the resolution and accuracy of genetic maps, facilitating the identification of candidate genes and genetic elements associated with important traits.

Applications in Crop Improvement

Marker-Assisted Breeding: Telocentric mapping provides detailed genetic maps that are essential for marker-assisted breeding. By precisely locating genes and markers, breeders can select individuals with desirable traits more effectively and accelerate the development of improved crop varieties. Marker-assisted breeding enhances the efficiency and precision of breeding programs, leading to the creation of high-performing crop varieties. The application of telocentric mapping in marker-assisted breeding involves the identification of genetic markers associated with key traits, such as yield, disease resistance, and stress tolerance. By incorporating these markers into breeding programs, breeders can select parent plants with the desired genetic profiles and predict the performance of their offspring. This targeted approach accelerates the breeding process and reduces the time required to develop new crop varieties, improving agricultural productivity and sustainability.

Identification of Quantitative Trait Loci (QTLs): Telocentric mapping aids in the identification of QTLs associated with important agronomic traits, such as yield, disease resistance, and stress tolerance. This information is crucial for understanding the genetic basis of complex traits and for developing targeted breeding strategies. Identifying QTLs provides insights into the genetic architecture of traits and helps breeders develop crop varieties with enhanced performance. QTL mapping using telocentric chromosomes involves analyzing the genetic variation and phenotypic traits of individuals to identify chromosomal regions associated with specific traits. Researchers use statistical tools and genetic models to correlate the presence of genetic markers with trait expression. This process enables the identification of QTLs and the dissection of the genetic basis of complex traits. By incorporating QTL information into breeding programs, breeders can develop crop varieties with improved yield, quality, and resilience to environmental stresses.

Gene Cloning and Functional Genomics: The precise localization of genes through telocentric mapping facilitates gene cloning and functional genomics studies. Researchers can isolate and characterize genes of interest, study their functions, and manipulate them for crop improvement. Functional genomics provides insights into gene expression, regulation, and interactions, enhancing our understanding of gene function. Telocentric mapping enables the precise identification of candidate genes associated with important traits. Researchers can clone these genes and study their functions using various molecular techniques, such as gene knockout, overexpression, and RNA interference. Functional genomics studies provide insights into the roles of specific genes in biological processes and their contributions to phenotypic traits. This knowledge is essential for developing genetic interventions and for improving crop performance through targeted gene manipulation.

Chapter - 8

Cytogenomics

A. Cytogenomics: Concept, Tools, and Techniques for Crop Improvement

Cytogenomics integrates cytogenetics and genomics to study the structure, function, and evolution of chromosomes in the context of the entire genome. This interdisciplinary field combines the detailed chromosome analysis provided by cytogenetics with the comprehensive genetic information offered by genomics, facilitating a deeper understanding of plant genetics and the development of improved crop varieties.

Concept of Cytogenomics

Cytogenomics is a scientific discipline that merges cytogenetic techniques with genomic technologies to investigate the chromosomal organization, genetic diversity, and functional genomics of organisms. This integration allows researchers to gain a comprehensive understanding of the genetic and chromosomal basis of traits, which is crucial for various applications, including crop improvement. The primary objectives of cytogenomics are to:

Analyze Chromosomal Structures

Study the Physical Organization and Structural Variations of Chromosomes: Cytogenomics involves the detailed examination of chromosomal structures to understand their physical organization and identify structural variations. Chromosomes can exhibit a variety of structural changes, including duplications, deletions, translocations, and inversions.

- **Duplications**: These occur when a segment of a chromosome is copied and inserted into the genome. Duplications can lead to gene redundancy, which might provide a buffer against deleterious mutations or create opportunities for evolutionary innovation through gene divergence.
- **Deletions**: Deletions involve the loss of a chromosomal segment, which can result in the absence of certain genes. This loss can have significant effects on an organism, especially if the deleted genes are essential for normal development or function.

DOI: 10.1201/9781003717775-8

- » **Translocations**: Translocations occur when a segment of one chromosome is moved to another chromosome. This can disrupt gene function and regulation, potentially leading to diseases such as cancer. In plants, translocations can be used to introduce new genetic material into a species, which is beneficial for breeding programs.
- » **Inversions**: Inversions involve the flipping of a chromosome segment, reversing the order of genes. This can affect gene expression and recombination rates. Inversions are useful in plant breeding for creating linkage blocks that maintain desirable gene combinations through generations.

Map Genes and Genetic Markers

Identify the Location of Genes and Genetic Markers on Chromosomes: Cytogenomics plays a crucial role in gene mapping by identifying the precise location of genes and genetic markers on chromosomes. This information is fundamental for understanding genetic linkage, trait inheritance, and the genetic basis of complex traits.

- » **Genetic Markers**: Markers such as single nucleotide polymorphisms (SNPs), simple sequence repeats (SSRs), and restriction fragment length polymorphisms (RFLPs) serve as reference points for mapping. They help identify regions of the genome associated with specific traits.
- » **Quantitative Trait Loci (QTL) Mapping**: QTL mapping involves linking genetic markers with phenotypic traits to identify chromosomal regions that control these traits. This approach is essential for understanding the genetic architecture of complex traits, such as yield, disease resistance, and stress tolerance in crops.
- » **Physical Mapping**: Physical mapping techniques, including radiation hybrid mapping and optical mapping, provide high-resolution maps by determining the physical distances between genes and markers. This is crucial for assembling genome sequences and identifying gene locations.

Genome Function and Regulation

Explore the Functional Roles of Genes and Regulatory Elements: Cytogenomics seeks to elucidate the functional roles of genes and regulatory elements, as well as their interactions within the genome. Understanding how genes are regulated and how they interact with each other is key to comprehending the complexities of gene expression and trait development.

- » **Gene Expression**: Techniques such as RNA sequencing (RNA-seq) and microarray analysis are used to study gene expression patterns. These methods help identify which genes are active under specific conditions and how their expression levels change in response to environmental stimuli or developmental stages.
- » **Regulatory Elements**: Regulatory elements, including promoters, enhancers, silencers, and insulators, control the timing, location, and intensity of gene expression. Cytogenomics involves mapping these elements to understand their roles in gene regulation. Chromatin immunoprecipitation sequencing (ChIP-seq) is a powerful technique for identifying DNA-protein interactions and mapping regulatory elements.

- **Epigenetics**: Epigenetic modifications, such as DNA methylation and histone modifications, play a crucial role in gene regulation without altering the DNA sequence. Cytogenomics integrates epigenetic studies to explore how these modifications influence gene expression and contribute to phenotypic diversity.
- **Gene Networks**: Genes do not function in isolation but interact in complex networks. Systems biology approaches, including network analysis and modeling, are used to understand these interactions. Cytogenomics helps identify key regulatory genes and pathways that control important traits, providing targets for genetic manipulation.

Facilitate Crop Improvement

Apply Cytogenomic Knowledge to Enhance Breeding Programs: Cytogenomics is a powerful tool for crop improvement, providing the genetic insights needed to develop disease-resistant, high-yielding, and resilient crop varieties. The integration of cytogenetic and genomic data allows for more precise and efficient breeding strategies.

- **Marker-Assisted Selection (MAS)**: MAS uses genetic markers linked to desirable traits to select individuals for breeding. Cytogenomics provides the detailed genetic maps and markers needed for MAS, enabling breeders to identify and select plants with the best genetic profiles more effectively.
- **Genomic Selection (GS)**: GS involves predicting the breeding value of individuals based on genome-wide markers. Cytogenomics enhances GS by providing high-resolution genetic data and identifying markers associated with complex traits. This approach improves the accuracy of selection and accelerates the breeding process.
- **Introgression of Desirable Traits**: Cytogenomics facilitates the introgression of beneficial traits from wild relatives or exotic germplasm into cultivated varieties. By identifying and mapping genes associated with desirable traits, breeders can incorporate these genes into new crop varieties, enhancing their performance and adaptability.
- **Hybrid Development**: Hybrid varieties often exhibit heterosis or hybrid vigor, resulting in improved yield and stress tolerance. Cytogenomics aids in the development of hybrid varieties by identifying and mapping genes associated with heterosis. This information helps breeders select parent lines that will produce superior hybrids.
- **Functional Genomics and Gene Editing**: Cytogenomics provides the foundation for functional genomics studies, enabling researchers to identify and manipulate genes associated with important traits. Techniques such as CRISPR/Cas9 gene editing allow precise modification of genes, providing powerful tools for developing improved crop varieties.
- **Resistance to Biotic and Abiotic Stresses**: Cytogenomics helps identify genes and regulatory elements involved in resistance to diseases, pests, and environmental stresses. By understanding the genetic basis of stress resistance, breeders can develop crop varieties that are more resilient to changing environmental conditions.

Tools and Techniques in Cytogenomics

Cytogenomics employs a variety of tools and techniques to analyze chromosomes and genomes. These methods range from classical cytogenetic techniques to advanced genomic technologies, providing a robust framework for studying plant genetics.

1. Fluorescence In Situ Hybridization (FISH):

Fluorescence In Situ Hybridization (FISH) is a potent cytogenetic method that enables the visualization of specific DNA sequences within chromosomes. This technique uses fluorescently labeled DNA probes that hybridize with their complementary DNA sequences on chromosomal spreads. The fluorescence emitted by these probes allows researchers to detect and map genes, repetitive sequences, and chromosomal anomalies with high precision. This method plays a crucial role in various genetic studies, including gene mapping, karyotyping, and the identification of chromosomal rearrangements. In crop improvement, FISH provides invaluable insights for developing targeted breeding strategies by identifying and locating genes associated with desirable traits such as disease resistance and stress tolerance.

Methodology of FISH

Probe Preparation: The first step in FISH involves creating DNA probes that are specific to the target sequences. These probes are labeled with fluorescent dyes, which emit light when exposed to certain wavelengths. The probes can be designed to target a wide range of DNA sequences, including unique genes, repetitive elements, and entire chromosomal regions. The preparation of DNA probes is a meticulous process that involves several steps to ensure specificity and efficiency. Researchers select DNA sequences of interest, which may involve cloning specific gene segments or synthesizing oligonucleotides. The selected sequences are then labeled with fluorescent dyes, which are chosen based on their spectral properties to avoid overlap when multiple probes are used. The labeled probes must be purified to remove any unincorporated dyes or contaminants that could affect hybridization efficiency. This step is critical for achieving clear and specific fluorescent signals during the hybridization process.

Chromosome Preparation: Chromosomes are typically isolated from actively dividing cells, such as those in metaphase. The cells are treated to arrest them in metaphase, where chromosomes are most condensed and visible. These cells are then lysed, and the chromosomes are spread on a microscope slide. Preparing chromosomes for FISH involves several steps to ensure that the chromosomes are well-preserved and suitable for hybridization. Cells are often cultured and synchronized to maximize the number of cells in metaphase. Chemical treatments, such as colchicine, are used to arrest cells in metaphase by inhibiting spindle fiber formation. The arrested cells are then subjected to a hypotonic solution to swell them, making the chromosomes more accessible. Following this, cells are fixed using a mixture of methanol and acetic acid, which preserves the chromosome structure and spreads them onto slides. The slides are then dried and treated to make the DNA accessible for hybridization with the probes.

Hybridization: The prepared fluorescently labeled probes are applied to the chromosomal spreads on the slides. The slide is then incubated under conditions that allow the DNA probes to hybridize with their complementary sequences on the chromosomes. This process involves denaturing the DNA (heating to separate the strands) and then allowing the probes to anneal to their targets as the temperature is gradually reduced. Hybridization is a critical step in FISH that requires precise control of environmental conditions to ensure specific and efficient binding of the probes to the target sequences. The slide containing the chromosomal spreads is first denatured, usually by heating, to separate the DNA strands. The denatured DNA probes are then applied to the slide and allowed to hybridize by cooling slowly, which encourages the formation of stable DNA-DNA duplexes between the probes and their complementary sequences on the chromosomes. This step is typically carried out in a humid chamber to prevent evaporation and maintain optimal hybridization conditions. Hybridization buffers containing formamide, dextran sulfate, and salts are used to enhance the specificity and stability of the probe-target interactions.

Detection: After hybridization, the slides are washed to remove any unbound probes. The hybridized probes, which are now bound to their target sequences on the chromosomes, are then visualized using a fluorescence microscope. The specific fluorescent signals indicate the location of the target sequences on the chromosomes, which can be captured and analyzed using imaging software. The detection step in FISH involves the visualization and analysis of the fluorescent signals emitted by the bound probes. After hybridization, the slides undergo a series of washes to remove non-specifically bound probes, reducing background fluorescence and enhancing signal clarity. The washed slides are then counterstained, often with a DNA-specific stain such as DAPI, to provide a contrast background that highlights the chromosomes. Using a fluorescence microscope equipped with appropriate filter sets, researchers can visualize the specific fluorescent signals corresponding to the bound probes. These signals are captured using high-resolution cameras, and specialized imaging software is used to analyze and interpret the data. The software allows for precise measurement of signal intensity, localization, and colocalization of multiple probes, providing detailed insights into the chromosomal arrangement and gene locations.

Applications of FISH in Crop Improvement

Gene Mapping: FISH is extensively used in gene mapping to pinpoint the exact locations of genes on chromosomes. By using probes specific to genes of interest, researchers can determine their precise chromosomal positions. This information is crucial for understanding the genetic architecture of crops and identifying genes associated with important traits. Gene mapping using FISH involves designing probes that specifically bind to target genes. Once hybridized, the fluorescent signals from these probes reveal the precise chromosomal locations of the genes. This high-resolution mapping allows researchers to construct detailed genetic maps, which are essential for identifying gene-trait associations. For instance, in crops like rice and maize, FISH can be used to map genes related to yield, drought tolerance, and disease resistance. These maps provide a framework for further genetic analysis and breeding efforts, enabling the development of improved crop varieties with enhanced traits.

Karyotyping: Karyotyping involves the visualization of the complete set of chromosomes in a cell to identify structural features and any anomalies. FISH enhances karyotyping by allowing the detection of specific chromosomal regions and identifying abnormalities such as duplications, deletions, and translocations. This application is vital for diagnosing genetic disorders and breeding programs. In crop improvement, FISH-based karyotyping helps in identifying chromosomal rearrangements that may impact plant development and performance. By using probes that bind to specific chromosomal regions, researchers can visualize and compare the karyotypes of different plant varieties. This comparative analysis reveals structural variations, such as translocations or inversions, that may contribute to phenotypic differences. For example, in wheat, FISH can identify chromosomal translocations that confer resistance to certain pathogens. By understanding these karyotypic changes, breeders can select parent plants with favorable chromosomal arrangements, improving the genetic stability and performance of new crop varieties.

Detection of Chromosomal Rearrangements: FISH is an effective tool for identifying chromosomal rearrangements, such as translocations, inversions, and other structural variations. By using probes that span the breakpoints of rearrangements, researchers can detect these changes and understand their impacts on gene function and expression. Chromosomal rearrangements can significantly affect gene expression and trait inheritance. For example, a translocation that disrupts a gene involved in disease resistance can render a plant susceptible to infections. FISH allows researchers to identify these rearrangements by using probes that bind to regions flanking the breakpoints. By analyzing the patterns of fluorescent signals, researchers can determine the presence and extent of rearrangements. This information is crucial for breeding programs, as it helps identify and eliminate deleterious rearrangements that may negatively impact crop performance. Conversely, beneficial rearrangements that enhance traits such as yield or stress tolerance can be identified and incorporated into breeding strategies.

Identification of Repetitive Elements: Repetitive DNA sequences, such as satellite DNA, telomeres, and centromeres, play significant roles in chromosomal structure and function. FISH can be used to map these repetitive elements, providing insights into their organization and contribution to genome stability and evolution. Repetitive elements are often involved in maintaining chromosome integrity and regulating gene expression. FISH probes designed to target these sequences allow researchers to visualize their distribution across the genome. For instance, telomeric probes can highlight the ends of chromosomes, while centromeric probes can identify the primary constriction points. Mapping these elements is crucial for understanding their roles in chromosomal behavior during cell division and their impact on genome stability. In crop improvement, knowledge of repetitive element organization can aid in identifying genetic markers linked to important traits, facilitating marker-assisted selection and improving breeding efficiency.

Locating Genes Associated with Desirable Traits: In crop improvement, FISH is instrumental in locating genes linked to beneficial traits such as disease resistance, stress tolerance, and yield improvement. By identifying the chromosomal locations of these genes, breeders can develop targeted breeding strategies. For example, if a gene associated with drought resistance is mapped to a specific chromosomal region, FISH can help confirm its location, allowing breeders to incorporate this trait into new varieties through marker-assisted selection. FISH provides a high-resolution method for mapping genes associated with desirable traits. By using probes that specifically target these genes, researchers can determine their precise locations on the chromosomes. This information is crucial for breeding programs, as it enables breeders to select parent plants with the desired genetic traits. For instance, if a gene conferring resistance to a particular pest is mapped using FISH, breeders can use this information to develop crop varieties that are less susceptible to pest damage. Similarly, mapping genes associated with abiotic stress tolerance, such as drought or salinity, allows breeders to create crops that can thrive in challenging environmental conditions.

Studying Polyploidy and Chromosomal Aberrations: Polyploidy, where organisms have more than two sets of chromosomes, is common in many crops. FISH is used to study polyploidy by identifying and differentiating between multiple sets of chromosomes. This is particularly useful in understanding the genetic composition and evolution of polyploid crops like wheat and sugarcane. Polyploidy plays a significant role in the evolution and diversification of plant species. FISH allows researchers to visualize and distinguish between different sets of chromosomes in polyploid organisms. By using probes that bind to specific chromosomal regions, researchers can identify the origin and arrangement of multiple chromosome sets. This information is essential for understanding the genetic complexity of polyploid crops and for developing strategies to harness their genetic potential. For example, in wheat, FISH can help identify homoeologous chromosomes (chromosomes derived from different ancestral species) and study their interactions, providing insights into the genetic basis of important traits and aiding in the development of improved wheat varieties.

Comparative Genomics: FISH enables comparative genomics studies by allowing the comparison of chromosomal structures and gene locations across different species or varieties. This can reveal evolutionary relationships and the conservation of important genes. For instance, comparing the chromosomal locations of disease-resistance genes in wild and cultivated varieties can provide insights into the domestication process and help in the transfer of beneficial traits from wild relatives to cultivated crops. Comparative genomics involves analyzing the similarities and differences in the genomic organization of different species or varieties. FISH provides a powerful tool for visualizing these differences at the chromosomal level. By using probes that target conserved gene sequences or chromosomal regions, researchers can compare the genomic landscapes of different organisms. This approach reveals evolutionary relationships and helps identify conserved genetic elements that play crucial roles in trait development. For example, in crop improvement, FISH can be used to identify disease-resistance genes in wild relatives of cultivated crops. By understanding the chromosomal organization and conservation of these genes, breeders can develop strategies to introgress beneficial traits from wild species into cultivated varieties, enhancing their resilience and performance.

Functional Genomics: FISH can be used to investigate the spatial organization of the genome and the relationship between gene location and gene expression. By mapping regulatory elements and genes involved in specific pathways, researchers can better understand how genes are regulated and how they interact within the genome. This knowledge is crucial for manipulating gene expression and developing crops with enhanced traits.

Functional genomics aims to understand the roles and interactions of genes within the genome. FISH contributes to this field by providing a visual representation of gene locations and their spatial relationships. By mapping regulatory elements, such as enhancers and silencers, researchers can study how these elements influence gene expression. FISH also allows for the visualization of gene clusters and their interactions, providing insights into the regulation of complex traits. For example, in crop improvement, understanding the spatial organization of genes involved in photosynthesis or stress response can help develop strategies to enhance these pathways, leading to crops with improved productivity and resilience.

Advantages of FISH

» **High Resolution:** FISH provides high-resolution mapping of specific DNA sequences, enabling precise localization of genes and other genomic elements. This precision is crucial for accurate genetic analysis and for understanding the genetic basis of complex traits.

» **Versatility:** The technique can be applied to a wide range of organisms, from plants to animals, making it a versatile tool in genetic research. This versatility allows researchers to study diverse biological systems and apply findings across different species.

» **Real-Time Analysis:** FISH allows for the direct observation of chromosomes and genetic material in their native state, providing real-time insights into chromosomal organization and anomalies. This capability is essential for studying dynamic processes such as cell division and gene expression.

» **Combining with Other Techniques:** FISH can be combined with other cytogenetic and molecular techniques, such as comparative genomic hybridization (CGH) and karyotyping, to provide comprehensive genomic analysis. By integrating multiple methods, researchers can obtain a more complete understanding of the genome and its functions.

Challenges and Considerations

Complexity and Technical Expertise: FISH is a complex technique that requires significant technical expertise and precision in probe design, hybridization conditions, and fluorescence microscopy. Proper training and experience are essential to achieve accurate and reliable results.

The technical complexity of FISH involves multiple steps, each requiring careful optimization and control. Designing specific probes, preparing high-quality chromosomal spreads, and optimizing hybridization conditions are critical for successful outcomes. Researchers must be proficient in using fluorescence microscopes and imaging software to accurately detect and analyze fluorescent signals. Additionally, interpreting FISH results requires a deep understanding of cytogenetics and genomic organization, highlighting the need for specialized training and expertise.

Fluorescence Overlap: When multiple fluorescent dyes are used, their spectra can overlap, complicating the interpretation of results. This challenge requires careful selection of dyes and the use of advanced imaging techniques to distinguish between different signals.

Fluorescence overlap, or spectral overlap, occurs when the emission spectra of different fluorescent dyes used in FISH overlap, making it difficult to distinguish between signals. This issue can lead to inaccurate interpretation of results, especially when multiple probes are used simultaneously. To address this challenge, researchers select fluorescent dyes with minimal spectral overlap and use advanced imaging techniques, such as spectral imaging and deconvolution, to separate overlapping signals. Proper calibration of the fluorescence microscope and careful selection of filter sets are also essential to reduce fluorescence overlap and enhance signal clarity.

Resolution Limits: While FISH offers high resolution compared to many other techniques, its resolution is still limited by the size of the probes and the chromosomal spreads. For very fine-scale mapping, additional techniques such as high-throughput sequencing may be needed.

The resolution of FISH is influenced by factors such as probe size, hybridization efficiency, and the quality of chromosomal spreads. Although FISH provides high-resolution mapping, it may not achieve the fine-scale resolution required for detailed analysis of closely spaced genes or small chromosomal regions. To overcome these limitations, researchers can complement FISH with high-throughput sequencing techniques, such as whole-genome sequencing or targeted sequencing, to achieve finer resolution and obtain more comprehensive genetic information.

Sample Preparation: The quality of chromosomal spreads is crucial for successful FISH analysis. Poor preparation can lead to suboptimal hybridization and signal detection, requiring meticulous sample preparation and handling.

Preparing high-quality chromosomal spreads is a critical step in FISH that affects the accuracy and reliability of the results. Cells must be carefully synchronized to maximize the number of cells in metaphase, and chromosomal spreads must be prepared to ensure optimal preservation and accessibility of the chromosomes. Proper fixation and drying of the slides are essential to maintain chromosomal integrity and prevent artifacts. Additionally, handling and storage conditions must be carefully controlled to prevent degradation and contamination, ensuring that the chromosomal spreads are suitable for hybridization and signal detection.

Fluorescence In Situ Hybridization (FISH) is a powerful cytogenetic technique that enables the precise localization of specific DNA sequences on chromosomes. By using fluorescently labeled probes, FISH allows researchers to visualize and map genes, repetitive elements, and chromosomal rearrangements with high resolution. This technique is widely used in gene mapping, karyotyping, and the identification of chromosomal anomalies. In crop improvement, FISH provides invaluable insights for developing targeted breeding strategies by identifying and locating genes associated with desirable traits such as disease resistance and stress tolerance. The versatility, precision, and real-time analysis capabilities of FISH make it a vital tool in genetic research and crop improvement. Despite its technical complexity and challenges, FISH remains a fundamental technique for advancing our understanding of genomic organization and enhancing crop performance.

2. Comparative Genomic Hybridization (CGH):

Comparative Genomic Hybridization (CGH) is a robust genomic technique utilized to detect and map chromosomal imbalances, including duplications and deletions. This method is pivotal for identifying structural changes within the genome that may impact gene function and trait expression. CGH involves the hybridization of differentially labeled DNA from test and reference samples to a microarray or chromosomal spread. By comparing the hybridization signals between these samples, researchers can pinpoint the presence and precise location of chromosomal aberrations. This technique provides a comprehensive overview of genomic variations and is particularly useful in crop improvement for identifying chromosomal regions linked to essential agronomic traits such as yield, quality, and stress tolerance.

Methodology of CGH

Sample Preparation: The first step in CGH involves preparing DNA samples from both the test and reference specimens. The test sample typically comes from a plant showing a particular trait or condition of interest, while the reference sample is taken from a standard or wild-type plant. The DNA from these samples is extracted using standard molecular biology techniques to ensure high purity and integrity. The quality of the DNA samples is crucial for successful CGH analysis. The DNA must be free of contaminants and degradation to ensure accurate hybridization and signal detection.

Researchers often use spectrophotometric and electrophoretic methods to assess DNA quality and quantity, ensuring that the samples meet the required standards for CGH.

Labeling: The extracted DNA from both the test and reference samples is differentially labeled with fluorescent dyes. Typically, the test sample is labeled with one fluorescent dye (e.g., green), and the reference sample is labeled with another (e.g., red). This differential labeling allows for the comparison of hybridization signals and the identification of chromosomal imbalances. The labeling process involves incorporating fluorescent nucleotides into the DNA during a labeling reaction. Enzymes such as DNA polymerase or terminal deoxynucleotidyl transferase (TdT) are used to facilitate the incorporation of the fluorescent nucleotides into the DNA strands. The choice of fluorescent dyes is critical, as they must have distinct emission spectra to ensure clear differentiation of signals during analysis.

Hybridization: The labeled DNA samples are mixed and co-hybridized to a microarray or a metaphase chromosomal spread. In a microarray-based CGH, the DNA samples are hybridized into an array containing thousands of immobilized DNA probes representing the entire genome or specific regions of interest. In a chromosomal spread-based CGH, the DNA samples are hybridized directly to chromosomes spread on a slide. The hybridization step requires precise control of environmental conditions, including temperature, humidity, and hybridization buffer composition. These conditions ensure specific and efficient binding of the labeled DNA to the complementary sequences on the microarray or chromosomes. The hybridization process typically takes several hours to overnight, allowing sufficient time for the DNA samples to anneal to their targets.

Detection: After hybridization, the microarray or chromosomal spread is washed to remove unbound and nonspecifically bound DNA. The hybridization signals are then detected using a fluorescence scanner or microscope. The intensity of the fluorescent signals reflects the relative copy number of the DNA sequences in the test and reference samples. In a microarray-based CGH, the fluorescence scanner captures the hybridization signals from each probe on the array. The resulting images are analyzed using specialized software to quantify the signal intensities and identify regions of chromosomal imbalances. In a chromosomal spread-based CGH, the fluorescent signals are visualized using a fluorescence microscope, and the images are captured for analysis.

Data Analysis: The fluorescent signals from the test and reference samples are compared to identify regions of chromosomal imbalance. Areas with increased test sample signals indicate duplications, while regions with decreased signals suggest deletions. The data analysis involves normalizing the signal intensities, calculating the log ratios of test to reference signals, and identifying significant deviations from the expected values. Advanced bioinformatics tools and statistical algorithms are used to analyze the CGH data and identify chromosomal regions with significant imbalances. The results are often visualized as log ratio plots or ideograms, showing the relative copy number variations along the chromosomes. These visualizations help researchers pinpoint the exact locations of duplications and deletions, providing insights into the genetic basis of the observed traits.

Applications of CGH in Crop Improvement

Identifying Chromosomal Regions Associated with Yield: CGH is instrumental in identifying chromosomal regions that influence yield traits in crops. By comparing the genomic DNA of high-yielding and low-yielding plants, researchers can identify duplications and deletions associated with yield differences. These regions often contain genes or regulatory elements that contribute to increased productivity, making them valuable targets for breeding programs. For example, in rice, CGH can be used to compare the genomes of high-yielding and low-yielding varieties. Regions with increased copy numbers in high-yielding varieties may contain genes associated with enhanced photosynthesis, nutrient uptake, or stress resistance. By mapping these regions, breeders can develop markers for selecting high-yielding plants and incorporate these beneficial traits into new rice varieties.

Quality Traits: CGH helps identify genomic regions linked to quality traits such as fruit size, color, flavor, and nutritional content. By detecting chromosomal imbalances in plants with superior quality traits, researchers can pinpoint the genetic variations responsible for these characteristics. This information is crucial for developing crop varieties with improved quality attributes. In tomato, for instance, CGH can be used to identify chromosomal regions associated with fruit quality traits. By comparing the genomes of tomato varieties with different fruit sizes, colors, and flavors, researchers can identify duplications and deletions associated with these traits. These findings can guide breeding programs aimed at developing tomatoes with enhanced quality, improving consumer satisfaction and market value.

Stress Tolerance: CGH is valuable for identifying chromosomal regions that confer tolerance to biotic and abiotic stresses, such as diseases, pests, drought, and salinity. By comparing the genomes of stress-tolerant and susceptible plants, researchers can identify genetic variations that contribute to stress resistance. This information aids in the development of resilient crop varieties capable of thriving under challenging environmental conditions. For example, in wheat, CGH can be used to identify genomic regions associated with drought tolerance. By comparing the genomes of drought-tolerant and susceptible wheat varieties, researchers can pinpoint duplications and deletions linked to drought resistance. These regions may contain genes involved in water uptake, root development, or stress response pathways. By incorporating these genetic variations into breeding programs, breeders can develop wheat varieties with enhanced drought tolerance, ensuring stable yields under water-limited conditions.

Genome Integrity and Stability: CGH can assess the genomic integrity and stability of crop varieties, particularly those developed through genetic engineering or hybridization. By detecting and mapping chromosomal imbalances, researchers can ensure that new varieties do not carry unintended genetic alterations that could impact their performance or safety. In genetically modified crops, CGH can be used to verify the integrity of the inserted transgenes and the surrounding genomic regions. By comparing the genomes of genetically modified and non-modified plants, researchers can detect any unintended duplications or deletions resulting from the genetic engineering process. This ensures the safety and stability of genetically modified crops, facilitating their regulatory approval and commercialization.

Introgression Breeding: CGH aids in introgression breeding by identifying and tracking the transfer of desirable traits from wild relatives or exotic germplasm into cultivated varieties. By comparing the genomes of introgressed lines and their parents, researchers can monitor the incorporation of beneficial traits and ensure that the desired genetic variations are retained in the new varieties. In maize, for example, CGH can be used to monitor the introgression of disease-resistance genes from wild relatives into cultivated lines. By comparing the genomes of introgressed lines and their parents, researchers can track the transfer of resistance genes and ensure that the desired traits are successfully incorporated. This approach enhances the efficiency of introgression breeding and accelerates the development of disease-resistant maize varieties.

Advantages of CGH

High-Resolution Detection: CGH provides high-resolution detection of chromosomal imbalances, allowing for precise mapping of duplications and deletions. This high resolution is essential for identifying small genetic variations that may have significant impacts on trait expression and crop performance. The high-resolution detection capability of CGH enables researchers to pinpoint the exact locations of chromosomal imbalances, even those involving small genomic regions. This precision is crucial for identifying genes and regulatory elements associated with important traits. By detecting small duplications and deletions, CGH provides detailed insights into the genetic architecture of crops, facilitating the development of targeted breeding strategies.

Comprehensive Genome Analysis: CGH offers a comprehensive overview of the entire genome, enabling the detection of multiple chromosomal imbalances simultaneously. This comprehensive analysis is particularly valuable for studying complex traits that involve interactions between multiple genomic regions. The ability of CGH to provide a genome-wide overview of chromosomal imbalances allows researchers to identify multiple regions associated with a single trait. This comprehensive analysis is essential for understanding the genetic basis of complex traits, such as yield and stress tolerance, which may involve the combined effects of several genetic variations. By detecting multiple imbalances simultaneously, CGH enhances the efficiency of genetic analysis and breeding programs.

Applicability to Diverse Species: CGH applies to a wide range of plant species, making it a versatile tool in crop improvement. This versatility allows researchers to study genomic variations across different crops and apply the findings to diverse breeding programs. The applicability of CGH to diverse plant species makes it a valuable tool for comparative genomics and cross-species genetic studies. Researchers can use CGH to analyze chromosomal imbalances in various crops, providing insights into the genetic basis of important traits across different species. This versatility facilitates the transfer of knowledge and genetic resources between crops, enhancing the overall efficiency of breeding programs.

Non-Requirement for Prior Sequence Information: Unlike some genomic techniques, CGH does not require prior sequence information for the test and reference samples. This makes it an ideal tool for studying non-model organisms or newly sequenced genomes where detailed sequence data may not be available. The ability of CGH to detect chromosomal imbalances without requiring prior sequence information is a significant advantage for studying non-model organisms and newly sequenced genomes. This feature allows researchers to analyze genetic variations in crops with limited genomic resources, expanding the scope of genetic studies and breeding programs. By providing a versatile and accessible tool for genome analysis, CGH enhances the ability to improve a wide range of crops.

Challenges and Considerations

Resolution and Sensitivity: While CGH provides high-resolution detection, its sensitivity may vary depending on the size of the chromosomal imbalances and the quality of the DNA samples. Small duplications and deletions may be challenging to detect, requiring optimized protocols and high-quality samples. The sensitivity of CGH to detect small chromosomal imbalances is influenced by factors such as probe design, hybridization efficiency, and signal-to-noise ratio. Researchers must carefully optimize these factors to ensure the accurate detection of small duplications and deletions. Additionally, the quality of the DNA samples is critical for successful CGH analysis. High-quality, intact DNA is essential for achieving reliable and reproducible results.

Technical Expertise: CGH is a technically complex technique that requires specialized expertise and equipment. Proper training and experience are essential for conducting CGH experiments and interpreting the results accurately.

The technical complexity of CGH involves multiple steps, each requiring specialized knowledge and skills. Researchers must be proficient in DNA extraction, labeling, hybridization, and fluorescence detection techniques. Additionally, interpreting CGH results requires expertise in bioinformatics and statistical analysis to identify and quantify chromosomal imbalances accurately. Proper training and access to specialized equipment, such as fluorescence scanners and imaging software, are essential for successful CGH experiments.

Fluorescence Overlap: Similar to FISH, fluorescence overlap can occur when multiple fluorescent dyes are used in CGH. This overlap can complicate the interpretation of hybridization signals, necessitating careful dye selection and advanced imaging techniques. Fluorescence overlap, or spectral overlap, occurs when the emission spectra of different fluorescent dyes used in CGH overlap, making it difficult to distinguish between signals. This issue can lead to inaccurate interpretation of results, especially when multiple probes are used simultaneously. To address this challenge, researchers select fluorescent dyes with minimal spectral overlap and use advanced imaging techniques, such as spectral imaging and deconvolution, to separate overlapping signals. Proper calibration of the fluorescence scanner and careful selection of filter sets are also essential to reduce fluorescence overlap and enhance signal clarity.

Sample Preparation: The quality of the DNA samples is critical for successful CGH analysis. Poorly prepared samples can lead to suboptimal hybridization and inaccurate detection of chromosomal imbalances. Meticulous sample preparation and handling are essential to ensure reliable results.

Preparing high-quality DNA samples is a critical step in CGH that affects the accuracy and reliability of the results. The DNA must be free of contaminants and degradation to ensure accurate hybridization and signal detection. Researchers often use spectrophotometric and electrophoretic methods to assess DNA quality and quantity, ensuring that the samples meet the required standards for CGH. Proper handling and storage of the DNA samples are also essential to prevent degradation and contamination, ensuring that the samples are suitable for hybridization and signal detection.

Comparative Genomic Hybridization (CGH) is a powerful genomic technique that enables the detection and mapping of chromosomal imbalances, including duplications and deletions. By comparing the hybridization signals of differentially labeled DNA from test and reference samples, CGH provides a comprehensive overview of genomic variations. This technique is particularly useful in crop improvement for identifying chromosomal regions associated with yield, quality, and stress tolerance. By detecting and mapping genomic imbalances, researchers can pinpoint genetic variations that contribute to important agronomic traits, facilitating the development of superior crop varieties. Despite its technical complexity and challenges, CGH remains a fundamental tool for advancing our understanding of genomic organization and enhancing crop performance.

3. Genome Sequencing:

Genome sequencing is a transformative genomic technique that involves determining the complete DNA sequence of an organism's genome. This comprehensive approach allows researchers to obtain a detailed view of the genetic blueprint of an organism, providing invaluable insights into its genetic composition, structural variations, and functional elements. High-throughput sequencing technologies, such as next-generation sequencing (NGS), have revolutionized genome sequencing by enabling rapid and cost-effective analysis of entire genomes.

Methodology of Genome Sequencing

High-Throughput Sequencing Technologies: High-throughput sequencing technologies, also known as next-generation sequencing (NGS), have significantly advanced the field of genome sequencing. NGS platforms, such as Illumina, PacBio, and Oxford Nanopore, enable the parallel sequencing of millions of DNA fragments, generating vast amounts of genetic data quickly and cost-effectively.

Illumina Sequencing: Illumina sequencing is one of the most widely used NGS platforms. It involves fragmenting the DNA, attaching adapters, and amplifying the fragments through bridge PCR. The amplified fragments are then sequenced using a process called sequencing-by-synthesis, where fluorescently labeled nucleotides are incorporated into the growing DNA strand. Each nucleotide incorporation is detected by a camera, producing a high-resolution sequence read.

PacBio Sequencing: PacBio sequencing, also known as single-molecule real-time (SMRT) sequencing, provides long-read sequencing capabilities. It involves sequencing single DNA molecules in real-time, allowing for the detection of long contiguous sequences and structural variations. This technology is particularly useful for resolving complex genomic regions and repetitive sequences.

Oxford Nanopore Sequencing: Oxford Nanopore sequencing uses nanopores embedded in a membrane to sequence DNA molecules as they pass through the pores. The changes in electrical current as the DNA translocates through the nanopores are used to determine the sequence. This technology provides ultra-long reads and is highly portable, making it suitable for field-based genomic studies.

Whole-Genome Sequencing (WGS): Whole-genome sequencing (WGS) involves sequencing the entire genome of an organism. This approach provides a comprehensive view of the genetic makeup, including coding and non-coding regions, regulatory elements, and structural variations. WGS is essential for de novo genome assembly, comparative genomics, and evolutionary studies.

RNA Sequencing (RNA-seq): RNA sequencing (RNA-seq) is a powerful technique used to study gene expression and transcriptomic profiles. It involves sequencing the RNA molecules in a sample to determine the abundance and diversity of transcripts. RNA-seq provides insights into gene expression patterns, alternative splicing events, and regulatory networks.

Applications of Genome Sequencing in Crop Improvement

Identification of Candidate Genes: Genome sequencing enables the identification of candidate genes associated with important traits such as yield, disease resistance, and stress tolerance. By comparing the genomes of different plant varieties, researchers can pinpoint genetic variations that contribute to these traits. This information is crucial for developing targeted breeding strategies. For example, in rice, genome sequencing can be used to identify genes associated with drought tolerance. By comparing the genomes of drought-tolerant and susceptible varieties, researchers can identify single nucleotide polymorphisms (SNPs), insertions, deletions, and other genetic variations linked to drought resistance. These candidate genes can then be validated through functional studies and used in breeding programs to develop drought-tolerant rice varieties.

Genetic Marker Development: Genome sequencing provides a wealth of genetic information that can be used to develop molecular markers for marker-assisted selection (MAS) and genomic selection (GS). These markers, such as SNPs and simple sequence repeats (SSRs), are essential for selecting individuals with desirable traits in breeding programs. Marker-assisted selection involves using genetic markers linked to specific traits to select parent plants with the desired genetic profile. For instance, in maize, genome sequencing can identify SNPs associated with resistance to a particular pest. These SNPs can be used as markers to select maize plants with the resistance trait, accelerating the breeding process and improving crop performance. Genomic selection involves predicting the breeding value of individuals based on genome-wide markers. Genome sequencing provides the necessary genetic data to build predictive models that accurately estimate the genetic potential of plants. This approach enhances the efficiency and precision of breeding programs, leading to the rapid development of improved crop varieties.

Structural Variation Analysis: Genome sequencing allows for the detection and analysis of structural variations such as duplications, deletions, inversions, and translocations. These variations can significantly impact gene function and trait expression. The role of structural variations in crop genomes is crucial for improving crop performance. For example, in wheat, genome sequencing can identify structural variations associated with disease resistance. Researchers can compare the genomes of resistant and susceptible wheat varieties to detect duplications or deletions in regions containing resistance genes. By understanding these structural variations, breeders can develop strategies to incorporate resistance traits into new wheat varieties, enhancing their ability to withstand diseases.

Functional Genomics: Genome sequencing provides a foundation for functional genomics studies, enabling researchers to explore the roles and interactions of genes within the genome. By integrating genomic data with transcriptomic, proteomic, and metabolomic data, researchers can gain insights into gene function and regulation. RNA sequencing (RNA-seq) is a key tool in functional genomics, allowing researchers to study gene expression patterns under different conditions. For example, in soybean, RNA-seq can be used to compare gene expression profiles between drought-stressed and well-watered plants. This analysis can identify genes that are upregulated or downregulated in response to drought, providing targets for genetic manipulation to improve drought tolerance.

Genome-Wide Association Studies (GWAS): Genome-wide association studies (GWAS) leverage genome sequencing data to identify genetic loci associated with specific traits. By analyzing the genetic variation across a population, GWAS can identify regions of the genome that contribute to trait variation. In tomatoes, for example, GWAS can be used to identify genetic loci associated with fruit size and shape. By sequencing the genomes of a diverse population of tomato plants and correlating genetic variations with fruit traits, researchers can pinpoint specific regions of the genome that influence fruit morphology. These findings can inform breeding programs aimed at developing tomatoes with desirable fruit characteristics.

Introgression Breeding: Genome sequencing facilitates introgression breeding by enabling the precise tracking of genetic material from wild relatives or exotic germplasm into cultivated varieties. This approach enhances the genetic diversity of crops and introduces beneficial traits from wild species. In barley, for example, genome sequencing can be used to track the introgression of disease-resistance genes from wild barley into cultivated varieties. By sequencing the genomes of introgressed lines and their parents, researchers can monitor the transfer of resistance genes and ensure that the desired traits are successfully incorporated. This approach accelerates the development of disease-resistant barley varieties, improving crop resilience and performance.

Comparative Genomics: Comparative genomics involves comparing the genomes of different species or varieties to understand evolutionary relationships, gene conservation, and genetic diversity. Genome sequencing provides the data necessary for comparative genomics studies, revealing insights into the genetic basis of important traits. For instance, in Brassica crops, comparative genomics can be used to study the genetic diversity and evolution of disease-resistance genes. By sequencing the genomes of different Brassica species and varieties, researchers can identify conserved and divergent regions of the genome. This information can guide breeding programs aimed at transferring resistance genes between species, enhancing the disease resistance of Brassica crops.

Advantages of Genome Sequencing

Comprehensive Genomic Analysis: Genome sequencing provides a complete view of an organism's genetic makeup, including coding and non-coding regions, regulatory elements, and structural variations. This comprehensive analysis is essential for understanding the genetic basis of complex traits and for developing targeted breeding strategies. The comprehensive nature of genome sequencing allows researchers to explore the entire genome, uncovering genetic variations that may not be detected using other techniques. This holistic approach provides a deeper understanding of the genetic architecture of crops, facilitating the identification of candidate genes and regulatory elements that contribute to important traits.

High Resolution and Accuracy: High-throughput sequencing technologies offer high resolution and accuracy, enabling the detection of even small genetic variations. This precision is crucial for identifying SNPs, indels, and other variations that may have significant impacts on trait expression and crop performance. The high resolution and accuracy of NGS platforms allow researchers to generate detailed genetic maps and identify fine-scale genetic variations. This level of precision is essential for studying complex traits, where small genetic changes can have significant phenotypic effects. By detecting subtle variations, genome sequencing provides the necessary data for accurate genetic analysis and breeding programs.

Rapid and Cost-Effective: Advancements in NGS technologies have significantly reduced the time and cost of genome sequencing. This accessibility allows for large-scale genomic studies and the sequencing of multiple individuals or populations, enhancing the power and scope of genetic research. The rapid and cost-effective nature of NGS technologies enables researchers to sequence the genomes of numerous plant varieties and populations, generating large datasets for genetic analysis. This scalability is essential for conducting genome-wide association studies, population genetics, and comparative genomics. By reducing the barriers to genome sequencing, NGS technologies facilitate the advancement of crop improvement efforts.

Integration with Other Omics Technologies: Genome sequencing can be integrated with other omics technologies, such as transcriptomics, proteomics, and metabolomics, to provide a comprehensive understanding of gene function and regulation. This integrative approach enhances our knowledge of the molecular mechanisms underlying important traits. The integration of genome sequencing with other omics technologies allows researchers to study the interactions between genes, proteins, and metabolites, providing a holistic view of biological processes. This systems biology approach enhances our understanding of the molecular networks that control trait expression, facilitating the development of targeted breeding strategies and genetic interventions.

Challenges and Considerations

Data Management and Analysis: Genome sequencing generates vast amounts of data that require advanced computational tools and bioinformatics expertise for analysis. Managing, storing, and analyzing these large datasets can be challenging, necessitating robust data infrastructure and skilled personnel. The sheer volume of data generated by genome sequencing necessitates the use of high-performance computing resources and sophisticated bioinformatics tools for data processing and analysis. Researchers must be proficient in data management, storage, and analysis techniques to handle the complexities of genome sequencing data. Additionally, the interpretation of sequencing data requires expertise in genomics, genetics, and molecular biology, highlighting the need for interdisciplinary collaboration.

Sequencing Errors and Biases: Despite advancements in NGS technologies, sequencing errors and biases can occur, affecting the accuracy of the data. These errors can arise from various sources, including library preparation, sequencing chemistry, and data processing. Sequencing errors and biases can impact the accuracy of variant calling and genome assembly, leading to incorrect conclusions. To mitigate these issues, researchers use quality control measures, such as duplicate sequencing, error correction algorithms, and validation with independent methods. Additionally, optimizing library preparation protocols and sequencing conditions can reduce the occurrence of errors and biases.

Interpretation of Variants: Interpreting the functional significance of genetic variants can be challenging, particularly for non-coding regions and structural variations. Functional validation through experimental studies is often required to determine the impact of these variants on gene function and trait expression. The interpretation of genetic variants involves linking sequence changes to phenotypic effects, which can be complex and context-dependent. Functional validation using techniques such as gene editing, gene expression analysis, and phenotypic assays is essential for confirming the roles of variants in trait development. Additionally, integrating genomic data with transcriptomic, proteomic, and metabolomic data can provide insights into the functional impact of variants, enhancing the accuracy of variant interpretation.

Genome sequencing is a transformative genomic technique that provides comprehensive insights into the genetic composition, structural variations, and functional elements of an organism's genome. High-throughput sequencing technologies, such as next-generation sequencing (NGS), have revolutionized genome sequencing by enabling rapid and cost-effective analysis of entire genomes. In crop improvement, genome sequencing enables the identification of candidate genes and genetic markers associated with important traits, facilitating the development of molecular markers for marker-assisted selection (MAS) and genomic selection (GS). By analyzing the genetic information, breeders can accelerate the breeding process, improve crop performance, and develop superior crop varieties.

Despite challenges such as data management, sequencing errors, and variant interpretation, genome sequencing remains a fundamental tool for advancing our understanding of genomic organization and enhancing crop improvement efforts.

4. Chromosome Painting:

Chromosome painting is a sophisticated cytogenetic technique that employs fluorescently labeled DNA probes to selectively stain specific chromosomes or chromosomal regions. This method facilitates the visualization and differentiation of individual chromosomes, enabling researchers to study chromosomal behavior and identify structural variations with high precision. Chromosome painting is especially valuable for karyotyping and detecting chromosomal rearrangements, providing critical insights into genome organization and evolution.

Methodology of Chromosome Painting

Probe Preparation: The first step in chromosome painting involves preparing DNA probes specific to the target chromosomes or regions. These probes are labeled with fluorescent dyes, which emit light when exposed to specific wavelengths. The DNA probes are designed to hybridize with complementary sequences on the chromosomes, ensuring specific and efficient binding. Probes for chromosome painting are typically generated from whole chromosomes or specific chromosomal regions using techniques such as flow sorting, microdissection, or amplification. These probes are labeled with different fluorescent dyes to distinguish between multiple chromosomes or regions in a single experiment. The choice of fluorescent dyes is critical to avoid spectral overlap and ensure clear differentiation of signals during analysis.

Chromosome Preparation: Chromosomes are prepared from dividing cells, usually at the metaphase stage of mitosis when they are most condensed and visible. Cells are treated to arrest them in metaphase and then lysed to release the chromosomes, which are subsequently spread on a microscope slide. The preparation of metaphase chromosomes involves several steps to ensure high-quality spreads suitable for hybridization. Cells are cultured and synchronized to maximize the number of metaphase cells. Chemicals such as colchicine or nocodazole are used to arrest cells in metaphase by disrupting spindle formation. The arrested cells are then treated with a hypotonic solution to swell them, making the chromosomes more accessible. The cells are fixed with a methanol-acetic acid mixture and spread onto slides, where they are dried and treated to denature the DNA, making it accessible for hybridization with the fluorescent probes.

Hybridization: The prepared fluorescently labeled probes are applied to the chromosomal spreads on the slides. The slides are incubated under conditions that allow the DNA probes to hybridize with their complementary sequences on the chromosomes. This process involves denaturing the DNA (heating to separate the strands) and then allowing the probes to anneal to their targets as the temperature is gradually reduced. Hybridization is a critical step in chromosome painting, requiring precise control of environmental conditions to ensure specific and efficient binding of the probes to the target sequences. The hybridization buffer typically contains formamide, dextran sulfate, and salts to enhance the specificity and stability of the DNA-DNA interactions. The slides are incubated in a humid chamber to prevent evaporation and maintain optimal hybridization conditions. Hybridization usually takes several hours to overnight, allowing sufficient time for the probes to anneal to their target sequences.

Detection: After hybridization, the slides are washed to remove any unbound or nonspecifically bound probes. The hybridized probes, which are now bound to their target sequences on the chromosomes, are visualized using a fluorescence microscope. The specific fluorescent signals indicate the location of the target sequences on the chromosomes, which can be captured and analyzed using imaging software. Detection involves visualizing the fluorescent signals emitted by the bound probes using a fluorescence microscope equipped with appropriate filter sets. The specific fluorescent dyes used in the probes emit light at different wavelengths, allowing researchers to distinguish between multiple probes in a single experiment. High-resolution cameras capture the fluorescent signals, and specialized imaging software is used to analyze and interpret the data. The software allows for precise measurement of signal intensity, localization, and colocalization of multiple probes, providing detailed insights into chromosomal structure and organization.

Applications of Chromosome Painting in Crop Improvement

Karyotyping: Chromosome painting enhances karyotyping by allowing researchers to visualize and differentiate individual chromosomes. This technique is particularly useful for identifying numerical and structural chromosomal abnormalities, such as aneuploidy, translocations, inversions, and deletions. In crop improvement, karyotyping provides valuable information about the chromosomal composition of different plant varieties. By comparing the karyotypes of different crops, researchers can identify chromosomal changes associated with desirable traits. For example, in wheat, chromosome painting can be used to identify and characterize translocations that confer disease resistance. This information is essential for selecting parent plants with favorable chromosomal compositions and developing new varieties with improved traits.

Identifying Chromosomal Rearrangements: Chromosome painting is an effective tool for detecting chromosomal rearrangements, such as translocations, inversions, and other structural variations. By using probes that span the breakpoints of rearrangements, researchers can detect these changes and understand their impacts on gene function and expression. Chromosomal rearrangements can significantly affect gene expression and trait inheritance. For instance, a translocation that disrupts a gene involved in disease resistance can render a plant susceptible to infections. Chromosome painting allows researchers to identify these rearrangements by using probes that bind to regions flanking the breakpoints. By analyzing the patterns of fluorescent signals, researchers can determine the presence and extent of rearrangements. This information is crucial for breeding programs, as it helps identify and eliminate deleterious rearrangements that may negatively impact crop performance.

Studying Chromosomal Behavior: Chromosome painting provides insights into chromosomal behavior during cell division and differentiation. By visualizing individual chromosomes, researchers can study processes such as homologous recombination, chromosome segregation, and chromosomal stability. Chromosomal behavior is essential for improving crop breeding strategies. For example, in maize, chromosome painting can be used to study homologous recombination during meiosis. By visualizing the pairing and segregation of chromosomes, researchers can identify factors that influence recombination rates and genetic diversity. This knowledge can be applied to develop breeding strategies that maximize genetic variation and improve crop performance.

Analyzing Hybridization: Chromosome painting helps identify and characterize chromosomal changes associated with hybridization. By comparing the chromosomes of hybrid plants to those of their parent species, researchers can detect genomic rearrangements and introgressions that occur during hybridization. Hybridization is a common technique in crop breeding, where two different plant species or varieties are crossed to combine desirable traits. Chromosome painting allows researchers to visualize the chromosomal composition of hybrid plants and identify any structural changes that result from hybridization. For example, in rice, chromosome painting can be used to study the introgression of disease resistance genes from wild relatives into cultivated varieties. By analyzing the chromosomal changes associated with hybridization, breeders can develop new crop varieties with enhanced traits and improved genetic diversity.

Polyploidy Analysis: Polyploidy, where organisms have more than two sets of chromosomes, is common in many crops. Chromosome painting is used to study polyploidy by identifying and differentiating between multiple sets of chromosomes. This is particularly useful in understanding the genetic composition and evolution of polyploid crops like wheat and sugarcane.

Polyploidy plays a significant role in the evolution and diversification of plant species. Chromosome painting allows researchers to visualize and distinguish between different sets of chromosomes in polyploid organisms. By using probes that bind to specific chromosomal regions, researchers can identify the origin and arrangement of multiple chromosome sets. This information is essential for understanding the genetic complexity of polyploid crops and for developing strategies to harness their genetic potential. For example, in wheat, chromosome painting can help identify homoeologous chromosomes (chromosomes derived from different ancestral species) and study their interactions, providing insights into the genetic basis of important traits and aiding in the development of improved wheat varieties.

Introgression Breeding: Chromosome painting facilitates introgression breeding by enabling the precise tracking of genetic material from wild relatives or exotic germplasm into cultivated varieties. This approach enhances the genetic diversity of crops and introduces beneficial traits from wild species.

Introgression breeding involves transferring beneficial traits from wild relatives or exotic germplasm into cultivated varieties. Chromosome painting allows researchers to monitor the incorporation of genetic material from donor species into the recipient genome. By using probes specific to the donor chromosomes, researchers can visualize the introgressed segments and track their inheritance in subsequent generations. For example, in maize, chromosome painting can be used to track the introgression of disease resistance genes from wild relatives into cultivated lines. By ensuring the successful transfer of beneficial traits, chromosome painting enhances the efficiency of introgression breeding and accelerates the development of improved crop varieties.

Advantages of Chromosome Painting

High Resolution and Specificity: Chromosome painting provides high-resolution visualization of specific chromosomes or chromosomal regions. This specificity allows researchers to identify and characterize structural variations with great precision, facilitating detailed genetic analysis and breeding strategies. The high resolution and specificity of chromosome painting enable researchers to detect even small chromosomal changes that may not be visible using traditional cytogenetic techniques. This precision is crucial for identifying genes and regulatory elements associated with important traits. By providing detailed insights into chromosomal structure and organization, chromosome painting enhances the accuracy of genetic analysis and breeding programs.

Visualization of Individual Chromosomes: The ability to visualize individual chromosomes allows for the study of chromosomal behavior during cell division and differentiation. This visualization provides valuable information about chromosomal stability, segregation, and recombination, which are essential for understanding genetic inheritance and improving breeding strategies. Visualizing individual chromosomes allows researchers to study the dynamic processes that occur during cell division, such as chromosome pairing, recombination, and segregation. This information is essential for understanding the mechanisms that drive genetic variation and inheritance. By studying chromosomal behavior, researchers can develop strategies to enhance genetic diversity and improve crop performance.

Detection of Structural Variations: Chromosome painting is an effective tool for detecting structural variations, such as translocations, inversions, duplications, and deletions. These variations can have significant impacts on gene function and trait expression, making their detection crucial for crop improvement. Structural variations can significantly affect gene expression and trait inheritance. Chromosome painting allows researchers to identify these variations with high precision, providing detailed insights into their impacts on gene function and trait development. By detecting structural variations, chromosome painting enhances the ability to develop targeted breeding strategies and improve crop performance.

Versatility and Applicability: Chromosome painting is a versatile technique that can be applied to a wide range of plant species, making it a valuable tool in crop improvement. This versatility allows researchers to study chromosomal behavior and structural variations across different crops and apply the findings to diverse breeding programs. The versatility of chromosome painting makes it a valuable tool for studying a wide range of plant species. By applying chromosome painting to different crops, researchers can gain insights into the genetic basis of important traits and develop strategies to improve crop performance. This versatility facilitates the transfer of knowledge and genetic resources between crops, enhancing the overall efficiency of breeding programs.

Challenges and Considerations

Technical Expertise: Chromosome painting is a technically complex technique that requires specialized expertise and equipment. Proper training and experience are essential for conducting chromosome painting experiments and interpreting the results accurately. The technical complexity of chromosome painting involves multiple steps, each requiring specialized knowledge and skills. Researchers must be proficient in probe design, chromosome preparation, hybridization, and fluorescence detection techniques. Additionally, interpreting the results requires expertise in cytogenetics and genomic analysis. Proper training and access to specialized equipment, such as fluorescence microscopes and imaging software, are essential for successful chromosome painting experiments.

Fluorescence Overlap: When multiple fluorescent dyes are used, their spectra can overlap, complicating the interpretation of results. This challenge requires careful selection of dyes and the use of advanced imaging techniques to distinguish between different signals. Fluorescence overlap, or spectral overlap, occurs when the emission spectra of different fluorescent dyes used in chromosome painting overlap, making it difficult to distinguish between signals. This issue can lead to inaccurate interpretation of results, especially when multiple probes are used simultaneously. To address this challenge, researchers select fluorescent dyes with minimal spectral overlap and use advanced imaging techniques, such as spectral imaging and deconvolution, to separate overlapping signals. Proper calibration of the fluorescence microscope and careful selection of filter sets are also essential to reduce fluorescence overlap and enhance signal clarity.

Sample Preparation: The quality of chromosomal spreads is crucial for successful chromosome painting. Poor preparation can lead to suboptimal hybridization and signal detection, requiring meticulous sample preparation and handling. Preparing high-quality chromosomal spreads is a critical step in chromosome painting that affects the accuracy and reliability of the results. The cells must be carefully synchronized to maximize the number of metaphase cells, and the chromosomal spreads must be prepared to ensure optimal preservation and accessibility of the chromosomes. Proper fixation and drying of the slides are essential to maintain chromosomal integrity and prevent artifacts. Additionally, handling and storage conditions must be carefully controlled to prevent degradation and contamination, ensuring that the chromosomal spreads are suitable for hybridization and signal detection.

Resolution Limits: While chromosome painting offers high resolution compared to many other techniques, its resolution is still limited by the size of the probes and the chromosomal spreads. For very fine-scale mapping, additional techniques such as high-throughput sequencing may be needed. The resolution of chromosome painting is influenced by factors such as probe size, hybridization efficiency, and the quality of chromosomal spreads. Although chromosome painting provides high-resolution mapping, it may not achieve the fine-scale resolution required for detailed analysis of closely spaced genes or small chromosomal regions. To overcome these limitations, researchers can complement chromosome painting with high-throughput sequencing techniques, such as whole-genome sequencing or targeted sequencing, to achieve finer resolution and obtain more comprehensive genetic information.

Chromosome painting is a powerful cytogenetic technique that allows for the precise visualization and differentiation of individual chromosomes or chromosomal regions. By using fluorescently labeled DNA probes, researchers can study chromosomal behavior, detect structural variations, and identify chromosomal rearrangements with high precision. This technique is particularly valuable for karyotyping and analyzing chromosomal changes associated with hybridization, polyploidy, and introgression in crop improvement. Chromosome painting provides detailed insights into genome organization and structural variations, facilitating the development of targeted breeding strategies and the improvement of crop performance. Despite its technical complexity and challenges, chromosome painting remains a fundamental tool for advancing our understanding of genomic organization and enhancing crop improvement efforts.

5. Single Nucleotide Polymorphism (SNP) Genotyping:

Single Nucleotide Polymorphism (SNP) genotyping is a genomic technique used to detect and analyze single nucleotide variations within the genome. These single nucleotide polymorphisms (SNPs) are among the most common types of genetic variation and are widely distributed across the genome. SNPs serve as valuable genetic markers for mapping genes, studying genetic diversity, and understanding the genetic basis of complex traits. High-throughput SNP genotyping platforms, such as SNP arrays and next-generation sequencing (NGS)-based methods, provide detailed and comprehensive information about genetic variations, enabling significant advancements in various fields, particularly crop improvement.

Methodology of SNP Genotyping

DNA Extraction: The first step in SNP genotyping involves extracting high-quality DNA from the organism of interest. The quality and purity of the DNA are critical for accurate genotyping. Various methods, such as CTAB extraction, phenol-chloroform extraction, and commercial DNA extraction kits, can be used to obtain high-quality genomic DNA. Extracted DNA must be assessed for concentration and purity using spectrophotometric methods (e.g., Nanodrop) and electrophoretic methods (e.g., agarose gel electrophoresis) to ensure it is suitable for subsequent genotyping procedures. High-quality DNA minimizes the risk of errors and ensures reliable genotyping results.

SNP Detection and Genotyping Platforms: SNP genotyping can be performed using several high-throughput platforms, each offering unique advantages depending on the study's goals and the available resources. Two primary methods are SNP arrays and NGS-based approaches.

SNP Arrays: SNP arrays, also known as SNP chips, are widely used for genotyping large numbers of SNPs simultaneously. This method involves hybridizing fragmented DNA to a microarray containing immobilized oligonucleotide probes specific to known SNPs. The hybridization results in fluorescent signals that are detected and analyzed to determine the genotype at each SNP locus. Commercially available SNP arrays, such as those from Illumina and Affymetrix, can genotype hundreds of thousands to millions of SNPs in a single experiment. SNP arrays are cost-effective, provide high throughput, and are suitable for large-scale genotyping projects. However, they are limited to detecting known SNPs and require prior knowledge of the genome.

NGS-Based SNP Genotyping: NGS-based methods, such as genotyping-by-sequencing (GBS) and whole-genome resequencing (WGR), offer greater flexibility and the ability to discover and genotype novel SNPs. GBS involves sequencing reduced representation libraries generated by digesting genomic DNA with restriction enzymes and ligating adapters. Sequencing the libraries on NGS platforms (e.g., Illumina) generates millions of short reads, which are aligned to a reference genome to identify SNPs. WGR involves sequencing the entire genome of individuals and identifying SNPs by comparing the sequences to a reference genome. NGS-based methods provide comprehensive and unbiased SNP detection, including rare and novel variants. These methods are particularly useful for species with limited genomic resources or for discovering SNPs in previously uncharacterized genomes.

Data Analysis: The data generated from SNP arrays or NGS platforms are processed and analyzed using bioinformatics tools. For SNP arrays, the fluorescent signals are quantified, and genotype calls are made based on signal intensities. For NGS-based methods, raw sequence reads are aligned to a reference genome, and variant calling algorithms identify SNPs. Bioinformatics pipelines, such as GATK (Genome Analysis Toolkit), SAMtools, and PLINK, are commonly used for variant calling, quality control, and downstream analysis. These tools enable researchers to filter low-quality SNPs, impute missing genotypes, and perform statistical analyses to identify associations between SNPs and traits of interest.

Applications of SNP Genotyping in Crop Improvement

Genome-Wide Association Studies (GWAS): GWAS leverages SNP genotyping data to identify genetic loci associated with specific traits across a population. By genotyping a large number of SNPs in diverse individuals and correlating genetic variations with phenotypic traits, researchers can pinpoint SNPs that influence important agronomic traits. In maize, for example, GWAS can identify SNPs associated with yield, drought tolerance, and disease resistance. By analyzing the genetic and phenotypic data from a diverse maize population, researchers can identify significant SNP-trait associations. These findings provide insights into the genetic architecture of complex traits and guide the development of improved maize varieties through marker-assisted selection (MAS).

Quantitative Trait Loci (QTL) Mapping: QTL mapping involves identifying chromosomal regions (QTLs) that control the variation of quantitative traits. SNP genotyping provides the dense genetic markers needed to construct high-resolution genetic maps and identify QTLs associated with traits such as yield, quality, and stress tolerance. In rice, QTL mapping can identify genomic regions associated with grain quality traits, such as amylose content and grain length. By genotyping a mapping population with SNP markers and analyzing the phenotypic data, researchers can detect QTLs that contribute to these traits. This information is crucial for breeding programs aiming to develop rice varieties with superior grain quality.

Genomic Selection (GS): Genomic selection uses genome-wide SNP data to predict the breeding value of individuals and select the best candidates for breeding. By incorporating SNP genotyping data into predictive models, researchers can estimate the genetic potential of plants more accurately and select individuals with superior genetic profiles. In wheat, genomic selection can be used to improve traits such as grain yield, disease resistance, and drought tolerance. By genotyping a training population with SNP markers and using phenotypic data to build predictive models, researchers can predict the breeding values of untested individuals. This approach accelerates the breeding process, increases selection accuracy, and enhances the efficiency of breeding programs.

Marker-Assisted Selection (MAS): MAS involves using genetic markers linked to specific traits to select individuals with desirable traits in breeding programs. SNP genotyping provides the markers needed for MAS, enabling breeders to select plants with favorable genetic profiles more efficiently. In tomato, SNP markers linked to disease-resistance genes can be used for MAS. By genotyping breeding lines with SNP markers associated with resistance to diseases such as late blight or fusarium wilt, breeders can identify and select resistant plants. This targeted approach reduces the time and resources needed for developing disease-resistant tomato varieties.

Genetic Diversity Studies: SNP genotyping is used to assess genetic diversity within and between populations. By analyzing the distribution and frequency of SNPs, researchers can study population structure, genetic relationships, and evolutionary history. In soybean, SNP genotyping can be used to analyze the genetic diversity of different landraces and cultivars. By genotyping a diverse collection of soybean accessions, researchers can assess the genetic variation and identify distinct genetic groups. This information is valuable for conserving genetic resources, managing germplasm collections, and designing breeding strategies to maximize genetic diversity.

Introgression Breeding: SNP genotyping facilitates introgression breeding by enabling the precise tracking of genetic material from wild relatives or exotic germplasm into cultivated varieties. This approach enhances the genetic diversity of crops and introduces beneficial traits from wild species. In barley, SNP genotyping can be used to track the introgression of disease-resistance genes from wild barley into cultivated varieties. By genotyping the introgressed lines and comparing them to the parent lines, researchers can monitor the transfer of resistance genes and ensure that the desired traits are successfully incorporated. This approach accelerates the development of disease-resistant barley varieties and improves crop performance.

Advantages of SNP Genotyping

High Resolution and Density: SNPs are abundant and evenly distributed throughout the genome, providing high-resolution markers for genetic studies. The high density of SNPs allows for fine-scale mapping of genetic loci and precise identification of associations between SNPs and traits. The high resolution and density of SNP markers enable researchers to detect small genetic variations that may significantly impact trait expression. This precision is crucial for studying complex traits, where multiple genes and regulatory elements contribute to phenotypic variation. By providing detailed genetic information, SNP genotyping enhances the accuracy and power of genetic studies.

Cost-Effectiveness: High-throughput SNP genotyping platforms, such as SNP arrays and NGS-based methods, offer cost-effective solutions for genotyping large numbers of samples and SNPs simultaneously. This cost-effectiveness makes SNP genotyping accessible for large-scale genetic studies and breeding programs. The cost-effectiveness of SNP genotyping platforms allows researchers to genotype large populations and conduct comprehensive genetic analyses. This scalability is essential for genome-wide association studies, genomic selection, and population genetics. By reducing the cost per sample, SNP genotyping platforms enable researchers to generate large datasets and gain valuable insights into the genetic basis of important traits.

High Throughput: SNP genotyping platforms can process large numbers of samples in parallel, enabling high-throughput genotyping and rapid data generation. This high throughput is essential for genome-wide studies and breeding programs that require genotyping thousands of individuals. The high throughput of SNP genotyping platforms accelerates the pace of genetic research and breeding programs. By processing large numbers of samples simultaneously, researchers can generate genetic data quickly and efficiently. This rapid data generation is crucial for timely decision-making in breeding programs and for advancing our understanding of the genetic architecture of crops.

Reproducibility and Accuracy: SNP genotyping platforms provide highly reproducible and accurate genotype calls, ensuring reliable data for genetic studies. The consistency and precision of SNP genotyping enhance the quality and reliability of genetic analyses. The reproducibility and accuracy of SNP genotyping platforms are critical for generating reliable genetic data. By providing consistent genotype calls, SNP genotyping platforms ensure that researchers can trust their results and make informed decisions based on accurate genetic information. This reliability is essential for breeding programs, where accurate genotype data is crucial for selecting the best candidates and improving crop performance.

Applicability to Diverse Species: SNP genotyping is applicable to a wide range of plant species, making it a versatile tool for genetic research and crop improvement. This versatility allows researchers to study genetic variation and trait associations across different crops and apply the findings to diverse breeding programs. The applicability of SNP genotyping to diverse plant species enables researchers to conduct genetic studies in a wide range of crops. By applying SNP genotyping to different species, researchers can gain insights into the genetic basis of important traits and develop strategies to improve crop performance. This versatility facilitates the transfer of knowledge and genetic resources between crops, enhancing the overall efficiency of breeding programs.

Challenges and Considerations

Bioinformatics Expertise: SNP genotyping generates large amounts of data that require advanced bioinformatics tools and expertise for analysis. Managing, storing, and analyzing these data can be challenging, necessitating robust data infrastructure and skilled personnel. The large datasets generated by SNP genotyping require sophisticated bioinformatics tools and expertise for data processing, analysis, and interpretation. Researchers must be proficient in using bioinformatics pipelines and software to handle the complexities of SNP genotyping data. Additionally, the storage and management of large datasets require robust data infrastructure, including high-performance computing resources and secure data storage solutions.

Sequencing Depth and Coverage: For NGS-based SNP genotyping methods, sequencing depth and coverage are critical factors that influence the accuracy and completeness of SNP detection. Insufficient coverage can lead to missing data and inaccurate genotype calls. Sequencing depth and coverage are important considerations for NGS-based SNP genotyping methods. Adequate sequencing depth ensures that each SNP is covered by multiple reads, reducing the risk of errors and increasing the accuracy of genotype calls. Researchers must carefully design their sequencing experiments to achieve sufficient coverage for reliable SNP detection and analysis.

Interpretation of SNPs: Interpreting the functional significance of SNPs can be challenging, particularly for non-coding regions. Functional validation through experimental studies is often required to determine the impact of SNPs on gene function and trait expression. The interpretation of SNPs involves linking genetic variations to phenotypic effects, which can be complex and context-dependent. Functional validation using techniques such as gene editing, gene expression analysis, and phenotypic assays is essential for confirming the roles of SNPs in trait development. Additionally, integrating SNP data with other omics data, such as transcriptomics and proteomics, can provide insights into the functional impact of SNPs and enhance their interpretation.

Population Structure and Stratification: Population structure and stratification can confound the results of SNP-based genetic studies. Proper statistical methods and experimental designs are necessary to account for population structure and avoid false associations. Population structure and stratification refer to the presence of subpopulations within a study population, which can lead to spurious associations between SNPs and traits. Researchers must use appropriate statistical methods, such as principal component analysis (PCA) and mixed models, to account for population structure and reduce the risk of false-positive associations. Proper experimental design, including the use of diverse and representative populations, is also essential for minimizing the impact of population structure on genetic studies.

Single Nucleotide Polymorphism (SNP) genotyping is a powerful genomic technique that enables the detection and analysis of single nucleotide variations across the genome. SNPs serve as valuable genetic markers for mapping genes, studying genetic diversity, and understanding the genetic basis of complex traits. High-throughput SNP genotyping platforms, such as SNP arrays and NGS-based methods, provide detailed and comprehensive genetic information, facilitating significant advancements in crop improvement. By leveraging SNP genotyping for genome-wide association studies (GWAS), QTL mapping, and genomic selection, breeders can identify SNPs associated with desirable traits and develop molecular markers for selecting individuals with superior genetic profiles.

This enhances the efficiency and precision of breeding programs, accelerating the development of improved crop varieties. Despite challenges such as data management, sequencing depth, and the interpretation of SNPs, SNP genotyping remains a fundamental tool for advancing genetic research and crop improvement efforts.

6. Optical Mapping: Optical Mapping

Optical mapping is an advanced genomic technique that produces high-resolution, physical maps of DNA molecules. This method is instrumental in identifying structural variations within the genome, such as insertions, deletions, inversions, and translocations. Optical mapping complements other genomic techniques by providing a detailed view of the genome's structural organization, which is crucial for understanding genetic variations and their implications in various biological processes. In crop improvement, optical mapping aids in identifying and characterizing structural variations associated with important agronomic traits. By integrating optical maps with sequence data, researchers can develop comprehensive genomic resources for breeding programs, facilitating the creation of high-yielding and resilient crop varieties.

Methodology of Optical Mapping

Sample Preparation: The first step in optical mapping involves isolating high molecular weight (HMW) DNA from the organism of interest. The integrity and purity of the DNA are critical for successful optical mapping, as fragmented or degraded DNA can compromise the quality of the maps. Isolating HMW DNA requires gentle extraction methods to minimize shearing. Common techniques include the use of agarose plugs to encapsulate cells or nuclei, followed by enzymatic lysis and purification. The extracted DNA is then embedded in agarose to maintain its integrity during subsequent processing steps.

DNA Fragmentation and Labeling: The HMW DNA is subjected to restriction enzyme digestion to create ordered restriction fragments. Specific restriction enzymes are chosen based on their recognition sequences and the desired resolution of the optical map. The restriction fragments are then labeled with fluorescent tags to visualize the fragment ends. The choice of restriction enzymes is crucial for generating informative and high-resolution maps. Enzymes with longer recognition sequences produce fewer, larger fragments, while those with shorter recognition sequences generate more, smaller fragments. The labeled fragments are subsequently elongated and immobilized on a specialized surface, such as a microfluidic chip or glass slide, to prepare for imaging.

Imaging and Map Construction: The immobilized and labeled DNA fragments are imaged using advanced microscopy techniques. High-resolution fluorescence microscopy captures the positions of the fluorescent labels along the DNA molecules, generating images that represent the physical structure of the DNA. The imaging process involves capturing multiple high-resolution images of the elongated DNA molecules to ensure accurate detection of the fluorescent labels. Image processing algorithms are then used to analyze the positions of the labels and construct a physical map of the DNA. This map, known as an optical map, represents the ordered restriction fragments along the DNA molecule.

Data Analysis and Integration: The optical maps are analyzed to identify structural variations, such as insertions, deletions, inversions, and translocations. These maps are then integrated with sequence data to provide a comprehensive view of the genome's structural organization. Data analysis involves aligning the optical maps with reference genomes or other sequence data to identify discrepancies and structural variations. Advanced bioinformatics tools and algorithms are used to detect and characterize these variations, providing detailed insights into the genome's architecture. Integrating optical maps with sequence data enhances the resolution and accuracy of structural variation analysis, enabling researchers to correlate genetic variations with phenotypic traits.

Applications of Optical Mapping in Crop Improvement

Identification of Structural Variations: Optical mapping is highly effective in identifying structural variations within the genome, such as insertions, deletions, inversions, and translocations. These structural variations can have significant impacts on gene function and trait expression, making their identification crucial for crop improvement. For example, in maize, optical mapping can be used to identify large insertions and deletions associated with disease resistance. By comparing the optical maps of resistant and susceptible varieties, researchers can pinpoint structural variations that contribute to resistance. These findings provide valuable targets for breeding programs aiming to develop disease-resistant maize varieties.

Characterization of Complex Genomes: Optical mapping is particularly useful for characterizing complex genomes, such as those of polyploid crops. The high resolution and detailed structural information provided by optical maps enable researchers to study the organization and evolution of polyploid genomes. In wheat, which is a hexaploid crop with three sets of homologous chromosomes, optical mapping can help disentangle the complex genomic structure. By generating optical maps for each set of chromosomes, researchers can identify homoeologous regions, structural variations, and chromosomal rearrangements. This information is essential for understanding the genetic basis of important traits and for developing improved wheat varieties.

Genome Assembly and Validation: Optical mapping complements whole-genome sequencing by providing physical maps that can be used to validate and improve genome assemblies. The high-resolution optical maps help resolve ambiguities and gaps in sequence data, leading to more accurate and complete genome assemblies. In rice, for instance, optical mapping can be used to validate and refine genome assemblies generated by next-generation sequencing (NGS). By aligning the optical maps with the sequence data, researchers can identify misassemblies, correct errors, and close gaps in the genome. This integrated approach enhances the quality of the genome assembly and provides a reliable reference for genetic studies and breeding programs.

Marker Development for Breeding: Optical mapping aids in the development of molecular markers for breeding programs. By identifying structural variations associated with desirable traits, researchers can develop markers that are linked to these variations, facilitating marker-assisted selection (MAS). In soybean, optical mapping can identify structural variations linked to traits such as drought tolerance and pest resistance. These variations can be used to develop SNP markers or other types of molecular markers for MAS. By genotyping breeding lines with these markers, breeders can select plants with the desired structural variations, accelerating the development of improved soybean varieties.

Introgression Breeding: Optical mapping facilitates introgression breeding by enabling the precise tracking of genetic material from wild relatives or exotic germplasm into cultivated varieties. This approach enhances genetic diversity and introduces beneficial traits from wild species into cultivated crops. In barley, optical mapping can be used to track the introgression of disease-resistance genes from wild barley into cultivated varieties. By generating optical maps for the introgressed lines and comparing them to the parent lines, researchers can monitor the transfer of resistance genes and ensure that the desired traits are successfully incorporated. This approach accelerates the development of disease-resistant barley varieties and improves crop performance.

Comparative Genomics: Optical mapping allows for comparative genomics studies by providing detailed structural information that can be compared across different species or varieties. This comparison reveals evolutionary relationships, gene conservation, and structural diversity. In Brassica crops, comparative optical mapping can be used to study the genetic diversity and evolution of disease resistance genes. By generating optical maps for different Brassica species and varieties, researchers can identify conserved and divergent structural regions. This information guides breeding programs aimed at transferring resistance genes between species, enhancing the disease resistance of Brassica crops.

Advantages of Optical Mapping

High-Resolution Structural Information: Optical mapping provides high-resolution physical maps that reveal detailed structural information about the genome. This high resolution is essential for identifying and characterizing structural variations, such as insertions, deletions, and inversions, that may significantly impact gene function and trait expression. The high-resolution structural information obtained from optical mapping enables researchers to detect even small structural variations that may not be visible using other techniques. This precision is crucial for studying complex traits, where multiple genes and regulatory elements contribute to phenotypic variation. By providing detailed insights into the genome's architecture, optical mapping enhances the accuracy and power of genetic studies.

Complementary to Sequencing Data: Optical mapping complements sequencing data by providing physical maps that can validate and refine genome assemblies. The integration of optical maps with sequence data enhances the resolution and accuracy of structural variation analysis. The complementary nature of optical mapping and sequencing data allows researchers to leverage the strengths of both techniques. While sequencing provides detailed nucleotide-level information, optical mapping offers high-resolution structural information that can resolve ambiguities and correct errors in genome assemblies. This integrated approach enhances the quality of genetic data and provides a comprehensive view of the genome.

Non-Biased Detection of Structural Variations: Optical mapping allows for the unbiased detection of structural variations, including large insertions, deletions, and complex rearrangements. Unlike sequencing-based methods, which may have biases due to library preparation and coverage, optical mapping provides a direct visualization of DNA molecules. The non-biased detection of structural variations is a significant advantage of optical mapping. By providing a direct and unbiased view of the genome, optical mapping ensures that researchers can accurately detect and characterize structural variations. This reliability is crucial for understanding the genetic basis of important traits and for developing targeted breeding strategies.

Versatility and Applicability: Optical mapping is a versatile technique that can be applied to a wide range of organisms, including plants, animals, and microorganisms. This versatility makes it a valuable tool for genetic research and crop improvement. The applicability of optical mapping to diverse organisms enables researchers to conduct genetic studies on a wide range of species. By applying optical mapping to different crops, researchers can gain insights into the genetic basis of important traits and develop strategies to improve crop performance. This versatility facilitates the transfer of knowledge and genetic resources between crops, enhancing the overall efficiency of breeding programs.

Challenges and Considerations

Technical Expertise and Equipment: Optical mapping requires specialized expertise and equipment, including high-resolution microscopes and advanced imaging software. Proper training and experience are essential for conducting optical mapping experiments and interpreting the results accurately. The technical complexity of optical mapping involves multiple steps, each requiring specialized knowledge and skills. Researchers must be proficient in DNA extraction, labeling, imaging, and data analysis techniques. Additionally, interpreting the results requires expertise in genomics and structural variation analysis. Proper training and access to specialized equipment, such as high-resolution microscopes and imaging software, are essential for successful optical mapping experiments.

Sample Quality: The quality of the DNA samples is critical for successful optical mapping. Poorly prepared or degraded DNA can compromise the quality of the optical maps, leading to inaccurate results. Preparing high-quality DNA samples is a critical step in optical mapping that affects the accuracy and reliability of the results. The DNA must be free of contaminants and degradation to ensure accurate imaging and map construction. Researchers must use gentle extraction methods and careful handling to maintain the integrity of the DNA samples. Proper storage and handling conditions are also essential to prevent degradation and contamination, ensuring that the DNA samples are suitable for optical mapping.

Resolution Limits: While optical mapping offers high resolution compared to many other techniques, its resolution is still limited by the size of the DNA fragments and the imaging technology. For very fine-scale mapping, additional techniques such as high-throughput sequencing may be needed. The resolution of optical mapping is influenced by factors such as the size of the DNA fragments, the choice of restriction enzymes, and the quality of the imaging system. Although optical mapping provides high-resolution structural information, it may not achieve the fine-scale resolution required for detailed analysis of closely spaced genes or small structural variations. To overcome these limitations, researchers can complement optical mapping with high-throughput sequencing techniques, such as whole-genome sequencing or targeted sequencing, to achieve finer resolution and obtain more comprehensive genetic information.

Optical mapping is a powerful genomic technique that provides high-resolution, physical maps of DNA molecules. By creating ordered restriction maps, optical mapping allows researchers to identify and characterize structural variations, such as insertions, deletions, inversions, and translocations. This technique complements other genomic methods by providing detailed insights into the genome's structural organization, which is crucial for understanding genetic variations and their implications in various biological processes. In crop improvement, optical mapping aids in identifying and characterizing structural variations associated with important traits. By integrating optical maps with sequence data, researchers can develop comprehensive genomic resources for breeding programs, facilitating the development of high-yielding and resilient crop varieties. Despite challenges such as technical expertise and sample quality, optical mapping remains a fundamental tool for advancing genetic research and enhancing crop improvement efforts.

Applications of Cytogenomics in Crop Improvement

1. Marker-Assisted Selection (MAS): Cytogenomics provides detailed genetic maps and molecular markers for MAS, enabling breeders to select individuals with desirable traits more effectively. By identifying markers linked to important agronomic traits, such as yield, disease resistance, and stress tolerance, breeders can accelerate the breeding process and develop superior crop varieties.

2. Genomic Selection (GS): GS involves using genome-wide markers to predict the breeding value of individuals. Cytogenomics enhances GS by providing high-resolution genetic information and identifying markers associated with complex traits. This approach allows breeders to select individuals with the highest genetic potential, improving the efficiency and accuracy of breeding programs.

3. Identification of Quantitative Trait Loci (QTLs): Cytogenomics facilitates the identification of QTLs associated with important agronomic traits. By integrating cytogenetic and genomic data, researchers can pinpoint chromosomal regions contributing to trait variation. This information is crucial for understanding the genetic basis of complex traits and for developing targeted breeding strategies.

4. Gene Cloning and Functional Genomics: The precise localization of genes through cytogenomics enables gene cloning and functional genomics studies. Researchers can isolate and characterize genes associated with desirable traits, study their functions, and manipulate them for crop improvement. Functional genomics provides insights into gene expression, regulation, and interactions, enhancing our understanding of gene function.

5. Development of Hybrid Varieties: Cytogenomics aids in the development of hybrid varieties with enhanced traits. By identifying and mapping genes associated with heterosis, breeders can develop hybrids with improved yield, quality, and stress tolerance. Cytogenomics also helps in the identification of genetic markers for maintaining hybrid vigor and fertility.

6. Introgression of Desirable Traits: Cytogenomics facilitates the introgression of desirable traits from wild relatives or exotic germplasm into cultivated varieties. By identifying and mapping genes associated with beneficial traits, breeders can incorporate these genes into new crop varieties, enhancing their performance and adaptability.

B. Chromosome sorting

Chromosome sorting is a crucial cytogenetic technique that allows for the separation and isolation of individual chromosomes from a complex mixture of chromosomal material. This method is essential for detailed genetic analysis, gene mapping, and genome sequencing. By sorting chromosomes, researchers can obtain pure chromosome preparations for various applications in genomics and crop improvement.

Concept of Chromosome Sorting

Chromosome sorting is a sophisticated technique that utilizes flow cytometry to separate and isolate individual chromosomes based on their size, shape, and fluorescence characteristics. This method is exceptionally valuable for plants with large and complex genomes, where isolating specific chromosomes can substantially enhance the precision of genetic studies. By enabling the isolation of pure chromosome preparations, chromosome sorting provides a means to conduct detailed genetic analyses, including gene mapping, structural variation studies, and the identification of specific genetic traits.

Methodology of Chromosome Sorting

The process of chromosome sorting typically involves several critical steps to ensure the accurate separation and collection of individual chromosomes. These steps include chromosome preparation, staining, flow cytometry, and collection.

1. Chromosome Preparation: Chromosome sorting begins with the isolation of chromosomes from dividing cells, typically at the metaphase stage of mitosis when chromosomes are most condensed and distinguishable. Cells are harvested from tissues that are actively dividing, such as root tips or meristems in plants. The cells are treated with a mitotic inhibitor, such as colchicine or nocodazole, to arrest them in metaphase by preventing the formation of the mitotic spindle. This step ensures that a large proportion of cells are in the desired stage of the cell cycle. The cells are then treated with a hypotonic solution to swell them, making the chromosomes more accessible for isolation. Following this, the cells are lysed to release the chromosomes, which are then fixed using a mixture of methanol and acetic acid to preserve their structure and prevent degradation.

2. Staining: The isolated chromosomes are stained with fluorescent dyes that bind specifically to DNA or particular chromosomal regions, allowing for their identification based on fluorescence. Commonly used dyes include DAPI (4',6-diamidino-2-phenylindole), which binds to AT-rich regions of DNA, and propidium iodide, which intercalates into the DNA helix. In addition to general DNA stains, fluorescently labeled probes that bind to specific chromosomal regions or sequences can be used. These probes, such as those used in fluorescence in situ hybridization (FISH), can target repetitive elements, centromeres, telomeres, or specific gene loci. This targeted staining allows for the differentiation of chromosomes based on their unique fluorescence patterns.

3. Flow Cytometry: The stained chromosomes are then passed through a flow cytometer, a device that detects and sorts particles based on their optical properties. As the chromosomes flow through the cytometer, they are illuminated by laser beams, causing the fluorescent dyes to emit light. The emitted fluorescence and the light-scattering properties of each chromosome are detected by the flow cytometer's sensors. These signals are analyzed to determine the size, shape, and fluorescence characteristics of each chromosome. Based on these properties, the flow cytometer sorts the chromosomes, separating them into distinct populations. Flow cytometry allows for the rapid and precise sorting of thousands of chromosomes per second, making it a highly efficient method for chromosome isolation.

4. Collection: Sorted chromosomes are collected into separate tubes or multi-well plates for further analysis. The purity and accuracy of the sorted chromosomes are verified using various techniques, such as fluorescence microscopy or PCR. Once collected, the isolated chromosomes can be used for a wide range of downstream applications. These applications include the construction of chromosome-specific libraries, high-resolution gene mapping, the study of structural variations, and the analysis of chromosomal behavior during cell division. The ability to isolate individual chromosomes with high purity and accuracy significantly enhances the precision of genetic studies and facilitates the detailed investigation of complex genomes.

Applications of Chromosome Sorting

Gene Mapping: Chromosome sorting is instrumental in gene mapping by providing pure chromosome preparations for detailed genetic analysis. By isolating specific chromosomes, researchers can map genes and genetic markers more accurately, facilitating the identification of quantitative trait loci (QTLs) and other genetic elements associated with important traits. For example, in wheat, chromosome sorting can be used to isolate specific chromosomes for detailed genetic mapping. By combining chromosome sorting with high-throughput sequencing, researchers can identify genes associated with traits such as disease resistance, yield, and stress tolerance. This information is crucial for developing marker-assisted selection (MAS) strategies and improving wheat breeding programs.

Structural Variation Analysis: Chromosome sorting allows for the detection and analysis of structural variations, such as duplications, deletions, inversions, and translocations. By isolating specific chromosomes, researchers can analyze these variations in detail, providing insights into their impact on gene function and trait expression. In maize, for example, chromosome sorting can be used to study structural variations associated with hybrid vigor. By isolating and analyzing specific chromosomes from hybrid and parental lines, researchers can identify structural changes that contribute to the enhanced performance of hybrids. This information is valuable for developing strategies to exploit hybrid vigor in maize breeding programs.

Comparative Genomics: Chromosome sorting facilitates comparative genomics studies by enabling the comparison of specific chromosomes from different species or varieties. This approach helps identify conserved and divergent regions, providing insights into evolutionary relationships and the genetic basis of important traits. In Brassica crops, chromosome sorting can be used to compare the genomes of different species, such as Brassica rapa, Brassica oleracea, and Brassica napus. By isolating and comparing specific chromosomes, researchers can identify genetic elements that are conserved across species and those that are unique to each species. This information is essential for understanding the evolutionary history of Brassica crops and for developing strategies to transfer beneficial traits between species.

Functional Genomics: Isolated chromosomes can be used for functional genomics studies, including gene expression analysis and epigenetic profiling. By studying the genes and regulatory elements on specific chromosomes, researchers can gain a deeper understanding of their roles in biological processes and trait development. In rice, for example, chromosome sorting can be used to isolate chromosomes carrying genes involved in stress response. By analyzing the gene expression patterns and epigenetic modifications on these chromosomes, researchers can identify key regulatory elements that control stress tolerance. This information can be used to develop strategies for enhancing stress tolerance in rice through genetic and epigenetic modifications.

Introgression Breeding: Chromosome sorting is valuable for introgression breeding, which involves transferring beneficial traits from wild relatives or exotic germplasm into cultivated varieties. By isolating and analyzing specific chromosomes carrying the desired traits, researchers can monitor the introgression process and ensure the successful transfer of these traits. In barley, chromosome sorting can be used to track the introgression of disease-resistance genes from wild barley into cultivated lines. By isolating the chromosomes carrying the resistance genes and analyzing their integration into the cultivated genome, researchers can develop barley varieties with enhanced disease resistance. This approach accelerates the introgression process and improves the efficiency of breeding programs.

Advantages of Chromosome Sorting

High Precision and Accuracy: Chromosome sorting provides high precision and accuracy in isolating specific chromosomes, enabling detailed genetic analysis and gene mapping. The ability to sort individual chromosomes reduces the complexity of genetic studies and enhances the resolution of genetic maps. The precision and accuracy of chromosome sorting are crucial for detecting subtle genetic variations and for studying complex traits. By isolating pure chromosome preparations, researchers can conduct high-resolution genetic analyses that are not possible with whole-genome approaches. This precision enhances the reliability and reproducibility of genetic studies, providing valuable insights into the genetic architecture of crops.

Enhanced Genetic Studies: By providing pure chromosome preparations, chromosome sorting enhances genetic studies, including genome sequencing, structural variation analysis, and functional genomics. This approach facilitates the identification of genetic elements associated with important traits and provides insights into their roles in trait development. Chromosome sorting enhances the efficiency and resolution of genetic studies by providing pure and specific chromosome preparations. This approach allows researchers to focus on individual chromosomes and their genetic content, reducing the complexity of genetic analyses. The ability to isolate and study specific chromosomes enhances our understanding of the genetic basis of important traits and facilitates the development of targeted breeding strategies.

Versatility: Chromosome sorting is a versatile technique that can be applied to a wide range of organisms, including plants, animals, and microorganisms. This versatility makes it a valuable tool for genetic research across different species and enables cross-species comparisons. The versatility of chromosome sorting allows researchers to study the genetic architecture of diverse organisms and to conduct comparative genomics studies. By applying chromosome sorting to different species, researchers can gain insights into the conservation and divergence of genetic elements and their roles in trait development. This versatility facilitates the transfer of knowledge and genetic resources between species, enhancing the overall efficiency of genetic research and breeding programs.

Complementary to Other Techniques: Chromosome sorting can be used in conjunction with other genetic techniques, such as genome sequencing, fluorescence in situ hybridization (FISH), and comparative genomic hybridization (CGH), to provide comprehensive genetic information. By integrating data from multiple approaches, researchers can obtain a more complete understanding of the genome's structure and function. The complementary nature of chromosome sorting and other genetic techniques allows researchers to leverage the strengths of each approach. For example, combining chromosome sorting with genome sequencing provides high-resolution genetic maps and detailed sequence information, enhancing the accuracy and resolution of genetic analyses. This integrated approach provides a comprehensive view of the genome, facilitating the identification of genetic elements associated with important traits.

Challenges and Considerations

Chromosomal Integrity: Maintaining the integrity of chromosomes during isolation and sorting is crucial for obtaining accurate results. Chromosomes can be fragile and prone to breakage, so careful handling and optimized protocols are necessary to preserve their structure.

The integrity of chromosomes is critical for successful chromosome sorting and subsequent genetic analyses. Researchers must use gentle extraction and handling methods to minimize the risk of chromosomal breakage and degradation. Optimized protocols for chromosome isolation, staining, and sorting are essential to maintain chromosomal integrity and ensure accurate results.

Resolution and Specificity: Achieving high resolution and specificity in chromosome sorting can be challenging, especially for small or similar-sized chromosomes. Advanced flow cytometry techniques and optimized staining protocols are essential to enhance resolution and ensure the accurate identification and sorting of chromosomes. The resolution and specificity of chromosome sorting are influenced by factors such as the choice of fluorescent dyes, staining protocols, and flow cytometry settings. Researchers must carefully optimize these factors to achieve high resolution and specificity in chromosome sorting. Advanced flow cytometry techniques, such as multicolor flow cytometry and spectral flow cytometry, can enhance the resolution and specificity of chromosome sorting, enabling the accurate identification and separation of small or similar-sized chromosomes.

Sample Preparation: The quality of chromosome preparations is critical for successful sorting. Isolating chromosomes from dividing cells and obtaining sufficient quantities of high-quality chromosomes require optimized sample preparation protocols and the use of synchronized cell cultures. Preparing high-quality chromosome preparations involves several steps, including cell synchronization, mitotic arrest, hypotonic treatment, and fixation. Each step must be carefully optimized to ensure the accurate isolation and preservation of chromosomes. The use of synchronized cell cultures and optimized protocols for chromosome isolation and staining are essential for obtaining high-quality chromosome preparations suitable for sorting.

Fluorescence Overlap: Fluorescence overlap can occur when multiple fluorescent dyes are used, leading to interference and reduced accuracy in chromosome sorting. Careful selection of dyes with minimal spectral overlap and the use of compensation techniques in flow cytometry can mitigate this issue. Fluorescence overlap, or spectral overlap, occurs when the emission spectra of different fluorescent dyes used in chromosome sorting overlap, making it difficult to distinguish between signals. To address this challenge, researchers must select fluorescent dyes with minimal spectral overlap and use compensation techniques in flow cytometry to separate overlapping signals. Proper calibration of the flow cytometer and careful selection of filter sets are also essential to reduce fluorescence overlap and enhance signal clarity.

Applications of Chromosome Sorting

1. Gene Mapping: Chromosome sorting allows for high-resolution gene mapping by providing pure chromosome preparations for detailed genetic analysis. By isolating specific chromosomes, researchers can map genes and genetic markers more accurately, facilitating the identification of quantitative trait loci (QTLs) and other genetic elements associated with important traits.

2. Genome Sequencing: Chromosome sorting enhances genome sequencing efforts by enabling the isolation of individual chromosomes for sequencing. This approach reduces the complexity of the sequencing process and allows for the assembly of chromosome-specific sequences. It is particularly useful for organisms with large and complex genomes, where whole-genome sequencing can be challenging.

3. Structural Variation Analysis: Chromosome sorting helps identify and characterize structural variations, such as duplications, deletions, inversions, and translocations. By isolating specific chromosomes, researchers can analyze these variations in detail, providing insights into their impact on gene function and trait expression.

4. Comparative Genomics: Chromosome sorting facilitates comparative genomics studies by enabling the comparison of specific chromosomes from different species or varieties. This approach helps identify conserved and divergent regions, providing insights into evolutionary relationships and the genetic basis of important traits.

5. Functional Genomics: Isolated chromosomes can be used for functional genomics studies, including gene expression analysis and epigenetic profiling. By studying the genes and regulatory elements on specific chromosomes, researchers can gain a deeper understanding of their roles in biological processes and trait development.

6. Crop Improvement: Chromosome sorting is a valuable tool for crop improvement, allowing breeders to isolate and analyze chromosomes carrying desirable traits. This approach facilitates the development of new crop varieties with enhanced yield, quality, and stress tolerance.

Techniques and Tools for Chromosome Sorting

Flow Cytometry: Flow cytometry is the primary technique used for chromosome sorting. This technology uses laser-based detection to analyze and sort particles, including chromosomes, based on their size, shape, and fluorescence characteristics. Flow cytometers equipped with multiple lasers and detectors can simultaneously analyze multiple parameters, enabling the precise sorting of chromosomes.

Fluorescent Staining: Fluorescent dyes are used to stain chromosomes, allowing for their identification based on fluorescence. Common dyes include DAPI (4',6-diamidino-2-phenylindole), which binds to AT-rich regions of DNA, and propidium iodide, which binds to DNA in general. Fluorescently labeled probes can also be used to target specific chromosomal regions or genes.

Chromosome Microdissection: Chromosome microdissection involves the physical isolation of specific chromosomal regions using fine needles or laser beams. This technique can be combined with flow cytometry to enhance the resolution of chromosome sorting and enable the analysis of specific chromosomal segments.

Single Chromosome Sequencing: Single chromosome sequencing involves the isolation and sequencing of individual chromosomes. This approach provides high-resolution genetic information and facilitates the assembly of chromosome-specific sequences. Techniques such as single-cell sequencing and long-read sequencing technologies are used to achieve this.

Challenges and Considerations

Chromosomal Integrity: Maintaining the integrity of chromosomes during isolation and sorting is crucial for obtaining accurate results. Chromosomes can be fragile and prone to breakage, so careful handling and optimized protocols are necessary to preserve their structure.

Resolution and Specificity: Achieving high resolution and specificity in chromosome sorting can be challenging, especially for small or similar-sized chromosomes. Advanced flow cytometry techniques and optimized staining protocols are essential to enhance resolution and ensure the accurate identification and sorting of chromosomes.

Sample Preparation: The quality of chromosome preparations is critical for successful sorting. Isolating chromosomes from dividing cells and obtaining sufficient quantities of high-quality chromosomes require optimized sample preparation protocols and the use of synchronized cell cultures.

Fluorescence Overlap: Fluorescence overlap can occur when multiple fluorescent dyes are used, leading to interference and reduced accuracy in chromosome sorting. Careful selection of dyes with minimal spectral overlap and the use of compensation techniques in flow cytometry can mitigate this issue.

Advantages of Chromosome Sorting

Precision and Accuracy: Chromosome sorting offers high precision and accuracy in isolating specific chromosomes, enabling detailed genetic analysis and gene mapping. The ability to sort individual chromosomes reduces the complexity of genetic studies and enhances the resolution of genetic maps.

Enhanced Genetic Studies: By providing pure chromosome preparations, chromosome sorting enhances genetic studies, including genome sequencing, structural variation analysis, and functional genomics. This approach facilitates the identification of genetic elements associated with important traits and provides insights into their roles in trait development.

Versatility: Chromosome sorting is a versatile technique that can be applied to a wide range of organisms, including plants, animals, and microorganisms. This versatility makes it a valuable tool for genetic research across different species and enables cross-species comparisons.

Complementary to Other Techniques: Chromosome sorting can be used in conjunction with other genetic techniques, such as genome sequencing, FISH, and CGH, to provide comprehensive genetic information. By integrating data from multiple approaches, researchers can obtain a more complete understanding of the genetic architecture of traits.

Applications in Crop Improvement

Chromosome sorting is a powerful tool that provides precise and detailed genetic information, significantly enhancing various aspects of crop improvement. This technique aids breeders in selecting, developing, and improving crop varieties by offering high-resolution genetic maps and molecular markers. The following are key applications of chromosome sorting in crop improvement:

Marker-Assisted Selection (MAS)

Marker-assisted selection (MAS) is a breeding technique that uses molecular markers linked to desirable traits to select individuals with superior genetic profiles. Chromosome sorting enhances MAS by providing high-resolution genetic maps and identifying markers associated with important agronomic traits.

Process:

1. **Isolation and Analysis:** Chromosome sorting isolates specific chromosomes from individuals in a breeding population. These chromosomes are analyzed to identify genetic markers linked to traits such as disease resistance, yield, or stress tolerance. This involves detailed mapping of single nucleotide polymorphisms (SNPs), microsatellites, and other molecular markers that

are closely associated with desired traits. High-resolution mapping allows breeders to pinpoint the exact location of these markers on the chromosomes, enhancing the precision of MAS.

2. **Marker Development:** High-resolution genetic maps are created by identifying SNPs and other molecular markers on the isolated chromosomes. These maps serve as a reference for developing markers that can be used in breeding programs. The markers identified through chromosome sorting are highly specific and tightly linked to the target traits, making them reliable tools for MAS.
3. **Selection:** Breeders use these markers to screen breeding populations, selecting individuals that possess the desired traits based on their genetic profiles. This process significantly reduces the time and resources required for traditional phenotypic selection, as breeders can select plants with the desired genetic makeup at an early stage.

Benefits:

» **Efficiency:** Chromosome sorting streamlines the process of marker identification, making MAS more efficient and accurate. By providing high-resolution genetic maps, breeders can identify and utilize markers that are closely linked to the target traits, ensuring that the selected individuals possess the desired characteristics.

» **Precision:** High-resolution genetic maps ensure that selected markers are closely linked to the target traits, reducing the likelihood of selecting undesirable traits. This precision enhances the reliability of MAS and improves the overall effectiveness of breeding programs.

Example: In wheat breeding, chromosome sorting can be used to identify markers linked to rust resistance. By isolating and analyzing chromosomes from resistant and susceptible varieties, breeders can develop markers for MAS, allowing them to select resistant individuals more effectively and accelerate the development of rust-resistant wheat varieties. This targeted approach not only improves the efficiency of breeding programs but also ensures the development of varieties with enhanced disease resistance.

Genomic Selection (GS)

Genomic selection (GS) predicts the breeding value of individuals based on genome-wide markers. Chromosome sorting enhances GS by providing detailed genetic data and identifying markers associated with complex traits, improving the accuracy of selection and accelerating the breeding process.

Process:

1. **Genetic Data Collection:** Chromosome sorting isolates chromosomes from a training population with known phenotypes. These chromosomes are analyzed to identify genome-wide markers, including SNPs and other genetic variations. This comprehensive analysis provides a detailed genetic profile of the training population, which is essential for developing accurate predictive models.
2. **Model Development:** Statistical models are developed to predict breeding values based on the identified markers and the phenotypic data of the training population. These models

use advanced algorithms to integrate the genetic data and predict the genetic potential of individuals in the breeding population. The accuracy of these models is enhanced by the high-resolution genetic data obtained through chromosome sorting.

3. **Selection:** The models are used to predict the breeding values of individuals in a breeding population, allowing breeders to select those with the highest potential for desired traits. This predictive approach enables breeders to make informed decisions and select individuals that are likely to produce superior offspring.

Benefits:

- **Accuracy:** Detailed genetic data from chromosome sorting improve the accuracy of breeding value predictions. The high-resolution markers provide a comprehensive view of the genetic makeup of individuals, allowing for more precise predictions of their breeding potential.
- **Speed:** GS accelerates the breeding process by enabling early selection of superior individuals based on genetic potential rather than phenotypic performance alone. This approach reduces the time required for traditional breeding cycles and accelerates the development of improved crop varieties.

Example: In maize breeding, chromosome sorting can enhance GS by identifying markers associated with yield and drought tolerance. By integrating these markers into predictive models, breeders can select high-yielding, drought-tolerant individuals more accurately and efficiently. This approach not only improves the accuracy of selection but also accelerates the development of maize varieties with enhanced performance under drought conditions.

Introgression of Desirable Traits

Introgression breeding involves transferring beneficial traits from wild relatives or exotic germplasm into cultivated varieties. Chromosome sorting facilitates this process by isolating and analyzing chromosomes carrying desirable traits, ensuring their successful incorporation into new crop varieties.

Process:

1. **Identification:** Chromosomes from wild relatives or exotic germplasm with desirable traits are isolated and analyzed to identify genetic regions associated with those traits. This involves detailed mapping of the genetic regions that contribute to the desired characteristics, such as disease resistance or stress tolerance.
2. **Crossing:** These chromosomes are then introgressed into cultivated varieties through controlled crossing. The integration of these desirable traits is monitored through successive generations to ensure their stable incorporation into the cultivated genome.
3. **Verification:** The presence and integration of the desirable traits are verified by isolating and analyzing the chromosomes of the resulting progeny. This step ensures that the introgressed traits are effectively transferred and expressed in the new varieties.

Benefits:

- **Precision:** Chromosome sorting ensures that specific genetic regions carrying desirable traits are successfully transferred. This precision enhances the efficiency of introgression breeding and reduces the risk of unwanted genetic changes.
- **Enhanced Performance:** Introgression of beneficial traits enhances the performance and adaptability of cultivated varieties. This approach leverages the genetic diversity of wild relatives to improve the resilience and productivity of crop varieties.

Example: In rice breeding, chromosome sorting can facilitate the introgression of drought tolerance from wild rice species into cultivated varieties. By isolating and analyzing chromosomes carrying drought tolerance genes, breeders can ensure the successful transfer of these genes and develop drought-tolerant rice varieties. This targeted approach enhances the resilience of rice crops to water-limited conditions, improving yield stability in drought-prone regions.

Development of Hybrid Varieties

Hybrid varieties are developed by crossing two genetically distinct parents, resulting in offspring with enhanced traits due to heterosis (hybrid vigor). Chromosome sorting aids in the development of hybrid varieties by identifying and mapping genes associated with heterosis and maintaining hybrid vigor and fertility.

Process:

1. **Gene Identification:** Chromosomes from parent lines are isolated and analyzed to identify genes associated with heterosis. This involves detailed mapping of genetic regions that contribute to hybrid vigor, such as genes involved in growth, yield, and stress tolerance.
2. **Crossing:** Identified genes are used to select parent lines for crossing to produce hybrid offspring with enhanced traits. The selected parent lines are crossed to create hybrid varieties that exhibit superior performance compared to their parent lines.
3. **Maintenance:** Chromosome sorting helps monitor and maintain hybrid vigor and fertility by ensuring the stability and integrity of the hybrid genome. This involves regular analysis of the chromosomes to detect any changes or instabilities that may affect hybrid performance.

Benefits:

- **Enhanced Traits:** Hybrid varieties often exhibit superior yield, quality, and stress tolerance compared to their parent lines. Chromosome sorting helps identify the genetic basis of these enhanced traits, enabling the development of high-performing hybrids.
- **Fertility Maintenance:** Chromosome sorting ensures the stability and fertility of hybrid varieties, maintaining their performance over successive generations. This approach reduces the risk of hybrid breakdown and ensures consistent productivity.

Example: In tomato breeding, chromosome sorting can be used to identify genes associated with fruit size and quality. By selecting parent lines based on these genes, breeders can develop hybrid tomatoes with larger, higher-quality fruits, improving marketability and consumer satisfaction. This targeted approach not only enhances the traits of hybrid varieties but also ensures their consistent performance across different growing conditions.

Resistance to Biotic and Abiotic Stresses

Chromosome sorting helps identify genes and regulatory elements involved in resistance to biotic (diseases, pests) and abiotic (drought, salinity) stresses. By isolating and analyzing specific chromosomes, breeders can develop crop varieties that are more resilient to changing environmental conditions and reduce crop losses due to these stresses.

Process:

1. **Gene Identification:** Chromosomes from stress-tolerant individuals are isolated and analyzed to identify genes and regulatory elements associated with resistance. This involves detailed mapping of genetic regions that contribute to stress resistance, including genes involved in pathogen defense, drought response, and salinity tolerance.
2. **Marker Development:** Molecular markers linked to these genes are developed for use in MAS and GS. These markers enable the precise selection of individuals with enhanced stress resistance, improving the efficiency of breeding programs.
3. **Breeding:** Breeders use these markers to select individuals with enhanced stress resistance, incorporating these traits into new crop varieties. This targeted approach ensures the development of resilient crops that can withstand various biotic and abiotic stresses.

Benefits:

» **Resilience:** Crop varieties with enhanced resistance to biotic and abiotic stresses are more resilient to environmental changes and pests. This resilience improves yield stability and reduces the risk of crop losses.

» **Reduced Losses:** Improved stress resistance reduces crop losses, ensuring stable yields and food security. This approach enhances the sustainability of agricultural production and contributes to global food security.

Example: In soybean breeding, chromosome sorting can be used to identify genes associated with resistance to soybean cyst nematode (SCN). By isolating and analyzing chromosomes from resistant varieties, breeders can develop markers for MAS, allowing them to select SCN-resistant individuals more effectively and develop soybean varieties with improved resistance to this pest. This targeted approach enhances the resilience of soybean crops to nematode infestations, reducing yield losses and improving overall productivity.

C. Isolation of specific chromosomes for development of molecular maps and gene location

The isolation of specific chromosomes is a crucial technique in genomics that significantly aids in the development of molecular maps and the precise location of genes. This technique leverages advanced cytogenetic and molecular biology methods to separate individual chromosomes from a cell's genome, allowing detailed genetic analysis. The resulting data provide valuable insights into the genetic architecture of organisms, facilitating the identification of genetic markers and the mapping of genes associated with important traits. This section elaborates on the methodology, applications, and benefits of isolating specific chromosomes for developing molecular maps and locating genes.

Methodology of Chromosome Isolation

The process of isolating specific chromosomes involves several meticulous steps to ensure the accurate separation and integrity of the chromosomes. These steps include chromosome preparation, staining, flow cytometry, and collection.

Chromosome Preparation: Chromosome isolation begins with the preparation of chromosomes from actively dividing cells, typically at the metaphase stage of mitosis when chromosomes are most condensed and distinguishable. This stage is optimal for isolation due to the compact and well-defined structure of chromosomes.

1. **Cell Harvesting:** Cells are harvested from tissues with high mitotic activity, such as root tips or meristems in plants. The choice of tissue is critical to ensure a sufficient number of metaphase cells. Typically, young, actively growing tissues are selected because they exhibit a high rate of cell division. The cells are harvested and placed in a suitable medium to maintain their viability during the preparation process.
2. **Mitotic Arrest:** Cells are treated with a mitotic inhibitor (e.g., colchicine or nocodazole) to arrest them in metaphase, ensuring a high proportion of cells are in the desired stage. The duration and concentration of the mitotic inhibitor treatment must be optimized to maximize the number of metaphase cells while minimizing cellular stress. This step is crucial for obtaining a high yield of metaphase chromosomes.
3. **Hypotonic Treatment:** Cells are treated with a hypotonic solution to swell them, making the chromosomes more accessible. The hypotonic treatment causes water to enter the cells, increasing their volume and spreading the chromosomes apart. This step facilitates the subsequent lysis of the cells and the release of chromosomes. The treatment duration and concentration of the hypotonic solution must be carefully controlled to achieve optimal swelling without causing cell lysis.
4. **Fixation:** The swollen cells are lysed to release the chromosomes, which are then fixed using a methanol-acetic acid mixture to preserve their structure and prevent degradation. The fixation process hardens the chromosomes and prevents enzymatic degradation, ensuring that they remain intact during subsequent handling and analysis. The ratio of methanol to acetic acid and the duration of fixation must be optimized to preserve chromosome morphology.

Staining: The isolated chromosomes are stained with fluorescent dyes that bind specifically to DNA or particular chromosomal regions. This staining allows for the identification and differentiation of chromosomes based on their fluorescence characteristics.

1. **DNA Stains:** Common dyes include DAPI (4',6-diamidino-2-phenylindole), which binds to AT-rich regions, and propidium iodide, which intercalates into the DNA helix. These dyes provide strong fluorescence signals, facilitating the visualization of chromosomes under a fluorescence microscope. The concentration of the dye and the staining duration must be optimized to achieve uniform and specific staining.

2. **Fluorescent Probes:** Fluorescently labeled probes can target specific chromosomal regions or sequences (e.g., centromeres, telomeres) for more precise identification. These probes can be designed to hybridize to unique sequences within the genome, allowing for the identification of specific chromosomes or chromosomal regions. The design and synthesis of fluorescent probes require careful consideration of the target sequence and the fluorophore used to ensure specific and strong hybridization signals.

Flow Cytometry: The stained chromosomes are passed through a flow cytometer, which detects and sorts them based on their fluorescence and light-scattering properties.

1. **Detection:** The flow cytometer uses laser beams to excite the fluorescent dyes, and the emitted light is detected by sensors. The intensity and wavelength of the emitted fluorescence are used to distinguish between different chromosomes. The flow cytometer can analyze thousands of chromosomes per second, providing high-throughput and high-resolution analysis. The settings of the flow cytometer, including laser power, detector sensitivity, and flow rate, must be optimized to achieve accurate detection and sorting.
2. **Sorting:** Based on the detected signals, the flow cytometer sorts the chromosomes into distinct populations, separating them by size, shape, and fluorescence. This sorting process allows for the isolation of specific chromosomes with high purity. The sorted chromosomes can be collected into different containers for subsequent analysis. The efficiency and accuracy of sorting depend on the optimization of flow cytometry parameters and the quality of the initial chromosome preparation.

Collection: Sorted chromosomes are collected into separate tubes or multi-well plates for further analysis. The purity and accuracy of the sorted chromosomes are verified using techniques such as fluorescence microscopy or polymerase chain reaction (PCR).

1. **Verification:** The purity of the sorted chromosomes is verified by reanalyzing a small fraction of the sorted material using flow cytometry or fluorescence microscopy. The absence of contamination by other chromosomes is confirmed by visual inspection or quantitative analysis of fluorescence signals. The accuracy of sorting can also be verified by PCR amplification of chromosome-specific sequences, confirming the identity of the isolated chromosomes.
2. **Storage:** Sorted chromosomes can be stored at low temperatures to preserve their integrity for long-term analysis. Proper storage conditions, such as low temperature and protection from light, are essential to prevent degradation and loss of fluorescence signals. The sorted chromosomes can be used for various downstream applications, including sequencing, hybridization, and functional analysis.

Applications in Developing Molecular Maps and Gene Location

The isolation of specific chromosomes facilitates the development of high-resolution molecular maps and precise gene location, which are essential for understanding genetic architecture and improving crop varieties.

High-Resolution Molecular Maps: Isolated chromosomes enable the creation of detailed molecular maps, which provide a comprehensive overview of the genetic landscape of specific chromosomes.

These maps are crucial for identifying genetic markers and understanding the organization of genetic material.

1. **Marker Identification:** By analyzing isolated chromosomes, researchers can identify single nucleotide polymorphisms (SNPs), simple sequence repeats (SSRs), and other molecular markers. These markers are essential for constructing high-resolution genetic maps that detail the distribution and organization of genetic elements. High-resolution maps enable the precise localization of genetic markers, facilitating the identification of genes associated with important traits.
2. **Gene Mapping:** Molecular maps facilitate the precise mapping of genes associated with important traits. Researchers can locate genes on specific chromosomes and determine their position relative to other genetic elements, enhancing the understanding of gene function and regulation. This information is crucial for identifying candidate genes for breeding programs and for understanding the genetic basis of complex traits.

Gene Location and Characterization: The isolation of specific chromosomes allows for the precise location and characterization of genes, providing insights into their roles in trait expression and genetic inheritance.

1. **Positional Cloning:** Isolated chromosomes are used in positional cloning to identify and clone genes associated with specific traits. This technique involves mapping the gene to a precise chromosomal location and then isolating and characterizing the gene. Positional cloning provides a detailed understanding of gene structure, function, and regulation, enabling the development of targeted breeding strategies.
2. **Functional Analysis:** By locating genes on specific chromosomes, researchers can perform functional analyses to understand the role of these genes in biological processes and trait development. This information is critical for identifying candidate genes for crop improvement and for developing strategies to manipulate gene expression to enhance desirable traits.

Structural Variation Analysis: Chromosome isolation allows for the detailed analysis of structural variations, such as insertions, deletions, inversions, and translocations. Understanding these variations is essential for studying their impact on gene function and trait expression.

1. **Detection of Variations:** By isolating and analyzing specific chromosomes, researchers can detect and characterize structural variations that may affect gene function. These variations can have significant implications for trait expression and genetic stability. Detailed analysis of structural variations provides insights into the mechanisms of genetic diversity and evolution, facilitating the development of strategies to exploit beneficial variations in breeding programs.
2. **Implications for Breeding:** Knowledge of structural variations helps breeders develop strategies to minimize negative effects and exploit beneficial variations in breeding programs. By identifying and characterizing structural variations, breeders can select individuals with desirable genetic profiles, improving the efficiency and effectiveness of breeding programs.

Comparative Genomics: Isolating specific chromosomes enables comparative genomics studies, allowing researchers to compare the genetic content and organization of chromosomes across different species or varieties.

1. **Evolutionary Insights:** Comparative analysis of isolated chromosomes provides insights into the evolutionary relationships between species and the conservation of genetic elements. This information is valuable for understanding the genetic basis of traits and the evolutionary processes shaping genomes. Comparative genomics studies can identify conserved genetic regions that play crucial roles in trait development, providing targets for genetic manipulation and breeding programs.
2. **Transfer of Traits:** By comparing chromosomes from different species, researchers can identify and transfer beneficial traits from wild relatives or exotic germplasm into cultivated varieties, enhancing genetic diversity and crop performance. This approach leverages the genetic diversity of wild species to improve the resilience and productivity of cultivated crops.

Benefits of Chromosome Isolation

Precision and Accuracy: The isolation of specific chromosomes provides high precision and accuracy in genetic analysis. This precision is crucial for developing detailed molecular maps and accurately locating genes. The ability to isolate individual chromosomes reduces the complexity of genetic analyses, enabling the identification and characterization of specific genetic elements with high resolution.

Enhanced Genetic Studies: Isolated chromosomes enhance genetic studies by reducing the complexity of the genome, allowing for more focused and detailed analysis. This reduction in complexity is particularly beneficial for organisms with large and complex genomes. Isolated chromosomes provide a simplified and manageable system for studying genetic variation, facilitating the identification of genetic markers and the mapping of genes.

Improved Breeding Programs: The detailed genetic information obtained from isolated chromosomes aids in the development of improved breeding strategies. By identifying genetic markers and locating genes, breeders can make informed decisions to enhance crop performance and resilience. The integration of chromosome isolation with breeding programs accelerates the development of superior crop varieties, improving yield, quality, and stress tolerance.

Versatility: Chromosome isolation is a versatile technique applicable to a wide range of organisms, making it a valuable tool for genetic research and crop improvement across diverse species. The ability to isolate specific chromosomes from different organisms enables comparative genomics studies, facilitating the transfer of knowledge and genetic resources between species.

Chapter - 9

Role of Polyploidy in Crop Evolution and Breeding

A. Role of polyploidy in crop evolution and breeding

Introduction

Polyploidy, the condition of having more than two complete sets of chromosomes, is a fundamental aspect of plant biology that has profoundly influenced the evolution and breeding of many crops. This genetic phenomenon has contributed significantly to the diversification, adaptability, and improvement of numerous plant species, playing a pivotal role in the development of agriculture. Polyploidy can occur naturally or be induced artificially, and it is classified into two main types: autopolyploidy and allopolyploidy. The mechanisms, historical significance, and applications of polyploidy is essential for advancing modern agricultural practices and improving crop yields and quality.

Polyploidy is prevalent in the plant kingdom and is particularly common among angiosperms, or flowering plants. It is estimated that a significant proportion of flowering plants are polyploid, including many of the world's most important crops such as wheat, cotton, and potatoes. The additional sets of chromosomes in polyploids can arise through various mechanisms, including genome duplication within a species (autopolyploidy) or the combination of genomes from different species (allopolyploidy). These processes lead to increased genetic diversity, which can enhance the ability of plants to adapt to changing environmental conditions and resist pests and diseases. The study of polyploidy has a rich history, with early observations dating back to the 19th century. Researchers have long been fascinated by the occurrence of polyploidy and its implications for plant evolution and breeding. The first documented evidence of polyploidy was provided by botanists who observed variations in chromosome numbers among different plant species. These early studies laid the groundwork for understanding the role of polyploidy in creating new plant species and driving evolutionary change.

In the early 20th century, the concept of polyploidy was further elucidated by the work of several prominent scientists. In 1907, the German botanist Hugo de Vries introduced the term "polyploidy" to describe plants with multiple sets of chromosomes. His pioneering research demonstrated that polyploidy could result in the formation of new species with distinct morphological and genetic characteristics. Around the same time, botanists such as Winge and Winkler conducted cytological

DOI: 10.1201/9781003717775-9

studies that revealed the prevalence of polyploidy in various plant families, further highlighting its significance in plant evolution. One of the most notable early discoveries related to polyploidy was the identification of polyploid species in the genus Brassica, which includes economically important crops such as cabbage, broccoli, and mustard. In the 1930s, botanist Woo Jang-choon (also known as Nagaharu U) developed the "triangle of U" theory, which explained the relationships between the diploid and polyploid species of Brassica. This groundbreaking work demonstrated that polyploidy played a crucial role in the evolution and diversification of the Brassica species complex.

Polyploidy has also been a driving force in the evolution of many other important crop species. For example, wheat (Triticum spp.) is a well-known polyploid crop that has undergone multiple rounds of polyploidization throughout its evolutionary history. Modern bread wheat (Triticum aestivum) is a hexaploid species with three sets of homologous chromosomes derived from different ancestral species. This complex genomic structure has contributed to the adaptability and high yield potential of wheat, making it one of the world's most widely cultivated and consumed crops.

The significance of polyploidy extends beyond natural evolution to include its applications in plant breeding and crop improvement. Polyploidy can be induced artificially through various techniques, such as colchicine treatment, which disrupts chromosome segregation during cell division and results in genome doubling. Induced polyploidy has been used to create new plant varieties with desirable traits, such as increased size, improved nutritional content, and enhanced resistance to environmental stresses. In addition to its applications in creating new crop varieties, polyploidy has been instrumental in improving the quality and yield of existing crops. For example, the induction of polyploidy in certain fruit crops, such as strawberries and bananas, has led to the development of seedless varieties with enhanced fruit size and flavor. Similarly, polyploidy has been used to improve the fiber quality of cotton and the starch content of potatoes.

The impact of polyploidy on plant breeding is not limited to the creation of new varieties and the enhancement of specific traits. Polyploidy also facilitates the transfer of genetic material between species, a process known as introgression. This is particularly important for incorporating beneficial traits from wild relatives into cultivated crops. By combining the genomes of different species through allopolyploidy, breeders can introduce traits such as disease resistance and drought tolerance into commercial varieties, thereby improving their resilience and performance. The role of polyploidy in plant breeding is further exemplified by the development of hybrid crops. Hybrids are created by crossing two genetically distinct parent plants, often resulting in offspring with superior vigor and productivity, a phenomenon known as heterosis or hybrid vigor. Polyploid hybrids, in particular, benefit from the combined genetic resources of both parent species, leading to enhanced growth, yield, and stress resistance. The success of polyploid hybrids in agriculture underscores the importance of polyploidy as a tool for crop improvement.

Despite its many benefits, polyploidy also presents challenges in plant breeding and genetic research. The increased complexity of polyploid genomes can complicate genetic analysis and the identification of specific genes associated with desirable traits. Additionally, polyploid plants may exhibit reduced fertility due to irregular chromosome pairing and segregation during meiosis. These challenges necessitate the development of advanced breeding techniques and molecular tools to effectively utilize polyploidy in crop improvement programs.

Advancements in genomics and biotechnology have greatly facilitated the study and application of polyploidy in plant breeding. High-throughput sequencing technologies, such as next-generation sequencing (NGS), allow researchers to analyze polyploid genomes in unprecedented detail. These technologies enable the identification of genetic markers, quantitative trait loci (QTL), and candidate genes associated with important traits in polyploid crops. Additionally, genome editing tools, such as CRISPR/Cas9, offer new opportunities for manipulating polyploid genomes to enhance crop performance and resilience. The ongoing research into polyploidy and its applications in crop breeding continues to yield valuable insights and innovations. For instance, the study of polyploidy in soybean (Glycine max) has led to the identification of genes involved in nitrogen fixation and stress tolerance, providing targets for genetic improvement. Similarly, research on polyploidy in maize (Zea mays) has revealed the genetic basis of hybrid vigor, guiding the development of more productive and resilient hybrid varieties.

History of Polyploidy in Crop Evolution and Breeding

The concept of polyploidy and its significance in plant evolution and breeding has a rich and fascinating history. Over the past century, numerous researchers have explored and documented the prevalence and impact of polyploidy in plants. These efforts have laid the foundation for our current understanding of how polyploidy contributes to genetic diversity, adaptability, and crop improvement.

Early Observations and Discoveries

The early study of polyploidy can be traced back to the late 19th and early 20th centuries, a period when cytology, the study of cells and their structures, was rapidly advancing. One of the first notable observations of polyploidy was made by the botanist Hugo de Vries. In the early 1900s, de Vries observed that some plants exhibited variations in chromosome numbers, which he theorized could lead to the formation of new species. His pioneering work on chromosome doubling laid the groundwork for future studies on polyploidy.

Around the same time, Swedish botanist Hans Winkler, who is credited with coining the term "polyploidy," conducted studies on plant chromosomes. Winkler's research demonstrated that polyploidy was a common phenomenon in plants and could result in significant morphological changes. These early observations were crucial in establishing the concept of polyploidy as an important factor in plant evolution and diversity.

In 1907, botanist E. W. Sinnott provided further evidence of polyploidy in his studies on the evening primrose (*Oenothera spp.*). Sinnott's work showed that polyploidy could result in increased size and vigor of plant organs, a phenomenon that would later be recognized as a key advantage in crop breeding. These early 20th-century studies collectively highlighted the potential of polyploidy to drive evolutionary change and enhance plant traits.

The Role of Polyploidy in Speciation

One of the most significant contributions to the study of polyploidy came from the German botanist Hugo de Vries. In his seminal work, "Species and Varieties: Their Origin by Mutation," published in 1905, de Vries proposed that polyploidy could lead to the sudden formation of new species. He observed that chromosome doubling in plants could create reproductive barriers, preventing hybrid offspring from backcrossing with their parent species. This idea, known as the polyploid speciation hypothesis, was a revolutionary concept that challenged the gradualist view of species formation.

In the 1920s and 1930s, botanists Winge and Winkler further investigated the role of polyploidy in plant speciation. Their cytological studies on various plant families revealed that polyploidy was prevalent in many species and often accompanied by significant genetic and phenotypic changes. They demonstrated that polyploidy could lead to the rapid establishment of new species with distinct characteristics, thus contributing to the diversity of plant life. One of the most famous examples of polyploid speciation is the formation of the wheat genus (Triticum). Early cytological studies by Kihara and Sears in the 1930s and 1940s revealed that modern bread wheat (Triticum aestivum) is a hexaploid species, containing three sets of homologous chromosomes derived from different ancestral species. This complex genomic structure resulted from two rounds of polyploidization and hybridization events, highlighting the role of polyploidy in the evolutionary history of wheat.

The Triangle of U and Brassica Evolution

A significant breakthrough in understanding polyploidy in crop evolution came from the work of the Japanese botanist Woo Jang-choon, also known as Nagaharu U. In 1935, U proposed the "triangle of U" theory, which explained the relationships between the diploid and polyploid species within the Brassica genus. The triangle of U illustrated how three diploid species (Brassica rapa, Brassica oleracea, and Brassica nigra) hybridized and underwent polyploidization to form three allopolyploid species (*Brassica juncea, Brassica napus*, and *Brassica carinata*). U's theory provided a clear example of how polyploidy and hybridization could generate new species with unique genetic and phenotypic traits. The Brassica species are economically important crops, including vegetables such as cabbage, broccoli, and kale, as well as oilseed crops like rapeseed (canola). The triangle of U remains a classic example in plant genetics, demonstrating the power of polyploidy to drive crop evolution and diversification.

Cytogenetic Techniques and Advancements

The mid-20th century saw significant advancements in cytogenetic techniques, which allowed for more detailed studies of polyploidy. The development of chromosome banding techniques, such as Giemsa staining and fluorescence in situ hybridization (FISH), enabled researchers to visualize and differentiate chromosomes with greater precision. These techniques revealed the intricate structures and arrangements of chromosomes in polyploid plants, providing insights into their genetic organization. In the 1950s and 1960s, the work of geneticists such as Barbara McClintock and Norman Borlaug further advanced the understanding of polyploidy. McClintock's discovery of transposable elements in maize demonstrated how genetic elements could move within the genome, contributing to genetic diversity and evolution. Borlaug's work on wheat breeding and the development of high-yielding, disease-resistant wheat varieties highlighted the practical applications of polyploidy in crop improvement. The use of colchicine, a chemical that induces polyploidy by disrupting spindle fiber formation during cell division, became a standard technique for creating polyploid plants. Colchicine treatment allowed breeders to artificially induce polyploidy in various crops, leading to the development of new varieties with enhanced traits. This method was particularly successful in fruit crops, such as strawberries and bananas, where polyploidy resulted in larger fruit size and improved quality.

Molecular Techniques and Genomic Insights

The advent of molecular biology in the latter half of the 20th century revolutionized the study of polyploidy. The development of DNA sequencing technologies enabled researchers to analyze the genetic composition of polyploid plants in unprecedented detail. Early molecular studies focused on comparing the DNA sequences of diploid and polyploid species to understand the genetic changes associated with polyploidy. One of the landmark achievements in this field was the sequencing of the Arabidopsis thaliana genome in 2000. Arabidopsis, a model organism for plant genetics, provided a reference genome for studying polyploidy in related species. Comparative genomic studies revealed that many polyploid species had undergone genome duplication events, resulting in the retention and diversification of duplicated genes. The development of high-throughput sequencing technologies, such as next-generation sequencing (NGS), further accelerated the study of polyploidy. NGS allowed researchers to sequence entire genomes of polyploid plants, uncovering the complexity of their genetic architecture. These studies provided insights into the mechanisms of genome duplication, gene loss, and chromosomal rearrangements that accompany polyploidy.

Modern Applications of Polyploidy in Crop Breeding

In recent decades, the understanding of polyploidy has been harnessed for practical applications in crop breeding. Polyploidy has been used to develop new crop varieties with improved traits, such as increased yield, enhanced nutritional content, and greater resistance to pests and diseases.

Wheat Breeding: Wheat (Triticum spp.) is one of the most significant polyploid crops. Modern bread wheat (Triticum aestivum) is a hexaploid species that has benefited from multiple rounds of polyploidization and hybridization. The complex genome of wheat has been extensively studied, and breeders have used this knowledge to develop high-yielding, disease-resistant varieties. The introduction of dwarfing genes, identified through the study of wheat's polyploid genome, led to the Green Revolution, significantly increasing wheat production worldwide.

Cotton Improvement: Cotton (*Gossypium spp.*) is another important polyploid crop. Modern cotton varieties are primarily allopolyploids, resulting from the hybridization of diploid species followed by chromosome doubling. The polyploid nature of cotton has contributed to its fiber quality, yield, and adaptability. Advances in genomic sequencing have allowed researchers to identify genes associated with fiber traits, leading to the development of improved cotton varieties.

Fruit Crops: Many fruit crops have benefited from the induction of polyploidy. For example, triploid watermelons are seedless and have better texture and taste than their diploid counterparts. Similarly, polyploidy in bananas has led to the development of seedless varieties that are more palatable and commercially viable. The use of colchicine and other chemicals to induce polyploidy has been instrumental in these advancements.

Ornamental Plants: Polyploidy is also widely used in the breeding of ornamental plants to enhance flower size, color, and other aesthetic traits. For instance, polyploidy induction in chrysanthemums and lilies has resulted in varieties with larger and more vibrant flowers. These improvements have significant commercial value in the horticultural industry.

Stress Resistance: Polyploidy has been exploited to improve the stress resistance of crops. Polyploid plants often exhibit greater tolerance to environmental stresses such as drought, salinity, and cold.

For example, polyploidy in canola (*Brassica napus*) has been associated with improved tolerance to cold and frost, making it more suitable for cultivation in temperate regions.

Hybrid Crops: The development of hybrid crops has also benefited from the understanding of polyploidy. Hybrid varieties, which are created by crossing two genetically distinct parent plants, often exhibit hybrid vigor, resulting in superior growth, yield, and stress resistance. Polyploid hybrids, in particular, leverage the combined genetic resources of both parent species, leading to enhanced performance. Examples include hybrid maize (*Zea mays*) and hybrid rice (*Oryza sativa*), which have revolutionized agricultural production.

Challenges and Future Directions

While polyploidy offers numerous benefits for crop evolution and breeding, it also presents challenges. The increased complexity of polyploid genomes can complicate genetic analysis and breeding efforts. Identifying and manipulating specific genes in polyploid plants can be difficult due to the presence of multiple homologous gene copies.

Advances in genomic technologies and molecular tools are helping to address these challenges. Techniques such as CRISPR/Cas9 genome editing allow for precise modification of polyploid genomes, enabling the targeted manipulation of specific genes. High-throughput sequencing and bioinformatics tools are improving the ability to analyze polyploid genomes and identify genetic markers associated with important traits. The future of polyploidy research holds great promise for further advancing crop breeding and agricultural productivity. Continued exploration of polyploidy in lesser-studied crops and wild relatives can uncover new genetic resources for crop improvement.

B. Auto- and allopolyploids

Polyploidy, the condition of possessing more than two complete sets of chromosomes, can be classified into two main types: autopolyploidy and allopolyploidy. Both types have significant implications for plant evolution and breeding, contributing to genetic diversity, adaptation, and the development of new crop varieties. The differences between these types and their roles in crop improvement is essential for advancing agricultural practices.

1. Autopolyploidy

Autopolyploidy, a form of polyploidy, occurs when an organism has multiple sets of chromosomes that originate from a single species. This type of polyploidy results from genome duplication events within the same species, leading to increased genetic redundancy. Autopolyploids have identical or nearly identical sets of chromosomes, which can contribute to increased cell size, larger plant organs, and greater overall vigor. The role of autopolyploidy in plant evolution and breeding is significant, providing opportunities for enhanced genetic diversity, adaptation, and the development of superior crop varieties.

Definition and Mechanisms

Autopolyploidy is characterized by the presence of multiple chromosome sets derived from a single species. This phenomenon typically arises through genome duplication events that can occur naturally or be induced artificially. The resulting autopolyploids possess increased genetic material, leading to enhanced genetic variability and potential evolutionary advantages.

Mechanisms of Formation: Autopolyploidy can arise through various mechanisms, including errors in cell division, specifically nondisjunction during meiosis or mitosis. Nondisjunction refers to the failure of chromosomes to separate properly during cell division, resulting in the formation of gametes or somatic cells with doubled chromosome numbers. There are several key mechanisms through which autopolyploidy can occur:

1. **Nondisjunction during Meiosis:** Nondisjunction during meiosis can lead to the formation of unreduced gametes, which contain twice the normal number of chromosomes. When such unreduced gametes fuse with normal gametes during fertilization, the resulting zygote will be an autopolyploid. For example, if a diploid (2n) organism produces unreduced diploid gametes (2n) and these gametes fuse with haploid (n) gametes, the resulting offspring will be triploid (3n). If both gametes are unreduced diploids (2n), the offspring will be tetraploid (4n).
2. **Nondisjunction during Mitosis:** Nondisjunction can also occur during mitosis, leading to somatic cell polyploidy. This process results in cells with doubled chromosome numbers within the same organism. If a diploid (2n) cell undergoes chromosome doubling without subsequent cell division, it can produce a tetraploid (4n) cell. This type of mitotic nondisjunction can occur in developing tissues, leading to the formation of polyploid sectors within an otherwise diploid organism.
3. **Induced Polyploidy:** Autopolyploidy can be induced artificially using chemicals such as colchicine. Colchicine disrupts the formation of spindle fibers during cell division, preventing chromosome separation and resulting in genome duplication. This technique is widely used in plant breeding to create polyploid varieties with desirable traits.

Characteristics of Autopolyploids: Autopolyploids exhibit several distinctive characteristics due to their increased genetic material and redundancy. These characteristics can influence their growth, development, and overall performance:

1. **Increased Cell Size:** One of the most notable characteristics of autopolyploids is increased cell size. The presence of additional sets of chromosomes leads to larger nuclei and cell volumes. This increase in cell size can contribute to the overall robustness and vigor of the plant.
2. **Larger Plant Organs:** Autopolyploids often exhibit larger plant organs, such as leaves, flowers, and fruits. The increased genetic material allows for the expression of traits associated with enhanced growth and development. For example, autotetraploid plants may produce larger leaves and flowers compared to their diploid counterparts.
3. **Greater Vigor:** Autopolyploids tend to exhibit greater overall vigor, characterized by increased biomass production, improved stress tolerance, and enhanced reproductive success. The genetic redundancy in autopolyploids provides a buffering effect against deleterious mutations, allowing for the expression of advantageous traits and improved adaptability.

Examples in Crop Plants

Autopolyploidy has been observed in various crop plants, contributing to their genetic diversity and adaptation. The presence of multiple chromosome sets in these crops allows for the expression of a wide range of phenotypic traits and enhances their resilience to environmental stresses and diseases.

1. **Potato (*Solanum tuberosum*):** The cultivated potato is an autotetraploid species with four sets of homologous chromosomes. This polyploid nature contributes to the potato's genetic diversity, allowing for the expression of a wide range of phenotypic traits. The increased genetic redundancy in potatoes also provides a buffer against deleterious mutations, enhancing the plant's resilience to environmental stresses and diseases.

   » **Genetic Diversity:** The autotetraploid nature of potatoes allows for the accumulation of a diverse array of alleles within the genome. This genetic diversity provides a broad genetic base for selection, enabling the development of potato varieties with improved traits such as disease resistance, drought tolerance, and enhanced nutritional content.

   » **Resilience to Environmental Stresses:** The increased genetic redundancy in autotetraploid potatoes provides a buffering effect against environmental stresses. The presence of multiple copies of each gene allows for greater flexibility in gene expression, enhancing the plant's ability to adapt to changing environmental conditions. This resilience is particularly important for ensuring stable potato yields in diverse and challenging growing environments.

2. **Alfalfa (*Medicago sativa*):** Alfalfa is another example of an autotetraploid crop. The tetraploid nature of alfalfa contributes to its high forage yield and quality, making it an essential crop for livestock feed. The genetic redundancy in autotetraploid alfalfa allows for greater genetic variation and the potential for selecting superior varieties with improved traits.

   » **High Forage Yield:** The autotetraploid nature of alfalfa contributes to its high biomass production and forage yield. The increased cell size and vigor of autotetraploid alfalfa plants result in greater overall biomass accumulation, making alfalfa an important forage crop for livestock production.

   » **Improved Forage Quality:** The genetic diversity and redundancy in autotetraploid alfalfa allow for the selection of varieties with improved forage quality traits, such as increased protein content, better digestibility, and enhanced palatability. These improvements are crucial for optimizing livestock nutrition and productivity.

Applications in Breeding: Autopolyploidy has practical applications in plant breeding, particularly for developing new crop varieties with enhanced traits.

1. **Induction of Autopolyploidy:** Breeders can induce autopolyploidy artificially using chemicals like colchicine, which disrupts chromosome segregation during cell division. This technique is used to create polyploid varieties with larger fruits, flowers, and improved yield. For example, colchicine treatment has been used to produce tetraploid watermelon varieties with larger and seedless fruits.
2. **Genetic Variation and Selection:** Autopolyploidy increases genetic variation within a species, providing a broader genetic base for selection. Breeders can exploit this variation to select individuals with desirable traits, such as disease resistance, drought tolerance, and improved

nutritional content. The increased genetic diversity in autopolyploids also facilitates the development of hybrids with enhanced vigor and productivity.

Characteristics

Autopolyploids tend to exhibit increased cell size, larger plant organs, and greater overall vigor. These characteristics can result from the increased genetic material and the buffering effect of having multiple copies of each gene.

1. **Increased Cell Size:** One of the most notable characteristics of autopolyploids is increased cell size. The presence of additional sets of chromosomes leads to larger nuclei and cell volumes. This increase in cell size can contribute to the overall robustness and vigor of the plant. In agricultural crops, larger cell size can translate to more substantial plant structures, including thicker stems, broader leaves, and larger fruits, enhancing their commercial value and productivity.
2. **Larger Plant Organs:** Autopolyploids often exhibit larger plant organs, such as leaves, flowers, and fruits. The increased genetic material allows for the expression of traits associated with enhanced growth and development. For example, autotetraploid plants may produce larger leaves and flowers compared to their diploid counterparts. This characteristic is particularly advantageous in horticultural crops, where larger flowers and fruits are desirable traits for ornamental and edible purposes.
3. **Greater Vigor:** Autopolyploids tend to exhibit greater overall vigor, characterized by increased biomass production, improved stress tolerance, and enhanced reproductive success. The genetic redundancy in autopolyploids provides a buffering effect against deleterious mutations, allowing for the expression of advantageous traits and improved adaptability. This vigor can lead to higher yields and better performance under suboptimal growing conditions, making autopolyploid crops more resilient and productive in diverse agricultural environments.

2. Allopolyploidy

Allopolyploidy, a type of polyploidy, occurs when an organism has multiple sets of chromosomes originating from different species. This genetic phenomenon is particularly significant in plant evolution and breeding because it combines the genomes of different species, leading to increased genetic diversity and the potential for new traits. Allopolyploidy typically arises from hybridization events followed by chromosome doubling, which restores fertility and allows for stable inheritance of the combined genomes. The distinct sets of chromosomes from each parent species contribute to unique genetic combinations, enhancing the adaptability and performance of allopolyploid plants.

Definition and Mechanisms

Allopolyploidy is defined by the presence of multiple sets of chromosomes derived from different species. This type of polyploidy results from the hybridization of two distinct species, followed by chromosome doubling, which allows for the successful meiotic pairing and segregation of chromosomes. Allopolyploids possess a unique combination of genetic material from their parent species, leading to novel phenotypic traits and increased genetic variation.

Mechanisms of Formation: Allopolyploidy typically involves the hybridization of two diploid species, followed by chromosome doubling to restore fertility. The process can occur naturally or be induced artificially through various breeding techniques. The primary mechanisms through which allopolyploidy can occur include:

1. **Hybridization:** Hybridization between two diploid species results in the formation of a hybrid organism with one set of chromosomes from each parent species. For example, if species A (2n) hybridizes with species B (2n), the resulting hybrid (AB) will have a diploid set of chromosomes from each parent. However, this initial hybrid is often sterile due to irregular chromosome pairing during meiosis. The hybrid chromosomes cannot pair properly, leading to meiotic arrest and sterility.
2. **Chromosome Doubling:** Chromosome doubling can occur naturally through errors in cell division or be induced artificially using chemicals like colchicine. The doubling of chromosomes in the hybrid organism results in the formation of an allopolyploid with two complete sets of chromosomes from each parent species. For instance, if the hybrid (AB) undergoes chromosome doubling, it will become an allotetraploid (AABB), restoring fertility and allowing for normal meiotic division. The presence of homologous pairs of chromosomes enables proper segregation during meiosis, leading to viable gametes and successful reproduction.

Characteristics of Allopolyploids: Allopolyploids exhibit several distinctive characteristics due to the combination of different genomes and the presence of homeologous chromosomes. These characteristics can influence their growth, development, and overall performance:

1. **Increased Genetic Diversity:** The combination of different genomes in allopolyploids results in increased genetic diversity. This diversity provides a broader genetic base for selection and enhances the ability of allopolyploid organisms to adapt to various environmental conditions. The presence of homeologous chromosomes allows for genetic recombination, introducing new variations and facilitating the expression of advantageous traits. This genetic diversity is particularly valuable in breeding programs, as it increases the potential for selecting superior individuals with desirable characteristics.
2. **Heterosis (Hybrid Vigor):** Allopolyploids often exhibit heterosis, characterized by increased growth, vigor, and productivity compared to their parent species. The combination of genetic material from different species can lead to the expression of superior traits, such as enhanced yield, improved stress tolerance, and greater resistance to pests and diseases. This hybrid vigor is a significant advantage in crop breeding, contributing to the development of high-performing varieties. The phenomenon of heterosis is particularly pronounced in allopolyploids due to the combination of diverse genetic backgrounds, which can result in enhanced metabolic and physiological efficiency.
3. **Novel Traits:** The unique combination of genomes in allopolyploids can result in the expression of novel traits not present in either parent species. These novel traits can include improved nutritional content, enhanced flavor, and better adaptability to specific environmental conditions. The ability to combine desirable traits from different species makes allopolyploidy a powerful tool in plant breeding and crop improvement. The presence of

homeologous chromosomes allows for the introgression of beneficial alleles, creating new genetic combinations that can enhance crop performance and resilience.

Examples in Crop Plants

Allopolyploidy is common in many important crop species, contributing to their genetic complexity and agronomic performance. The presence of multiple sets of chromosomes from different species allows for the expression of a wide range of phenotypic traits and enhances their resilience to environmental stresses and diseases.

1. Wheat (*Triticum aestivum*): Modern bread wheat is a hexaploid species (AABBDD) resulting from two rounds of allopolyploidization. The first hybridization event involved the merging of diploid species *Triticum urartu* (AA) and a species related to *Aegilops speltoides* (BB), forming tetraploid wheat (AABB). A subsequent hybridization with *Aegilops tauschii* (DD) resulted in hexaploid wheat. This complex genomic structure has contributed to wheat's adaptability and high yield potential, making it one of the world's most important crops.

» **Adaptability and Yield Potential:** The allopolyploid nature of wheat provides a broad genetic base for adaptation to diverse environmental conditions. The presence of multiple sets of chromosomes allows for the expression of a wide range of traits, enhancing the plant's ability to thrive in different climates and soils. The genetic diversity within hexaploid wheat also contributes to its high yield potential, making it a staple crop for global food security.

» **Disease Resistance:** The combination of different genomes in wheat has contributed to its resistance to various diseases, such as rusts and blights. The presence of multiple sets of chromosomes allows for the introgression of resistance genes from different species, providing a robust defense against pathogens. This genetic resilience is critical for maintaining stable wheat production and ensuring food security.

2. Cotton (*Gossypium hirsutum*): Upland cotton is an allotetraploid species (AADD) derived from the hybridization of an African diploid species (*Gossypium arboreum*, AA) and an American diploid species (*Gossypium raimondii*, DD). The allopolyploid nature of cotton has led to the development of high-quality fiber traits and improved stress tolerance, making it a crucial crop for the textile industry.

» **Fiber Quality:** The allopolyploid nature of cotton contributes to the development of high-quality fibers, characterized by longer, stronger, and finer threads. These fiber traits are essential for producing high-quality textiles, making cotton one of the most valuable crops for the textile industry. The genetic diversity within allopolyploid cotton allows for the selection of superior fiber traits, enhancing the crop's commercial value.

» **Stress Tolerance:** The genetic complexity of allopolyploid cotton enhances its ability to tolerate various environmental stresses, such as drought, salinity, and pests. The presence of multiple sets of chromosomes provides a genetic buffer against stress conditions, allowing the plant to maintain productivity under suboptimal growing conditions. This resilience is critical for ensuring stable cotton yields and supporting the global textile industry.

Applications of Allopolyploidy in Plant Breeding

Allopolyploidy, which involves the combination of multiple sets of chromosomes from different species, has substantial implications for plant breeding. It enables the creation of new crop varieties with enhanced traits, such as improved yield, quality, and stress resistance. By combining beneficial characteristics from different species, allopolyploidy enhances genetic diversity and adaptability in crops.

Creation of Allopolyploid Hybrids

The creation of allopolyploid hybrids involves crossing different species and inducing chromosome doubling. This technique allows breeders to combine desirable traits from both parent species, resulting in new crop varieties with enhanced performance.

1. Hybridization and Chromosome Doubling: Breeders can create allopolyploid hybrids by crossing different species and inducing chromosome doubling. This approach combines the genetic material from both parent species, leading to hybrids with improved yield, quality, and stress resistance. For example, triticale (× *Triticosecale*), an allopolyploid hybrid between wheat (*Triticum*) and rye (*Secale*), combines the high yield and quality of wheat with the disease resistance and hardiness of rye.

» **Development of New Varieties:** The creation of allopolyploid hybrids enables breeders to develop new crop varieties with a combination of traits from different species. This approach allows for the introduction of desirable characteristics, such as improved yield, disease resistance, and environmental adaptability, into commercial crop varieties. The use of allopolyploid hybrids in breeding programs enhances the genetic diversity and performance of crops, contributing to agricultural sustainability and food security.

» **Enhancement of Crop Traits:** Allopolyploid hybrids often exhibit hybrid vigor, resulting in enhanced growth, productivity, and stress tolerance. The combination of genetic material from different species creates new genetic combinations that can improve crop performance and resilience. Breeders can exploit these traits to develop superior crop varieties that meet the demands of modern agriculture.

Genetic Diversity and Adaptation

Allopolyploids benefit from increased genetic diversity due to the combination of different genomes. This diversity enhances the ability of allopolyploid crops to adapt to various environmental conditions and resist pests and diseases.

2. Genetic Diversity and Adaptation: The genetic diversity within allopolyploid crops enhances their ability to adapt to diverse environmental conditions. The presence of multiple sets of chromosomes allows for the expression of a wide range of traits, providing a genetic buffer against environmental stresses.

» **Adaptation to Environmental Conditions:** The genetic diversity within allopolyploid crops enhances their ability to adapt to diverse environmental conditions. The presence of multiple sets of chromosomes allows for the expression of a wide range of traits, providing a genetic buffer against environmental stresses. This adaptability is particularly valuable in the context of climate change, as it enhances the resilience of crops to changing environmental conditions.

- **Resistance to Pests and Diseases:** The genetic complexity of allopolyploid crops contributes to their resistance to pests and diseases. The combination of different genomes allows for the introgression of resistance genes from various species, enhancing the plant's ability to defend against pathogens. This genetic resilience is critical for maintaining stable crop yields and ensuring food security.

Genomic Insights and Advances

Recent advances in genomic technologies have provided deeper insights into the complexities of allopolyploid genomes and their implications for crop breeding. High-throughput sequencing and genome editing tools have revolutionized the study and manipulation of allopolyploid genomes, offering new opportunities for crop improvement.

1. Sequencing Technologies: High-throughput sequencing technologies, such as next-generation sequencing (NGS), have enabled researchers to sequence the entire genomes of allopolyploid plants. These studies have revealed the intricate structures and arrangements of chromosomes in allopolyploid species, providing valuable information for breeding and genetic improvement.

- **Genome Sequencing:** The sequencing of allopolyploid genomes provides a comprehensive understanding of their genetic architecture. These studies reveal the presence of homeologous chromosomes, gene duplications, and structural variations, offering insights into the evolutionary history and genetic complexity of allopolyploid species. The detailed genetic information obtained from genome sequencing is invaluable for identifying genetic markers, quantitative trait loci (QTL), and candidate genes associated with important traits.
- **Comparative Genomics:** Comparative genomic studies between diploid and allopolyploid species reveal the impact of polyploidy on genome structure and function. These studies provide insights into the mechanisms of genome duplication, gene loss, and chromosomal rearrangements that accompany allopolyploidy. The knowledge gained from comparative genomics enhances our understanding of the evolutionary processes and genetic consequences of allopolyploidy.

2. Genome Editing: Genome editing tools, such as CRISPR/Cas9, offer new opportunities for manipulating allopolyploid genomes. These tools allow for precise modification of specific genes, enabling breeders to enhance desirable traits and improve crop performance. The ability to target and edit multiple homologous genes in allopolyploid plants opens new avenues for crop improvement.

- **Precision Breeding:** Genome editing tools enable the precise modification of specific genes within allopolyploid genomes, allowing for the targeted enhancement of desirable traits. For example, CRISPR/Cas9 can be used to knock out genes associated with undesirable traits or to introduce beneficial alleles that enhance crop performance. The precision and efficiency of genome editing make it a powerful tool for accelerating the breeding process and developing superior crop varieties.
- **Functional Genomics:** Genome editing facilitates the functional analysis of genes within allopolyploid genomes, providing insights into their roles in growth, development, and stress response. By selectively modifying specific genes, researchers can study their effects on plant phenotype and identify key regulatory pathways. The knowledge gained from

functional genomics informs breeding strategies and guides the development of crops with improved traits.

Practical Applications in Crop Breeding

Allopolyploidy plays a crucial role in practical plant breeding, particularly for the development of new crop varieties with enhanced traits. This section explores the specific applications and the challenges associated with allopolyploidy in crop improvement.

Development of Enhanced Crop Varieties

The creation of allopolyploid hybrids enables breeders to combine desirable traits from different species, resulting in the development of new crop varieties with improved yield, quality, and stress resistance.

1. Creation of High-Yielding Varieties: Allopolyploid hybrids often exhibit higher yields due to hybrid vigor. By combining the high yield potential of one parent species with the hardiness and disease resistance of another, breeders can develop crops that perform better under various environmental conditions.

- » **Triticale Development:** Triticale (× *Triticosecale*), a hybrid between wheat (*Triticum*) and rye (*Secale*), is a prime example of an allopolyploid crop developed for high yield and quality. Triticale combines the high yield and nutritional quality of wheat with the disease resistance and environmental adaptability of rye. This hybrid is particularly valuable in regions where wheat cultivation is challenging due to soil and climatic conditions.

2. Enhancing Crop Quality: Allopolyploid hybrids can improve the quality of crops, including nutritional content, flavor, and storage properties. By combining traits from different species, breeders can develop crops with superior quality attributes.

- » **Nutritional Enhancement:** Allopolyploid hybrids can be developed to enhance the nutritional content of crops. For example, hybridization between different species of rice (*Oryza*) has led to the development of varieties with higher levels of essential nutrients such as iron and zinc. These nutritionally enhanced varieties are crucial for addressing micronutrient deficiencies in regions where rice is a staple food.

Addressing Agricultural Challenges

Allopolyploid crops are valuable for addressing various agricultural challenges, including pest and disease resistance, environmental adaptability, and sustainability.

1. Pest and Disease Resistance: The genetic complexity of allopolyploid crops contributes to their resistance to pests and diseases. By combining resistance genes from different species, breeders can develop crops that are more resilient to biotic stresses.

- » **Disease-Resistant Varieties:** Allopolyploid hybrids often inherit disease resistance traits from their parent species, making them more resilient to pathogens. For example, hybridization between different species of Brassica has led to the development of varieties with improved resistance to clubroot, a soil-borne disease affecting many cruciferous crops. These disease-resistant varieties reduce the need for chemical pesticides and promote sustainable agriculture.

2. Environmental Adaptability: Allopolyploid crops are often more adaptable to diverse environmental conditions due to their genetic diversity. This adaptability is particularly valuable in the context of climate change, as it helps ensure stable crop production despite changing environmental conditions.

» **Climate-Resilient Crops:** The genetic diversity within allopolyploid crops enhances their resilience to environmental stresses such as drought, salinity, and extreme temperatures. For instance, hybridization between different species of cotton has led to the development of varieties that are more tolerant to drought and heat, ensuring stable cotton production in arid and semi-arid regions.

Challenges and Solutions in Allopolyploid Breeding

While allopolyploid breeding offers significant advantages, it also presents challenges that need to be addressed to achieve successful outcomes.

1. Overcoming Genetic Incompatibilities: Genetic incompatibilities between the parent species can result in hybrid inviability, sterility, or reduced fitness. Overcoming these incompatibilities requires advanced breeding techniques.

» **Embryo Rescue:** Embryo rescue is a technique used to overcome post-zygotic barriers by culturing hybrid embryos in vitro. This method allows the development of viable hybrid plants that would otherwise not survive. For example, embryo rescue has been used to produce hybrids between wheat and rye, resulting in the development of triticale.

2. Ensuring Genetic Stability: The stability of genetic traits in hybrid populations is crucial for the long-term success of allopolyploid crops. Chromosomal rearrangements and genetic segregation can impact the stability and performance of hybrids.

» **Chromosome Doubling:** Chromosome doubling using chemicals like colchicine is often employed to stabilize hybrids by creating polyploid offspring with balanced chromosome sets. This technique ensures normal meiotic division and successful reproduction, leading to stable and fertile hybrid populations.

C. Distant hybridization

Distant Hybridization

Distant hybridization, also known as wide hybridization or interspecific hybridization, refers to the crossing of genetically distinct species or genera to produce hybrid offspring. This technique has been extensively used in plant breeding to combine desirable traits from different species, enhance genetic diversity, and develop new crop varieties with improved characteristics. Distant hybridization often involves overcoming significant reproductive barriers, but it can result in hybrids with unique and valuable traits that are not present in either parent species.

Definition and Mechanisms

Distant hybridization involves the crossbreeding of two genetically distant species or genera to produce hybrid progeny. This process can be challenging due to pre- and post-zygotic reproductive barriers that prevent successful hybridization and seed development. However, with advanced techniques and careful manipulation, breeders can overcome these barriers and create viable hybrids.

Mechanisms of Distant Hybridization: Distant hybridization can occur naturally or be induced artificially through various techniques. The process involves several key steps, including pollination, fertilization, embryo rescue, and chromosome doubling.

1. **Pollination and Fertilization:** The first step in distant hybridization is the transfer of pollen from the male parent (donor species) to the female parent (recipient species). This step requires careful selection of compatible species and precise timing to ensure successful pollination. In some cases, manual pollination techniques, such as hand pollination or artificial pollination, are used to facilitate crossbreeding. Successful fertilization results in the formation of hybrid embryos.
2. **Embryo Rescue:** Due to post-zygotic barriers, hybrid embryos may fail to develop properly or may abort at an early stage. To overcome this challenge, embryo rescue techniques are employed. This involves the extraction of hybrid embryos at an early developmental stage and their cultivation in artificial nutrient media under controlled conditions. Embryo rescue allows for the recovery and growth of viable hybrid plants that would otherwise not survive.
3. **Chromosome Doubling:** Hybrids resulting from distant hybridization often exhibit reduced fertility due to irregular chromosome pairing during meiosis. To restore fertility and produce stable hybrids, chromosome doubling is induced using chemicals like colchicine. Chromosome doubling results in the formation of polyploid hybrids with balanced chromosome sets, enabling normal meiotic division and successful reproduction.

Characteristics of Distant Hybrids

Distant hybrids exhibit a unique combination of traits from both parent species. These hybrids often show increased genetic diversity, novel phenotypic traits, and enhanced vigor. The characteristics of distant hybrids can vary widely depending on the genetic distance between the parent species and the specific traits being combined.

1. **Increased Genetic Diversity:** Distant hybrids possess a broader genetic base compared to hybrids derived from closely related species. This increased genetic diversity provides a wider range of phenotypic variation and enhances the potential for selecting superior individuals with desirable traits. The introduction of new genetic material from distant species can also introduce novel alleles and genetic combinations that are not present in the parent species.
2. **Novel Phenotypic Traits:** Distant hybrids often exhibit unique phenotypic traits that are not found in either parent species. These traits can include improved stress tolerance, enhanced nutritional content, novel flavors, and unique morphological characteristics. The combination of genetic material from different species can result in the expression of new and desirable traits that are valuable for crop improvement.
3. **Hybrid Vigor (Heterosis):** Distant hybrids frequently exhibit hybrid vigor, characterized by increased growth, vigor, and productivity compared to their parent species. The combination of diverse genetic backgrounds can lead to enhanced metabolic efficiency, improved stress tolerance, and greater resistance to pests and diseases. This hybrid vigor is a significant advantage

in breeding programs, contributing to the development of high-performing crop varieties.

Examples in Crop Plants

Distant hybridization has been successfully used in various crop plants to develop new varieties with improved traits. The following examples highlight the applications and benefits of distant hybridization in crop improvement.

1. Triticale (× *Triticosecale*): Triticale is an allopolyploid hybrid derived from the distant hybridization of wheat (*Triticum*) and rye (*Secale*). This hybrid combines the high yield and quality of wheat with the disease resistance and hardiness of rye.

» **Improved Yield and Quality:** Triticale exhibits high yield potential and good grain quality, making it a valuable crop for food and feed production. The hybrid nature of triticale allows for the expression of desirable traits from both parent species, enhancing its agronomic performance.

» **Disease Resistance and Hardiness:** The introduction of rye's genetic material into triticale provides increased resistance to diseases such as rusts and blights. Additionally, triticale exhibits greater tolerance to environmental stresses, such as drought and cold, compared to wheat, making it suitable for cultivation in diverse and challenging environments.

2. Strawberries (*Fragaria × ananassa*): The cultivated strawberry is a hybrid species resulting from the distant hybridization of two wild species, *Fragaria virginiana* and *Fragaria chiloensis*. This hybridization event led to the development of the modern cultivated strawberry with improved fruit size, flavor, and yield.

» **Enhanced Fruit Characteristics:** The hybridization of *F. virginiana* and *F. chiloensis* resulted in strawberries with larger fruit size, improved flavor, and better texture compared to their wild ancestors. These enhanced fruit characteristics have made strawberries a popular and commercially valuable fruit crop.

» **Increased Yield:** The hybrid nature of cultivated strawberries contributes to their high yield potential. The combination of genetic material from different species allows for the expression of traits associated with increased fruit production, making strawberries a profitable crop for growers.

3. Broccoli (*Brassica oleracea* var. *italica*): Broccoli is a member of the Brassica genus and is a result of distant hybridization events within the Brassica species complex. The development of modern broccoli involved the hybridization of different Brassica species, leading to the selection of desirable traits such as improved head size and nutritional content.

» **Improved Head Size and Quality:** The hybridization of different Brassica species contributed to the development of broccoli with larger and more uniform heads. These traits are important for commercial production and marketability, enhancing the economic value of broccoli as a vegetable crop.

» **Enhanced Nutritional Content:** Distant hybridization in broccoli has also led to improvements in nutritional content, including higher levels of vitamins and antioxidants. These nutritional enhancements make broccoli a valuable crop for promoting human health and nutrition.

Applications of Distant Hybridization in Plant Breeding

Distant hybridization, or the crossing of genetically distant species or genera, has been a powerful tool in plant breeding, allowing for the creation of new crop varieties with enhanced traits. This technique enables the combination of beneficial characteristics from different species, thereby increasing genetic diversity and improving the adaptability and performance of crops. The following sections explore the significant implications of distant hybridization in plant breeding, highlighting its role in creating new varieties, enhancing crop traits, and increasing genetic diversity and adaptability.

Hybridization and Chromosome Doubling

One of the primary applications of distant hybridization is the development of new crop varieties through hybridization and chromosome doubling. By crossing different species and inducing chromosome doubling, breeders can create hybrids that combine the desirable traits of both parent species. This process can lead to significant improvements in yield, quality, and stress resistance.

Development of New Varieties:

The creation of distant hybrids through hybridization and chromosome doubling allows breeders to develop new crop varieties that combine the traits of different species. This approach enables the introduction of desirable characteristics, such as improved yield, disease resistance, and environmental adaptability, into commercial crop varieties.

- **Enhanced Yield:** One of the primary goals of distant hybridization is to improve crop yield. By combining the genetic material of different species, breeders can create hybrids that exhibit higher productivity. For instance, triticale (× *Triticosecale*), a hybrid between wheat (*Triticum*) and rye (*Secale*), combines the high yield potential of wheat with the hardiness and disease resistance of rye. This hybrid has become an important crop in regions where traditional wheat cultivation is challenging due to environmental conditions.
- **Disease Resistance:** Distant hybrids often inherit disease resistance genes from their parent species, making them more resilient to various pathogens. This genetic resistance is crucial for maintaining healthy crops and reducing the reliance on chemical pesticides. For example, hybridization between different species of Brassica has led to the development of varieties with improved resistance to clubroot, a soil-borne disease that affects many cruciferous crops.
- **Environmental Adaptability:** The ability to adapt to different environmental conditions is another significant advantage of distant hybrids. By combining the genetic traits of species adapted to diverse climates and soils, breeders can develop crops that thrive in a range of environments. This adaptability is particularly valuable in the context of climate change, as it helps ensure stable crop production despite changing environmental conditions.

Enhancement of Crop Traits:

Distant hybrids often exhibit hybrid vigor, or heterosis, which results in enhanced growth, productivity, and stress tolerance. This hybrid vigor is a result of the genetic diversity introduced by combining different species, leading to new genetic combinations that can significantly improve crop performance and resilience.

3. **Improved Growth and Productivity:** Hybrid vigor often results in faster growth rates and higher biomass production. This increased productivity is beneficial for both food crops and forage crops. For instance, hybrids of sorghum (*Sorghum bicolor*) and Sudan grass (*Sorghum sudanense*) are widely grown as forage crops due to their rapid growth and high biomass yield, providing valuable feed for livestock.

4. **Stress Tolerance:** Distant hybrids frequently show enhanced tolerance to various abiotic stresses, such as drought, salinity, and extreme temperatures. This stress tolerance is critical for maintaining crop yields in challenging growing conditions. For example, hybridization between different species of cotton has led to the development of varieties that are more tolerant to drought and heat, ensuring stable cotton production in arid and semi-arid regions.

5. **Nutritional Quality:** Distant hybridization can also improve the nutritional quality of crops by combining traits related to nutrient content from different species. For instance, hybrids between different species of rice (*Oryza*) have been developed to enhance the levels of essential nutrients, such as iron and zinc, addressing nutritional deficiencies in regions where rice is a staple food.

Genetic Diversity and Adaptation

Distant hybrids benefit from increased genetic diversity due to the combination of different genomes. This genetic diversity enhances the ability of distant hybrid crops to adapt to various environmental conditions and resist pests and diseases. The following sections explore the benefits of genetic diversity and its role in crop adaptation and pest and disease resistance.

Adaptation to Environmental Conditions:

The genetic diversity within distant hybrid crops enhances their ability to adapt to diverse environmental conditions. The presence of multiple sets of chromosomes allows for the expression of a wide range of traits, providing a genetic buffer against environmental stresses.

- » **Environmental Resilience:** The increased genetic diversity in distant hybrids contributes to their resilience to environmental stresses. For instance, hybrid maize varieties developed through distant hybridization have shown improved tolerance to drought, enabling farmers to maintain yields even during periods of water scarcity. Similarly, hybrid rice varieties with enhanced tolerance to salinity have been developed to ensure stable rice production in coastal areas affected by soil salinization.
- » **Climate Change Adaptation:** The adaptability of distant hybrids is particularly valuable in the context of climate change. As global temperatures rise and weather patterns become more unpredictable, the ability of crops to withstand extreme conditions is crucial for food security. Distant hybrids with enhanced tolerance to heat, cold, and other climatic extremes are essential for sustaining agricultural productivity in the face of climate change.
- » **Soil Adaptation:** Distant hybrids can also exhibit improved adaptation to different soil types and conditions. For example, hybridization between different species of legumes has led to the development of varieties that can thrive in soils with varying pH levels and nutrient availability. This adaptability is important for expanding crop cultivation to marginal lands and improving soil health.

Resistance to Pests and Diseases:

The genetic complexity of distant hybrid crops contributes to their resistance to pests and diseases. The combination of different genomes allows for the introgression of resistance genes from various species, enhancing the plant's ability to defend against pathogens and pests.

» **Pest Resistance:** Distant hybridization can introduce pest resistance traits from wild or less commonly cultivated species into commercial crops. For instance, hybridization between cultivated tomato (*Solanum lycopersicum*) and its wild relatives has led to the development of tomato varieties with improved resistance to insect pests such as whiteflies and aphids. These resistant varieties reduce the need for chemical pesticides, promoting more sustainable agricultural practices.

» **Disease Resistance:** The introduction of disease-resistance genes through distant hybridization is a critical strategy for managing crop diseases. For example, hybridization between different species of wheat has resulted in varieties with improved resistance to rust diseases, which are major threats to global wheat production. The incorporation of multiple resistance genes from different species provides a durable defense against evolving pathogen populations.

» **Pathogen Diversity Management:** Distant hybrids also help manage the diversity of pathogens by creating a genetic mosaic within crop populations. This genetic mosaic can slow the spread of diseases and reduce the likelihood of epidemics. For example, hybrid sorghum varieties with diverse genetic backgrounds exhibit a range of resistance traits, making it difficult for pathogens to adapt and overcome the plant's defenses.

Genomic Insights and Advances

Recent advances in genomic technologies have provided deeper insights into the complexities of distant hybrid genomes and their implications for crop breeding. High-throughput sequencing and genome editing tools have revolutionized the study and manipulation of distant hybrid genomes, offering new opportunities for crop improvement.

Sequencing Technologies:

High-throughput sequencing technologies, such as next-generation sequencing (NGS), have enabled researchers to sequence the entire genomes of distant hybrid plants. These studies have revealed the intricate structures and arrangements of chromosomes in distant hybrid species, providing valuable information for breeding and genetic improvement.

» **Genome Sequencing:** The sequencing of distant hybrid genomes provides a comprehensive understanding of their genetic architecture. These studies reveal the presence of homeologous chromosomes, gene duplications, and structural variations, offering insights into the evolutionary history and genetic complexity of distant hybrid species. The detailed genetic information obtained from genome sequencing is invaluable for identifying genetic markers, quantitative trait loci (QTL), and candidate genes associated with important traits.

» **Comparative Genomics:** Comparative genomic studies between parent species and distant hybrids reveal the impact of hybridization on genome structure and function. These studies provide insights into the mechanisms of genome duplication, gene loss, and chromosomal rearrangements that accompany distant hybridization. The knowledge gained from comparative genomics enhances our understanding of the evolutionary processes and genetic consequences of distant hybridization.

Genome Editing:

Genome editing tools, such as CRISPR/Cas9, offer new opportunities for manipulating distant hybrid genomes. These tools allow for precise modification of specific genes, enabling breeders to enhance desirable traits and improve crop performance. The ability to target and edit multiple homologous genes in distant hybrid plants opens new avenues for crop improvement.

» **Precision Breeding:** Genome editing tools enable the precise modification of specific genes within distant hybrid genomes, allowing for the targeted enhancement of desirable traits. For example, CRISPR/Cas9 can be used to knock out genes associated with undesirable traits or to introduce beneficial alleles that enhance crop performance. The precision and efficiency of genome editing make it a powerful tool for accelerating the breeding process and developing superior crop varieties.

» **Functional Genomics:** Genome editing facilitates the functional analysis of genes within distant hybrid genomes, providing insights into their roles in growth, development, and stress response. By selectively modifying specific genes, researchers can study their effects on plant phenotype and identify key regulatory pathways. The knowledge gained from functional genomics informs breeding strategies and guides the development of crops with improved traits.

D. Barriers to interspecific and intergeneric hybridization

Interspecific and intergeneric hybridization involves the crossing of different species or genera to produce hybrid offspring. These techniques hold great potential for introducing beneficial traits and enhancing genetic diversity, which is crucial for the development of new crop varieties with improved characteristics. However, these processes are often hindered by significant reproductive barriers. These barriers can be classified into pre-zygotic and post-zygotic barriers, each presenting unique challenges that must be overcome to achieve successful hybridization. These barriers and the techniques to overcome them are essential for advancing plant breeding programs and achieving agricultural sustainability.

Pre-zygotic Barriers

Pre-zygotic barriers prevent fertilization and the formation of hybrid embryos. These barriers operate before the zygote is formed and include a range of mechanisms that impede successful mating or fertilization between different species or genera.

1. Temporal Isolation: Temporal isolation occurs when species or genera have different mating or flowering periods, preventing them from interbreeding. This difference in timing can be due to seasonal variations or daily activity patterns. When the reproductive periods do not overlap, it becomes impossible for these species to interbreed.

Example: Different species of oak trees (*Quercus* spp.) may flower at different times of the year, preventing cross-pollination between them. For instance, *Quercus robur* and *Quercus petraea* have different flowering periods, which reduces the chances of hybridization between these two species. This temporal separation ensures that each species remains reproductively isolated, maintaining their genetic integrity.

2. Spatial Isolation: Spatial isolation involves the physical separation of species or genera, preventing them from coming into contact. This can occur due to geographical barriers such as mountains, rivers, or distance, or it can result from habitat specialization, where species occupy different ecological niches. Spatial isolation ensures that even if the species are capable of interbreeding, they do not encounter each other to do so.

Example: Wild and cultivated rice species are often geographically separated, preventing natural hybridization. Wild rice (*Oryza rufipogon*) and cultivated rice (*Oryza sativa*) grow in different habitats or regions, reducing the chances of cross-pollination. This geographical separation helps in maintaining the distinct characteristics of wild and cultivated rice species.

3. Mechanical Isolation: Mechanical isolation refers to differences in the reproductive structures of species or genera that prevent successful mating or fertilization. This can include differences in flower morphology, size, or the structure of reproductive organs, making it difficult for pollen to reach the ovules. The physical incompatibility of reproductive organs acts as a barrier to hybridization.

Example: In some flowering plants, the shape and size of the flowers may be adapted to specific pollinators, preventing cross-pollination with other species that have different floral structures. For instance, the floral morphology of certain orchids is specialized for specific pollinators, such as bees or butterflies, which limits hybridization with other orchid species that attract different pollinators.

4. Behavioral Isolation: Behavioral isolation involves differences in mating behaviors or signals that prevent interbreeding between species or genera. This can include differences in pollinator preferences, mating rituals, or chemical cues that attract mates. Behavioral differences ensure that even if species come into contact, they do not recognize each other as potential mates.

Example: In flowering plants, different species may attract different pollinators due to variations in flower color, scent, or nectar composition. For example, red flowers may attract hummingbirds, while white flowers may attract moths, leading to reproductive isolation between plant species. This selective attraction of pollinators ensures that cross-pollination does not occur between species with different pollination strategies.

5. Gametic Isolation: Gametic isolation occurs when the gametes (sperm and egg) of different species or genera are incompatible and cannot fuse to form a zygote. This can result from differences in the biochemical properties of the gametes or the presence of species-specific recognition molecules. Gametic incompatibility ensures that even if the gametes come into contact, fertilization does not occur.

Example: In many flowering plants, the pollen tube may fail to grow or penetrate the ovule of a different species, preventing fertilization. This gametic incompatibility ensures that hybridization does not occur even if pollen from one species reaches the stigma of another species. For example, pollen from *Brassica oleracea* (cabbage) may not be able to fertilize the ovules of *Brassica rapa* (turnip) due to gametic incompatibility.

Post-zygotic Barriers

Post-zygotic barriers operate after fertilization has occurred and prevent the development, survival, or reproduction of hybrid offspring. These barriers can result in the inviability, sterility, or reduced fitness of hybrids, limiting the success of interspecific and intergeneric hybridization.

1. Hybrid Inviability: Hybrid inviability refers to the failure of hybrid embryos to develop properly or survive to reproductive maturity. This can result from genetic incompatibilities that disrupt normal development, leading to embryo abortion or early death of hybrid individuals. Hybrid inviability prevents the establishment of hybrid populations.

Example: In some interspecific crosses between species of the genus *Nicotiana* (tobacco), hybrid embryos may fail to develop due to genetic imbalances, leading to inviable offspring. These genetic incompatibilities can prevent the formation of viable hybrids even if fertilization occurs. For instance, crosses between *Nicotiana tabacum* and *Nicotiana glauca* often result in hybrid embryos that do not survive.

2. Hybrid Sterility: Hybrid sterility occurs when hybrid individuals are unable to produce functional gametes, preventing them from reproducing. This sterility often results from chromosomal mismatches that disrupt meiotic division, leading to the production of nonviable or unbalanced gametes. Hybrid sterility ensures that even if hybrids reach maturity, they cannot contribute to the gene pool.

Example: Mules, which are hybrids between horses (*Equus ferus caballus*) and donkeys (*Equus africanus asinus*), are sterile due to differing chromosome numbers of the parent species. Horses have 64 chromosomes, while donkeys have 62, resulting in mules with 63 chromosomes. This chromosomal mismatch prevents proper pairing during meiosis, leading to sterility. As a result, mules cannot produce offspring.

3. Hybrid Breakdown: Hybrid breakdown occurs when the first-generation hybrids are viable and fertile, but their offspring (second generation) exhibit reduced fitness, sterility, or inviability. This can result from the segregation of incompatible genetic combinations in subsequent generations. Hybrid breakdown limits the long-term success of hybridization.

Example: In some crosses between different species of sunflowers (*Helianthus*), the first-generation hybrids may be vigorous and fertile, but the second-generation hybrids often show reduced vigor, sterility, or developmental abnormalities. This breakdown of hybrid fitness prevents the establishment of stable hybrid populations over multiple generations.

4. Reduced Hybrid Fitness: Reduced hybrid fitness refers to the lower survival or reproductive success of hybrids compared to their parent species. This can result from ecological, physiological, or behavioral disadvantages that reduce the competitiveness of hybrids in their environment. Reduced hybrid fitness prevents hybrids from thriving in natural environments.

Example: In hybrid zones where closely related species of toads (*Bufo* spp.) interbreed, the hybrids may have reduced fitness due to intermediate phenotypes that are less well-adapted to specific ecological niches. This reduced fitness can prevent the establishment and spread of hybrid populations. For instance, hybrids between *Bufo americanus* and *Bufo fowleri* may exhibit lower survival rates in their respective habitats.

Overcoming Hybridization Barriers

Despite the challenges posed by pre-zygotic and post-zygotic barriers, various techniques have been developed to overcome these obstacles and achieve successful interspecific and intergeneric hybridization. These techniques include embryo rescue, somatic hybridization, chromosome doubling, and genetic engineering.

1. Embryo Rescue: Embryo rescue involves the extraction of hybrid embryos at an early developmental stage and their cultivation in artificial nutrient media. This technique bypasses the barriers to normal embryo development, allowing for the growth of viable hybrid plants. Embryo rescue is particularly useful in overcoming post-zygotic barriers such as hybrid inviability.

Example: In crosses between wheat (*Triticum aestivum*) and rye (*Secale cereale*), embryo rescue has been used to recover hybrid embryos that would otherwise abort. The rescued embryos are cultured in vitro, resulting in the development of triticale, a valuable hybrid crop that combines the high yield and quality of wheat with the disease resistance and hardiness of rye.

2. Somatic Hybridization: Somatic hybridization involves the fusion of protoplasts (cells without cell walls) from different species or genera to create hybrid cells. These hybrid cells can be regenerated into whole plants, bypassing the reproductive barriers associated with sexual hybridization. Somatic hybridization allows for the combination of genetic material from species that are otherwise reproductively isolated.

Example: Somatic hybridization has been used to create hybrids between potato (*Solanum tuberosum*) and tomato (*Solanum lycopersicum*), combining desirable traits from both species. The resulting somatic hybrids exhibit a range of novel traits, including improved disease resistance and stress tolerance. These hybrids are valuable for breeding programs aimed at improving the performance of both crops.

3. Chromosome Doubling: Chromosome doubling using chemicals like colchicine is employed to restore fertility in hybrids by producing polyploid offspring with balanced chromosome sets. This technique is particularly useful for overcoming hybrid sterility. Chromosome doubling allows hybrids to undergo normal meiotic division and produce viable gametes.

Example: In the creation of triticale, chromosome doubling is used to convert sterile hybrids between wheat and rye into fertile allotetraploids. The resulting triticale plants combine the desirable traits of both parent species and are capable of normal reproduction. Chromosome doubling ensures that triticale can be cultivated and propagated as a stable crop.

4. Genetic Engineering: Genetic engineering techniques, such as CRISPR/Cas9 genome editing, allow for the precise manipulation of specific genes to overcome hybridization barriers. These tools can be used to introduce or modify genes associated with fertility, development, and stress resistance. Genetic engineering offers a powerful approach to overcoming both pre-zygotic and post-zygotic barriers.

Example: Genome editing has been used to modify genes involved in pollen-pistil interactions, enhancing the compatibility of interspecific crosses in crops like maize (*Zea mays*) and sorghum (*Sorghum bicolor*). These modifications improve the success rate of hybridization and facilitate the development of new hybrid varieties. For instance, CRISPR/Cas9 can be used to knock out genes that cause incompatibility, allowing for successful fertilization and hybrid development.

E. Behaviour of interspecific and intergeneric crosses

Interspecific and intergeneric crosses involve the hybridization of different species or genera, respectively, to produce hybrid offspring. These types of crosses are significant in plant breeding as they allow the introduction of beneficial traits and the enhancement of genetic diversity. However, the behavior of these crosses is influenced by various biological and genetic factors that determine the success and stability of the hybrids.

Genetic Compatibility

The genetic compatibility between the parent species or genera plays a crucial role in the success of interspecific and intergeneric crosses. Compatibility is determined by the similarity of the genetic material and the ability of the chromosomes to pair and segregate properly during meiosis.

1. Chromosome Pairing: In successful interspecific and intergeneric crosses, chromosomes from the different species or genera must be able to pair and segregate properly during meiosis. This pairing is essential for the formation of viable gametes and the production of fertile offspring.

» **Homologous Pairing:** In some cases, the chromosomes of the parent species are sufficiently similar to pair correctly during meiosis. This homologous pairing leads to the production of viable and fertile hybrids. For example, in the case of triticale (× *Triticosecale*), the chromosomes of wheat (*Triticum*) and rye (*Secale*) are able to pair properly, resulting in fertile hybrids. Homologous pairing ensures that the genetic material from both parent species can be successfully inherited by the hybrid offspring, leading to stable and fertile populations.

» **Homeologous Pairing:** In other cases, the chromosomes are only partially homologous, leading to homeologous pairing. This can result in reduced fertility or sterility of the hybrids due to improper chromosome segregation. Homeologous pairing is often observed in distant hybridization, where the genetic divergence between the parent species is greater. The partial homology of the chromosomes can lead to irregularities during meiosis, resulting in unbalanced gametes and reduced fertility in the hybrids.

2. Chromosome Number: The difference in chromosome numbers between the parent species or genera can affect the success of hybridization. Discrepancies in chromosome numbers can lead to unbalanced gametes and reduced fertility.

» **Polyploidy:** Polyploidy, or the presence of multiple sets of chromosomes, can help overcome issues related to chromosome number differences. Chromosome doubling can restore fertility by creating balanced chromosome sets. For example, in the creation of triticale, chromosome doubling is used to convert sterile hybrids between wheat and rye into fertile allotetraploids. Polyploid hybrids often exhibit greater genetic stability and fertility compared to diploid hybrids, making them valuable in breeding programs.

» **Chromosome Doubling:** Artificial induction of chromosome doubling using chemicals like colchicine is a common technique to overcome sterility in hybrids. By doubling the chromosome number, breeders can produce polyploid hybrids with balanced chromosome sets, enabling normal meiotic division and successful reproduction. Chromosome doubling is a crucial step in stabilizing hybrids and ensuring their fertility and viability.

Hybrid Vigor and Phenotypic Expression

Hybrids resulting from interspecific and intergeneric crosses often exhibit hybrid vigor, also known as heterosis. This phenomenon is characterized by enhanced growth, vigor, and productivity compared to the parent species. The expression of phenotypic traits in hybrids can vary widely depending on the genetic makeup of the parent species.

1. Hybrid Vigor (Heterosis): Hybrid vigor is a common outcome of interspecific and intergeneric crosses, leading to improved performance of the hybrids. This enhanced vigor results from the combination of diverse genetic material, which can lead to superior traits.

- **Growth and Biomass:** Hybrids often show increased growth rates and higher biomass production compared to their parent species. This trait is particularly valuable for crops intended for food, feed, or biomass production. For instance, hybrids of sorghum (*Sorghum bicolor*) and Sudan grass (*Sorghum sudanense*) are widely grown for their rapid growth and high biomass yield. The increased growth and biomass production of these hybrids make them suitable for various agricultural applications, including forage and bioenergy production.
- **Yield and Productivity:** Hybrid vigor can result in higher yield and productivity in hybrid crops. The combination of beneficial traits from both parent species can lead to increased crop yield and better quality produce. For example, hybrid maize varieties developed through interspecific crosses exhibit higher yields and improved resistance to environmental stresses. The enhanced yield and productivity of these hybrids contribute to increased food production and agricultural sustainability.

2. Phenotypic Expression: The phenotypic traits expressed in hybrids can vary depending on the genetic background of the parent species. The combination of genetic material from different species can lead to novel phenotypic traits that are not present in either parent.

- **Novel Traits:** Hybrids often exhibit unique traits that result from the combination of genetic material from different species. These novel traits can include improved nutritional content, enhanced flavor, and better adaptability to specific environmental conditions. For example, hybrids between different species of rice (*Oryza*) have been developed to enhance the levels of essential nutrients, such as iron and zinc. The expression of novel traits in hybrids provides new opportunities for improving crop quality and nutritional value.
- **Intermediate Traits:** In some cases, hybrids may exhibit intermediate traits that are a blend of the characteristics of both parent species. This blending of traits can result in hybrids with desirable attributes that are advantageous for specific agricultural applications. For example, hybrids between different species of Brassica may exhibit a combination of traits such as disease resistance from one parent and high yield from the other parent. The expression of intermediate traits in hybrids allows breeders to combine the best attributes of both parent species.

Fertility and Reproductive Success

The fertility and reproductive success of hybrids resulting from interspecific and intergeneric crosses are influenced by various genetic and biological factors. Understanding these factors is essential for achieving stable hybrid populations and successful breeding programs.

1. Hybrid Fertility: Hybrid fertility is a critical factor that determines the success of interspecific and intergeneric crosses. Fertility can be affected by genetic incompatibilities, chromosomal mismatches, and meiotic irregularities.

» **Sterility:** Many hybrids resulting from distant hybridization are sterile due to chromosomal mismatches that disrupt meiotic division. Sterility can prevent hybrids from producing viable gametes, limiting their reproductive success. For example, mules, which are hybrids between horses (*Equus ferus caballus*) and donkeys (*Equus africanus asinus*), are sterile due to differing chromosome numbers of the parent species. The sterility of mules illustrates the challenges associated with hybrid fertility in distant crosses.

» **Fertility Restoration:** Chromosome doubling is often used to restore fertility in sterile hybrids. By creating polyploid hybrids with balanced chromosome sets, breeders can enable normal meiotic division and successful reproduction. This technique is commonly used in the development of hybrid crops such as triticale. Fertility restoration through chromosome doubling ensures that hybrids can produce viable offspring and contribute to the gene pool.

2. Reproductive Barriers: Reproductive barriers can affect the success of hybridization and the stability of hybrid populations. These barriers can be pre-zygotic (before fertilization) or post-zygotic (after fertilization).

» **Pre-zygotic Barriers:** Pre-zygotic barriers, such as temporal isolation, spatial isolation, mechanical isolation, and gametic isolation, prevent fertilization and the formation of hybrid embryos. Overcoming these barriers often requires careful manipulation and advanced breeding techniques. For example, manual pollination and controlled environmental conditions can help overcome temporal and spatial isolation in hybridization experiments.

» **Post-zygotic Barriers:** Post-zygotic barriers, such as hybrid inviability, hybrid sterility, hybrid breakdown, and reduced hybrid fitness, operate after fertilization and prevent the development, survival, or reproduction of hybrid offspring. Techniques such as embryo rescue and genetic engineering can help overcome these barriers. For example, embryo rescue involves culturing hybrid embryos in vitro to overcome developmental barriers and produce viable hybrid plants.

Genetic Stability and Segregation

The genetic stability and segregation of traits in hybrid populations are important considerations in interspecific and intergeneric crosses. The behavior of hybrid genomes during meiosis and subsequent generations can impact the stability and performance of hybrid crops.

1. Genetic Stability: Genetic stability refers to the consistency of genetic traits across generations. In stable hybrid populations, desirable traits are reliably passed on to subsequent generations without significant variation.

» **Polyploid Stability:** Polyploid hybrids often exhibit greater genetic stability compared to diploid hybrids. The presence of multiple sets of chromosomes provides a buffering effect against genetic variations, ensuring the consistent expression of desirable traits. Polyploid hybrids are commonly used in breeding programs to achieve stable crop performance. For example, polyploid wheat varieties exhibit stable yields and disease resistance, making them valuable for commercial agriculture.

» **Chromosomal Stability:** The stability of chromosomal arrangements is crucial for the long-term success of hybrid populations. Chromosomal rearrangements, such as translocations and inversions, can impact the stability and fertility of hybrids. Monitoring and selecting for chromosomal stability are important steps in hybrid breeding. Techniques such as cytogenetic analysis and molecular markers are used to assess and ensure chromosomal stability in hybrid populations.

2. Segregation of Traits: The segregation of genetic traits in hybrid populations can result in the expression of a wide range of phenotypes. The segregation patterns are influenced by the genetic makeup of the parent species and the behavior of chromosomes during meiosis.

» **Mendelian Segregation:** In some cases, traits in hybrids follow Mendelian inheritance patterns, where traits segregate according to predictable ratios. Understanding these patterns helps breeders select for desirable traits and develop stable hybrid varieties. For example, simple Mendelian traits such as flower color and seed shape can be easily tracked and selected in hybrid populations.

» **Non-Mendelian Segregation:** In complex hybrids, traits may segregate in non-Mendelian patterns due to interactions between multiple genes and chromosomal rearrangements. These complex segregation patterns require careful analysis and selection to achieve desired outcomes. For example, quantitative traits such as yield and disease resistance often exhibit non-Mendelian inheritance and require advanced statistical and genomic tools for selection.

Practical Applications and Challenges

The practical applications of interspecific and intergeneric crosses in plant breeding are vast, but they also come with challenges that need to be addressed for successful implementation.

1. Crop Improvement: Interspecific and intergeneric crosses are used to introduce beneficial traits such as disease resistance, stress tolerance, and improved yield into commercial crops. These crosses expand the genetic base of crops, providing new opportunities for crop improvement.

» **Disease Resistance:** Breeding for disease resistance is a primary goal of interspecific and intergeneric hybridization. By introducing resistance genes from wild or less commonly cultivated species, breeders can develop crops that are more resilient to pathogens and reduce the reliance on chemical pesticides. For example, hybridization between cultivated tomato (*Solanum lycopersicum*) and wild relatives has led to the development of tomato varieties with improved resistance to diseases such as late blight and bacterial wilt.

» **Stress Tolerance:** Enhancing stress tolerance in crops is critical for ensuring stable production in diverse and challenging environments. Interspecific and intergeneric crosses can introduce traits related to drought tolerance, salinity tolerance, and temperature resilience, improving the adaptability of crops to changing environmental conditions. For example, hybridization between different species of rice (*Oryza*) has led to the development of rice varieties with enhanced tolerance to drought and salinity, ensuring stable yields in stress-prone areas.

» **Yield Improvement:** The introduction of hybrid vigor through interspecific and intergeneric crosses can lead to significant yield improvements. High-yielding hybrids contribute to food security by increasing crop productivity and reducing the land area required for cultivation. For example, hybrid maize varieties exhibit increased yield potential and improved resistance to environmental stresses, contributing to higher food production and agricultural sustainability.

2. Challenges: Despite the benefits, there are several challenges associated with interspecific and intergeneric hybridization that need to be addressed for successful breeding programs.

» **Genetic Incompatibilities:** Genetic incompatibilities between the parent species can result in hybrid inviability, sterility, or reduced fitness. Overcoming these incompatibilities requires advanced techniques such as embryo rescue, chromosome doubling, and genetic engineering. For example, embryo rescue techniques are used to overcome post-zygotic barriers and produce viable hybrid plants from otherwise incompatible crosses.

» **Regulatory and Biosafety Concerns:** The use of genetic engineering and other advanced breeding techniques in hybridization raises regulatory and biosafety concerns. Ensuring the safe and responsible use of these technologies is essential for gaining public acceptance and achieving successful implementation. Regulatory frameworks and biosafety guidelines are necessary to address potential risks associated with genetically modified hybrids and ensure their safe use in agriculture.

» **Resource Intensity:** Interspecific and intergeneric hybridization can be resource-intensive, requiring significant time, effort, and financial investment. The development of viable and stable hybrids often involves multiple generations of breeding, selection, and evaluation. Long-term breeding programs and comprehensive field trials are necessary to ensure the success and stability of hybrid crops. The resource-intensive nature of hybrid breeding highlights the need for adequate funding and support for research and development in plant breeding.

The behavior of interspecific and intergeneric crosses is influenced by various genetic, biological, and environmental factors. These behaviors is crucial for effective plant breeding programs and the development of new crop varieties with enhanced traits. The success of interspecific and intergeneric hybridization depends on overcoming reproductive barriers, achieving genetic compatibility, ensuring hybrid fertility and stability, and leveraging hybrid vigor and phenotypic expression. Despite the challenges, the potential benefits of distant hybridization in crop improvement are immense, offering new opportunities for enhancing crop performance, resilience, and sustainability in agriculture. As research continues to advance, the application of interspecific and intergeneric crosses will play an increasingly important role in meeting the demands of modern agriculture and ensuring food security for future generations.

References

Abdurashidova, G., et al. (2000). Start sites of bi-directional DNA synthesis at the human lamin B2 origin. *Science, 287*, 2023–2026.

Adams, D., et al. (1992). The role of topoisomerase IV in partitioning bacterial replicons and the structure of catenated intermediates in DNA replication. *Cell, 71*, 277–288.

Adams, M. D., Kelley, J. M., Gocayne, J. D., Dubnick, M., Polymeropoulos, M. H., Xiao, H., & Merril, C. R. (1991). Complementary DNA sequencing: Expressed sequence tags and human genome project. *Science, 252*, 1651–1656.

Akira, S., Okazaki, K., & Sakano, H. (1987). Two pairs of recombination signals are sufficient to cause immunoglobulin V-(D)-J joining. *Science, 238*, 1134–1138.

Alberts, B., et al. (2002). *Molecular Biology of the Cell* (4th ed.). Garland Science.

Albright, R., & Matthews, B. (1998). How Cro and λ-repressor distinguish between operators: The structural basis underlying a genetic switch. *Proceedings of the National Academy of Sciences, USA, 95*, 3431–3436.

Allard, R. W. (1999). *Principles of Plant Breeding* (2nd ed.). John Wiley & Sons.

Amedeo, P., et al. (2000). Disruption of the plant gene MOM releases transcriptional silencing of methylated genes. *Nature, 405*, 203–206.

Amit, A., et al. (1986). Three-dimensional structure of an antigen-antibody complex at 2.8 Å resolution. *Science, 233*, 747–753.

Anthony, J. F., Griffiths, J. H. M., Suzubi, D., Lewontin, R. C., & Gilbert, W. M. (2000). *An Introduction to Genetic Analysis*. W. H. Freeman and Company.

Antson, A., et al. (1999). Structure of the trp RNA-binding attenuation protein, TRAP, bound to RNA. *Nature, 401*, 235–242.

Arkhipova, I. (1995). Promoter elements in *Drosophila melanogaster* revealed by sequence analysis. *Genetics, 139*, 1359–1369.

Autexier, C., & Greider, C. (1996). Telomerase and cancer: Revisiting the telomere hypothesis. *Trends in Biochemical Sciences, 21*, 387–391.

Avery, O. T., MacLeod, C. M., & McCarty, M. (1944). Studies on the chemical nature of the substance inducing transformation of Pneumococcal types. *Journal of Experimental Medicine, 79*, 137–158.

B.D. Singh (2014). *Fundamentals of Genetics* (332nd ed.). Kalyani Publishers.

Bachmair, A., Finley, D., & Varshavsky, A. (1986). In vivo half-life of a protein is a function of its amino-terminal residue. *Science, 234*, 179–186.

Basak, J., Kundagrami, S., Ghoose, T. K., & Pal, A. (2004). Development of yellow mosaic virus

(YMV) resistance-linked DNA marker in *Vigna mungo* from populations segregating for YMV reaction. *Molecular Breeding, 14*, 375–382.

Beadle, G. W., & Tatum, E. (1941). Genetic control of biochemical reactions in Neurospora. *Proceedings of the National Academy of Sciences, USA, 27*, 499–506.

Bell, S., et al. (1999). Orientation of the transcription preinitiation complex in Archaea. *Proceedings of the National Academy of Sciences, USA, 96*, 13662–13667.

Benko-Iseppon, A., Winter, M., Huettel, P., Staginnus, B., Muehlbauer, C., & Kahl, F. J. G. (2003). Molecular markers closely linked to fusarium resistance genes in chickpea show significant alignments to pathogenesis-related genes located on *Arabidopsis* chromosomes 1 and 5. *Theoretical and Applied Genetics, 107*, 379–386.

Bentley, D. (1999). Coupling RNA polymerase II transcription with pre-mRNA processing. *Current Opinion in Cell Biology, 11*, 347–351.

Bharadwaj, S. K., Chauhan, S. K., Yadav, S., Satyavati, C. T., Singh, R., Kumar, J., Srivastava, R., & Rajguru, G. (2010). *Indian Journal of Agricultural Sciences, 81*, 116–118.

Bhattacharyya, M., et al. (1990). The wrinkled-seed character of peas described by Mendel is caused by a transposon-like insertion in a gene encoding starch-branching enzyme. *Cell, 60*, 115–122.

Bird, A., & Wolffe, A. (1999). Methylation-induced repression: Belts, braces, and chromatin. *Cell, 99*, 451–454.

Bischoff, J., et al. (1996). An adenovirus mutant that replicates selectively in p53-deficient human tumor cells. *Science, 274*, 373–376.

Blattner, F. R., Plunkett, G. 3rd, Bloch, C. A., Perna, N. T., Burland, V., Riley, M., Collado-Vides, J., et al. (1997). The complete genome sequence of *Escherichia coli* K-12. *Science, 277*, 1453–1474.

Blum, A. (1988). *Plant Breeding for Stress Environments*. CRC Press.

Bradshaw, J. E. (2016). *Plant Breeding: Past, Present and Future*. CBS Publishers & Distributors Pvt. Ltd.

Brody, E., & Abelson, J. (1985). The spliceosome: Yeast pre-messenger RNA associates with a 40S complex in a splicing-dependent reaction. *Science, 228*, 963–967.

Brown, J., et al. (1993). Three-dimensional structure of the human class II histocompatibility antigen HLA-DR1. *Nature, 364*, 33–39.

Brown, T. (2012). *Introduction to Genetics: A Molecular Approach*. Garland Science, Taylor & Francis Group.

Burton, D., & Moore, J. (1998). Why do we not have an HIV vaccine and how can we make one? *Nature Medicine Vaccine Supplement, 4*, 495–498.

Busby, S., & Ebright, R. (1994). Promoter structure, promoter recognition, and transcription activation in prokaryotes. *Cell, 79*, 743–746.

Cao, J., Zhao, J. Z., Tang, D., Shelton, M., & Earle, D. (2002). Broccoli plants with pyramided cry1Ac and cry1C Bt genes control diamondback moths resistant to Cry1A and Cry1C proteins. Theoretical and Applied Genetics, 105, 258–264.

Carroll, S. (1995). Homeotic genes and the evolution of arthropods and chordates. *Nature, 376*, 479–485.

Casares, F., & Mann, R. (1998). Control of antennal versus leg development in *Drosophila. Nature, 392*, 723–726.

Caspari, T. (2000). Checkpoints: How to activate p53. *Current Biology, 10*, R315–R317.

Cattaneo, R. (1991). Different types of messenger RNA editing. *Annual Review of Genetics, 28*, 71–88.

Cavanee, W., & White, R. (1995). The genetic basis of cancer. *Scientific American, March*, 72–79.

Cech, T. (1986). RNA as an enzyme. *Scientific American, November*, 64–75.

Cech, T. (1990). Self-splicing of group I introns. *Annual Review of Biochemistry, 59*, 543–568.

Cern, F. J., Cianzio, S. R., Rafalski, A., Tingey, S., & Dyer, D. (1997). Relationship between seed yield heterosis and molecular marker heterozygosity in soybean. Theoretical and Applied Genetics, 95, 460–467.

Chahal, G. S., & Gosal, S. S. (2002). *Principles and Procedures of Plant Breeding: Biotechnological and Conventional Approaches*. Narosa Publishing House.

Chaitieng, B., Kaga, A., Han, O. K., Wang, Y. W., Wongkaew, S., Laosuwan, P., Tomooka, N., & Vaughan, D. A. (2002). Mapping a new source of resistance to powdery mildew in mungbean. Plant Breeding, 121, 521–525.

Chaudhary, H. K. (1996). *Elementary Principles of Plant Breeding* (2nd ed.). Oxford & IBH Publishing.

Cheng, S. W., et al. (1991). Functional importance of sequence in the stem-loop of a transcription terminator. *Science, 254*, 1205–1207.

Chopra, V. L., & Paroda, R. S. (1986). *Approaches for Incorporating Drought and Salinity Resistance in Crop Plants*. Oxford and IBH Publishing.

Conaway, J., & Conaway, R. (1999). Transcription elongation and human disease. *Annual Review of Biochemistry, 68*, 301–319.

Cramer, P., et al. (2000). Architecture of RNA polymerase II and implications for the transcription mechanism. *Science, 288*, 640–649.

Crick, F. (1970). Central dogma of molecular biology. *Nature, 227*, 561–563.

Crow, J. (1993). N. I. Vavilov, martyr to genetic truth. *Genetics, 134*, 1–4.

Das, A. (1993). Control of transcription termination by RNA-binding proteins. *Annual Review of Biochemistry, 62*, 893–930.

David, F. (n.d.). *Molecular Biology*. Narosa Publishing House.

Dhanasekar, P., Dhumal, K. N., & Reddy, K. S. (2010). Identification of RAPD markers linked to plant type gene in pigeonpea. Indian Journal of Biotechnology, 9, 58–63.

Dickerson, R., et al. (1982). The anatomy of A-, B-, and Z-DNA. *Science, 216*, 475–485.

Dietrich, W. F., Miller, J., Steen, R., Merchant, M. A., Damron-Boles, D., Husain, Z., & Dredge, R. (1996). A comprehensive genetic map of the mouse genome. *Nature, 380*, 149–152.

Donis-Keller, H., Green, P., Helms, C., Cartinhour, S., Weiffenbach, B., Stephens, K., & Keith, T. P. (1987). A genetic linkage map of the human genome. *Cell, 51*, 319–337.

Donovan, S., & Diffley, J. (1996). Replication origins in eukaryotes. *Current Opinion in Genetics and Development, 6*, 203–207.

Driscoll, R., Youngquist, M., & Baldeschwieler, D. (1990). Atomic-scale imaging of DNA using scanning tunneling microscopy. *Nature, 346*, 294–296.

Dunham, I., Shimizu, N., Roe, B. A., Chissoe, S., Hunt, A. R., Collins, J. E., & Bruskiewich, R. (1999). The DNA sequence of human chromosome 22. *Nature, 402*, 489–495.

Duran, Y., & Vega, P. M. (2004). Assessment of genetic variation and species relationships in a collection of Lens using RAPD and ISSR. Spanish Journal of Agricultural Research, 2, 538–544.

Duvick, D. N. (1990). The romance of plant breeding and other myths. In J. P. Gustafson (Ed.), *Gene Manipulation and Plant Improvement II* (pp. 143–157). Plenum Press.

Ek, M., Eklund, M., Von Post, R., Dayteg, C., Henriksson, T., Weibull, P., Ceplitis, A., Isaac, P., & Tuvesson, S. (2005). Microsatellite markers for powdery mildew resistance in pea (Pisum sativum L.). Hereditas, 142, 86–91.

Elliot, M. M. (n.d.). Prehistory and history of Arabidopsis research. *California Institute of Technology*.

Falconer, D. S., & Mackay, T. F. C. (1986). *Introduction to Quantitative Genetics* (4th ed.). Longman.

Falkenberg, M., Lehman, I., & Elias, P. (2000). Leading- and lagging-strand DNA synthesis in vitro by a reconstituted herpes simplex virus type 1 replisome. *Proceedings of the National Academy of Sciences, USA, 97*, 3896–3900.

Fisher, R. A. (1936). Has Mendel's work been rediscovered? *Annals of Science, 1*, 115–137.

Fleischmann, R. D., Adams, M. D., White, O., Clayton, R. A., Kirkness, E. F., Kerlavage, A. R., & Bult, C. J. (1995). Whole-genome random sequencing and assembly of *Haemophilus influenzae* Rd. *Science, 269*, 496–512.

Fraenkel-Conrat, H., & Singer, B. (1957). Virus reconstitution. II: Combination of protein and nucleic acid from different strains. *Biochimica et Biophysica Acta, 24*, 540–548.

Frank-Kamenetskii, M., & Mirkin, S. (1995). Triplex DNA structures. *Annual Review of Biochemistry, 64*, 65–95.

Fraser, C. M., Gocayne, J. D., White, O., Adams, M. D., Clayton, R. A., Fleischmann, R. D., & Bult, C. J. (1995). The minimal gene complement of *Mycoplasma genitalium*. *Science, 270*, 397–403.

Ganapathy, K. N., Gnanesh, B., Gowda, N., Byre, M., Venkatesha, S. C., Gomashe, S. S., & Mallikarjuna, V. C. (2011). AFLP analysis in pigeonpea (Cajanus cajan (L.) Millsp.) revealed close relationships of cultivated genotypes with some of its wild relatives. Genetic Resources and Crop Evolution, 58, 837–847.

Gardener, E. J., Simmons, M. J., & Snustad, D. P. (1998). *Principles of Genetics*. John Wiley & Sons.

Gaskill, M. (2016). First DNA sequencing in space: A game changer. *International Space Station Program Office*.

Gaur, R., Sethy, N. J. K., Choudhary, S., Shokeen, B., Gupta, V., & Bhatia, S. (2011). Advancing the STMS genomic resources for defining new locations on the intraspecific genetic linkage map of chickpea (Cicer arietinum L.). BMC Genomics, 12, 117.

Gerald, K. (2002). *Cell and Molecular Biology: Concepts and Experiments*. John Wiley & Sons.

Goff, S. A., et al. (2002). A draft sequence of the rice genome. *Science, 296*, 92–100.

Griffith, F. (1928). Significance of pneumococcal types. *Journal of Hygiene, 27*, 113–159.

Gupta, P. K. (1985). *Cytology, Genetics and Evolution: A Textbook for University Students*. Rastogi Publications.

Gupta, P. K. (2002). *Cytology, Genetics and Molecular Biology*. Rastogi Publications.

Hershey, A., & Chase, M. (1952). Independent functions of viral protein and nucleic acid in growth of bacteriophage. *Journal of General Physiology, 36*, 39–56.

Horowitz, N. H. (1990). George Wells Beadle (1903–89). *Genetics, 124*, 1–6.

Hull, D. (1996). A revolutionary philosopher of science. *Nature, 382*, 203–204.

James, D. W. (2017). *Molecular Biology of the Gene* (7th ed.). Pearson Education.

Jean, G. (2016). From Mendel to epigenetics: History of genetics. *Journal of Genetics*.

Kaeppler, S. M., Phillips, R. L., & Kim, T. S. (1993). Use of near-isogenic lines derived by backcrossing or selfing to map quantitative traits. *Theoretical and Applied Genetics, 87*, 233–237.

Keightley, P. D. (1996). A metabolic basis for dominance and recessivity. *Genetics, 143*, 621–625.

King, R., & Stansfield, W. D. (1996). *A Dictionary of Genetics* (5th ed.). Oxford University Press.

Kuhn, T. (1962). *The Structure of Scientific Revolutions*. University of Chicago Press.

Liu, Y. (2008). A new perspective on Darwin's pangenesis. *Biological Review of the Cambridge Philosophical Society*.

Maddox, J. (1999). The unexpected science to come. *Scientific American, 281*(6), 62–67.

Meyerowitz, E. M. (n.d.). Prehistory and history of Arabidopsis research. *California Institute of Technology*.

Moore, J. A. (1993). *Science as a Way of Knowing*. Harvard University Press.

Morgan, T. H. (1910). Sex-limited inheritance in Drosophila. *Science*.

Neel, J. V. (1994). *Physician to the Gene Pool: Genetic Lessons and Other Stories*. Wiley.

Niklas, K. J. (1997). *The Evolutionary Biology of Plants*. University of Chicago Press.

Obe, G., & Natarajan, A. T. (1990). *Chromosomal Aberrations: Basic and Applied Aspects*. Springer Verlag.

Olson, M., Hood, L., Cantor, C., & Botstein, D. (1989). A common language for physical mapping of the human genome. *Science, 245*, 1434–1435.

P. S. Verma & A. K. Agarwal (2009). *Genetics* (9th ed.). S. Chand & Company Pvt. Ltd.

Popper, K. (1962). *Conjectures and Refutations: The Growth of Scientific Knowledge*. Basic Books.

Ruse, M. (1989). *Philosophy of Biology*. Macmillan.

Saxena, R. K., Prathima, C., Saxena, K. B., Hoisington, D. A., & Varshney, R. K. (2010). Novel SSR markers for polymorphism detection in pigeonpea (*Cajanus spp.*). *Plant Breeding, 129*, 142–148.

Simmonds, N. W. (1979). *Principles of Crop Improvement*. Longman.

Singh, B. D. (2015). *Plant Breeding: Principles and Methods*. Kalyani Publishers.

Strickberger, M. W. (2015). *Genetics* (3rd ed.). Pearson Education India.

Sturtevant, A. H. (1965). *A History of Genetics*. Harper & Row.

Tamarin, R. H. (2002). *Principles of Genetics* (7th ed.). Tata McGraw-Hill.

Taran, B., Buchwald, L., Tullu, A., Banniza, S., Warkentin, T. D., & Vandenberg, A. (2003). Using molecular markers to pyramid genes for resistance to ascochyta blight and anthracnose in lentil (*Lens culinaris Medik*). *Euphytica, 134*, 223–230.

The C. elegans Sequencing Consortium. (1998). Genome sequence of the nematode *C. elegans*: A platform for investigating biology. *Science, 282*, 2012–2018.

Veer, B. R. (n.d.). *Genetics*. Kedar Nath Ram Nath Publishers, Meerut.

Watson, J. D., & Jordan, E. (1989). The Human Genome Program at the National Institute of Health. *Genomics, 5*, 654–656.

Weber, J. L., & May, P. E. (1989). Abundant class of human DNA polymorphisms which can be typed using the polymerase chain reaction. *American Journal of Human Genetics, 44*, 388–396.

Winter, P. C., Hickey, G. I., & Fletcher, H. L. (2003). *Instant Notes in Genetics*. Viva Books Pvt. Ltd.

Yu, J., et al. (2002). A draft sequence of the rice genome. *Science, 296*, 79–92.